Powered Flight

David R. Greatrix

Powered Flight

The Engineering of Aerospace Propulsion

 Springer

David R. Greatrix
Department of Aerospace Engineering
Ryerson University
Victoria Street 350
Toronto, ON M5B 2K3
Canada

ISBN 978-1-4471-2484-9   e-ISBN 978-1-4471-2485-6
DOI 10.1007/978-1-4471-2485-6
Springer London Heidelberg New York Dordrecht

British Library Cataloguing in Publication Data
A catalogue record for this book is available from the British Library

Library of Congress Control Number: 2011944772

Cover design: eStudio Calamar S.L.

Printed on acid-free paper

Springer is part of Springer Science+Business Media (www.springer.com)

Preface

This book documents the engineering behind most of today's aerospace propulsion systems. Some of the material coverage in this book for a number of relevant propulsion topics is more detailed, in the context of a university or technical college course textbook at the introductory or advanced level, so that the reader gets some appreciation of the mathematics (as well as the physics and history) involved in the science and engineering of propulsion. The wider coverage of topics in this book gives more context to the reader, and allows one to observe similarities and differences between systems, observations that otherwise would not be possible in a book dedicated to only one, or a few, propulsion systems. The book has been divided into two main parts, mostly based on the historical separation of the two well-established general categories: airplane propulsion and rocket propulsion. The roots of this book arise from several undergraduate and graduate aerospace engineering courses taught by the author at Ryerson University over the years, including *Aerospace Propulsion* and *Rocket Propulsion*, but also relevant elements of courses like *Flight Mechanics*, *Aircraft Performance* and *Gas Dynamics* that help bring additional background.

In addition to teaching, the author conducts research in the areas of solid and hybrid rocket propulsion, and the flight dynamics of vehicles. It is hoped that this research activity brings further background and insight in to this book. For example, in terms of historical context, the author can certainly understand the demands faced, and the need for persistence required, by researchers past and present, who struggle forward in the face of underfunding and lack of interest from the surrounding technical community.

Prior to joining Ryerson University in 1993, the author worked in the Canadian aerospace industry for a number of years, as well as a few years at a Canadian government laboratory. The author is presently an international member of the AIAA Solid Rockets Technical Committee, and over the years has been a frequent presenter of technical papers at various propulsion-related conferences in North America and Europe.

Acknowledgments

Assistance in the writing and publication of this book was provided by a number of people, and I wish to express my gratitude for their help. I wish to express my thanks to the external reviewers who took the time to read the manuscript, and provide constructive comments. I would like to acknowledge the various representatives of companies, technical/educational organizations, academic institutions and government agencies who arranged for the usage of material helpful towards understanding various topics under discussion in this literary effort. In terms of background preparation for bringing forth the material presented in this book, I am indebted to my professional colleagues past and present, close and far away, and to my undergraduate and graduate students past and present, who directly or indirectly have helped to raise my knowledge level from year to year to a point such that a comprehensive book on aerospace propulsion could ultimately be produced. In addition, some have laid the seeds of motivation that helped inspire the author to initiate and proceed to complete this venture. Whether intentional or not, they have all contributed to the realization of this book, and these individuals have my thanks.

Colleagues who in different ways have been helpful in support of this literary effort include my current Ryerson University Propulsion Research Facility colleagues, Jerry Karpynczyk and Peter Bradley, and from the past and present Aerospace Engineering program at Ryerson University, Profs. Cruchley, Downer, Mölder, McTavish, Liu, Kumar and Okouneva, and from the Mechanical Engineering program at Ryerson, Prof. Kawall. My thanks for the more recent assistance of Prof. Brian Cantwell of Stanford University, Prof. Frank Lu of the University of Texas at Arlington, and Prof. William Roberts of North Carolina State University, in graciously providing input to the present effort.

Professors who were instrumental in laying the earlier groundwork for the author's subsequent efforts include Prof. Jeffrey Tinkler of the University of Manitoba, and Prof. James Gottlieb of the University of Toronto. The exchange of educational information over the years on gas turbine and other engines by my long-time friend at GE Aviation, Greg Johnson, is acknowledged here with thanks. My father, John Greatrix, a long-time Royal Canadian Air Force fighter pilot and

Transport Canada aviation inspector, has over the years provided useful information about aircraft, aircraft propulsion and aircraft flight operations, and I am happy to acknowledge his underlying contribution to the present effort, and his longstanding support. Finally, for her longstanding general support, I should mention Mary Carmel Roberta (Fraser) Greatrix… thanks, Mom.

Contents

Part II   Rocket Propulsion

Symbols

A	Local core flow cross-sectional area, m^2
A_c	Mean chamber cross-sectional area, m^2
A_e	Nozzle exit plane cross-sectional area, m^2
A_{ef}	Fan nozzle exit plane cross-sectional area, m^2
A_{inj}	Injector orifice cross-sectional area, m^2
A_{line}	Propellant line cross-sectional area, m^2
A_p	Port cross-sectional area, m^2
A_t	Nozzle throat cross-sectional area, m^2
A_w	Wetted area, m^2
A_1	Propeller disk area, m^2
a	Gas sound speed, m/s; also, coefficient, mass-flux-dependent burning rate, m/s-(kg/s m^2)^n
a_ℓ	Longitudinal (or lateral) acceleration, m/s^2
a_n	Normal acceleration, m/s^2
B	Number of propeller blades; magnetic field strength, T (teslas)
$\mathcal{B}$	Bypass ratio [TF]
b	Nonequilibrium two-phase sound speed, m/s; also, wing span, m
C	De St. Robert coefficient, m/s-Pa^n
C_D	Drag coefficient
C_{Di}	Lift-induced drag coefficient
C_{Do}	Zero-lift drag coefficient
$C_{d,inj}$	Injector orifice flow discharge coefficient
C_F	Thrust coefficient [rocket engine]
$C_{F,v}$	Vacuum thrust coefficient
C_h	Stanton number
C_L	Lift coefficient

$C_{L,\text{me}}$	Lift coefficient for maximum endurance
$C_{L,\text{mr}}$	Lift coefficient for maximum air range
C_m	Particle or droplet specific heat, J/(kg K)
C_P	Propeller power coefficient
C_p	Constant-pressure specific heat, gas phase, J/(kg K)
C_S	Speed coefficient [propeller]
C_s	Specific heat, solid phase, J/(kg K)
C_T	Propeller thrust coefficient
C_v	Constant-volume specific heat, gas phase, J/(kg K)
c	Effective exhaust velocity, m/s; speed of light, m/s; airfoil chord, m
c^*	Characteristic exhaust velocity, m/s
D	Aerodynamic drag, N; diameter, m
D_H	Hydraulic diameter, m
DN	Bearing speed capability index, [(bearing bore) diameter, mm] × [speed (of shaft rotation), rpm)]
d	Hydraulic diameter of core flow, m; propeller diameter, m
d_e	Nozzle exit diameter, m
d_m	Particle mean diameter, m
d_p	Propeller diameter, m; port diameter [SRM]
d_t	Nozzle throat diameter, m
E	Total gas specific energy, J/kg; electric field strength, N/C or V/m
E_p	Total particle specific energy, J/kg
e	Oswald efficiency factor
e_s	Surface ablation rate, m/s
F	Thrust, N; also, F_{net}, Net thrust
F_{gross}	Gross thrust, N (before effect of air-intake momentum drag included)
$F_{\text{installed}}$	Installed thrust, N (net thrust produced from engine in place on aircraft)
F_Q	Force contributing to torque on propeller or rotor shaft, N
F_{net}	Net thrust (including momentum drag effect), N
F_o	Static thrust, N
f	Darcy-Weisbach friction factor; frequency, Hz; fuel-to-air ratio
f^*	Zero-transpiration friction factor
f_{AB}	Fuel-to-gas ratio, afterburner
f_{act}	Actual fuel-air ratio
f_{cyc}	Frequency of operational cycle, Hz
f_{lim}	Limit friction factor for negative erosive burning
f_{stoich}	Stoichiometric fuel-air ratio
f_{1L}	Fundamental [first] axial resonant [harmonic] acoustic frequency for chamber of length L, Hz
f_{1R}	Fundamental [first] radial resonant acoustic frequency for chamber of radius R, Hz

f_{1T}	Fundamental [first] tangential resonant acoustic frequency for chamber of width D, Hz
G	Axial mass flux, kg/(m^2 s)
G_a	Accelerative mass flux, kg/(m^2 s)
g	Gravitational acceleration, m/s^2
g_o	Reference [sea-level] gravitational acceleration, m/s^2
H	Headrise, m [turbopump]
ΔH_s	Net surface heat release, J/kg
h	Effective convective heat transfer coefficient, W/(m^2 K); altitude, m; enthalpy, J/kg; wall thickness, m
$h*$	Zero-transpiration convective heat transfer coefficient, W/(m^2 K)
h_{AGL}	Altitude above ground level, m
h_c	Convective heat transfer coefficient, W/(m^2 K)
h_G	Geometric altitude, m
h_p	Pressure altitude, m
h_{sc}	Screen [obstacle clearance] height, takeoff and landings, m
$h_{w,\,g}$	Mean height of wing above ground, in ground roll, m
Δh	Difference in pressure head, m
I	Electric current, A (amperes or C/s)
I_d	Density specific impulse [average liquid or solid propellant specific gravity $\times$ I_{sp}], s; also, I_ρ [average propellant density $\times$ I_{sp}], s·kg/m^3
I_{sp}	Specific impulse, s
I_{tot}	Total impulse, N·s
J	Propeller advance ratio
j	Electric current density, A/m^2
K	Lateral/longitudinal acceleration burning rate displacement orientation angle coefficient; lift-induced drag factor
K_b	Limiting coefficient on transient burning rate, s^{-1}
K_δ	Shear layer coefficient, m^{-1}
k	Gas thermal conductivity, W/(m K)
k_z	Thermal conductivity, solid phase, W/(m K)
L	Aerodynamic lift, N; length, m
$L*$	Characteristic length, m
L_c	Axial length of chamber, m
$\ell_{c/p}$	Effective length of PJ combustion tube [combustor + tailpipe], m
ℓ_f	Length of HRE solid fuel grain, m
ℓ_p	Length of SRM propellant grain, m; also L_p
$\mathfrak{M}$	Average molecular mass of gas, amu; also $\mathcal{M}$
M_{ac}	Pitching moment about the aerodynamic centre
Ma	Mach number
Ma$_1$	Mach number of flow upstream of oblique shock

Ma_2	Mach number of flow downstream of oblique shock
Ma_∞	Vehicle flight Mach number
Ma_{rel}	Relative flow Mach number
m	Vehicle mass, kg; unit mass of gas, kg
$\dot{m}_a$	Air mass flow [engine core intake], kg/s
$\dot{m}_{by}$	Air mass flow through fan [TF, bypass flow], kg/s
$\dot{m}_f$	Fuel mass flow, kg/s
$\dot{m}_o$	Oxidizer mass flow, kg/s
m_p	Mean mass of particle or droplet, kg; mass of propellant, kg
m_{pp}	Dry mass of EP powerplant, kg
N	Engine shaft speed, rpm
N_s	Pump specific speed, $rpm \cdot (\ell/min)^{1/2}/m^{3/4}$
N_1	Low-pressure shaft speed, rpm; also, N_{LP}
N_2	High-pressure shaft speed, rpm; also, N_{HP}
Nu	Nusselt number
n	Propeller shaft rotational speed, revolutions/sec; burning rate exponent [SRM, HRE]
P	Ideal power, W
P_{inp}	Input electric power, W
P_{jet}	Kinetic power of jet exhaust [EP], W
P_S	Shaft power, W
P_{So}	Static shaft power, W
Pr	Gas Prandtl number
p	Local gas static pressure, Pa
p_c	Chamber pressure, Pa
p_e	Nozzle exit static pressure, Pa
p_∞	Outside ambient air pressure, Pa
Δp_{inj}	Injector pressure drop, Pa
Q	Torque, N m; volumetric flow rate, m^3/s
Q_s	Net near-surface heat release, J/kg
q	Dynamic pressure, Pa; electric charge magnitude, C (+/-, coulombs)
q_R	Heat of reaction, J/kg, typically lower heating value
q_{rad}	Heat of radiation, J/kg
R	Specific gas constant, J/(kg K); propeller or rotor tip radius, m; air-referenced range, m
$^\circ R_c$	Degree of reaction, compressor
$^\circ R_t$	Degree of reaction, turbine
$\Re$	Universal gas constant, J/(kg K)
R_E	Mean earth radius, m
R_g	Ground-referenced range (distance), m
Re_d	Local gas Reynolds number based on core hydraulic diameter

Re_p	Relative flow Reynolds number about particle
r	Radial distance, m; oxidizer-to-fuel mixture ratio; compression ratio
r_a	Acceleration-dependent burning rate, m/s
r_b	Overall burning rate, m/s
$r_{b,qs}$	Quasi-steady burning rate, m/s
r_e	Erosive burning rate positive component, m/s
r_o	Base burning rate, m/s
r_p	Pressure-dependent burning rate, m/s
r_{st}	Stoichiometric mixture ratio
r_u	Velocity-dependent burning rate, m/s
r_v	Volumetric compression ratio
S	Wing reference area, m^2; entropy, J/K; cross-sectional area of missile body, m^2
S_s	Pump suction specific speed, rpm·$(\ell/\text{min})^{1/2}/\text{m}^{3/4}$
St	Strouhal number [combustion]
s	Distance, m; entropy per unit mass, J/(kg K)
s_{TO}	Takeoff distance, m
T	Local gas static temperature, K
T_{as}	Auto-ignition temperature, K
T_{ds}	Decomposition gas temperature resulting from ablation, K
T_f	Flame temperature, K
T_i	Initial temperature, solid phase, K
T_o	Stagnation temperature, K
T_p	Particle temperature, K
T_s	Burning surface temperature, K
T_∞	Outside ambient air temperature, K; local central core gas temperature, K
t	Time, s; airfoil thickness, m
t_B	Burning [motor firing] time, s
t_c	Residence (stay) time for combustion, s
Δt_c	Effective combustion time period for one operational cycle, s
Δt_{cyc}	Time period of operational cycle, s
U	Local rotor tangential velocity, m/s
U_t	Rotor tip speed, m/s
u	Local axial gas velocity, m/s; internal energy, J/kg
u_e	Nozzle exit gas or exhaust jet velocity, m/s; also V_e
u_p	Local axial particle velocity, m/s
u_∞	Bulk axial gas velocity, m/s
V	Velocity, m/s; true airspeed, m/s; voltage, V (volts)
ΔV	Incremental vehicle velocity gain from motor thrust, m/s
V_E	Overall resultant velocity, blade, m/s

$V_{e,\infty}$	Fully-expanded exhaust exit velocity, m/s
$V_{ef,\infty}$	Fully-expanded fan duct exhaust exit velocity, m/s
V_g	Ground speed, m/s
V_R	Resultant velocity [forward flight + rotation], blade, m/s
V_w	Wind speed, m/s
V_∞	True airspeed of flight vehicle, m/s
Ψ	Volume of element, m^3
Ψ_c	Volume of chamber, m^3
Ψ_p	Volume of propellant, m^3
Ψ_t	Overall volume of tank, m^3
Ψ_u	Useable volume of tank, m^3
$\Delta\Psi_{max}$	Auxiliary-usage volume of propellant, m^3
$\Delta\Psi_b$	Boil-off volume of propellant, m^3
$\Delta\Psi_c$	Cooling bleed-off volume of propellant, m^3
$\Delta\Psi_t$	Trapped (residual) volume of propellant, m^3
$\Delta\Psi_u$	Ullage volume of propellant, m^3
v	Gas velocity, m/s; specific volume, m^3/kg
v_f	Normal gas flow velocity of flame, m/s
v_{inj}	Hydraulic injection velocity, m/s
W	Vehicle weight, N
W_c	Work by compressor on air, J
W_E	Empty vehicle weight, N; dry engine weight, N
W_F	Fuel weight, N
W_O	Initial vehicle weight, N
W_{PL}	Payload weight, N
W_t	Work by gas on turbine, J
w	Induced velocity, m/s; swirl velocity, m/s
x	Axial distance, m
y	Distance from propellant surface, m
α	Angle of attack, deg; Lenoir-Robillard model coefficient
α_g	Gas phase void fraction
α_i	Induced angle of attack, rad
α_p	Particle mass loading fraction [SRM]
α_s	Thermal diffusivity, solid phase, m^2/s; specific power, W/kg [EP systems]
β	Propeller or rotor blade pitch angle, deg; angle of sideslip [aircraft], deg; Lenoir-Robillard model coefficient
β_g	Particle-to-gas mass flow ratio
β_{ref}	Reference propeller or rotor blade pitch angle at a specified radial position, deg
Γ	Dihedral angle (wing, blade), deg
γ	Ratio of specific heats of gas; flight path angle, degree or rad

γ_a	Ratio of specific heats of air
δ	Atmospheric air pressure ratio wedge angle, deg
$\delta_{\max}$	Wedge angle at point of shock detachment, deg
δ_o	Reference energy zone thickness, m
δ_r	Resultant energy zone thickness, m
ε	Effective propellant surface roughness height, m; emissivity coefficient [radiation]; impeller slip factor
ε_c	Combustion efficiency [HRE, LRE]
ζ	payload mass fraction [rocket vehicle]
η_b	Burner [combustor] adiabatic component efficiency
η_c	Compressor adiabatic component efficiency
η_d	Diffuser [intake] adiabatic component efficiency
η_f	Fan adiabatic component efficiency
η_{fn}	Fan nozzle adiabatic component efficiency
η_m	Turbomachinery adiabatic component efficiency
η_n	Core nozzle adiabatic component efficiency
η_o	Overall efficiency
η_p	Jet propulsive efficiency
η_{pr}	Propeller propulsive efficiency
η_t	Turbine adiabatic component efficiency; thruster efficiency [EP systems]
η_{th}	Thermal efficiency
θ	Atmospheric air temperature ratio pitch elevation angle, degree oblique shock angle, deg
θ_R	Runway upslope angle, deg
θ_r	Resultant angle of stretched energy zone, deg
κ	Dilatation term, s^{-1}
Λ	Sweep angle (wing, blade), deg
λ	Nondimensional velocity ratio [propeller]; wavelength [EM], m
μ	Absolute [dynamic] gas viscosity, kg/(m s); coefficient of rolling friction; main rotor advance ratio [helicopter]
υ	Kinematic gas viscosity, m^2/s
ξ	Damping ratio
π_b	Burner [combustor] stagnation pressure ratio
π_c	Compressor stagnation pressure ratio
π_d	Diffuser [intake] stagnation pressure ratio
π_f	Fan stagnation pressure ratio [TF]
ρ	Local gas density, $\mathrm{kg/m}^3$
ρ_m	Density, particle or droplet material, $\mathrm{kg/m}^3$
ρ_p	Density, particles within gas flow volume, $\mathrm{kg/m}^3$
ρ_q	Net electric charge density, $\mathrm{C/m}^3$
ρ_s	Density, solid phase, $\mathrm{kg/m}^3$
σ	Atmospheric air density ratio; propeller or rotor solidity

σ_p	Pressure-dependent burning-rate temperature sensitivity, K^{-1}
τ_s	Surface fluid shear stress, Pa
ϕ	Acceleration vector orientation angle, deg; ground effect factor; angle of resultant velocity from plane of blade rotation, deg; equivalence ratio [combustion]; flow coefficient [turbine]; roll angle, deg [vehicle]
ϕ_d	Acceleration vector displacement orientation angle, deg
ψ	Rotor blade azimuth angle, deg [helicopter]; turbine stage loading coefficient
ω	Angular frequency, rad/s
ω_n	Natural frequency, rad/s
ω_r	Resonant frequency, rad/s

Chapter 1
Introduction to Aerospace Propulsion

1.1 First Thoughts

Mankind's adventure of powered, controlled human flight officially began on a cool winter day, December 17, 1903, on the windswept sands of Kill Devil Hills near Kitty Hawk, North Carolina [1]. The heavier-than-air flying machine, the *Flyer I*, was able to lift off the ground with the aid of propulsive thrust delivered by two pusher propellers (mounted behind the main lift-producing wing) driven by a single in-line 4-cylinder 12-horsepower piston engine (see Fig. 1.1). Prior to this point in history, controlled human flight had been restricted to unpowered gliders and essentially unpowered (and in most cases, poorly controlled) lighter-than-air aerostats [predominantly unpowered balloons of various designs, and a few poorly powered and controlled airships (dirigibles, blimps) using propellers]. While Orville and Wilbur Wright did not invent the internal combustion engine or the propeller, their innovative engineering skills, bringing the needed technology together, enabled the first powered airplane to take flight. It took several more years, largely at Huffman Prairie airfield in Dayton, Ohio, for the two brothers to refine and further test various variants of their *Flyer*, before it could be called a practical flying aircraft that was useable to others.

Significant elements of our current aerospace propulsion knowledge predate the Wright Brothers. In some cases, one can go back hundreds, to literally thousands of years. The Chinese are generally credited with the invention and use [2] of the solid-propellant rocket at a time (1232) when bows, arrows, and catapult-based projectile launchers were the state of the art in Western Europe. The conventional liquid-propellant rocket took a little longer to enter our lives, as a contemporary of the Wrights, Robert Goddard, labored under isolation and little support to ultimately accomplish this feat in 1926 [2]. While the 17th century mathematician and scientist, Sir Isaac Newton, is rightly famous for establishing that "for every action there is an equal and opposite reaction" [2], essentially laying out the principle behind jet and rocket propulsion in 1687s *Principia Mathematica*, it must be acknowledged that the Greeks, in the time of Alexander the Great and the Caesars,

D. R. Greatrix, *Powered Flight*, DOI: 10.1007/978-1-4471-2485-6_1,
© Springer-Verlag London Limited 2012

Fig. 1.1 At *right*, a famous view of the *Wright Flyer I* of 1903 becoming airborne. Displayed at the Smithsonian Air and Space Museum, one can view the *gray-colored* in-line gasoline-fueled 4-cylinder piston engine that powered the *Flyer*. The engine is effectively lying on its side, so that the reciprocating pistons are moving parallel to the wing beneath it, rather than normal to it. The engine is positioned just off-center on the lower wing counterbalanced by the prone pilot on the other (*left*) side of the engine, powering the two propellers, in a pusher propeller configuration to allow for the positioning of the forward canards to give pitch control. Cutaway diagram of the engine from the Library of Congress (U.S). Photo at *right* from the U.S. National Archives

were demonstrating functional propulsive nozzles for various applications, like that demonstrated by Hero of Alexandria's *aeolipile* in 62 (Fig. 1.2; [2]).

Since 1903, advancements in aerospace propulsion have in relative terms been more frequent, to some degree in lockstep with advances in overall flight vehicle design for atmospheric and space flight. Indeed, it is commonly heard (in the aerospace propulsion community, anyways) that every notable advance in airplane and space vehicle capability in recent memory has been enabled first by a corresponding development in propulsion engineering. Advancements, however, do not always come easily or readily. As was the case for Robert Goddard in his earlier years of developing the liquid rocket engine (Fig. 1.3; [2]), Sir Frank Whittle demonstrated extraordinary persistence in pushing forward the development of the turbojet engine in the late 1920s and well into the 1930s, sometimes in the face of withering criticism and disbelief from the technical community. Progress by his team, and that of counterpart Hans von Ohain in Germany, ultimately led to the first appearance of "odd-sounding, propellerless" aircraft over the skies of Germany (first jet flight, August 27, 1939), and then England (first flight of the Gloster E.28/29 powered by a Whittle W-1X jet engine, May 15, 1941), in the early 1940s. A General Electric/Whittle engine powered a Bell aircraft on October 2, 1942, for America's first jet flight [3].

Fig. 1.2 Hero of Alexandria's steam-powered turbine, a.k.a., *aeolipile*, documented first around the year 62, at the Library of Alexandria. Diagram from Knight's American Mechanical Dictionary, Vol. 1, 1876, as copied from much older Arabic documents. Comparable versions of the aeolipile are credited to other authors a number of decades or more before Hero

1.2 "Seemed Like a Good Idea at the Time"

Among the good ideas that have come out over the years as relates to aerospace propulsion (see above) are the occasional bad ones. In my younger days as an aerospace engineer, I worked on the development and flight testing of a rocket-powered target drone. Target drones are a subset of the family of unmanned (uninhabited) air vehicles [UAVs] that are remotely piloted from a distant flight mission control center (ground or ship-based control station, or controlled from another aircraft), typically in parallel with some autonomous control via a flight control computer on-board the UAV itself. The principal purpose of a target drone is to provide training in terms of defending one's location against an incoming air-based threat. The end result usually involves the destruction of the drone, so basically a target drone has on average a one-flight lifetime. With this background, it would not be surprising to hear that target drones usually occupy the low end of the totem pole of UAVs, with considerable pressure from various parties to bring

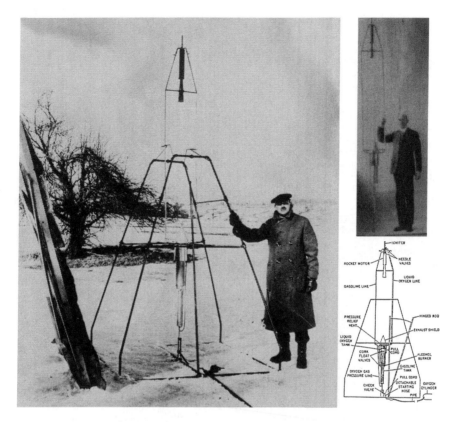

Fig. 1.3 Photos of Robert Goddard beside his first successful liquid-propellant rocket engine powered flight vehicle, launched March 16, 1926. The respective liquid oxygen and gasoline storage tanks are in the lower section of the vehicle (*next to Goddard*), while the combustion chamber and main exhaust nozzle are in the upper section (*nose*) of the vehicle, being fed by long propellant lines extending up from the tanks below. Photos and schematic diagram courtesy of NASA

the unit price down. The novel basis for our particular vehicle's thrust delivery was a sequenced firing of off-the-shelf 70 mm-diameter solid rocket motors packed into the rear end of our flight vehicle, to produce a subsonic boost-coast-boost flight profile from the drone's launch to the point of approaching the defended test location. As an aside, a comparable approach was in fact tried for powering gliders in pre-WWII Germany.

While there had been some early rationale that the set of rocket motors was cheaper than one small turbojet engine plus fuel tank, and thus a good design choice to proceed upon, it later came out that the rockets were cheaper initially because they were being donated to us by a supplier, at a loss, for the duration of our prototype test phase, and that their price status would change "significantly" upwards once multi-unit production was to begin. During flight testing, the

Fig. 1.4 Photo of Messerschmitt ME 163 Komet rocket plane on display at the Canadian Aviation and Space Museum

ongoing complaints from potential clients and interested spectators were: "your drone's range is way too short for our needs" commonly followed by "the boost-coast-boost profile does not really emulate existing threats." As discussed later in this book, a solid rocket motor operating near sea level will have a specific impulse of around 240 s (specific impulse, I_{sp}, is a measure of thrust efficiency, with the higher the value, the better). An air-breathing turbojet engine, even a small unsophisticated one designed for UAVs, has an effective specific impulse of around 3,000 s. In hindsight, a turbojet-powered version of our target drone would in fact have been cost-effective and ultimately cheaper than the solid-rocket-package approach, and would have met our potential clients' need for greater range and consistent threat flight characteristics.

A second example of hindsight being 20/20 is the novel, rocket-powered Messerschmitt ME 163 *Komet* of World War II fame (see Fig. 1.4), the world's first and still only production rocket-powered fighter aircraft. It underwent years of development and testing, and ultimately went into production (over 300 units being the final total), and entered into service with the Luftwaffe as an anti-bomber interceptor in 1944. The single tail-mounted Walter bipropellant rocket engine was powered by a volatile hypergolic fuel (hydrazine hydrate–methanol mixture, called C-Stoff) and oxidizer (hydrogen peroxide concentration, called T-Stoff) that was notoriously dangerous to handle. Once flight operations began with the *Komet*, several drawbacks became clear: (1) the short 7-minute life of the rocket engine usually meant a relatively slow, unpowered glide back to one's home airfield after engaging a number of Allied aircraft at altitude, which left the *Komet*'s pilot vulnerable to Allied aircraft seeking retribution, (2) the rocket plane was fast while under thrust, and for most pilots as it turned out, too fast, to allow for accurate targeting of Allied bombers upon engagement, and (3) the unpressurized cockpit of the *Komet* was an unfriendly environment for their pilots, at the high altitudes that the Allied bombers were flying at. In the end, only 16 Allied aircraft were downed by the *Komet*, a miserable result for the time and lives expended in bringing it to operation.

A further example of "good ideas" not panning out is the use of jet propulsors at the blade tips of a helicopter's main rotor (in place of a central engine driving the rotor), in order to eliminate the use of a tail rotor, and in some applications, improve the main rotor's lifting capability. Various means for driving these tip propulsion systems have been tried over the years, including channeled pressurized air delivered from a central compressor (Sud-Ouest's prototype helicopter, the S.O. 1100 *Ariel*, 1947), mini-turbojets positioned at each tip, etc. A main drawback noted for this approach to powering rotor rotation is the introduction of a counter-productive high drag component on the rotor blade when in an autorotation scenario, where the helicopter, having lost power, needs to make an unpowered descent and final transient lift recovery with application of the main rotor's collective pitch. Additionally, the use of miniature propulsion systems like turbojets or ramjets at the rotor blade tip has not proven to be robust or sufficiently long-life, to date. As a result, maybe not surprisingly given the proven track record of the high power-to-weight turboshaft engine, one has yet to see a production helicopter utilizing the above approach.

1.3 From Design to Certification

Let's briefly consider the present process of defining, designing, building and supporting a new engine for a commercial flight application. Initially, the aerospace engine manufacturer (original equipment manufacturer (OEM), like General Electric, Pratt & Whitney, Rolls-Royce, etc.) would want to define as best as possible the product, and they typically do this through creating a major top-level document commonly entitled something along the lines of *Marketing Requirements and Objectives* (MR and O). It is common that a new engine is designed as part of the introduction of a new flight vehicle, so there would likely be close cooperation between the engine and prospective vehicle manufacturers during the design and build process. The MR and O report would contain various items related to finance, partner/contractor support, training and guarantees. Of particular interest to this book, the report would also contain something along the lines of *Design Requirements and Objectives* (DR and O), which would outline the technical characteristics of the engine that the company's engineering team will have to address. Note that a requirement is a "must have", while an objective is a "nice to have". The DR and O defines technical requirements that fall roughly into three categories:

1. **Customer requirements:** required thrust at different throttle settings (maximum takeoff, maximum climb, maximum cruise), fuel consumption at various throttle settings (including idle), refueling times, dispatch and flight reliability, ease of maintenance, time between overhauls, etc.
2. **Certification requirements:** airworthiness (tied to engineering design of the engine, e.g., engine must be able to throttle up from idle to full thrust in an expeditious manner, i.e. less than 5 s), manufacturing and inspection (tied to the building or fabrication process), operations (tied to maintenance, flight and ground operations, troubleshooting, repairs), safety considerations (e.g. bird

strike, hail ingestion, de-icing, separated blade containment, fire containment and extinguishment, etc.)

3. **Company requirements:** standard materials and processes (to avoid excessive fabrication costs), documentation (including detailed drawings), partnership and shared work arrangements with supporting contractor companies providing engine components, industry standards (International Standards Organization (ISO), quality control, etc.), etc.

The DR and O essentially defines the "type design" for the new engine, which ultimately will be submitted for evaluation by the pertinent transport authorities like Transport Canada or the Federal Aviation Administration (FAA, in the United States). If successful, the engine will receive a "type certificate" that will permit production beyond the prototype building phase. Initial prototypes would be built first for flight, systems and structural testing to meet certification evaluation requests (provide "proof of compliance"). Note that the first prototype engine would only be built after the design has been cleared for advanced development, after a detailed preliminary design phase had shown that the performance R and O's would be met.

Certification of a complete airplane will be addressed through various parts or chapters of a transport authority's regulations, e.g. the National Aircraft Certification branch of Transport Canada produces the Canadian Aviation Regulations (CARs), which are similar to the FAA's Federal Aviation Regulations (FARs), and in Europe, regulations fall under the jurisdiction of the European Aviation Safety Agency (EASA). These regulations cover the aircraft's airframe, analysis and test of flight and structural performance, landing gear, systems and equipment safety and reliability (note famous FAR section/clause 1309, "… must perform under any operating condition… failure of a system must not endanger the aircraft…"), lightning protection, fuel storage, etc. Commercial transport aircraft over a 12,500 lb (5,700 kg) takeoff weight are covered in FAR Part 25 (CAR Part V, Subpart 25), while for smaller utility aircraft, one would refer to FAR Part 23 (CAR Part V, Subpart 23). Transport helicopters are covered in FAR Part 29. Engines are covered in FAR Part 33, covering such items as bird strike, vibration, fire containment, throttle control limits, etc. Propellers are covered in FAR Part 35, e.g., blade pitch control, durability, fatigue limits, etc. Chemical emissions are covered in FAR Part 34 (fuel venting, engine exhaust), while noise is covered in FAR Part 36 (required limits on acoustic emissions from the engine(s)).

Safety is a primary theme running through these and other government regulations. Design requirements for redundancy of systems or components in the event of failure are common ("fail-safe" approach for single-point failure, where such a failure leading to catastrophic loss of the flight vehicle would be extremely improbable). One also finds "fault-tolerant", "safe-life", or "damage-resistant" systems or components that can withstand local failure or damage for a minimum period of time before detection and repair can take place. When a failure does occur, the hazard to the aircraft and the people on-board must be minimized as much as possible. For example, when a pylon-mounted engine on a wing begins to

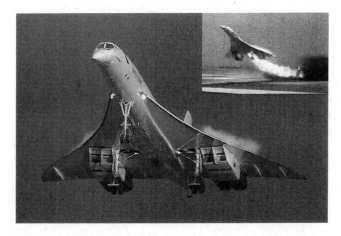

Fig. 1.5   Concorde supersonic airliner, in landing approach (*inset*, 2000 incident)

vibrate beyond the allowed acceleration limit (or abruptly encounters an obstacle, say during takeoff or landing), it must separate cleanly from the wing (typically via several shear pins or bolts giving way once their loading threshold is exceeded, etc.), before it can lead to the destruction of the wing and thus the aircraft itself. A tire burst on takeoff or landing should not lead to the destruction of the aircraft [e.g. burst tire material during takeoff punctured a wing fuel tank above a landing gear, the fuel leak in turn resulting in a fire in the landing gear bay and nearest engine, disabling various systems in the process as the fire spread, ultimately leading to the destruction of a Concorde airliner (Fig. 1.5) in 2000; loose titanium metal components inadvertently dropped from a previous airliner onto the runway led to the Concorde's initial tire puncture]. A compressor or turbine blade separation within a gas turbine engine, or a propeller blade separation, should not lead to the destruction of an aircraft, or lead to undue injuries to people on-board (see engine of Fig. 1.6; an Airbus A380 had its left inboard engine's intermediate pressure turbine break up in flight in November of 2010, causing some left wing structural damage as a result, but the flight crew was able to land the aircraft without incident). A bird or collection of birds striking an engine or cockpit windshield (see Fig. 1.7) should not lead to the destruction of an aircraft (e.g. the striking of a flock of Canada geese after takeoff led to a complete engine shutdown on an Airbus A320 in 2009, forcing a successful ditching of the airplane onto the Hudson river near New York). The initiation of fire or smoke on or within the aircraft should not lead to its destruction (e.g. in 1998, an MD-11 airliner was brought down by a fire spreading through the passenger cabin and cockpit wall structure, likely initiated by arcing of faulty wiring of the in-flight entertainment system).

   With respect to safety, cost is a major issue. The aircraft's manufacturer will typically want to minimize costs where possible, especially if the customer (e.g.

Fig. 1.6   Cutaway diagram of the Rolls-Royce Trent 900 turbofan engine. This engine powers the Airbus A380 airliner. *Lower right* photo of Rolls-Royce outdoor static engine test at NASA Stennis Space Center. Diagram and photo courtesy of Rolls-Royce plc

Fig. 1.7   Bird strike results on helicopter cockpit windshield and surrounding airframe (*actual incident*)

airline utilizing the aircraft for profit, or the passenger flying thereon) is unwilling to pay above a certain dollar limit. The engine manufacturer may feel that economic pressure in conjunction. The respective government is also aware that a suitable balance must be struck in stipulating minimum acceptable safety compliance that can be met at a reasonable cost, versus an excessive cost. Concluding

Fig. 1.8 Jet engine nacelle positioned beneath the wing (*left*) and at the rear fuselage (*right*)

on a positive note, new technology and processes continue to appear and be developed, helping to move the regulations toward greater safety.

1.4 Integration of the Propulsion System to the Flight Vehicle

Any discussion of the viability of propulsion systems will ultimately lead to how well the propulsion system will couple to the prospective flight vehicle, and by extension, how well the propulsion system will perform once installed and operating in flight. A number of items pertinent to the proper integration of an engine to a flight vehicle are governed by transport authority regulations, but ultimately the success of the integration is established by the success of the given flight vehicle's performance with the engine(s) installed. Differing categories of aircraft or spacecraft, in conjunction with differing propulsion system types, will likely have some differing requirements for systems' interface and overall integration. Several representative examples are provided below to give the reader a sense of the issues associated with propulsion system integration.

1.4.1 Engine-Airframe Integration for Airplanes

Aircraft engines for modern subsonic civil airliners are usually pod-mounted, for good aerodynamic intake performance ("clean" undisturbed air, away from airframe interference) and for ease of access for engine maintenance and repair [4]. Along these lines, the engine nacelle (pod) would commonly be positioned on a pylon (strut) beneath the wing (if there is sufficient nacelle-to-ground clearance; a bonus with this approach is the providing of some relief from the

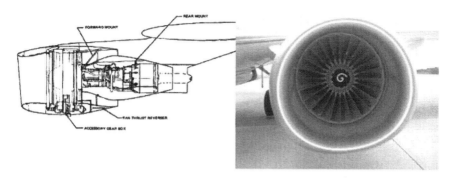

Fig. 1.9 Schematic diagram of pylon-mounted turbofan engine nacelle placement under the wing of a Boeing 737. The small forward angle (*from the vertical reference*) of the intake face compensates for the local airflow upwash component at that position upstream of the lifting wing (intake face perpendicular to the local incoming airflow at design cruise conditions). The distinctive non-circular "hamster pouch" shape of the front engine intake seen on later 737s (*evident in the photos*) is due to the low ground clearance of larger diameter, higher bypass turbofan engines installed on these newer airplanes; engine accessories are also by necessity placed to the side of the engine as a result of this design constraint

aerodynamic-lift-inducing upward bending moment acting on the wing), or extending from the rear (aft) side or upper fuselage (see Figs. 1.8, 1.9; [5]). Figure 1.10 provides an example of a less common choice for an airliner, a central engine placement in the tail of the airplane [5]. The effect of the engine's location relative to the airframe, with thrust on and off, on the aircraft's overall flight stability and control, needs to be established and accounted for in setting up the aircraft's flight control management. The hot exhaust gas and particulate matter from the engine will need to be ducted safely away from the surrounding airframe (avoid heating/erosion zones). Depending on the engine's location, the threat of foreign object damage (FOD) from runway debris being sucked into the engine's intake during takeoffs and

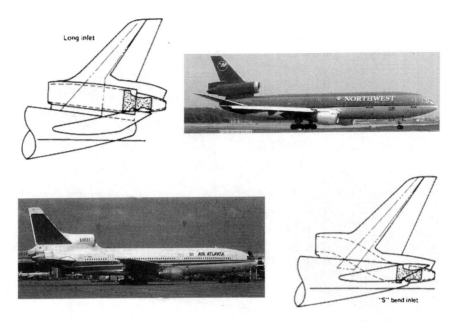

Fig. 1.10 Schematic diagrams of two possible central tail placements of a turbofan engine in a large widebody commercial airliner. The *upper right photo* of a McDonnell Douglas DC-10 shows the resulting choice for one case. A contemporary of the DC-10, the Lockheed L-1011 TriStar in the *lower left photo*, employed the S-bend option

landings needs to be addressed, e.g. one needs to set the appropriate limit on the use of engine thrust reversers (curtail usage below a certain airspeed in landing). In the special case of amphibious aircraft (see Fig. 1.11), one needs a healthy clearance of the engine from the water's surface, to keep water from entering the engine and affecting its operation. Military aircraft may also have unique requirements (see Figs. 1.12, 1.13; [6], and 1.14). The engine position relative to the cockpit and passenger locations needs to meet safety and noise requirements. Fuel delivery from the fuel storage tanks (tanks preferably placed in the wing, for greatest passenger safety in case of a crash) to the given engine needs to be established, controlled manually via the pilot's throttle setting, and monitored by the flight control computer. Thermal conditions in the nacelle environment need to be monitored and controlled properly. Similarly, as illustrated by the various systems and accessories in close proximity to the example engine of Fig. 1.15; [7], the electronic, hydraulic, pneumatic (typical for some pressurized air to be bled off from the engine compressor for various applications, including the aircraft's environmental control system; some interest exists in having a "more or all electric" engine for future aircraft, with less or no engine air bleed-off) and safety systems interface between the engine and the aircraft flight management computer needs to be properly in place for effective and safe engine operation.

Fig. 1.11   Photo of Consolidated PBY-5A Canso (*lower*: PBY-5 Catalina) amphibious reconnaissance (patrol) aircraft. Note the position of the radial piston engines high up on the wing, to avoid the intake of water when landing or taking off from the water's surface (engine: air-cooled 1200-hp 14-cylinder Pratt & Whitney Twin Wasp R-1830-92, two back-to-back rows [banks] of 7 cylinders in radial configuration)

1.4.2 Motor or Engine Integration into Rocket Vehicle

As an example of a larger design issue related to rockets, one on occasion has to decide on the number of motor stages to meet the needs of a given flight vehicle application. For example, a single motor stage to reach Earth orbit from the Earth's surface is a very demanding requirement with current technology (see the SSTO example of Fig. 1.16), as compared to the more common approach of using a number of stages that are consecutively jettisoned once expended. With respect to a more direct issue as relates to the integration of a rocket engine to its flight vehicle, one in some cases has to decide on the division of thrust delivery between

Fig. 1.12 Photo of Fairchild Republic A-10 Thunderbolt II (a.k.a., Warthog) low-level attack aircraft. The aft pod-mounted engines are positioned high-up and shielded to some degree by the airframe, to avoid FOD, help avoid detection by heat-seeking missiles, and also to help avoid damage from flying shrapnel and ground-based anti-aircraft fire (engine: General Electric TF34-GE-100 turbofan). Courtesy of USAF

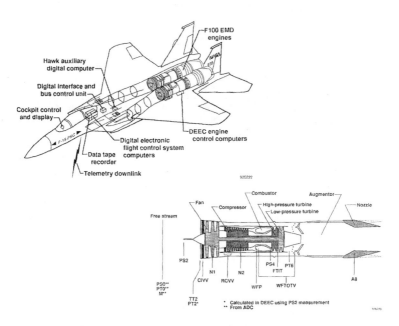

Fig. 1.13 *Upper diagram* illustrates the positioning of the two Pratt & Whitney F100 EMD low-bypass afterburning turbofan engines embedded within the fuselage of a McDonnell Douglas F-15 Eagle fighter aircraft fitted out for a NASA control study [6]. *Side schematic view* of an F100 EMD engine shown in the *lower diagram*. Diagrams reprinted with permission of the American Institute of Aeronautics and Astronautics

Fig. 1.14 *Bottom view* of the Boeing/Lockheed Martin F-22 Raptor fighter aircraft, *at left*. Note the faceted/angled geometry of the front air intakes and the exhaust nozzles for the two embedded engines (engine: Pratt & Whitney F119-PW-100 afterburning low-bypass turbofan with vectorable exhaust nozzle), meeting the needs for traditional engine operation as well as producing a low radar and heat signature (stealth requirement). Photos courtesy of USAF; *left* and *lower right photos*: John Karpynczyk

an integral engine (embedded to some degree within the rocket vehicle's body) and expendable, strap-on booster motors that in essence contribute to the first-stage thrust of a two- or multi-stage vehicle, and then are readily jettisoned when emptied of propellant.

As one might expect, allied to the above examples, volume and weight constraints can be quite tight on rockets and missiles. The given integral engine must of course fit within the specified confines of the vehicle's body, and allow for piping, wiring, ancillary equipment, etc., to be positioned about its periphery [8]. If the engine is an air-breather requiring an intake, there may be some constraint on the intake geometry (shape, size) in order to conform to the overall vehicle's space requirements for launch and post-launch flight. The propulsion system must allow for the launch platforms that are anticipated for the rocket vehicle, e.g. launch from a host aircraft at high altitude (cold, low-pressure environment). If there is some anticipation of an undesirable vibrational interaction between the propulsion system and the surrounding flight vehicle, some damping mechanisms may be necessitated. One must allow for piping between the engine and propellant storage tank(s) when applicable, with feed controlled by the vehicle's flight management computer via electronic, hydraulic and pneumatic means, or some combination thereof, depending on the engine and its application. Thermal management in the vicinity of the engine may be required, depending on the expected temperature range to be encountered internally and externally. One-only or multiple ignition of an engine will necessitate the proper electronic and safety interface between the vehicle's electric power source and the engine igniter. The ignition operation and

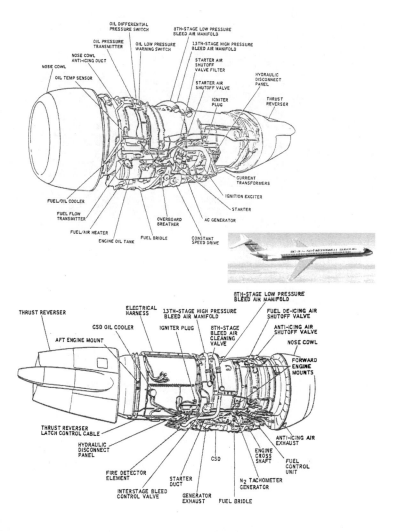

Fig. 1.15 Schematic diagrams of both sides of a Pratt & Whitney JT8D low-bypass turbofan engine, illustrating various systems and accessories associated with the operation of the engine and the airplane (McDonnell Douglas DC-9) [7]

other electronic systems on-board the flight vehicle must not be adversely affected by the local electromagnetic environment (EME). A command self-destruct capability is commonly required as a safety contingency, in the case of a post-launch vehicle flight trajectory that has gone awry; the self-destruct process usu-ally entails the rapid disabling of the propulsion system in some manner.

Fig. 1.16   Illustration of the proposed NASA/Lockheed Martin X-33 orbital spaceplane subscale technology demonstrator flight vehicle lifting off from the tarmac. The vehicle was to use two side-by-side Rocketdyne XRS-2200 aerospike-nozzled liquid-propellant rocket engines, as a single-stage-to-orbit craft. The project was canceled in 2001. Diagram courtesy of NASA

1.4.3 Thruster-Spacecraft System Integration

In this final example, we will look at the system integration issues surrounding the use of an attitude/position-control thruster on a spacecraft. See Fig. 1.17 for one well-known spacecraft example. Commonly, attitude control is shared between propulsive thrusters and non-propulsive inertial momentum devices (e.g. reaction wheels, magnetorquers, extendible gravity or aerodynamic-drag booms; Fig. 1.18; [9]), and this separation of duties needs to be established as part of the thruster integration to the space vehicle. A number of thrusters may be used for controlling the vehicle in terms of attitude and position, so their combined duties for successful maneuvering must also be clearly established [10]. The collection of thrusters, if chemical systems, may all be fed from a single propellant storage tank (if a monopropellant system; two tanks, for fuel and oxidizer storage, if a bipropellant system), so there must be allowance for proper valving, tubing, etc. to get the required amount of propellant to each thruster as needed, as managed by the spacecraft's flight management computer. Temperature levels within and around the various components will need to be monitored and controlled to ensure proper operation, bearing in mind the category of propellant being used (e.g. liquid at

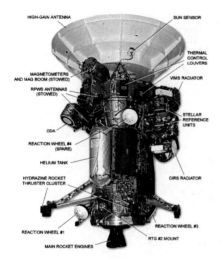
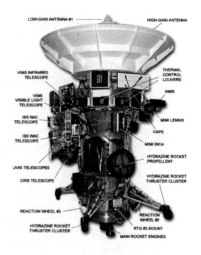

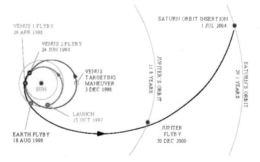

Fig. 1.17 Schematic diagram of the unmanned Cassini spacecraft for interplanetary exploration (launched in 1997, for investigation of Saturn and its moons; entered Saturn's orbit in July 2004). Note the use of reaction wheels in combination with hydrazine monopropellant thrusters for attitude control, positioned around the main body of the spacecraft. Electric power provided by radioisotope-decay thermoelectric generators (plutonium based), since solar arrays are insufficient at such a large distance from the Sun. The main bipropellant (nitrogen tetroxide/MMH) rocket engines are for larger corrections in the vehicle's flight path, e.g., in entering into an orbit around Saturn. For clarity of viewing, the thermal blankets normally in place have been removed in the above diagram. Courtesy of ASI/ESA/NASA

room temperature versus liquid only at very low temperatures). The propellant, if a chemical system, is likely pressure-fed, via a pressurized driver gas like helium or nitrogen, which will have its own storage tank and feed regulator system. Leaks from tanks, tubing, fittings and valves are a major concern, and need to be minimized, in order to enable a sufficient operational lifetime for the thruster in question. Electric thrusters, powered by electricity from solar cells, etc., will have their own special requirements, as will be discussed later in this book.

Fig. 1.18 Illustration of the use of thrusters and momentum devices for spacecraft attitude control [9]. Courtesy of NASA

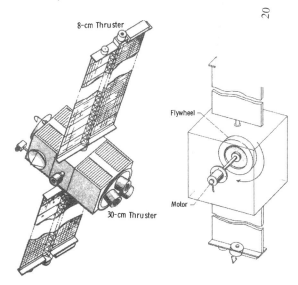

1.5 Review of Gasdynamics and Thermodynamics

Given that internal gasdynamic flow is an important component of the propulsion system analysis to be undertaken throughout this book, let us end this chapter with a review some of the more pertinent equations.

Recall the equation of state for a thermally perfect gas:

$$p = \frac{\rho \Re T}{M} = \rho R T = \frac{RT}{v} \tag{1.1}$$

where

p = absolute pressure (Pa or lb/ft^2)
T = absolute temperature (K or °R)
ρ = gas density (kg/m^3 or slug/ft^3)
v = specific volume (m^3/kg or ft^3/slug)
R = specific gas constant (J/kg-K or ft-lb/slug-°R)
$\mathcal{M}$ = gas molecular mass (a.m.u.)
$\Re$ = universal gas constant (8,314 J/kg-K or 49,709 ft-lb/slug-°R)
For air: M = 29 a.m.u. = 29 g per g-mole, 1 g-mole = 6.022 × 10^{23} molecules

We will see later how molecular mass $\mathcal{M}$ of the working gas will play a role in thrust delivered. For this course, the above equation of state for a given ideal gas will be a reasonable approximation in most cases for the range of temperatures and pressures of propulsion systems we will be examining. Having said that, one should be aware of real gas effects as temperatures become quite high, where molecular dissociation (breaking apart of larger molecules into smaller ones, or into individual atoms) and at higher temperatures, ionization (separation of electrons from molecules or atoms), can become significant in altering the behavior of the gas from the assumed ideal-gas framework. In the case of air, primarily made up of diatomic oxygen (O_2) and diatomic nitrogen (N_2), dissociation begins to become evident above around 1,500 K (thermodynamic properties such as C_p also begin becoming a function of pressure, as well as temperature), and both dissociation and ionization is quite significant upon reaching 5,000 K [11].

A gas will have a component of energy produced by molecular spin and vibration, defined as the specific internal or intrinsic energy u (J/kg). For our purposes, we can relate this energy component to temperature T (K) via

$$u = C_v T, \quad \text{J/kg} \tag{1.2}$$

where C_v is the constant-volume specific heat (J/kg-K) of the gas. We will also want to define a larger energy quantity known as enthalpy h, which is the sum of the specific internal energy and the pressure energy (per unit mass, due to molecular collisions) of the gas:

$$h = u + \frac{p}{\rho}, \quad \text{J/kg} \tag{1.3}$$

Substituting gives

$$h = C_v T + RT = (C_v + R)T = C_p T \tag{1.4}$$

where C_p is defined as the constant-pressure specific heat (J/kg-K). While C_v and C_p may vary somewhat with temperature, we will assume, where convenient, mean values (i.e., a calorically perfect gas).

We can define γ as the ratio of specific heats:

$$\gamma = \frac{C_p}{C_v} = \frac{C_p}{C_p - R} = \frac{1}{1 - \frac{R}{C_p}} \quad \text{or} \quad C_p = \frac{\gamma R}{\gamma - 1} \tag{1.5}$$

Example values for γ are 1.4 for diatomic gases like air or nitrogen (N_2), 1.3 for an intermediate mixture of complex molecules, and the highest possible value, 1.67, for monatomic gases like the inert gas, argon.

Recall for the speed of sound in a gas:

$$a = \sqrt{\gamma R T} = \sqrt{\frac{\gamma p}{\rho}} \tag{1.6}$$

Note the definition of Mach number Ma (current S.I. symbol; formerly, M):

$$\text{Ma} = \frac{V}{a} \tag{1.7}$$

where V is the local gas velocity (sometimes using u for gas velocity).

Recall the definition for Reynolds number:

Fig. 1.19 Schematic diagram of fluid flow through streamtube

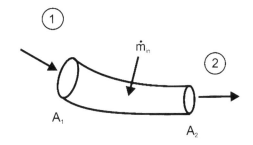

$$\text{Re}_\ell = \frac{\rho V \ell}{\mu} = \frac{V \ell}{\upsilon} \tag{1.8}$$

where ℓ is a characteristic length (like duct diameter), μ is absolute (dynamic) gas viscosity, and υ is the kinematic gas viscosity.

For an isentropic flow process ($\Delta s = 0$; [11]), as shown by the diagram of Fig. 1.19, moving from flow station ① to flow station ② in a duct:

$$s_2 - s_1 = C_p \cdot \ell n(\frac{T_2}{T_1}) - R \cdot \ell n(\frac{p_2}{p_1}) = 0,$$

$$\frac{T_2}{T_1} = \left(\frac{p_2}{p_1}\right)^{\frac{\gamma-1}{\gamma}} \tag{1.9}$$

and via the ideal equation of state:

$$\frac{\rho_2}{\rho_1} = \left(\frac{p_2}{p_1}\right)^{\frac{1}{\gamma}} = \frac{v_1}{v_2} \tag{1.10}$$

Conservation of mass (continuity) for a steady axial flow from station ① to ②, with mass inflow from the duct boundary, gives:

$$\rho_1 A_1 V_1 + \dot{m}_{in} = \rho_2 A_2 V_2 \tag{1.11}$$

where the cross-sectional area of the tube at a given location is given by A. As above, we will be restricting our analyses to steady-state cases (i.e. no traveling compression or rarefaction waves in the flow, with the exception of our pulsejet evaluation).

Conservation of momentum for a steady axial flow (as shown by Fig. 1.20a, with a duct holding force F_x being positive to the right, such that thrust F is positive to the left) gives:

$$F_x + (p_1 - p_\infty)A_1 + \dot{m}_1 V_1 = (p_2 - p_\infty)A_2 + \dot{m}_2 V_2 \tag{1.12}$$

Here, the net pressure force acting on the control volume is that above ambient surrounding pressure p_∞. Now, considering the rightward flow from the exit plane

Fig. 1.20 *Upper* diagram **a** showing schematic of simple streamtube having rightward-moving flow producing leftward thrust, and *lower* diagram **b** showing by analogy the leftward thrust production by a rocket engine (exhaust jet plume as control volume for thrust estimation)

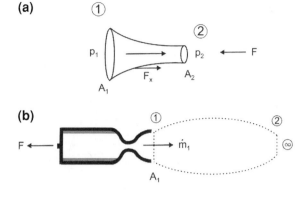

Fig. 1.21 Schematic diagram of fluid flow through streamtube (showing heat input, and shaft work output)

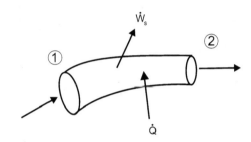

of a propulsive nozzle (Fig. 1.20b), such that p_1 and V_1 are at the exit plane of area A_1 and further downstream in the exhaust jet wake, p_2 becomes p_∞ and V_2 goes to zero, then gross thrust F (i.e., neglecting any mass intake at left of control volume; thrust in newtons (N) or pounds (lb$_f$)) becomes:

$$F = -F_x = (p_1 - p_\infty)A_1 + \dot{m}_1 V_1 \tag{1.13}$$

Later, we will be looking at various formats for the thrust equation, depending on the propulsion system being looked at.

Heat input in a conventional propulsion system is typically generated by a combustion process, where a fuel and an oxidizer are brought together, the resulting reaction producing reactant products and heat. Referring to Fig. 1.21, given that Q is the rate of heat input (J/s), and W_s is the rate of shear (mechanical, shaft) work output (J/s or W, watts), then conservation of energy in going from station ① to ② gives:

$$Q + \dot{m}_1 \left(h_1 + \frac{V_1^2}{2} \right) = W_s + \dot{m}_2 \left(h_2 + \frac{V_2^2}{2} \right) \tag{1.14}$$

Later, we will see how heat input via combustion correlates to work output performance (through mechanical rotation or thrust delivery).

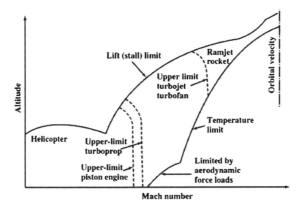

Fig. 1.22 Flight envelope for fixed- and rotary-winged flight vehicles using differing means for their propulsion [12]. Quantitatively, to give some scale to the above diagram, the upper flight Mach number for turboshaft powered helicopters in forward level flight is around 0.4, piston engine propellered airplanes in steady level flight is around 0.6, for turboprop airplanes around 0.7, for high-bypass turbofan powered airplanes around 0.9, low-bypass turbofan and turbojet-powered airplanes around 3.0, for ramjet-powered aircraft around 5.0, for scramjet-powered aircraft around 10.0, and for non-airbreathing chemical rocket-powered vehicles, up to and beyond orbital flight Mach numbers (20.0 and higher). Due to aerodynamic structural loading and aeroheating, one needs to go higher in altitude as one goes faster (conditions less severe at lower air density). Of course, on the other hand in going very high in altitude, one needs sufficient air intake if one is using air-breathing engines (combustion requirement), in addition to needing sufficient air density for aerodynamic lift at a given airspeed. Graph reprinted with permission of the American Institute of Aeronautics and Astronautics

Additional isentropic, adiabatic (no heat addition) steady flow relations will be of use in this course. For example, conservation of mass and energy provides the following useful relationship between duct cross-sectional area A and flow Mach number Ma:

$$\frac{A_1}{A_2} = \frac{\mathrm{Ma}_2}{\mathrm{Ma}_1}\left[\frac{2 + (\gamma - 1)\mathrm{Ma}_1^2}{2 + (\gamma - 1)\mathrm{Ma}_2^2}\right]^{\frac{\gamma+1}{2(\gamma-1)}} \qquad (1.15)$$

where the static temperature ratio is given by

$$\frac{T_2}{T_1} = \left[\frac{2 + (\gamma - 1)\mathrm{Ma}_1^2}{2 + (\gamma - 1)\mathrm{Ma}_2^2}\right] \qquad (1.16)$$

As the book progresses, we will see the usage of various forms of the above equations.

1.6 Closing Comments for this Chapter

From this point, with a little background now behind us, we will commence to take a look at various aerospace propulsion devices and systems. Referring to Fig. 1.22; [12], one has a variety of choices to select from. With the Wright brothers in mind, let's begin with atmospheric flight at the lower speed end of the spectrum, and the corresponding propulsion approaches that have been developed to bring power to flight above the Earth's surface.

1.7 Example Problems

1.1. Identify various technological contributions made by individuals prior to the Wright Brothers' 1903 flight of the *Flyer I*, that would be considered tangible advancements for the field of aerospace propulsion. *(Example: The demonstration of jet propulsion by exhausting a heated gas… the aeolipile, documented in 62 A.D., as shown in Fig. 1.2).*

1.2. Identify proposed ideas for improving aerospace propulsion, that did not quite turn out to be what one would have hoped (at least to date). *(Example: the use of a piston engine to deliver a compressed-gas jet exhaust, as a principal means for vehicle thrust delivery… not particularly efficient, in comparison to the conventional turbojet engine).*

1.3. *Project:* Identify a particular flight application requiring the integration of a propulsion system to a flight vehicle, and elaborate on the various factors that need to be addressed in order to successfully meet the requirements for that particular application to move ahead.

1.8 Solutions to Example Problems

1.1. Another example: inventor of the practical four-stroke gasoline-fuelled spark-ignition internal combustion engine, Nicolaus Otto, in 1876 (see Chap. 4).

1.2. Another example: piston rotary engines (see Chap. 4).

1.3. Make sure to consult the government certification requirements for your particular engine, and how it is to be safely mated to the given flight vehicle (regulations can usually be found online).

References

1. Graham I (1995) Aircraft. Andromeda Oxford, Abingdon
2. Gatland KW et al (1981) The illustrated encyclopedia of space technology—a comprehensive history of space exploration. Crown Publishers, New York

3. Anonymous (1990) Eight decades of progress—a heritage of aircraft turbine technology. General Electric Company, Cincinnati

4. Anderson JD Jr (1999) Aircraft performance and design. McGraw-Hill, New York

5. Kroo I, Shevell RS (2006) Aircraft design: synthesis and analysis. Desktop Aeronautics, Stanford

6. Connors T (1992) Thrust stand evaluation of engine performance improvement algorithms in an F-15 airplane. In: Proceedings of 28th AIAA/ASME/SAE/ASEE Joint Propulsion Conference, Nashville, July 6–8

7. Anonymous (1982) Alitalia AOM DC-9 training manual. Alitalia Airlines, Rome

8. Sutton GP, Biblarz O (2001) Rocket propulsion elements, 7th edn. Wiley, New York

9. Anonymous (1977) Ion propulsion for spacecraft. NASA Lewis research center, Cleveland

10. Turchi PJ (1995) Electric rocket propulsion systems. In: Humble RW, Henry GN, Larson WJ (eds) Space propulsion analysis and design. McGraw-Hill, New York

11. John JEA (1984) Gas dynamics, 2nd edn. Prentice-Hall, Upper Saddle River

12. Mattingly JD (2006) Elements of propulsion: gas turbines and rockets. AIAA, Reston

Part I
Airplane Propulsion

Chapter 2
Introduction to Atmospheric Flight

2.1 Introduction to Propulsion for Airplanes

In evaluating the design and performance of airplanes for various atmospheric flight applications, we see the need for generating thrust F in order to counter the atmospheric aerodynamic drag D that is acting to slow a flight vehicle at a given airspeed V, and for a fixed-wing aircraft, the need for generating aerodynamic lift L to counter the weight W of the vehicle, in order to maintain a given altitude h. We note the correlation on the performance parameter F/W (thrust-to-weight, ~ 0.2 to 0.4 for most conventional fixed-wing aircraft, where thrust F in this case is typically the maximum sea level static value, F_o, and weight is the nominal maximum takeoff weight at sea level of the airplane, W_o) in sizing airplanes. Thus, the bigger the airplane, the more thrust required from the propulsion system, and thus more fuel consumed. One will encounter takeoff requirements that further establish the influence of the above important performance ratio, in this case on meeting the required runway length. Further in regard to thrust, we note the use of various propulsion systems to deliver thrust for a conventional fixed-wing aircraft, namely the propeller [driven by a piston, rotary or turboprop (TP) engine], turbojet (TJ) and turbofan (TF). In the case of conventional rotary-winged aircraft (helicopters), thrust and lift are primarily delivered by one or more large main rotors [driven by a turboshaft (TS) engine in most higher power applications].

Brake or thrust specific fuel consumption, or specific impulse (essentially the inverse of thrust specific fuel consumption), are common measures of aircraft engine performance and flight economy. In comparing one propulsion system to another, one often sees the use of charts such as the example of Fig. 2.1. In that example, high-bypass-ratio turbofans produce the best (highest) specific impulse, but these systems only function economically up to the high subsonic flight region for the most part $(Ma_\infty < 0.9)$, up to about 13 km in altitude at best. On the other end of the spectrum, if one needs to fly quickly, as fast as Ma_∞ approaching 8 (at that speed, one would likely need to be cruising at a fairly high altitude, on the order of 31 km or higher) then a scramjet would be the only available option,

D. R. Greatrix, *Powered Flight*, DOI: 10.1007/978-1-4471-2485-6_2,
© Springer-Verlag London Limited 2012

Fig. 2 1 Performance comparison (in terms of specific impulse, I_{sp}) of various air-breathing engines that might be used for airplanes, at differing flight Mach numbers. In practice, altitude will also play a role. AFRL refers to the Air Force Research Laboratory (US), and HC PDE refers to pulse detonation engine employing a conventional hydrocarbon fuel. Graph courtesy of DARPA

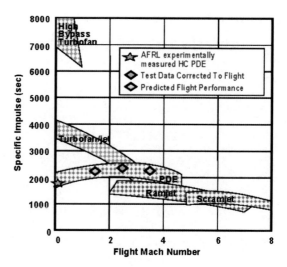

according to the example chart, unfortunately at the lowest peak specific impulse of the various air-breathing engines.

In order to provide some background to the larger picture that surrounds airplanes and the propulsion systems that are integrated to them to allow for a successful mission, an example set of mathematical equations describing airplane cruise performance reveals the close connection that exists between engine and airframe.

2.2 Example Mission Requirement: Range and Endurance of Fixed-Wing Airplanes

The actual overall range of a fixed-wing aircraft will include mission components such as takeoff, climb to altitude, cruise at altitude, descend, loiter, further descent and land. This accumulated distance, if including a substantial distance component for maneuvering and loitering, may be significantly greater than the nominal trip distance from points A to B. Cruising at altitude is a major component of a typical conventional flight mission profile, and performance assessment for this state is a good indicator of a given aircraft's range and endurance capability. We will focus on cruise performance for this example airplane mission requirement.

We note that turbojet and turbofan engine performance is largely ascertained as a function of thrust F, while propellered engine performance is determined as a function of power (shaft power delivered by engine, P_S; and occasionally, one might use thrust power, FV). Given this correlation, let us separate the aircraft's range and endurance performance assessment into two categories: jet aircraft and propellered aircraft.

Fig. 2.2 Free-body force
diagram for airplane in steady
level flight. The diagram
assumes that the airplane's
angle of attack α is relatively
small, for mathematical
convenience

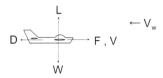

2.2.1 Jet Aircraft

One can define a pertinent performance parameter, thrust specific fuel consumption (TSFC), as follows:

$$\text{TSFC} = \text{kg/h of fuel/N of thrust delivered, kg/(h\,N)}$$
$$= \text{lb/h of fuel/lb of thrust delivered, lb/(h\,lb)}$$

The corresponding fuel consumption rate in terms of weight for a jet engine would thus be:

$$\dot{w}_f = \frac{\mathrm{d}w_f}{\mathrm{d}t} = -\frac{\mathrm{d}W}{\mathrm{d}t} = g \cdot \text{TSFC} \cdot F, \quad \text{N/h, where } F \text{ in newtons}$$
$$= \text{TSFC} \cdot F, \qquad\qquad \text{lb/h, where } F \text{ in pounds}$$

(2.1)

The aircraft's overall weight, W, would drop as fuel is consumed, so that for a small time increment, for the S.I. unit case:

$$\mathrm{d}t = -\frac{\mathrm{d}W}{g \cdot \text{TSFC} \cdot F}$$

(2.2)

noting that the weight increment above is negative in value, in order for the time increment to be positive. One can integrate, to get an estimate of the time elapsed over a cruise segment, or in other words, the endurance of the aircraft for a given amount of fuel burned (as reflected below by the difference between the aircraft's weight at the start of the given cruise segment, W_{ini}, and at the end of the flight segment, W_{fin}):

$$t_e = \int \mathrm{d}t = \frac{1}{g} \int_{W_{\text{fin}}}^{W_{\text{ini}}} \frac{1}{\text{TSFC} \cdot F} \cdot \mathrm{d}W$$

(2.3)

For the corresponding range of the cruise segment,

$$R = \int V \mathrm{d}t = \frac{1}{g} \int_{W_{\text{fin}}}^{W_{\text{ini}}} \frac{V}{\text{TSFC} \cdot F} \cdot \mathrm{d}W$$

(2.4)

One recalls that for steady level flight (SLF), as per Fig. 2.2, one can correlate to the airplane's overall lift and drag coefficients, via:

$$F = D = W\frac{C_D}{C_L} \tag{2.5}$$

where $L = C_L\frac{1}{2}\rho V^2 S$ and $D = C_D\frac{1}{2}\rho V^2 S$, S being the aircraft's reference wing area, ρ the local air density and V the aircraft's true airspeed.

Substituting into the integral,

$$R = \frac{1}{g}\int_{W_{fin}}^{W_{ini}}\frac{V\cdot\frac{C_L}{C_D}}{\text{TSFC}\cdot W}\cdot dW \tag{2.6}$$

If one assumes constant or mean values for V, C_L/C_D, and TSFC for the given cruise flight segment, integration produces the well-known Breguet range equation [1], in this case for jet aircraft and S.I. units:

$$R = \frac{1}{g}\frac{V}{\text{TSFC}}\frac{C_L}{C_D}\cdot\ln\left(\frac{W_{ini}}{W_{fin}}\right), \quad \text{km, where g = 9.81 m/s}^2,$$
$$\text{(2.7)}$$
$$V \text{ in km/h, TSFC in kg/(h N)}$$

In British (Imperial) units, the Breguet range equation for jets may appear as follows:

$$R = \frac{V}{\text{TSFC}}\frac{C_L}{C_D}\cdot\ln\left(\frac{W_{ini}}{W_{fin}}\right), \quad \text{nm (nautical miles)},$$
$$\text{(2.8)}$$
$$\text{where} \quad 1\text{ nm } = 1.1508 \text{ statute miles } = 6076 \text{ ft,}$$
$$V \text{ in kt (i.e., nm/h), and TSFC in lb/h}\cdot\text{lb}$$

The best estimate of mean weight for a given cruise segment (possibly needed for calculating a mean C_L/C_D or V) may be found from inspection of the Breguet equation:

$$\overline{\ln(W)} = \frac{\ln(W_{ini}) + \ln(W_{fin})}{2} = \ln\sqrt{W_{ini}} + \ln\sqrt{W_{fin}} = \ln\sqrt{W_{ini}\cdot W_{fin}}$$

which in essence says:

$$\overline{W} = \sqrt{W_{ini}\cdot W_{fin}} \tag{2.9}$$

One can in turn establish that the corresponding endurance for a jet aircraft, for the S.I. unit case described for range above, may be found via:

$$t_e = \frac{R}{V} = \frac{1}{g}\frac{C_L/C_D}{\text{TSFC}}\cdot\ln\left(\frac{W_{ini}}{W_{fin}}\right), \text{ h} \tag{2.10}$$

Here, V would be the mean or constant airspeed for the cruise flight segment, in km/h. Inspection of the above endurance equation for jets suggests that one should maximize C_L/C_D for maximum endurance (time airborne). This requirement conforms with flying for minimum drag:

$$C_{L,\text{me}} = C_{L,\text{md}} = \sqrt{\frac{C_{Do}}{K}}, \quad C_{D,\text{me}} = C_{D,\text{md}} = 2C_{Do}, \quad V_{\text{md}} = \sqrt{\frac{2\overline{W}}{\rho S C_{L,\text{md}}}} \qquad (2.11)$$

where one has the standard subsonic flight (subtransonic) drag polar expression, summing zero-lift drag and lift-induced drag components, as follows:

$$C_D = C_{Do} + KC_L^2 \qquad (2.12)$$

Inspection of the range equation above suggests that for jets, one would want to maximize $V \cdot C_L/C_D$, or correspondingly, maximize the following:

$$V \cdot \frac{C_L}{C_D} = \sqrt{\frac{2\overline{W}}{\rho S C_L}} \cdot \frac{C_L}{C_D} = \sqrt{\frac{2\overline{W}}{S}} \cdot \frac{1}{\rho^{1/2}} \cdot \frac{C_L^{1/2}}{C_D} \qquad (2.13)$$

While one cannot do much as a pilot about the aircraft weight W or reference wing area S, according to the above, a lower air density ρ (higher altitude h) would be helpful to increase range, in addition to maximizing $C_L^{1/2}/C_D$. Let us now consider the latter term, but for convenience, look at minimizing the inverse of that term:

$$\frac{C_D}{C_L^{1/2}} = \frac{C_{Do} + KC_L^2}{C_L^{1/2}} = C_{Do}C_L^{-1/2} + KC_L^{3/2} \qquad (2.14)$$

Differentiate to find the trough point:

$$\frac{d}{dC_L}[C_{Do}C_L^{-1/2} + KC_L^{3/2}] = 0 = -\frac{1}{2}C_{Do}C_L^{-3/2} + \frac{3}{2}KC_L^{1/2} \qquad (2.15)$$

Rearranging, one arrives at the following for maximum jet air range:

$$C_{L,\text{mr}} = \sqrt{\frac{C_{Do}}{3K}}, \quad C_{D,\text{mr}} = C_{Do} + \frac{KC_{Do}}{3K} = \frac{4}{3}C_{Do} \qquad (2.16)$$

The corresponding airspeed for best air range by a jet aircraft would in turn be:

$$V_{\text{mr}} = \sqrt{\frac{2\overline{W}}{\rho S C_{L,\text{mr}}}} = \frac{1}{\rho^{1/2}} \cdot \sqrt{\frac{2\overline{W}}{S}} \cdot \left(\frac{3K}{C_{Do}}\right)^{1/4} = 1.32 V_{\text{md}} \qquad (2.17)$$

The above estimate holds until the airplane begins to appreciably enter *drag divergence*, where C_{Do} begins to rise with further increases in flight Mach number Ma_∞. Of course, if such is the case, one is violating the original premise that produced the above result (i.e., the assumption of a constant C_{Do}), and thus are in a transonic domain where the above guideline is invalid. One can also note at this juncture that TSFC for turbofan and turbojet engines also tends to worsen as the airplane moves above a certain threshold altitude. This could be in the neighborhood of 40,000–50,000 ft (12–15 km) for an airplane employing a conventional commercial jet engine.

A tailwind (V_w negative in value) is beneficial for gaining extra distance, while a headwind is detrimental, as per the correlation between ground range R_g as a function of air or still-air range R:

$$R_g = R - V_w t_e \tag{2.18}$$

where groundspeed V_g correlates to airspeed V via:

$$V_g = V - V_w \tag{2.19}$$

The endurance t_e will remain the same value, regardless of the magnitude of the wind. From pilot observation, it has been ascertained that one should fly at a bit faster airspeed than the still-air criterion when in a headwind to attain the maximum possible ground range, and the reverse (fly a bit slower, airspeed wise) when a tailwind is present. One can demonstrate this mathematically. For maximum ground range, observation of the Breguet range equation suggests that one needs to maximize the following:

$$(V - V_w)\frac{C_L}{C_D} = \sqrt{\frac{2W}{\rho S}} \cdot \frac{C_L^{1/2}}{C_D} - V_w \cdot \frac{C_L}{C_D} \tag{2.20}$$

Differentiation of the above allows one to find a peak point:

$$\frac{d}{dC_L}\left[\sqrt{\frac{2W}{\rho S}} \cdot \frac{C_L^{1/2}}{C_D} - V_w \cdot \frac{C_L}{C_D}\right] = \frac{d}{dC_L}\left[\frac{1}{C_{Do} + KC_L^2}\left\{\sqrt{\frac{2W}{\rho S}} \cdot C_L^{1/2} - V_w C_L\right\}\right] = 0 \tag{2.21}$$

which ultimately reduces to the following for best C_L to fly at in a wind:

$$0 = C_{Do} - 3KC_L^2 - 2\sqrt{\frac{\rho S}{2W}} \cdot V_w C_{Do} C_L^{1/2} + 2\sqrt{\frac{\rho S}{2W}} \cdot V_w KC_L^{5/2} \tag{2.22}$$

2.2.2 Propellered Aircraft

In the case of propellered airplanes, fuel consumption by the engine driving the propeller is proportional to power, rather than thrust. As a result, one defines brake specific fuel consumption (BSFC) as a suitable performance parameter, as follows:

$$\text{BSFC} = \text{kg/h of fuel/W of power delivered, } \text{kg}/(\text{h W})$$
$$= \text{lb/h of fuel/bhp of power delivered, } \text{lb}/(\text{h hp})$$

where bhp refers to brake horsepower (1 hp = 550 ft lb/s) as measured from the drive shaft of the engine. The corresponding fuel consumption rate in terms of weight for a shaft power (P_S) producing engine would thus be:

$$\dot{w}_f = \frac{dw_f}{dt} = -\frac{dW}{dt} = g \cdot \text{BSFC} \cdot P_S, \qquad \text{N/h, where } P_S \text{ in watts} \tag{2.23}$$
$$= \text{BSFC} \cdot P_S, \qquad \text{lb/h, where } P_S \text{ in horsepower}$$

The aircraft's overall weight, W, would drop as fuel is consumed, so that for a small time increment, for the S.I. unit case:

$$dt = -\frac{dW}{g \cdot \text{BSFC} \cdot P_S} \tag{2.24}$$

One can integrate to arrive at the propellered aircraft's endurance for a given flight segment:

$$t_e = \int dt = \frac{1}{g} \int_{W_{\text{fin}}}^{W_{\text{ini}}} \frac{1}{\text{BSFC} \cdot P_S} \cdot dW \tag{2.25}$$

Similarly, for air range, one has:

$$R = \int V dt = \frac{1}{g} \int_{W_{\text{fin}}}^{W_{\text{ini}}} \frac{V}{\text{BSFC} \cdot P_S} \cdot dW \tag{2.26}$$

One notes the following correlations for shaft power, for the S.I. unit case, for an aircraft in level flight:

$$P_S = \frac{FV}{3.6\eta_{\text{pr}}} = \frac{DV}{3.6\eta_{\text{pr}}} = \frac{WC_D}{C_L} \cdot \frac{V}{3.6\eta_{\text{pr}}}, \qquad \text{watts, where } V \text{ in km/h} \tag{2.27}$$

The symbol η_{pr} is the propeller propulsive efficiency, and the 3.6 factor arises from there being 3,600 s in one hour. Substituting,

$$R = \frac{1}{g} \int_{W_{\text{fin}}}^{W_{\text{ini}}} \frac{V}{\text{BSFC} \cdot \dfrac{WC_D}{C_L} \cdot \dfrac{V}{3.6\eta_{\text{pr}}}} \cdot dW = \frac{1}{g} \int_{W_{\text{fin}}}^{W_{\text{ini}}} \frac{3.6\eta_{\text{pr}} C_L/C_D}{\text{BSFC} \cdot W} \cdot dW \tag{2.28}$$

Assuming constant or mean values for η_{pr}, BSFC and C_L/C_D over a given cruise flight segment, integration produces the Breguet range equation for propellered aircraft, with S.I. units:

$$R = \frac{1}{g} \cdot \frac{3.6 \eta_{\mathrm{pr}} C_L / C_D}{\mathrm{BSFC}} \cdot \ln\left(\frac{W_{\mathrm{ini}}}{W_{\mathrm{fin}}}\right), \quad \mathrm{km}, \ \mathrm{where} \ g = 9.81 \ \mathrm{m/s^2}, \quad (2.29)$$
$$\mathrm{BSFC \ in \ kg/h \cdot W}$$

In British (Imperial) units, the Breguet range equation for propellered aircraft may appear as follows:

$$R = \frac{326 \eta_{\mathrm{pr}} C_L / C_D}{\mathrm{BSFC}} \cdot \ln\left(\frac{W_{\mathrm{ini}}}{W_{\mathrm{fin}}}\right), \quad \mathrm{nm}, \ \mathrm{where \ BSFC \ in \ lb/h \cdot hp} \quad (2.30)$$

The 326 factor arises from the combination of conversion factors for horsepower, distance and time as they appear in the above equation, to produce distance in nautical miles (nm). Note that the above is for air range (we will account for wind later).

Similarly, for endurance of a propellered aircraft over a given flight segment, S.I. units as for the corresponding range equation above,

$$t_e = \frac{R}{V} = \frac{1}{g} \cdot \frac{3.6 \eta_{\mathrm{pr}} C_L / C_D}{V \cdot \mathrm{BSFC}} \cdot \ln\left(\frac{W_{\mathrm{ini}}}{W_{\mathrm{fin}}}\right), \quad \mathrm{h} \quad (2.31)$$

noting that V is the constant or mean airspeed in km/h over the cruise flight segment.

By inspection of the above, maximum endurance for a propellered airplane would require the maximization of:

$$\frac{1}{V} \frac{C_L}{C_D} = \sqrt{\frac{\rho S}{2W}} \cdot \frac{C_L^{3/2}}{C_D} \quad (2.32)$$

One can observe that the above requirement corresponds to flying at minimum power:

$$C_{L,\mathrm{me}} = C_{L,\mathrm{mp}} = \sqrt{\frac{3 C_{Do}}{K}} \quad (2.33\mathrm{a})$$

$$= C_{L\max}, \quad \mathrm{if} \ \sqrt{\frac{3 C_{Do}}{K}} \geq C_{L\max} \quad (2.33\mathrm{b})$$

Note the requirement that the aircraft cannot fly at a C_L value above the aerodynamic stall limit, $C_{L\max}$.

For maximum air range, inspection of the Breguet range equation suggests a propellered aircraft should fly at a maximum value for C_L / C_D:

$$C_{L,\mathrm{mr}} = C_{L,\mathrm{md}} = \sqrt{\frac{C_{Do}}{K}}, \quad V_{\mathrm{mr}} = \sqrt{\frac{2\overline{W}}{\rho S C_{L,\mathrm{mr}}}} = \frac{1}{\rho^{1/2}} \cdot \sqrt{\frac{2\overline{W}}{S}} \cdot \left(\frac{K}{C_{Do}}\right)^{1/4}, \quad (2.34)$$

The above equations suggest that there is some benefit in going to a higher altitude (lower ρ), to increase cruise airspeed and thus shorten the trip time in going from point A to point B. The practical economic cruise altitude for a conventional turboprop aircraft is in the neighborhood of 15,000–25,000 ft (4.5–7.6 km), a bit lower for an airplane with a turbo-supercharged piston engine, and substantially lower still for an aircraft employing a normally-aspirated (i.e., non-supercharged) piston engine to drive the propeller (10,000 ft or less). Shaft power for turboprop engines does decrease with increasing altitude above a certain altitude, and propeller performance degrades as air density gets too low; these are factors that come into play with respect to practical economic cruise airspeeds and altitudes.

As was the case for jet aircraft, propellered aircraft endurance is independent of the wind levels present. Ground range, as before, is ascertained via

$$R_g = R - V_w t_e \tag{2.35}$$

In addition, it can be shown that airspeed must be increased above the still-air criterion above, in the presence of a headwind, for maximum ground range attainment. Conversely, in a tailwind, the airspeed must be decreased below the still-air case. For maximum ground range, observation of the Breguet range equation suggests that one needs to maximize the following:

$$\frac{C_L}{C_D} - \frac{V_w}{V} \cdot \frac{C_L}{C_D} = \frac{C_L}{C_{Do} + KC_L^2} - \frac{V_w}{\sqrt{\frac{2W}{\rho S}}} \cdot \frac{C_L^{3/2}}{(C_{Do} + KC_L^2)} \tag{2.36}$$

Differentiating the above with respect to C_L, and setting the resulting equation to zero, one ultimately arrives at the following:

$$0 = C_{Do} - KC_L^2 + \frac{V_w}{\sqrt{\frac{2W}{\rho S}}} \left[\frac{1}{2} KC_L^{5/2} - \frac{3}{2} C_{Do} C_L^{1/2} \right] \tag{2.37}$$

It is hoped that the above example of airplane cruise performance analysis, allied to engine performance (here, largely as influenced by BSFC or TSFC), gives the reader a sense of importance that the propulsion system plays in making a particular aircraft viable. Please refer to the solution for Example Problem 2.1 at the end of this chapter, for a completed sample analysis of the cruise performance of a particular propellered aircraft. For students learning this material as part of a course, I would encourage them to attempt the example problem first, and then check the solution to see if any mistakes were made.

2.3 Example Mission Requirement: Takeoff of Fixed-Wing Airplanes

In order to provide further background to the larger picture that surrounds airplanes and the propulsion systems that are integrated to them to allow for a successful mission, an example set of mathematical equations describing airplane

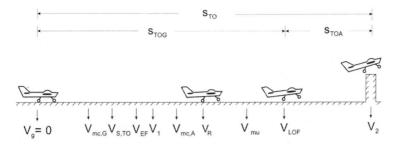

Fig. 2.3 Takeoff profile for fixed-wing aircraft

takeoff performance reveals further the close connection that exists between engine and airframe.

2.3.1 Introduction to Takeoff Performance

Government airworthiness regulations, like the Canadian Aviation Regulations (CARs) by Transport Canada, the Federal Aviation Regulations (FARs) by the US Federal Aviation Administration (see www.faa.gov for online reading of FARs), and the Joint Aviation Regulations (JARs) by the European Aviation Safety Agency (EASA), stipulate the minimum standards for safe aircraft operation, including those for takeoff and landing. Here, reference will be made to FAR Part 23 for small (light) airplanes, and FAR Part 25 for heavier (>12,500 lb, 5,670 kg) transport aircraft (typically multi-engined; see FAR 25.105–25.115 for takeoff info).

Referring to Fig. 2.3 for a fixed-wing aircraft's takeoff profile, one can define a number of pertinent parameters describing the aircraft's takeoff performance (some parameters applicable only to multi-engined FAR 25 airplanes):

s_{TO} overall takeoff distance, from a standing start (groundspeed $V_g = 0$, airspeed $V > 0$ if a headwind present) to clearing a specified obstacle (screen) height (FAR 25, $h_{sc} = 35$ ft (10.67 m); FAR 23, 50 ft (15.25 m)) at airspeed V_2

s_{TOG} takeoff ground roll distance

s_{TOA} takeoff airborne distance; also called climb segment #1 distance

$V_{mc, G}$ minimum control airspeed on ground; should a critical engine fail for a multi-engined airplane, aircraft will not yaw or bank appreciably given adequate application of rudder and aileron by the pilot

$V_{S,TO}$ stall airspeed, aircraft in takeoff configuration (partial flaps typical)

V_{EF} critical engine failure airspeed, $>V_{mc, G}$ (otherwise, brake to full stop on runway)

V_1 critical engine failure recognition airspeed, $>V_{mc, G}$ and $<V_{LOF}$, sufficient field length for pilot to continue takeoff with critical engine inoperative

(multi-engined aircraft); if engine fails at $V < V_1$, brake to full stop on runway; $V_1 < V_{1,\ MBE}$, i.e., do not exceed *maximum brake energy* limit (above which, catastrophic failure of overheated brakes or tires may be anticipated)

$V_{mc,\ A}$ minimum control airspeed in air, $<1.2V_{S,TO}$; should a critical engine fail for a multi-engined aircraft, the pilot will have sufficient rudder, aileron and elevator so as not to yaw or bank appreciably

V_R takeoff rotation airspeed, $>1.05V_{mc,\ A}$; pilot begins rotation of aircraft in pitch to desired angle-of-attack (α) and corresponding C_L in preparation for liftoff; at pilot's discretion, the nose landing gear may begin to be retracted beyond this speed (to lower drag)

V_{mu} minimum unstick airspeed; with critical engine inoperative, aircraft could still liftoff under adequate control

V_{LOF} liftoff airspeed, $>1.1V_{mu}$ with all engines up, or $>1.05V_{mu}$ with critical engine inoperative; $\approx 1.2V_{S,\ TO}$

V_2 takeoff climb airspeed, $>1.2V_{S,\ TO}$ and $>1.1V_{mc,\ A}$; must be sufficiently high to satisfy climb segment #2 (35 ft (10.67 m) to 400 ft (122 m) AGL, FAR 25) minimum climb gradient requirement with critical engine inoperative; main landing gear can begin to be retracted (to lower drag)

A headwind in takeoff reduces the aircraft's groundspeed while maintaining the specified airspeed $(V = V_g + V_w)$, thus producing a shorter takeoff distance. Conversely, a tailwind will result in a longer takeoff distance. Note, FAR 25 specifies the use of only 50% of the predicted headwind in preparations/calculations undertaken by the pilot prior to takeoff, and 150% of the predicted tailwind magnitude, to be conservative and allow for some safety margin.

Below a height h_{AGL} equivalent to the aircraft's wing span b, the airplane will be, to some degree, in ground effect. The ground imposes a reflecting boundary condition that inhibits downward flow of air (downwash from wings, primarily), associated with the production of lift. With respect to takeoff and landing, the primary effects of interest are an increase in the lift-curve slope, $C_{L\alpha}$, and lift coefficient, C_L and a decrease in the lift-induced drag coefficient, C_{Di}. Recall for the aerodynamics of airplanes, the following:

$$\text{wing aspect ratio, } AR = \frac{b^2}{S} \tag{2.38}$$

$$\text{lift-induced drag factor, } K = \frac{1}{\pi \cdot AR \cdot e} \tag{2.39}$$

and the drag polar expression for overall drag coefficient,

$$C_D = C_{Do} + C_{Di} = C_{Do} + \phi K C_L^2 \tag{2.40}$$

Note that S in this case is the reference wing area, e is the Oswald efficiency factor, C_{Do} is the base (zero-lift) drag coefficient and ϕ is the ground effect factor on lift-induced drag $(0 < \phi < 1)$. Once the airplane is out of ground effect as it gains altitude, ϕ will go to unity. One expression for ϕ, proposed by Lan and Roskam [2], compares reasonably well with experimental data (h is the mean height of the wing over the ground, for this formula):

$$\phi = 1 - \frac{\left(1 - 1.32\frac{h}{b}\right)}{\left(1.05 + 7.4\frac{h}{b}\right)}, \quad 0.033 < h/b < 0.25 \tag{2.41}$$

Unless indicated otherwise, the airplane would be assumed to be out of ground effect for calculation purposes, once the screen height is reached or exceeded (again, to be conservative). At a height ratio of 0.1, the value for ϕ is about 0.52 according to the above, a result that is similar to experimental data.

One understands that the thrust generated from propellers, turbojets (at lower altitudes) or turbofans typically drops substantially with increasing airspeed from a standing start. We will have to account for this in any takeoff and early climb-out analysis. Do not make the mistake of using maximum static thrust throughout the ground roll and initial climb segment (this is not realistic). Also, for propellered airplanes, do not assume peak propeller propulsive efficiency η_{pr} applies at low speeds. In practice, more propeller thrust is available at lower speeds (a good thing for shortening takeoff distance), but at the price of a lower η_{pr}. A rough estimate of η_{pr} would be closer to 0.3–0.5 for the takeoff ground run's mean value, rather than the peak value seen at cruise conditions (0.8–0.85).

With a critical engine failure, note that there will be a drag increment due to propeller drag (at least until the propeller can be feathered) or turbine windmilling, and due to the asymmetric flight condition (rudder + aileron/spoiler deflection required to correct). Some yawing of the aircraft, as well as some accompanying roll, can occur with a thrust imbalance on a multi-engined airplane, and an additional roll component can occur due to engine torque imbalance.

2.3.2 Takeoff Analysis: Ground Roll

Consider the free-body force diagram of an airplane undertaking a ground roll, as the first part of a takeoff run, as shown in Fig. 2.4. Newton's second law of motion ($F = m \cdot a$) stipulates the following with respect to the airplane moving in the longitudinal s-direction:

$$m \cdot a_{\text{long}} = m\frac{dV_g}{dt} = \sum F_s = F - D - \mu(W - L) - W \cdot \sin\theta_R \tag{2.42}$$

Since the angle of incline of the runway θ_R is typically small (usually well within $\pm 5°$), one can apply small-angle assumptions in that regard (e.g., $\cos\theta_R \approx 1$). The coefficient of rolling friction μ with the brakes off is usually a small value, e.g., about 0.02 for dry asphalt. One can establish distance travelled s by the airplane via the following integration of groundspeed V_g with time:

Fig. 2.4 Forces acting on fixed-wing aircraft in ground roll

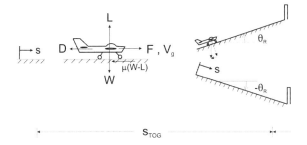

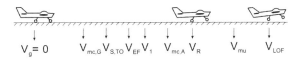

$$s = \int_0^t V_g \, dt, \quad \text{m} \tag{2.43}$$

Given aircraft mass $m = W/g$, and noting that longitudinal acceleration of the aircraft in the ground roll is given by:

$$a_{\text{long}} = \frac{dV_g}{dt} = \frac{g}{W}[F - D - \mu(W - L) - W \sin \theta_R] \tag{2.44}$$

then integrating the above provides

$$V_g = \int_0^t \frac{g}{W}[F - D - \mu(W - L) - W \sin \theta_R] dt \tag{2.45}$$

One would use airspeed V (recalling $V = V_g + V_w$) for estimating thrust F, lift L and drag D as follows:

F, determined from empirical equations or charts, as a function of V or flight
  Mach number Ma
$L = C_{L,\text{TOG}} \frac{1}{2} \rho V^2 S$, where the lift coefficient is for the ground roll only
$D = C_{D,\text{TOG}} \frac{1}{2} \rho V^2 S = \frac{1}{2} \rho V^2 S (C_{Do} + \phi K C_{L,\text{TOG}}^2)$, where one finds the ground
effect factor ϕ value at the wing's mean height above ground.

One can numerically integrate the appropriate equations above to arrive at an estimate of the airplane's ground roll distance s_{TOG}. For preliminary calculations, one could assume no aircraft rotation (in pitch) until the aircraft actually lifts off ($V = V_{\text{LOF}}$), rather than the actual case ($V = V_R$). Otherwise, one could allow for a transition phase from V_R to V_{LOF}, where the aircraft's lift coefficient is also

Fig. 2.5 Forces acting on
fixed-wing aircraft in climb

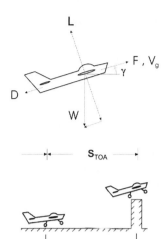

changing from $C_{L,\text{TOG}}$ to $C_{L,\text{TOA}}$, in effect when the pilot is pulling back on the control column, to apply elevator in bringing the aircraft's nose up in preparation for liftoff and becoming airborne. The default conservative assumption is that the airplane is assumed to be out of ground effect, upon liftoff ($\phi = 1$).

The effective ground roll lift coefficient $C_{L,\text{TOG}}$ is set by the pitch attitude of the aircraft as it rolls along the ground, and by the flap setting (typically, a partial deflection of the flaps, held constant during the roll and early climb out). In order to maximize the longitudinal acceleration of the airplane in the ground roll (and thus shorten the ground roll distance), one would like to minimize the force sum of the drag and rolling resistance linked to $C_{L,\text{TOG}}$. Applying differential calculus (differentiation with respect to $C_{L,\text{TOG}}$) for estimating a trough point,

$$\frac{d}{dC_{L,\text{TOG}}}(C_{Do} + \phi K C_{L,\text{TOG}}^2 - \mu C_{L,\text{TOG}}) = 0 \tag{2.46}$$

$$2\phi K C_{L,\text{TOG}} - \mu = 0 \tag{2.47}$$

so that one finally arrives at an expression for the nominal best choice:

$$C_{L,\text{TOG}}\Big|_{\text{max.accel.}} = \frac{\mu}{2\phi K} \tag{2.48}$$

The flap deflection, and the resulting actual $C_{L,\text{TOG}}$, may have to be a bit higher as a compromise, in order to get sufficient lift as the pilot commences applying elevator at rotation speed V_R. The influence of zero-lift drag coefficient C_{Do} should not be ignored in this process, with respect to its value during the ground roll (landing gear fully extended; investigate means for reducing drag where possible) and its value as the nose and main gear are retracted as the airplane becomes airborne.

2.3.3 Takeoff Analysis: Climb Segment #1

At and beyond V_{LOF}, as the aircraft becomes airborne and gains height through Climb Segment #1 (see Fig. 2.5), the main landing gear can be retracted at the discretion of the pilot (thus lowering C_{Do}). Along the flight path at a given climb angle γ, Newton's law of motion stipulates:

$$F - D - W \sin \gamma = \frac{W}{g} \frac{dV_g}{dt} \tag{2.49}$$

such that

$$\frac{dV_g}{dt} = \frac{g}{W}[F - D - W \sin \gamma] = a_{\text{long}} \tag{2.50}$$

In this airborne transition, there is a curvilinear path being followed, such that the aircraft's pitch rotation rate $d\gamma/dt$ may be assumed to be governed by the pertinent forces normal to the flight path contributing to centripetal acceleration of the airplane:

$$L - W \cos \gamma = \frac{W}{g} V_g \frac{d\gamma}{dt} \tag{2.51}$$

such that

$$\frac{d\gamma}{dt} = \frac{g}{WV_g}[L - W \cos \gamma], \quad \text{rad/s} \tag{2.52}$$

Note that the airplane's angle-of-attack α is assumed to be small in these calculations, and that pitch rotation acceleration $d^2\gamma/dt^2$ due to pitching moment contributions is implicit in this simplified representation. We will also need the vertical distance travelled (height gained), via

$$\frac{dh}{dt} = V \sin \gamma, \quad \text{m/s} \tag{2.53}$$

so via integration with time,

$$h = \int_{t_{LOF}}^{t} V \sin \gamma \cdot dt, \quad \text{m} \tag{2.54}$$

One can also numerically integrate for groundspeed V_g:

$$V_g = V_{g, \text{LOF}} + \int_{t_{LOF}}^{t} \frac{g}{W}[F - D - W \sin \gamma]dt \tag{2.55}$$

and flight path angle γ:

$$\gamma = \gamma_o + \int_{t_{LOF}}^{t} \frac{g}{WV_g}[L - W\cos\gamma]dt, \quad \text{rad} \tag{2.56}$$

The initial flight path angle at liftoff, γ_o, may have a positive or negative value, depending on the incline angle θ_R of the runway. Here, V_g is assumed to be following the airplane's flight path, so ground distance covered is established via

$$s = s_{TOG} + \int_{t_{LOF}}^{t} V_g \cos\gamma \cdot dt \tag{2.57}$$

Accounting for the aircraft's pitch attitude relative to the Earth's surface below, airspeed V for thrust, lift and drag calculations is found from:

$$V = V_g + V_w \cos\gamma \tag{2.58}$$

where one is assuming the wind is horizontal, and V_w is the head or tailwind component acting on the airplane (i.e., not accounting for a crosswind component here). Unless indicated otherwise, simplifying assumptions for preliminary calculations of airborne ground distance covered can include assuming $V_R \approx V_{LOF}$ (i.e., if V_R is known, but not V_{LOF}), ground effect factor $\phi \approx 1$ and that the liftoff lift coefficient $C_{L,TOA}$ remains unchanged through the first climb segment (to reach the screen height h_{sc}). In practice, one may have a decrease in $C_{L,TOA}$ while one accelerates from V_{LOF} to V_2, and in conjunction to reach and hold a desired initial climb angle.

For Climb Segment #1, one would continue the above integrations with time from liftoff ($t = t_{LOF}, h_{AGL} = 0$) up to reaching the screen height, $h_{AGL} = h_{sc}$. Upon reaching the screen height, one arrives at an airplane airspeed V_2 that one needs to confirm is acceptable under government guidelines (e.g., FAR 25, $V_2 > 1.2V_{S,TO}$), but also acceptable for further climbing out from the runway (segments #2 [h_{sc} to an altitude of 400 ft (122 m)] and #3 [to an altitude of 1,500 ft (457 m)]). If it turns out V_2 is too low in value, one would possibly need to increase V_{LOF} to get the value of V_2 up as well (alternatively, a lower initial climb angle might help in this respect, but will potentially extend the airborne transition ground distance). Note that one would have an upper limit value on V_{LOF}, with regards to the tire friction/heating threshold for the landing gear.

In completing the integrations up to the end of Climb Segment #1, one thus arrives at the overall takeoff ground distance, namely

$$s_{TO} = s_{TOG} + s_{TOA} \tag{2.59}$$

If the value for s_{TO} is higher than desired, one will need to revisit a number of parameters that contribute to that value, including $V_{S,TO}$, a critical variable that a number of parameters depend upon. One would not necessarily assume immediately that more powerful engines (higher thrust) are the only means of resolution to address such a difficulty.

2.3.4 Approximation Model for Ground Roll

Recall from the earlier discussion on takeoff ground roll:

$$\frac{dV_g}{dt} = \frac{g}{W}[F - D - \mu(W - L) - W\sin\theta_R] \qquad (2.60)$$

and

$$s = \int_0^t V_g\,dt \qquad (2.61)$$

We can also consider the following relationships between groundspeed and longitudinal ground distance covered:

$$V_g = \frac{ds}{dt}, \quad V_g dV_g = \frac{ds}{dt}dV_g, \quad ds = \frac{V_g}{\dfrac{dV_g}{dt}}dV_g \qquad (2.62)$$

Integrating for distance, one can make some substitutions:

$$s = \int ds = \int \frac{V_g}{\dfrac{dV_g}{dt}} \cdot dV_g = \int \frac{V_g \cdot dV_g}{\dfrac{g}{W}[F - D - \mu(W - L) - W\sin\theta_R]} \qquad (2.63)$$

The above expression for distance covered during the ground roll portion of the takeoff is not, in general, readily solved analytically (i.e., more likely to be solved via application of a numerical method, e.g., the finite difference approach). However, for preliminary design or quick estimation, one can introduce a few convenient assumptions to in turn produce a quick and relatively reasonable approximation for s_{TOG}.

A common approach for ground roll distance estimation is the so-called mean kinetic energy approximation. Equating work completed with kinetic energy developed during the ground roll, one can show that the following is approximately true for the ground distance covered:

$$s_{TOG} \approx \frac{V_{g,LOF}^2}{2\left(\dfrac{dV}{dt}\right)} = \frac{V_{g,LOF}^2/2}{\dfrac{g}{W}[\overline{F} - \overline{D} - \mu(W - \overline{L}) - W\sin\theta_R]} \qquad (2.64)$$

where

$$V_{g,LOF} = V_{LOF} - V_w \qquad (2.65)$$

noting that a headwind is treated as positive. As noted by the overbar on several parameters above, one needs to estimate thrust, drag and lift at a mean kinetic energy airspeed. Allowing for a headwind or tailwind, the airspeed at the mean kinetic energy point in the ground roll is:

$$\overline{V} = \sqrt{\frac{V_{\text{LOF}}^2 + V_w|V_w|}{2}} \tag{2.66}$$

In the case of a normal takeoff, the time elapsed for the ground roll can be estimated via

$$\Delta t_{\text{TOG}} = \frac{V_{g,\text{LOF}}}{\dfrac{dV}{dt}} = \frac{V_{g,\text{LOF}}}{\dfrac{g}{W}\left[\overline{F} - \overline{D} - \mu(W - \overline{L}) - W\sin\theta_R\right]} \tag{2.67}$$

With one engine inoperative at some point in the ground roll,

$$\Delta t_{\text{TOG}} = \frac{V_{g,\text{EF}}}{\left(\dfrac{dV}{dt}\right)_1} + \frac{V_{g,\text{LOF}} - V_{g,\text{EF}}}{\left(\dfrac{dV}{dt}\right)_2} = \frac{V_{g,\text{EF}}}{\dfrac{g}{W}\left[\overline{F} - \overline{D} - \mu(W - \overline{L}) - W\sin\theta_R\right]_1}$$

$$+ \frac{V_{g,\text{LOF}} - V_{g,\text{EF}}}{\dfrac{g}{W}\left[\overline{F} - \overline{D} - \mu(W - \overline{L}) - W\sin\theta_R\right]_2} \tag{2.68}$$

The airspeed-based parameters at state 1 (period prior to engine failure) would be determined at:

$$\overline{V}_{1,\text{OEI}} = \sqrt{\frac{V_{\text{EF}}^2 + V_w|V_w|}{2}} \tag{2.69}$$

while at the state 2 condition (period after engine failure), they would be determined via:

$$\overline{V}_{2,\text{OEI}} = \sqrt{\frac{V_{\text{EF}}^2 + V_{\text{LOF}}^2}{2}} \tag{2.70}$$

The corresponding ground distance travelled over the 2-part ground roll would be:

$$s_{\text{TOG}} = \frac{V_{g,\text{EF}}^2}{2\left(\dfrac{dV}{dt}\right)_1} + \frac{V_{g,\text{LOF}}^2 - V_{g,\text{EF}}^2}{2\left(\dfrac{dV}{dt}\right)_2} = \frac{V_{g,\text{EF}}^2/2}{\dfrac{g}{W}\left[\overline{F} - \overline{D} - \mu(W - \overline{L}) - W\sin\theta_R\right]_1}$$

$$+ \frac{(V_{g,\text{LOF}}^2 - V_{g,\text{EF}}^2)/2}{\dfrac{g}{W}\left[\overline{F} - \overline{D} - \mu(W - \overline{L}) - W\sin\theta_R\right]_2} \tag{2.71}$$

2.3.5 Approximation Model for Climb Segment #1

Let us now look at how we can estimate the distance covered over the latter part of the takeoff, when the aircraft is now off the ground. One can separate the airborne transition component of the takeoff run (Climb Segment #1) into two parts, for mathematical convenience:

I. V increasing from V_{LOF} to V_2, with wheels off the ground but little height gained
II. Altitude above ground level, $h_{AGL,}$ increasing, but with little velocity gained

In summary then, for Part I, assume acceleration from V_{LOF} to V_2 occurs at h_{AGL} near zero. For Part II, assume climb at constant airspeed of approximately V_2, with a constant flight path angle γ, from zero height above ground to screen height h_{sc} . The conservative default assumption is that the airplane is out of ground effect for both parts, such that $\phi = 1$, unless indicated otherwise.

Part I:

The first part can be described by:

$$\frac{dV}{dt} \approx \frac{g}{W}(\overline{F} - \overline{D}) \tag{2.72}$$

For this flight phase, mean thrust and drag are estimated at:

$$\overline{V} = \sqrt{\frac{V_{LOF}^2 + V_2^2}{2}} \tag{2.73}$$

Distance covered for Part I is found via:

$$s_{TOA,I} = \frac{V_2^2 - V_{LOF}^2}{2\left(\dfrac{dV}{dt}\right)} = \frac{V_2^2 - V_{LOF}^2}{2\dfrac{g}{W}(\overline{F} - \overline{D})} \tag{2.74}$$

Note that this is still-air distance, or air distance, for Part I (wind not considered on ground distance covered; we'll look at ground distance a bit later).

Time elapsed for Part I is:

$$\Delta t_{TOA,I} = \frac{V_2 - V_{LOF}}{\left(\dfrac{dV}{dt}\right)} = \frac{V_2 - V_{LOF}}{\dfrac{g}{W}(\overline{F} - \overline{D})} \tag{2.75}$$

Part II:

For the second part, the airplane is assumed to be climbing at a constant angle γ In this case, assuming the climb angle is not overly steep, one can estimate its value via the following:

$$\sin\gamma = \frac{F - D}{W} \approx \frac{F - D_{SLF}}{W} \approx \frac{F}{W} - \frac{C_D}{C_L}\Big|_{SLF} \tag{2.76}$$

The acronym SLF refers to steady level flight ($\gamma = 0°$), noting that

$$C_{L, \text{SLF}} = \frac{2W}{\rho S V^2} \tag{2.77}$$

and

$$C_D = C_{Do} + K C_L^2 \tag{2.78}$$

For Part II, thrust and drag are ascertained at V_2. The climb rate is found from:

$$\frac{dh}{dt} = V_2 \sin \gamma \tag{2.79}$$

Considering the triangle formed by the parameters associated with a climb, one can observe that

$$\Delta h = \Delta s \cdot \tan \gamma \tag{2.80}$$

which indicates that

$$s_{\text{TOA, II}} = \frac{h_{sc}}{\tan \gamma} \tag{2.81}$$

As for Part I, note that this is air distance for Part II (wind not considered). A bit later, we will account for a headwind or tailwind on distance covered.

Also from the climb's trigonometry,

$$\Delta t \cdot V_2 \cos \gamma = \Delta s \tag{2.82}$$

so that

$$\Delta t_{\text{TOA, II}} = \frac{s_{\text{TOA, II}}}{V_2 \cos \gamma} \tag{2.83}$$

Ground Distance with Wind Present:

Given the fact that the starting and ending criteria for Parts I and II are based on airspeeds (V_{LOF}, V_2), the times $\Delta t_{\text{TOA,I}}$ and $\Delta t_{\text{TOA,II}}$ noted above will remain the same regardless if there is a headwind or tailwind present. However, we know that air distance and ground distance tend to differ when V_w is finite, and that is certainly true here, as noted below (we will use the subscript g explicitly here for ground distance covered, to distinguish from the air distance estimate):

$$s_{g,\text{TOA,I}} = s_{\text{TOA,I}} - V_w \Delta t_{\text{TOA, I}} \tag{2.84}$$

and

$$s_{g,\text{TOA, II}} = s_{\text{TOA,II}} - V_w \Delta t_{\text{TOA, II}} \tag{2.85}$$

As a result, the overall ground distance covered during the airborne component of the takeoff is in the general case (allowing for wind):

$$s_{TOA} = s_{g,TOA, I} + s_{g,TOA, II} \qquad (2.86)$$

2.3.6 Approximated Overall Takeoff Distance

In summarizing then, one arrives at an estimate of overall takeoff distance from a standing start at the beginning of the runway to the screen height h_{sc}:

$$s_{TO} = s_{TOG} + s_{TOA} \qquad (2.87)$$

For some problems, one would continue estimating the distance and height covered as the airplane moves into Climb Segment #2 (h_{sc} to 400 ft AGL and then Climb Segment #3 (400–1,500 ft AGL) in climbing out and leaving the airport's control domain.

It is hoped that the above example of airplane takeoff performance analysis, allied to engine performance (here, largely as influenced by maximum takeoff thrust, that decreases with forward airspeed in the case of air-breathing engines), gives the reader a sense of importance that the propulsion system plays in making a particular aircraft viable. Please refer to the solution for Example Problem 2.5(b) at the end of this chapter, for a completed sample analysis of the takeoff performance of a particular jet aircraft.

In the next chapter, we will begin our study of propulsion systems with the propeller, a principal means of thrust delivery for lower speed airplanes. Later, we will discuss the various powerplants that can be used to drive the propeller's rotation.

2.4 Example Problems

2.1. A de Havilland DHC-8 (Dash 8) has a reference wing area S of 54.4 m^2, a zero-lift drag coefficient C_{Do} of 0.0255, and a value for the lift-induced drag factor K of 0.032. The regional transport airplane is powered by two Pratt and Whitney PW120A turboprops with propellers operating at 85% efficiency ($\eta_{pr} = 0.85$) at maximum cruise airspeed of 490 km/h, at an altitude of 4,600 m ASL(above sea level). Cruise BSFC is 2.2×10^{-4} kg/(h W). The airplane's maximum takeoff weight is 153.477 kN, maximum payload weight is 37.376 kN, maximum fuel weight is 25.261 kN and operating empty weight is 100.533 kN. For the current flight mission, the payload weight is *lower* than the maximum allowed, at 30.411 kN. Assume 1.374 kN of fuel is consumed in takeoff and climb to cruise, and 1.374 kN of fuel is consumed in descent from cruise and landing.

(a) Determine cruise range for the above flight at max. cruise speed and altitude under still-air conditions, for the maximum possible fuel that can be loaded into the airplane (assume no fuel reserve in your calculations).
(b) Repeat (a), but for a 56 km/h headwind.
(c) Determine the maximum-range (minimum-fuel) airspeed at the mean aircraft weight for the cruise segment, for the still-air case.
(d) Repeat (c), but for a 56 km/h headwind.

2.2. A Boeing 747-100 has a reference wing area of 511 m^2, a zero-lift C_{Do} of 0.02 and a lift-induced K of 0.065. The heavy airliner is powered by four Pratt and Whitney JT9D-7A turbofans with a cruise TSFC of 0.0694 kg/(h N) at a maximum cruise true Mach number of 0.8 at 9,150 m ASL. The airplane's maximum takeoff weight is 3,286 kN, maximum payload weight of 667.1 kN, maximum fuel weight of 1442.1 kN and operating empty weight of 1,687 kN (maximum zero-fuel weight of 2354.4 kN). For the current flight mission, the payload weight is *lower* than the maximum allowed at 402.2 kN. Assume 73.6 kN of fuel is consumed in takeoff and climb to cruise, and 49.1 kN of fuel in descent and landing.

(a) Determine cruise range for the above flight at max. cruise Mach number and altitude under still-air conditions, for the maximum possible fuel that can be loaded into the airplane (assume no fuel reserve in your calculations).
(b) Repeat (a), but for a 56 km/h tailwind.
(c) Determine the mean maximum-range airspeed and corresponding flight Mach number, for the still air case. Note that drag divergence is not being accounted for, when you see the result.
(d) Repeat (c), but for a 56 km/h tailwind.

2.3. A de Havilland DHC-5D (Buffalo) has a reference wing area S of 945 ft^2, a zero-lift drag coefficient C_{Do} of 0.024 and a value for the lift-induced drag factor K of 0.04. The transport airplane is powered by two General Electric CT64-820-4 turboprops with propellers operating at 85% efficiency ($\eta_{pr} = 0.85$) at cruise at an altitude of 10,000 ft ASL. Cruise BSFC is 0.5 lb/h·hp. The airplane's maximum takeoff weight is 43,000 lb, maximum payload weight is 12,000 lb, maximum fuel weight is 14,000 lb and operating empty weight is 25,000 lb. At the present time, the airplane is facing a 30-kt headwind in cruise at 10,000 ft, with the current aircraft weight being 41,000 lb. What would be the maximum-ground-range airspeed and flight Mach number at this juncture? How does this value compare to the still-air case?

2.4. Note the characteristics of the following aircraft:

Aircraft: Single-engined light recreational aircraft; wing span 32 ft; wing ref. area 180 sq ft; Oswald efficiency 0.75$C_{L\,max}$ 1.7 (partial flaps); aircraft weight 2,800 lb; $V_{LOF} = 1.15V_{S,L}$ C_{Do} 0.04, takeoff configuration; clearing 50-ft height, $V_2 = 1.3V_{S,L}$

Engines   One 4-cylinder normally aspirated air-cooled piston engine
Variable-pitch constant-speed propeller, $n = 2,700$ rpm, $d_p = 6$ ft
Forward thrust, at sea level, where $J = V/(nd_p)$,static $F_{o,S/L} = 500$ lbf, $F_{S/L}/F_{o,S/L} = 1$–$0.3\,J$, lbf;$F(h)/F_{S/L} = \sigma$(thrust vs. altitude, $\sigma = \rho/\rho_{S/L}$)

Airfield   ISA conditions at specified altitudes; mean wing height $h_{w,g} = 8$ ft, $\mu = 0.05$, free roll on short-grass runway; upslope of $0°$, $C_{L,\text{TOG}} = 0.9$; runway obstacle height 50 ft

Evaluate the following cases for takeoff distance, using an approximate method:

(a) Sea level, no wind
(b) Sea level, 10-kt headwind
(c) 3,000 ft altitude, no wind
(d) 3,000 ft altitude, 10-kt headwind.

2.5. Note the characteristics of the following aircraft:

Aircraft   Boeing 747-100 heavy wide-body commercial airliner; wing span 196 ft; wing ref. area 5,600 sq. ft.; Oswald efficiency $0.7C_{\text{Lmax}}$ 1.8 (partial flaps); aircraft weight 735,000 lb; $V_{\text{LOF}} = 1.18V_{S,L}C_{D_o}$ 0.036, takeoff configuration; clearing 35-ft screen height, $V_2 = 1.2V_{S,L}$

Engines   Four Pratt and Whitney JT9D-7A turbofan engines; for a single engine, engine, max. forward thrust for regular takeoff given by $F_{\text{eng, }S/L} = 46100 - 46.7\,V + 0.0467\,V^2$, lbf (thrust at S/L, V in ft/s), and $F(h)/F_{S/L} = \sigma^{0.8}$ (variation of thrust as function of altitude);

Airfield   ISA conditions at specified altitudes; mean wing height $h_{w,g} = 20$ ft; runway downslope of $0°$; free roll on asphalt, μ 0.02; $C_{L,\text{TOG}} = 1.0$ runway obstacle height 35 ft.

Evaluate the following cases for takeoff distance, using an approximate method:

(a) Sea level, no wind
(b) Sea level, 30-kt tailwind
(c) 5,000 ft altitude, no wind.

2.5 Solution for Problems

2.1. (a) Cruise range, no wind:

Air density at 4,600 m ISA alt. is $\rho = 0.77$ kg/m³. Sound speed at 4,600 m alt. is $a = 322$ m/s.

Given K = 0.032.

Breguet range equation for propellered aircraft:

$$R_{\text{cr}} = \frac{3.6}{g}\eta_{\text{pr}}\frac{C_L/C_D}{\text{BSFC}}\ln\left(\frac{W_{\text{ini}}}{W_{\text{fin}}}\right);$$ note: weights, if given in kg, should be converted to newtons.

$W_{\text{ini}} = W_o - W_{F,t/o+\text{climb}} = 153477 - 1374 = 152103\ \text{N} = 152.1\ \text{kN}$, weight at cruise's start

$W_{Fo} = W_o - W_E - W_{P/L} = 153477 - 100553 - 30411 = 22513$ N $= 22.5$ kN, initial fuel weight loaded into aircraft.

$W_{fin} = W_{ini} - (W_{Fo} - W_{F,t/o+climb} - W_{F,des+ldg} - W_{Fres})$, where W_{Fres} (reserve fuel) $= 0$ (not specified)

$W_{fin} = 152103 - (22513 - 1374 - 1374 - 0) = 132338$ N $= 132.3$ kN

$V_{cr} = 490$ km/h $= 136.5$ m/s (specified, so unless told otherwise, do not change this cruise speed)

$\overline{W} = \sqrt{W_{ini}W_{fin}} = (152100 \cdot 132300)^{1/2} = 141850$ N $= 141.9$ kN, mean cruise weight

$$\overline{C_L} = \frac{\overline{W}}{\frac{1}{2}\rho V_{cr}^2 S} = 141900/(0.5 \cdot 0.77 \cdot 136.5^2 \cdot 54.4) = 0.364$$

$\overline{C_D} = C_{Do} + K\,\overline{C_L^2} = 0.0255 + 0.032(0.364^2) = 0.03$

$R_{cr} = 3.6/9.81 \cdot 0.85 \cdot (0.364/0.03)/2.2 \times 10^{-4}$ $\ln(152.1/132.3) = 2400$ km, still-air cruise range

(b) Cruise ground range, 56 km/h headwind:

$R_{G,cr} = R_{cr} - V_w \cdot t_{cr}$, where $t_{cr} = R_{cr}/V_{cr} = 2400/490 = 4.9$ h

$R_{G,cr} = 2400 - 56(4.9) = 2126$ km, so shorter cruise range with a headwind.

(c) Mean maximum-range velocity at 4,600 m altitude:

Still air, propellered aircraft, $C_{L,mr} = \sqrt{\dfrac{C_{Do}}{K}} = (0.0255/0.032)^{1/2} = 0.89$,

$$V_{mr} = \left(\frac{\overline{W}}{\frac{1}{2}\rho C_{L,mr}S}\right)^{1/2} = (141900/(0.5 \cdot 0.77 \cdot 0.89 \cdot 54.4))^{1/2} = 87.3 \text{ m/s in still air.}$$

(d) With a 56 km/h (15.6 m/s) headwind, one knows to fly a bit faster than the still-air V_{mr}. Referring to the textbook equation:

$$0 = C_{Do} - KC_{L,mr}^2 + \frac{V_w}{\sqrt{\dfrac{2\,\overline{W}}{\rho S}}}\left[\frac{1}{2}KC_{L,mr}^{5/2} - \frac{3}{2}C_{Do}C_{L,mr}^{1/2}\right]$$

$$0 = 0.0255 - 0.032C_{L,mr}^2 + 0.189\left[0.016C_{L,mr}^{2.5} - 0.03825C_{L,mr}^{0.5}\right]$$

Solving iteratively, one arrives at $C_{L,mr} = 0.806$, which says:

$$V_{mr} = \left(\frac{\overline{W}}{\frac{1}{2}\rho C_{L,mr}S}\right)^{1/2} = (141900/(0.5 \cdot 0.77 \cdot 0.806 \cdot 54.4))^{1/2}$$
$$= 91.7 \text{ m/s in } 56 \text{ km/h headwind.}$$

2.2. (a) Cruise range, no wind:

Air density at 9,150 m ISA alt. is $\rho = 0.459$ kg/m^3. Sound speed at 9,150 m alt. is $a = 304$ m/s.

AR $= 59.8^2/511 = 7$, K $= 1/(\pi \cdot 7 \cdot 0.7) = 0.065$

Breguet range equation for jet aircraft:

$$R_{cr} = \frac{1}{g} \frac{V_{cr}}{\text{TSFC}} \frac{C_L}{C_D} \ln\left(\frac{W_{ini}}{W_{fin}}\right);$$ weights given in kg should be converted to newtons.

$W_{ini} = W_o - W_{F,t/o+climb} = 3.286 \times 10^6 - 73575 = 3.212 \times 10^6$ N $= 3212$ kN, weight at cruise's start

$W_E = W_{ZF} - W_{P/L,max} = 2.354 \times 10^6 - 667080 = 1.687 \times 10^6$ N

$W_{Fo} = W_o - W_E - W_{P/L} = 3.286 \times 10^6 - 1.687 \times 10^6 - 402210 = 1.197 \times 10^6$ N $= 1197$ kN, initial fuel weight loaded into aircraft.

$W_{fin} = W_E + W_{P/L} + W_{F,des+ldg} + W_{Fres}$, where W_{Fres} (reserve fuel) $= 0$ (not specified)

$W_{fin} = 1.687 \times 10^6 + 402210 + 49050 + 0 = 2.138 \times 10^6$ N $= 2138$ kN

$V_{cr} = a(Ma_{cr}) = 304 (0.8) = 243$ m/s $= 875$ km/h (specified, so unless told otherwise, do not change this cruise speed)

$\overline{W} = \sqrt{W_{ini} W_{fin}} = (3212 \cdot 2138)^{1/2} = 2621$ kN, mean cruise weight

$$\overline{C_L} = \frac{\overline{W}}{\frac{1}{2}\rho V_{cr}^2 S} = 2621000/(0.5 \cdot 0.459 \cdot 243^2 \cdot 511) = 0.379$$

$$\overline{C_D} = C_{Do} + K \overline{C_L^2} = 0.02 + 0.065(0.379^2) = 0.0293$$

$R_{cr} = 1/9.81 \cdot 875/0.0694 \cdot (0.379/0.0293)$ $\ln(3212/2138) = 6767$ km,   still-air cruise range.

(b) Cruise ground range, 56 km/h tailwind:

$R_{G,cr} = R_{cr} - V_w \cdot t_{cr}$, where $t_{cr} = R_{cr}/V_{cr} = 6767/875 = 7.73$ h

$R_{G, cr} = 6767 + 56(7.73) = 7200$ km, so greater cruise range with a tailwind.

(c) Mean maximum-range velocity at 9,150 m altitude:

Still air, jet aircraft, $C_{L,mr} = \sqrt{\dfrac{C_{Do}}{3K}} = (0.02/(3 \cdot 0.065))^{1/2} = 0.32$,

$$V_{mr} = \left(\frac{\overline{W}}{\frac{1}{2}\rho C_{L,mr} S}\right)^{1/2} = (2621000/(0.5 \cdot 0.459 \cdot 0.32 \cdot 511))^{1/2} = 264 \text{m/s in still air.}$$

$Ma_{mr} = V_{mr}/a = 264/304 = 0.868$, which for this older airplane, is likely bringing it significantly into drag divergence. Likely would lower the airplane's cruise Mach number closer to 0.8.

(d) With a 56 km/h tailwind ($V_w = -15.6$ m/s), one knows to fly a bit slower than the still-air V_{mr}. Referring to the textbook:

$$0 = C_{Do} - 3KC_{L,\,mr}^2 + 2\sqrt{\frac{\rho S}{2\,\overline{W}}} \cdot V_w \left[KC_{L,\,mr}^{5/2} - C_{Do}C_{L,\,mr}^{1/2} \right]$$

$0 = 0.02 - 3(0.065)C_{L,\,mr}^2 + 2(0.459 \cdot 511/(2 \cdot 2621000))^{0.5}(-15.6)[0.065C_{L,\,mr}^{2.5} - 0.02C_{L,\,mr}^{0.5}]$

$0 = 0.02 - 0.195C_{L,\,mr}^2 - 0.2087[0.065C_{L,\,mr}^{2.5} - 0.02C_{L,\,mr}^{0.5}]$

Solving iteratively, one arrives at $C_{L,mr} = 0.333$, which says:

$$V_{mr} = \left(\frac{\overline{W}}{\frac{1}{2}\rho C_{L,\,mr}S} \right)^{1/2} = (2621000/(0.5 \cdot 0.459 \cdot 0.333 \cdot 511))^{1/2} = 259 \text{ m/s, in}$$

56 km/h tailwind, giving a cruise Mach number of $0.852 < 0.868$.

2.3. Max. ground-range airspeed, 30-kt headwind:

Air density at 10,000 ft ISA alt. is $\rho = 0.001756$ slug/ft^3. Sound speed at 10,000 ft alt. is $a = 1077$ ft/s.

AR $= 96^2/945 = 9.75$, K $= 1/(\pi \cdot 9.75 \cdot 0.82) = 0.04$

Still air, prop aircraft, $C_{L,\,mr} = \sqrt{\frac{C_{Do}}{K}} = (0.024/(0.04))^{1/2} = 0.775$,

$$V_{mr} = \left(\frac{\overline{W}}{\frac{1}{2}\rho C_{L,\,mr}S} \right)^{1/2} = (41000/(0.5 \cdot 0.001756 \cdot 0.775 \cdot 945))^{1/2} = 253 \text{ ft/s in still}$$

air (150 kt).

$Ma_{mr} = V_{mr}/a = 253/1077 = 0.235$, a fairly low subsonic value.

With a 30-kt headwind ($V_w = 50.7$ ft/s), one knows to fly a bit faster than the still-air V_{mr}. Referring to the lecture notes:

$$0 = C_{Do} - KC_{L,\,mr}^2 + \frac{V_w}{\sqrt{\frac{2\,\overline{W}}{\rho S}}} \left[\frac{1}{2}KC_{L,\,mr}^{5/2} - \frac{3}{2}C_{Do}C_{L,\,mr}^{1/2} \right]$$

$0 = 0.024 - 0.04C_{L,\,mr}^2 + 0.22808[0.02C_{L,\,mr}^{2.5} - 0.036C_{L,\,mr}^{0.5}]$

Solving iteratively, one arrives at $C_{L,mr} = 0.69$, which says:

$$V_{mr} = \left(\frac{\overline{W}}{\frac{1}{2}\rho C_{L,\,mr}S} \right)^{1/2} = (41000/(0.5 \cdot 0.001756 \cdot 0.69 \cdot 945))^{1/2} = 268 \text{ ft/s, in 30-}$$

kt headwind (159 kt), giving a cruise Mach number of 0.267.

2.4. (a) Find the takeoff distance using an approximate method, in this case for sea level altitude ($\rho = 0.002378$ slugs/ft^3) and no wind ($V_w = 0$ ft/s).

General info:

AR $= b^2/S = 32^2/180 = 5.7$, K $= 1/(\pi AR \cdot e) = 1/(\pi \cdot 5.7 \cdot 0.75) = 0.0745$, $n = 2700/60 = 45$ revs./s

$V_{S,t/o} = (2W_o/(\rho SC_{Lmax}))^{1/2} = (2 \cdot 2800/(0.002378 \cdot 180 \cdot 1.7))^{1/2} = 87.7$ ft/s or 52 kt

$V_{LOF} = 1.15 \, V_{S,t/o} = 101$ ft/s, $V_2 = 1.30 \, V_{S,t/o} = 114$ ft/s, $\theta_R = 0°$ (no runway slope)

Ground roll:

Using textbook method, begin with $\overline{V} = V_{LOF}/\sqrt{2} = 71.4$ ft/s, $\overline{J} = \dfrac{\overline{V}}{nd_p} = 71.4/$

$(45 \cdot 6) = 0.264$

$\overline{T} = T_o(1 - 0.3\,\overline{J}) = 500(1 - 0.3(0.264)) = 460$ lbf

$\phi = 1 - \dfrac{(1 - 1.32h/b)}{(1.05 + 7.4h/b)} = 1 - (1 - 1.32(8/32))/(1.05 + 7.4(8/32)) = 0.77$

$C_{D,GE} = C_{Do} + \phi K C_{Lg}^2 = 0.04 + 0.77(0.0745)0.9^2 = 0.087$

$\overline{L} = \frac{1}{2}\rho\overline{V}^2 S C_{Lg} = 982$ lbf, $\overline{D} = \frac{1}{2}\rho\overline{V}^2 S C_{D,GE} = 94$ lbf, $V_{G,LOF} = V_{LOF} - V_w = 101$ ft/s

$$s_{TOG} = \dfrac{V_{G,LOF}^2/2}{\dfrac{g}{W}[\overline{T} - \overline{D} - \mu(W - \overline{L}) - W\sin\theta_R]}$$

$$= 101^2/2/(32.2/2800)/[460 - 94 - 0.05(2800 - 982) - 0]$$

$$= 1615 \text{ ft}$$

Early airborne transition (I):

$\overline{V} = \sqrt{\dfrac{V_{LOF}^2 + V_2^2}{2}} = [(101^2 + 114^2)/2]^{1/2} = 108$ ft/s,

$\overline{J} = \dfrac{108}{45 \cdot 6} = 0.4$, $\overline{T} = 500[1 - 0.3(0.4)] = 440$ lbf

$\overline{C_L} = \dfrac{W}{\frac{1}{2}\rho\overline{V}^2 S} = 2800/(0.5 \cdot 0.002378 \cdot 108^2 \cdot 180) = 1.12$

$\overline{C_D} = 0.04 + 1.0(0.0745)(1.12^2) = 0.133$, noting $\phi = 1.0$ is assumed once airplane is airborne

$\overline{D} = \dfrac{1}{2}\rho\,\overline{V}^2 S\,\overline{C_D} = 332$ lbf

$s_{TOA,I} = \dfrac{(V_2^2 - V_{LOF}^2)/2}{\frac{g}{W}[\overline{T} - \overline{D}]} = (114^2 - 101^2)/2/(32.2/2800)/[440 - 332] = 1125$ ft

Late airborne transition (II):

$V = V_2 = 114$ ft/s, $J = 114/(45 \times 6) = 0.42$, $T = 500(1 - 0.3(0.42)) = 437$ lbf,

$C_L = 2800/(0.5 \times 0.002378 \times 114^2 \times 180) = 1.0$,

$C_D = 0.04 + 0.0745(1.0^2) = 0.115$

$\gamma = \dfrac{T}{W} - \dfrac{C_D}{C_L} = 437/2800 - 0.115/1.0 = 0.041$ radians or $2.4°\,s_{TOA,}$

$II = \dfrac{h}{\tan\gamma} = 50/\tan(2.4°) = 1195$ ft

Overall: $s_{TO} = s_{TOG} + s_{TOA,I} + s_{TOA,II} = 1615 + 1125 + 1195 = 3935$ ft

Note: for a light aircraft, this is fairly long (around 2,000 ft would be more typical for an airplane of this size), suggesting the airplane is a bit underpowered

(b) Sea level, headwind V_w of 10 kt (17 ft/s): $V_{G,LOF} = V_{LOF} - V_W = 101 - 17 = 84$ ft/s

Ground roll, using textbook method:

$$\bar{V} = \sqrt{\frac{V_w^2 + V_{LOF}^2}{2}} = 72.4 \text{ ft/s}, \quad \bar{J} = 72.4/(45 \cdot 6) = 0.268,$$

$\bar{T} = 500(1 - 0.3(0.268)) = 460$ lbf

$\bar{L} = 0.5(0.002378)72.4^2(180)0.9 = 1010$ lbf,

$\bar{D} = 0.5(0.002378)72.4^2(180)0.087 = 98$ lbf

$s_{TOG} = 84^2/2/(32.2/2800)/[460 - 98 - 0.05(2800 - 1010) - 0] = 1126$ ft ... use this estimate for later.

Early airborne (I):

$s_{TOA,I} = s_{TOA,I}(0) - V_w \Delta t_{TOA,I}$, where $s_{TOA,I}(0)$, no wind case, is found in (a) above. Given the same $\bar{V}$ as (a), mean thrust and drag are same as (a):

$$\Delta t_{TOA,I} = \frac{V_2 - V_{LOF}}{\frac{g}{W}[\bar{T} - \bar{D}]} = (114 - 101)/(32.2/2800)/[440 - 332] = 10.5 \text{ s}$$

$s_{TOA,I} = 1125 - 17(10.5) = 947$ ft

Late airborne (II):

Same γ as (a), since calculated at same V_2.

$s_{TOA,II} = s_{TOA,II}(0) - V_w \Delta t_{TOA,II}$,

$$\Delta t_{TOA,II} = \frac{s_{TOA,II}(0)}{V_2 \cos \gamma} = 1195/(114 \cdot \cos 2.4°) = 10.5 \text{ s}$$

$s_{TOA,II} = 1195 - 17(10.5) = 1017$ ft

Overall:

$s_{TO} = 1126 + 947 + 1017 = 3090$ ft

(c) 3,000 ft alt., headwind V_w of 0 kt:

$\sigma = 0.915$ given $\rho = 0.00218$ slugs/ft^3 at 3000 ft ISA.

$V_{S,t/o} = (2W_o/(\rho S C_{L \text{ max}}))^{1/2} = (2 \cdot 2800/(0.00218 \cdot 180 \cdot 1.7))^{1/2} = 91.7$ ft/s or 54 kt

$V_{LOF} = 1.15 V_{S,t/o} = 105.5$ ft/s, $V_2 = 1.30 V_{S,t/o} = 119.2$ ft/s

$$T = \sigma T_{S/L}$$

Ground roll, using textbook method:

$$\bar{V} = \sqrt{\frac{V_w^2 + V_{LOF}^2}{2}} = 74.6 \text{ ft/s}, \quad \bar{J} = 74.6/(45 \cdot 6) = 0.276,$$

$\bar{T} = 458(1 - 0.3(0.276)) = 420$ lbf

$\overline{L} = 0.5(0.00218)74.6^2(180)0.9 = 983$ lbf,
$\overline{D} = 0.5(0.00218)74.6^2(180)0.087 = 95$ lbf
$s_{TOG} = 105.5^2/2/(32.2/2800)/[420-95-0.05(2800-983)-0] = 2067$ ft

Early airborne (I):

$$\overline{V} = \sqrt{\frac{V_{LOF}^2 + V_2^2}{2}} = [(105.5^2 + 119.2^2)/2]^{1/2} = 112.6 \text{ ft/s}, \ \overline{J} = \frac{112.6}{45 \cdot 6} = 0.42,$$
$\overline{T} = 458[1-0.3(0.42)] = 400$ lbf
$\overline{C_L} = \frac{W}{\frac{1}{2}\rho\overline{V}^2 S} = 2800/(0.5 \cdot 0.00218 \cdot 112.6^2 \cdot 180) = 1.126$
$\overline{C_D} = 0.04 + 1.0(0.0745)(1.126^2) = 0.134$
$\overline{D} = \frac{1}{2}\rho \overline{V}^2 S \overline{C_D} = 333$ lbf
$s_{TOA,I} = \frac{(V_2^2 - V_{LOF}^2)/2}{\frac{g}{W}[\overline{T} - \overline{D}]} = (119.2^2 - 105.5^2)/2/(32.2/2800)/[400-333] = 1998$ ft

Late airborne (II):

$V = V_2 = 119.2, \ J = 119.2/(45 \times 6) = 0.442, \ T = 458(1-0.3(0.442))$
$\quad = 397$ lbf,
$C_L = 2800/(0.5 \times 0.00218 \times 119.2^2 \times 180) = 1.0,$
$C_D = 0.04 + 0.0745(1.0^2) = 0.115$
$\gamma = \frac{T}{W} - \frac{C_D}{C_L} = 397/2800 - 0.115/1.0 = 0.027$ radians or $1.55°$
$s_{TOA,II} = \frac{h}{\tan\gamma} = 50/\tan(1.55°) = 1848$ ft
Overall: $s_{TO} = 2067 + 1998 + 1848 \approx 5913$ ft

(d) 3000 ft alt., headwind V_w of 10 kt (17 ft/s): $V_{G,LOF} = V_{LOF} - V_W = 105.5 - 17 = 88.5$ ft/s

Ground roll, using textbook method:

$$\overline{V} = \sqrt{\frac{V_w^2 + V_{LOF}^2}{2}} = 72.4 \text{ ft/s}, \ \overline{J} = 72.4/(45 \cdot 6) = 0.268, \ \overline{T} = 500(1-0.3(0.268)) =$$
460 lbf
$\overline{L} = 0.5(0.002378)72.4^2(180)0.9 = 1010$ lbf, $\overline{D} = 0.5(0.002378)72.4^2(180)$
$0.087 = 98$ lbf
$s_{TOG} = 88.5^2/2/(32.2/2800)/[419.5-97.6-0.05(2800-1009.2)-0] = 1466$ ft

Early airborne (I):

$s_{TOA,I} = s_{TOA,I}(0) - V_w \Delta t_{TOA,I}$, where $s_{TOA,I}(0)$, no wind case, is found in (a) above. Same thrust and drag as (c) above, given the same mean airspeed:
$$\Delta t_{TOA,I} = \frac{V_2 - V_{LOF}}{\frac{g}{W}[\overline{T} - \overline{D}]} = (119.2-105.5)/(32.2/2800)/[400-333] = 17.8 \text{ s}$$

$s_{\text{TOA,I}} = 1998 - 17(17.8) = 1695$ ft

Late airborne (II):

Same γ as (c), since at same V_2.

$s_{\text{TOA,II}} = s_{\text{TOA,II}}(0) - V_w\,\Delta t_{\text{TOA,II}}$,

$\Delta t_{\text{TOA,II}} = \dfrac{s_{\text{TOA,II}}(0)}{V_2 \cos \gamma} = 1848/(119.2{\cdot}\cos(1.55°)) = 15.5$ s

$s_{\text{TOA,II}} = 1848 - 17(15.5) = 1585$ ft

Overall:

$s_{TO} = 1466 + 1695 + 1585 \approx 4746$ ft

2.5. (a) Find the nominal takeoff distance using an approximate method, in this case for sea level altitude ($\rho = 0.002378$ slugs/ft^3) and no wind ($V_w = 0$ ft/s).

General info:

$AR = b^2/S = 196^2/5600 = 6.9$, $K = 1/(\pi{\cdot}AR \cdot e) = 1/(\pi{\cdot}6.9{\cdot}0.7) = 0.066$

$V_{S,t/o} = (2W_o/(\rho SC_{L\ \max}))^{1/2} = (2{\cdot}735000/(0.002378{\cdot}5600{\cdot}1.8))^{1/2} = 248$ ft/s

or 147 kt

$V_{LOF} = 1.18\,V_{S,t/o} = 293$ ft/s, $V_2 = 1.20\,V_{S,t/o} = 298$ ft/s, $\theta_R = 0°$ (no runway slope)

Ground roll: $\phi = 1 - \dfrac{(1 - 1.32h/b)}{(1.05 + 7.4h/b)} = 1 - (1 - 1.32(20/196))/(1.05 + 7.4(20/$

$196)) = 0.52$

$C_{D,GE} = C_{Do} + \phi KC_{Lg}^2 = 0.036 + 0.52(0.066)1.0^2 = 0.07$; normal t/o, $N = 4$

engines operating $\overline{V} = \sqrt{\dfrac{V_{LOF}^2 + V_w|V_w|}{2}} = \sqrt{\dfrac{293^2 + (0)(0)}{2}} = 207.2$ ft/s,

$\overline{T} = 46100 - 46.7\,\overline{V} + 0.0467\,\overline{V}^2 = 38429$ lbf for **one** engine, 153715 lbf for 4 engines.

$\overline{L} = \frac{1}{2}\rho\overline{V}^2 SC_{Lg} = 285857$ lbf, $\overline{D} = \frac{1}{2}\rho\overline{V}^2\,SC_{D,GE} = 20010$ lbf

$s_{TOG} = \dfrac{V_{G,LOF}^2/2}{\dfrac{g}{W}[\overline{T} - \overline{D} - \mu(W - \overline{L}) - W \sin \theta_R]}$

$= 293^2/2/(32.2/735000)/[153715 - 20010 - 0.02(735000 - 285857) - 0]$

$= 7856$ ft

Early airborne transition (I):

$\overline{V} = \sqrt{\dfrac{V_{LOF}^2 + V_2^2}{2}} = [(293^2 + 298^2)/2]^{1/2} = 296$ ft/s,

$\overline{T} = 4(46100 - 46.7(296) + 0.0467(296^2)) = 145474$ lbf

$\overline{C_L} = \dfrac{W}{\frac{1}{2}\rho\overline{V}^2 S} = 735000/(0.5{\cdot}0.002378{\cdot}296^2{\cdot}5600) = 1.26$

$\overline{C_D} = 0.036 + 1.0(0.066)(1.26^2) = 0.141$, noting $\phi = 1.0$ is assumed once airplane is airborne

$\overline{D} = \frac{1}{2}\rho \overline{V^2} S \overline{C_D} = 82257$ lbf

$s_{\text{TOA},I} = \frac{(V_2^2 - V_{\text{LOF}}^2)/2}{\frac{g}{W}[\overline{T} - \overline{D}]} = (298^2 - 293^2)/2/(32.2/735000)/[145474 - 82257] = 533$ ft

Late airborne transition (II):

$V = V_2 = 298$ ft/s, $T = 4(46100 - 46.7(298) + 0.0467(298^2)) = 145322$ lbf,
$C_L = 735000/(0.5 \times 0.002378 \times 298^2 \times 5600) = 1.243$,
$C_D = 0.036 + 0.066(1.243^2) = 0.138$
$\gamma = \frac{T}{W} - \frac{C_D}{C_L} = 145322/735000 - 0.138/1.243 = 0.087$ radians or $5°$

$s_{\text{TOA,II}} = \dfrac{h}{\tan \gamma} = 35/\tan(5°) = 400$ ft

Overall:

$s_{\text{TO}} = s_{\text{TOG}} + s_{\text{TOA,I}} + s_{\text{TOA,II}} = 7856 + 533 + 400 = 8789$ ft
Note: for a fully loaded Boeing 747, this is a reasonable estimate.

(b) Find the nominal takeoff distance using an approximate method, in this case for sea level altitude ($\rho = 0.002378$ slugs/ft^3) and a 30-kt tailwind ($V_w = -51$ ft/s). $V_{G,\text{LOF}} = V_{\text{LOF}} - V_W = 293 - (-51) = 344$ ft/s

Ground roll, using textbook method:

$\overline{V} = \sqrt{\dfrac{V_{\text{LOF}}^2 + V_w|V_w|}{2}} = \sqrt{\dfrac{293^2 + (-51)(51)}{2}} = 204$ ft/s,

$\overline{T} = 46100 - 46.7\,\overline{V} + 0.0467\,\overline{V^2} = 38517$ lbf for **one** engine, 154067 lbf for 4 engines.

$\phi = 1 - \dfrac{(1 - 1.32h/b)}{(1.05 + 7.4h/b)} = 1 - (1 - 1.32(20/196))/(1.05 + 7.4(20/196)) =$

0.52, from 2.5(a) $C_{D,GE} = C_{Do} + \phi K C_{Lg}^2 = 0.036 + 0.52(0.066)1.0^2 = 0.07$, from 2.5(a)

$\overline{L} = \frac{1}{2}\rho\overline{V^2}SC_{Lg} = 277096$ lbf, $\overline{D} = \frac{1}{2}\rho\overline{V^2}SC_{D,GE} = 19397$ lbf

$s_{\text{TOG}} = \dfrac{V_{G,\text{LOF}}^2/2}{\dfrac{g}{W}[\overline{T} - \overline{D} - \mu(W - \overline{L}) - W\sin\theta_R]}$

$= 344^2/2/(32.2/735000)/[154067 - 19397 - 0.02(735000 - 277096) - 0]$
$= 10761$ ft

Early airborne transition (I):

$s_{\text{TOA,I}} = s_{\text{TOA,I}}(0) - V_w \Delta t_{\text{TOA,I}}$, where $s_{\text{TOA,I}}(0)$, no wind case, is found in 2.5(a) solution:

$$\overline{V} = \sqrt{\frac{V_{LOF}^2 + V_2^2}{2}} = [(293^2 + 298^2)/2]^{1/2} = 296 \text{ ft/s},$$

$$\overline{T} = 4(46100 - 46.7(296) + 0.0467(296^2)) = 145474 \text{ lbf}$$

$$\overline{C_L} = \frac{W}{\frac{1}{2}\rho\overline{V}^2 S} = 735000/(0.5 \cdot 0.002378 \cdot 296^2 \cdot 5600) = 1.26$$

$\overline{C_D} = 0.036 + 1.0(0.066)(1.26^2) = 0.141$, noting $\phi = 1.0$ is assumed once airplane is airborne

$$\overline{D} = \frac{1}{2}\rho\overline{V}^2 S \overline{C_D} = 82257 \quad \text{lbf} \quad s_{TOA,I}(0) = \frac{(V_2^2 - V_{LOF}^2)/2}{\frac{g}{W}[\overline{T} - \overline{D}]} = (298^2 - 293^2)/2/$$

$(32.2/735000)/[145474 - 82257] = 533 \text{ ft}$

Given the same $\overline{V}$ as 2.5(a), mean thrust and drag are same as 2.5(a):

$$\Delta t_{TOA,I} = \frac{V_2 - V_{LOF}}{\frac{g}{W}[\overline{T} - \overline{D}]} = (298 - 293)/(32.2/735000)/[145474 - 82257] = 1.8 \text{ s}$$

$$s_{TOA,I} = 533 - (-51)(1.8) = 625 \text{ ft}$$

Late airborne (II):

Same γ as 2.5(a), since calculated at same V_2:
$V = V_2 = 298$ ft/s, $T = 4(46100 - 46.7(298) + 0.0467(298^2)) = 145322$ lbf,
$C_L = 735000/(0.5 \times 0.002378 \times 298^2 \times 5600) = 1.243$,
$C_D = 0.036 + 0.066(1.243^2) = 0.138$

$$\gamma = \frac{T}{W} - \frac{C_D}{C_L} = 145322/735000 - 0.138/1.243 = 0.087 \text{ radians or } 5°$$

$$s_{TOA,II}(0) = \frac{h}{\tan\gamma} = 35/\tan(5°) = 400 \text{ ft}$$

$$s_{TOA,II} = s_{TOA,II}(0) - V_w \Delta t_{TOA,II}, \quad \Delta t_{TOA,II} = \frac{s_{TOA,II}(0)}{V_2 \cos\gamma} = 400/(298 \cdot \cos 5°) =$$

1.35 s
$$s_{TOA,II} = 400 - (-51)(1.35) = 469 \text{ ft}$$

Overall:

$$s_{TO} = s_{TOG} + s_{TOA,I} + s_{TOA,II} = 10761 + 625 + 469 = 11855 \text{ ft}$$

(c) Find the nominal takeoff distance using an approximate method, in this case for 5,000 ft altitude ($\rho = 0.002$ slugs/ft^3) and no wind ($V_w = 0$ ft/s).

General info:

$\sigma = 0.86$ given $\rho = 0.002$ slugs/ft^3 at 5000 ft ISA.
$V_{S,t/o} = (2W_o/(\rho S C_{Lmax}))^{1/2} = (2 \cdot 735000/(0.002 \cdot 5600 \cdot 1.8))^{1/2} = 270$ ft/s or
147 kt
$V_{LOF} = 1.18 \, V_{S,t/o} = 319$ ft/s, $V_2 = 1.20 \, V_{S,t/o} = 324$ ft/s

Ground roll:

$$T = \sigma^{0.8} T_{S/L}$$

$$\overline{V} = \sqrt{\frac{V_{LOF}^2 + V_w|V_w|}{2}} = \sqrt{\frac{319^2 + (0)(0)}{2}} = 225.6 \text{ ft/s,}$$

$\overline{T} = 40860 - 41.4\,\overline{V} + 0.0414\,\overline{V^2} = 33627$ lbf for **one** engine, 134509 lbf for 4 engines.

$$\overline{L} = \tfrac{1}{2}\rho\overline{V}^2 S C_{Lg} = 285014 \text{ lbf, } \overline{D} = \tfrac{1}{2}\rho\overline{V}^2 S C_{D,GE} = 19951 \text{ lbf}$$

$$s_{TOG} = \frac{V_{G,LOF}^2/2}{\frac{g}{W}\left[\overline{T} - \overline{D} - \mu(W - \overline{L}) - W\sin\theta_R\right]}$$

$$= 319^2/2/(32.2/735000)/[134509 - 19951 - 0.02(735000 - 285014) - 0]$$

$$= 11002 \text{ ft}$$

Early airborne transition (I):

$$\overline{V} = \sqrt{\frac{V_{LOF}^2 + V_2^2}{2}} = [(319^2 + 324^2)/2]^{1/2} = 322 \text{ ft/s,}$$

$$\overline{T} = 4\cdot0.86^{0.8}\cdot(46100 - 46.7(322) + 0.0467(322^2)) = 127294 \text{ lbf}$$

$$\overline{C_L} = \frac{W}{\tfrac{1}{2}\rho\overline{V}^2 S} = 735000/(0.5\cdot0.002\cdot322^2\cdot5600) = 1.27$$

$\overline{C_D} = 0.036 + 1.0(0.066)(1.27^2) = 0.142$, noting $\phi = 1.0$ is assumed once airplane is airborne

$$\overline{D} = \frac{1}{2}\rho\,\overline{V^2}\,S\,\overline{C_D} = 0.5(0.002)322^2(5600)0.142 = 82450 \text{ lbf}$$

$$s_{TOA,I} = \frac{(V_2^2 - V_{LOF}^2)/2}{\frac{g}{W}[\overline{T} - \overline{D}]} = (324^2 - 319^2)/2/(32.2/735000)/[127294 - 82450] = 818 \text{ ft}$$

Late airborne transition (II):

$$V = V_2 = 324 \text{ ft/s,}$$
$$T = 4(0.886)(46100 - 46.7(324) + 0.0467(324^2)) = 127129 \text{ lbf,}$$
$$C_L = 735000/(0.5 \times 0.002 \times 324^2 \times 5600) = 1.25,$$
$$C_D = 0.036 + 0.066(1.25^2) = 0.139$$
$$\gamma = \tfrac{T}{W} - \tfrac{C_D}{C_L} = 127129/735000 \quad - \quad 0.139/1.25 = 0.062 \quad \text{radians} \quad \text{or} \quad 3.5°$$
$$s_{TOA,II} = \frac{h}{\tan\gamma} = 35/\tan(3.55°) = 564 \text{ ft}$$

Overall:

$$s_{TO} = s_{TOG} + s_{TOA,I} + s_{TOA,II} = 11002 + 818 + 564 = 12384 \text{ ft}$$

References

1. McCormick BW (1995) Aerodynamics, aeronautics and flight mechanics, 2nd edn. Wiley, New York
2. Lan CE, Roskam J (1997) Airplane aerodynamics and performance. DARcorporation, Lawrence (Kansas)

Chapter 3
The Propeller

3.1 Introduction

As may be viewed by the few examples of propellers used through the years, in Fig. 3.1, the rotating blade of a propeller shares similar characteristics to that of a wing passing through an airflow. Referring to Fig. 3.2, we know that a wing at a positive angle of attack α generates a lift force L perpendicular to the freestream direction, and a drag force D parallel to it, with a corresponding low-pressure zone above the wing, and a high-pressure zone beneath it. Above a certain angle of attack α_{max}, the wing may stall at a given airfoil section (considerable loss of lift). Similarly, a rotating propeller blade generates thrust F via a lift force component, demands a required engine torque Q to overcome a drag force component, and can stall locally if α_{max} is exceeded relative to the resultant freestream direction. There will be a static pressure increase in passing through the propeller plane from upstream (left) to downstream (right). There is also an increase in the air flow speed in moving through the propeller, and we shall see that this is a significant energy demand on the propeller/engine combination. Other factors, like trailing vortex generation from each blade (analogous to that seen for wings, i.e., vortex-induced downwash aft of the wing due to smaller aspect [span-to-chord] ratios), will introduce further demands on the engine.

3.2 Actuator Disk Theory

One approach for evaluating some aspects of propeller performance is via a simple application of the flow momentum conservation principle. Here, we will be treating the propeller as a thin actuator disk, through which there will be a step increase in static pressure (as observed for actual propellers). As per Fig. 3.3, the pertinent flow of air will be assumed to follow along a contracting stream tube, with the disk at mid-length.

D. R. Greatrix, *Powered Flight*, DOI: 10.1007/978-1-4471-2485-6_3,
© Springer-Verlag London Limited 2012

Fig. 3.1 Example propellers through the years

In going from station 0 to station 3, with the same ambient pressure conditions at each end, conservation of linear momentum gives thrust as follows:

$$F = \dot{m}(V_3 - V_0) = \rho A_3 V_3 (V_3 - V_0) \tag{3.1}$$

We are assuming a constant-density (incompressible) flow situation for mathematical convenience; strictly speaking, this assumption is more accurate at lower flow speeds, at flow Mach numbers below 0.5 typically. With continuous flow through the actuator disk of cross-sectional area A_1, one can see that thrust may also be ascertained via

$$F = A_1 (p_2 - p_1) \tag{3.2}$$

We can apply Bernoulli's equation (an incompressible form of the energy conservation equation) from station 0 to station 1:

$$p_0 + \frac{1}{2}\rho V_0^2 = p_1 + \frac{1}{2}\rho V_1^2 \tag{3.3}$$

and from station 2 to station 3:

Fig. 3.2 Forces acting on
wing airfoil section (*above*)
and propeller blade section
(*below*)

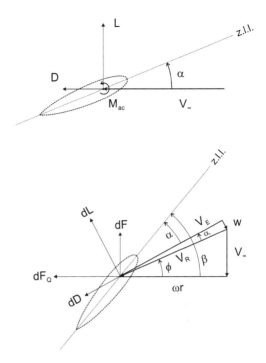

Fig. 3.3 Schematic diagram
of air flow moving along a
stream tube, passing through
an actuator disk
representation of a propeller

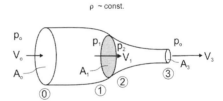

$$p_2 + \frac{1}{2}\rho V_2^2 = p_3 + \frac{1}{2}\rho V_3^2 \qquad (3.4)$$

We cannot apply Bernoulli's equation from station 1 to station 2, since energy is
being added in some quantity to the flow at the actuator disk (propeller). Subtracting Eq. 3.3 from Eq. 3.4, noting that $V_2 = V_1$,

$$p_2 - p_1 = \frac{1}{2}\rho(V_3^2 - V_0^2) = \frac{1}{2}\rho(V_3 + V_0)(V_3 - V_0) \qquad (3.5)$$

Given $A_3 V_3 = A_1 V_1$ from continuity (mass conservation), examination of
Eqs. 3.1, 3.2 and 3.5 gives:

$$F = A_1(p_2 - p_1) = \rho A_3 V_3 (V_3 - V_0) \qquad (3.6)$$

$$p_2 - p_1 = \rho \frac{A_3}{A_1} V_3 (V_3 - V_0) = \frac{1}{2} \rho (V_3 + V_0)(V_3 - V_0) \tag{3.7}$$

$$\rho \frac{A_3}{\left(\frac{A_3 V_3}{V_1}\right)} V_3 = \frac{1}{2} \rho (V_3 + V_0) \tag{3.8}$$

so that

$$V_1 = \frac{V_3 + V_0}{2} \tag{3.9}$$

This indicates that the air velocity through the propeller (disk) equals the average of the velocities far upstream and far downstream of the propeller. Following along from this result, one can define w as the propeller-induced velocity component, so that

$$V_1 = V_0 + w \tag{3.10}$$

and via Eq. 3.9,

$$V_3 = V_0 + 2w \tag{3.11}$$

As a side note, for a shrouded (ducted) propeller, the analysis is more convenient with the assumption that $V_3 = V_0 + w$. Returning to the analysis for an unducted propeller, thrust may now be found via

$$F = \rho A_1 V_1 (V_3 - V_0) = \rho A_1 (V_0 + w)(V_0 + 2w - V_0) = 2\rho A_1 (V_0 + w)w \tag{3.12}$$

Now, ideal power can be interpreted as the rate of change of kinetic energy of the flow, i.e., that being added by the propeller:

$$P = \frac{1}{2} \dot{m} V_3^2 - \frac{1}{2} \dot{m} V_0^2 = \frac{1}{2} \rho A_1 (V_0 + w)[(V_0 + 2w)^2 - V_0^2] = 2\rho A_1 w (V_0 + w)^2 \tag{3.13}$$

Substituting Eq. 3.12 into 3.13, one arrives at:

$$P = F(V_0 + w) \tag{3.14}$$

One can observe that this expression indicates that ideal power needed from the engine is in fact the thrust power (product of thrust and reference velocity) found locally at the actuator disk (propeller). Since power delivered from a piston or turboprop engine is relatively constant through the flight regime (from low to high speed, at a given altitude, with power overall dropping with altitude increasing above a certain altitude), this expression also suggests that one can expect thrust to drop as one picks up aircraft flight speed (V_0), a phenomenon that is observed in practice. The first part of Eq. 3.14 is referred to as the useful power:

$$P_{\text{use}} = F V_0 \tag{3.15}$$

while the second part is referred to as the induced power:

$$P_{\text{ind}} = Fw \tag{3.16}$$

Actuator disk theory also provides an estimate of the induced velocity w. Assuming thrust F is a known quantity, one can solve for w via Eq. 3.12:

$$(2\rho A_1)w^2 + (2\rho A_1 V_0)w - F = 0 \tag{3.17}$$

or

$$w = \frac{-V_0}{2} + \frac{1}{2}\sqrt{V_0^2 + \frac{2F}{\rho A_1}} \tag{3.18}$$

Here, we have chosen the positive root of the two-root quadratic equation solution, for a pertinent result.

For the static case, when V_0 is zero,

$$w_o = \sqrt{\frac{F_o}{2\rho A_1}} \tag{3.19}$$

and the ideal static power required from the powerplant is thus

$$P_o = P_{\text{ind},o} = F_o w_o = \frac{F_o^{3/2}}{\sqrt{2\rho A_1}} \tag{3.20}$$

Example: Consider a propeller having a 1-ft (30.5 cm) diameter, driven by a 1-hp (0.746 kW, 550 ft lb/s) piston engine. This is a case similar to the radio-controlled model airplane used for the SAE Aero Design student competition. What is the sea-level static thrust being delivered, approximately? Propeller disk area is found from:

$$A_1 = \frac{\pi d_p^2}{4} = \frac{\pi(1.0)^2}{4} = 0.79 \text{ ft}^2 \text{ or } 0.073 \text{ m}^2$$

Static thrust is estimated via:

$$F_o \approx P_{\text{ind},o}^{2/3}(2\rho A_1)^{1/3} = (1 \cdot 550)^{2/3}(2 \cdot 0.002378 \cdot 0.79)^{1/3} \approx 10.4 \text{ lbf (or 46 N)}$$

In practice, the delivered thrust is about 8.0 lbf. One can conclude that to some degree, the difference is due to the factors absent from actuator disk theory, namely blade profile drag and trailing vortex generation losses.

Returning to actuator disk theory for additional information, let us define an ideal propeller propulsive efficiency $\eta_{pr,i}$ as being the ratio of useful power to total ideal power (see Eqs. 3.14–3.16):

$$\eta_{pr,i} = \frac{FV_0}{F(V_0 + w)} = \frac{1}{1 + \frac{w}{V_0}} \tag{3.21}$$

The above suggests that we have 100% efficiency when we have no induced velocity w, i.e., negligible thrust delivery, so a balance must be struck between the two performance targets in practice. Defining the aircraft's reference dynamic pressure as

$$q = \frac{1}{2}\rho V_0^2 \tag{3.22}$$

then from Eq. 3.18,

$$\frac{w}{V_0} = \frac{1}{2}\left(-1 + \sqrt{\frac{F}{qA_1}}\right) \tag{3.23}$$

Returning to Eq. 3.21,

$$\eta_{pr,i} = \frac{2}{1 + \sqrt{1 + \frac{F}{qA_1}}} \tag{3.24}$$

This ideal expression from actuator disk theory suggests that as disk loading (F/A_p) goes to zero, propeller propulsive efficiency goes to one (or, 100%). In practice, a substantial disk loading is often desirable, so again, one notes that we have to be careful in trading off between two potential performance objectives (in this case, high thrust versus high efficiency).

The actual propeller propulsive efficiency η_{pr} is in terms of actual engine shaft power P_S needed, rather than ideal power:

$$\eta_{pr} = \frac{FV_\infty}{P_S} < \eta_{pr,i} \tag{3.25}$$

Here, the standard symbol for airplane forward airspeed is used (V_∞, rather than using the synonymous V_0 from actuator disk theory).

In summary, the actuator disk theory provides some interesting and useful information, for example, providing an estimate for induced velocity w that can be incorporated into more sophisticated analyses for propeller and helicopter rotor performance. As one might expect, the actuator disk theory overpredicts the amount of thrust delivered for a given engine power delivery, given the absence of some loss factors in the analysis. As a first guess in preliminary design, one could assume a correction factor of around 75 to 80% in the correlation between ideal power and actual engine shaft power:

$$P = F(V_0 + w) = (\text{correction factor}) \times P_S \tag{3.26}$$

A variable-pitch propeller (varying β as per Fig. 3.2, the angle between the blade section's zero lift line [z.l.l.] and the rotation plane), versus a fixed-pitch propeller, will be better able to approach the ideal actuator disk result, in compensating at least in part for flight at different airspeeds (from takeoff at low β, β_{ref} in the 20° range, to cruise conditions at high, β, β_{ref} in the 40° range, where β_{ref} is commonly measured at the 75% radius blade position [0.75R]). A variable-pitch propeller is commonly driven on a propeller shaft rotating at a relatively constant speed for peak engine efficiency, hence the descriptor as constant-speed propellers.

Please refer to the solutions for Example Problems 3.1 and 3.6 at the end of this chapter, for two completed sample analyses using actuator disk theory. For students learning this material as part of a course, I would encourage them to attempt the given example problem first, and then check the solution to see if any mistakes were made.

3.3 Momentum-Blade Element Theory

In order to design a new propeller for a given application, or to more accurately estimate existing propeller performance, one has to look at the aerodynamics in more detail than that done with actuator disk theory. One approach is to mathematically discretize the given blade into various sections or elements, and thereafter evaluate the aerodynamic properties locally in moving along the blade from hub ($r = r_h$) to tip ($r = R$), as per Fig. 3.4.

Referring to Fig. 3.2, the local pitch angle β of a propeller blade section is measured from the plane of rotation of the propeller to the section's zero lift line [1]. We see that there are three velocity components in forward flight as depicted in Fig. 3.2, namely V_∞ for forward airspeed (of the airplane), ωr for the local blade rotation speed, and w for the propeller-induced velocity component (α_i is the corresponding induced angle-of-attack component). The following equation applies for incremental forward thrust generated from the local blade section:

$$dF = dL\cos(\phi + \alpha_i) - dD\sin(\phi + \alpha_i) \tag{3.27}$$

where ϕ is the angle of the resultant velocity V_R (due to propeller rotation and forward flight speed) from the propeller plane of rotation, noting that

$$V_R = \sqrt{(\omega r)^2 + V_\infty^2} \tag{3.28}$$

One can now consider the sideways force components that contribute to incremental torque:

$$dQ = r dF_Q = r[dL\sin(\phi + \alpha_i) + dD\cos(\phi + \alpha_i)] \tag{3.29}$$

Incremental lift and drag of the blade section having a local chord (length) of c may be found as follows:

$$dL = \frac{1}{2}\rho V_E^2 c C_\ell dr \tag{3.30}$$

Fig. 3.4 Schematic diagram
of a three-bladed propeller,
and framework for
discretizing an individual
blade for analysis

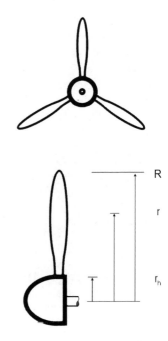

$$dD = \frac{1}{2}\rho V_E^2 c C_d dr \qquad (3.31)$$

Accounting for the induced velocity component, the overall resultant velocity V_E may be calculated via:

$$V_E = \sqrt{(\omega r - w\sin(\phi + \alpha_i))^2 + (w\cos(\phi + \alpha_i) + V_\infty)^2} \qquad (3.32)$$

while the induced angle of attack may be estimated from

$$\alpha_i = \sin^{-1}\left(\frac{w}{V_R}\right) \qquad (3.33)$$

Local lift coefficient C_l may be estimated, if in the linear aerodynamic regime (substantially away from stall conditions) for the lift-curve slope a_o, by the following:

$$C_\ell = C_{\ell_\alpha}(\beta - \alpha_i - \phi) = a_o(\beta - \alpha_i - \phi) \qquad (3.34)$$

Local drag coefficient C_d may be estimated via

$$C_d = C_{d,min}, \quad C_\ell < C_{\ell,min} \qquad (3.35a)$$

$$C_d = C_{d,min} + k(C_\ell - C_{\ell,min})^2, \quad C_{\ell,min} < C_\ell < C_{\ell,max} \qquad (3.35b)$$

$$C_d = C_{d,\alpha_{max}} + k_1(\alpha - \alpha_{max}), \quad \alpha > \alpha_{max} \tag{3.35c}$$

depending on which of the various aerodynamic regimes that the blade is presently within. The airfoil enters stall conditions when C_ℓ reaches $C_{\ell,max}$ and α correspondingly reaches α_{max}.

At this juncture, we need an equation that allows us to estimate α_i as a function of known properties. Referring to Eq. 3.27 for incremental thrust, one can simplify it with assumptions of small angles (note: ϕ is not necessarily small) and small w such that $V_E \approx V_R$, and drag being relatively small in comparison to lift, so that one is left with:

$$dF \approx dL \cdot \cos\phi = \frac{B}{2}\rho V_R^2 ca_o(\beta - \alpha_i - \phi)dr \cdot \cos\phi \tag{3.36}$$

Here, we have introduced the number of blades B, as we begin collecting various elements towards our final estimation of overall propeller performance parameters. Referring to actuator disk theory, one can introduce an alternative expression for incremental thrust as a function of effective V_∞ and w parallel to forward thrust, as produced by an inscribed annulus of thickness dr:

$$dF \approx 2\rho \cdot dA \cdot (V_0 + w)w \approx 2\rho \cdot dA \cdot (V_\infty + w\cos\phi)w\cos\phi$$
$$\approx 2\rho(2\pi r dr) \cdot (V_\infty + \alpha_i V_R \cos\phi)\alpha_i V_R \cos\phi \tag{3.37}$$

where α_i is in units of radians in the above expression. Equating Eqs. 3.36 and 3.37, one arrives at the following quadratic equation:

$$\alpha_i^2 + \left(\frac{V_\infty}{V_R \cos\phi} + \frac{ca_o B}{8\pi r \cos\phi}\right)\alpha_i - \frac{ca_o B}{8\pi r \cos\phi}(\beta - \phi) = 0 \tag{3.38}$$

Before proceeding further, let us introduce some other propeller performance-related parameters. One defines the propeller's overall or reference solidity as follows:

$$\sigma_{ref} = \frac{\text{blade area}}{\text{disk area}} = \frac{B \cdot c_{ref} R}{\pi R^2} = \frac{Bc_{ref}}{\pi R} \tag{3.39}$$

Local solidity is defined in terms of local chord c by:

$$\sigma = \frac{Bc}{\pi R} = x\frac{Bc}{\pi r} \tag{3.40}$$

where $x = r/R$ in this context (nondimensional radial location). Nondimensional flight airspeed may be given in terms of advance ratio J:

$$J = \frac{V_\infty}{nd_p} = \frac{V_\infty}{(\omega/(2\pi))(2R)} = \frac{\pi V_\infty}{\omega R} \tag{3.41}$$

Here, n is the propeller shaft rotation speed, in revolutions per second [rps] in fundamental units (although it may be quoted in revolutions per minute [rpm]). Do not confuse propeller shaft speed n with engine drive shaft speed N, since they may not coincide in value (this would almost surely be true for a turboprop engine, where the gas turbine's free turbine shaft for powering the propeller is likely rotating at at least 10,000 rpm, while the propeller rotation may be closer to 1,500 rpm; a speed reduction unit is required to gear down to the propeller's speed). Alternatively, for nondimensional flight airspeed relative to propeller rotation speed:

$$\lambda = \frac{V_\infty}{\omega R} = \frac{J}{\pi} \tag{3.42}$$

From the above result, one notes that

$$\phi = \tan^{-1}\left(\frac{V_\infty}{\omega r}\right) = \tan^{-1}\left(\frac{\lambda}{x}\right) \tag{3.43}$$

Given that static blade tip speed $V_T = \omega R$, then the following relationship exists:

$$V_R \cos \phi = \omega r = x V_T \tag{3.44}$$

Returning now to Eq. 3.38, substituting for various parameters above, one arrives at:

$$\alpha_i^2 + \left(\frac{\lambda}{x} + \frac{\sigma a_o V_R}{8x^2 V_T}\right)\alpha_i - \frac{\sigma a_o V_R}{8x^2 V_T}(\beta - \phi) = 0 \tag{3.45}$$

Solving the above quadratic equation for the appropriate root gives:

$$\alpha_i = \frac{1}{2}\left\{-\left(\frac{\lambda}{x} + \frac{\sigma a_o V_R}{8x^2 V_T}\right) + \left[\left(\frac{\lambda}{x} + \frac{\sigma a_o V_R}{8x^2 V_T}\right)^2 + \frac{\sigma a_o V_R}{2x^2 V_T}(\beta - \phi)\right]^{1/2}\right\} \tag{3.46}$$

Thus, we have shown that we can solve for the induced angle of attack as a function of the propeller's geometry and rotational speed, and the airplane's forward speed. By now knowing α_i via Eq. 3.46, one can then move to Eqs. 3.27 and 3.29 (with additional information from Eqs. 3.30 and 3.31), and integrate for overall propeller thrust F and propeller shaft torque Q.

Before leaving this section, let us consider two additional performance parameters, in addition to propeller propulsive efficiency η_{pr}, that are often charted with respect to advance ratio J, for a propeller with a given number of blades. Given that force in the propeller context is the product of dynamic pressure (referenced to propeller rotation speed) and area, one can produce the following correlation for thrust coefficient C_T:

$$C_T = \frac{F}{\rho n^2 d^4} \tag{3.47}$$

Similarly, if one considers power as the product of force and velocity, one can produce the following correlation for the propeller's power coefficient C_P :

$$C_P = \frac{P_S}{\rho n^3 d^5} \quad (3.48)$$

Note that shaft power required from the engine, P_S, may be determined via

$$P_S = Q\omega \quad (3.49)$$

with Q being the torque acting on the propeller shaft having an angular rotation speed ω.

These coefficients may be used for scaling up or down in evaluating a propeller's needed performance, in conjunction with advance ratio, which ties in the needed thrust and propeller torque for a given flight speed. Curves for various reference blade pitch values β_{ref} are generally displayed (see Figs. 3.5 and 3.6) [1, 2].

An additional performance parameter that is occasionally used is the speed or speed-power coefficient C_S, defined by

$$C_S = \left(\frac{\rho V_\infty^5}{P_S n^2}\right)^{1/5} = \frac{J}{C_P^{1/5}} \quad (3.50)$$

Depending on the available information (note that propeller diameter d is not appearing in the above expression), this parameter can be used for scaling and ultimately setting desired values for d or β at design-point conditions, e.g., at cruise flight speed and altitude. One can find charts providing η_{pr} as a function of C_S (with curves at increasing values of β_{ref}), or J as a function of C_S.

Returning to the momentum-blade element analysis, one may want to use this approach for generating plots of C_T or C_P as a function of J. From Eqs. 3.27, 3.30 and 3.31, incremental thrust from a blade section (times the number of blades B) is

$$dF = \frac{1}{2}\rho V_E^2 Bc[C_\ell \cos(\phi + \alpha_i) - C_d \sin(\phi + \alpha_i)]dr \quad (3.51)$$

The required shaft power, from Eqs. 3.29 and 3.49, dictates the following for incremental power from a given blade section (times the number of blades B):

$$dP_S = \omega r \frac{1}{2}\rho V_E^2 Bc[C_\ell \sin(\phi + \alpha_i) + C_d \cos(\phi + \alpha_i)]dr \quad (3.52)$$

For simplicity, one can assume the following, for usage later:

$$V_E^2 \approx V_R^2 = V_\infty^2 + \omega^2 r^2 = \frac{\omega^2 r^2}{\pi^2}(J^2 + \pi^2 x^2) \quad (3.53)$$

From earlier, we note the following applies for the thrust coefficient:

$$C_T = \frac{\pi^2}{4\rho\omega^2 R^4}F \quad (3.54)$$

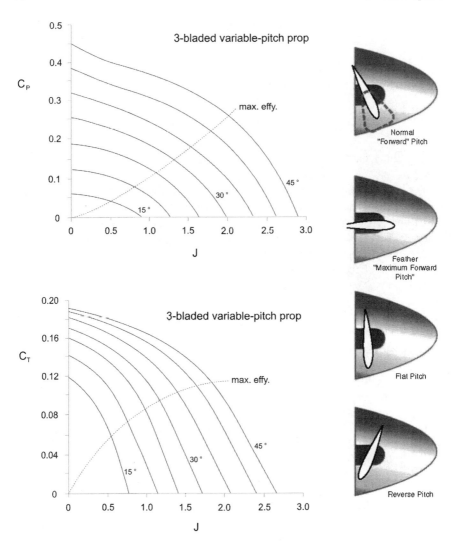

Fig. 3.5 Performance charts for a three-bladed, moderate-AF variable-pitch propeller, for curves at progressively higher values of β_{ref} (15–45°). *Right diagram* shows various positions that a variable-pitch propeller might take during the course of a flight, including reverse pitch for braking in the landing ground roll [2]

As a result, one can integrate to find:

$$C_T = \frac{\pi^2}{4\rho\omega^2 R^4} \int dF = \frac{\pi}{8} \int_{x_h}^{1} \sigma(J^2 + \pi^2 x^2)[C_\ell \cos(\phi + \alpha_i) - C_d \sin(\phi + \alpha_i)]dx$$

$$(3.55)$$

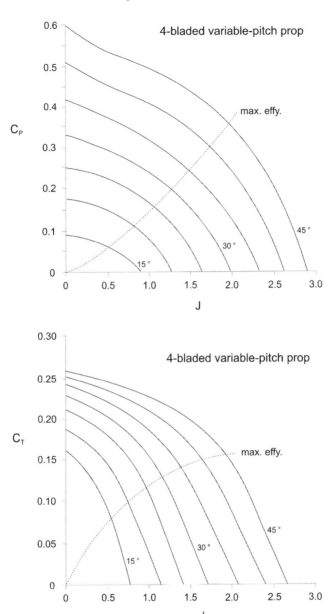

Fig. 3.6 Performance charts for a four-bladed, moderate-AF variable-pitch propeller

Similarly, for:

$$C_P = \frac{\pi^3}{4\rho\omega^3 R^5} P_S \qquad (3.56)$$

integrating gives:

$$C_P = \frac{\pi^3}{4\rho\omega^3 R^5} \int dP_S = \frac{\pi^2}{8} \int_{x_h}^{1} \sigma x (J^2 + \pi^2 x^2)[C_\ell \sin(\phi + \alpha_i) + C_d \cos(\phi + \alpha_i)] dx$$

(3.57)

The above equations for C_T and C_P can be integrated from the hub station $(x = x_h)$ to the blade tip $(x = 1)$ using a numerical approach as one moves along the blade of varying β and c, calculating the various pertinent parameters $(C_\ell, C_d, \alpha_i, \text{etc.})$ in conjunction. Please refer to the solution for Example Problem 3.4 at the end of this chapter, for a completed sample analysis using the momentum-blade element approach.

In this analysis, we did not account for tip losses (note: blade loading must vanish at the blade tip, in a continuous manner, i.e., not a step) which apply at low and high speed. The analysis does not account for energy losses associated with trailing vortex generation, which occurs with blades of finite aspect ratio (analogous to wings). Vortex theory using bound and trailing line vortices can better model the above effects, as well as producing a better estimate of induced velocity w [1]. However, as one might expect, the mathematics become a bit more involved, and an iterative solution is required.

3.4 Propeller Propulsive Efficiency

Earlier, we defined propeller propulsive efficiency as the useful (thrust) power over overall shaft power delivered from the engine:

$$\eta_{pr} = \frac{FV_\infty}{P_S}$$

(3.58)

Substituting from earlier results, one can show that:

$$\eta_{pr} = \frac{C_T \rho n^2 d^4 V_\infty}{C_P \rho n^3 d^5} = \frac{C_T}{C_P} J$$

(3.59)

One can view Fig. 3.7 for a sample chart of η_{pr} as a function of advance ratio J, with curves at various values of the reference blade pitch angle (β at $0.75R$, typically). One should be aware that as J goes to zero at low speeds, efficiency goes to zero as well, while C_T and C_P are finite, and indeed have their highest values near $J = 0$. As a result, one must be wary of using η_{pr} as a guideline in the low-speed regime of a given flight mission. Variable-pitch propellers are designed for a high η_{pr} in cruise, to minimize fuel consumption over long periods. This may be done at the expense of a lower η_{pr} for lower-speed, lower-altitude portions of a typical flight mission (e.g., takeoff, early climb). However, one must not forget that

Fig. 3.7 Chart illustrating propeller propulsive efficiency for an example propeller

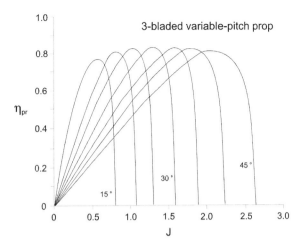

a high C_T, for high thrust delivery, is a critical requirement for the takeoff part of the flight mission.

3.5 Compressibility Tip Loss

Depending on the blade airfoil section design, drag divergence (compressibility) effects will become evident when the propeller blade's resultant tip speed $V_{R,tip}$ exceeds a local flow Mach number Ma_{tip} of around 0.85 (critical value, Ma_{cr}). As a result, one would not typically be cruising at much greater speed than a flight Mach number Ma_∞ of around 0.6, where

$$Ma_{tip} = \sqrt{\left(\frac{\pi n d}{a_\infty}\right)^2 + Ma_\infty^2} \qquad (3.60)$$

noting a_∞ is the local sound speed of the outside ambient air. Dommasch [3] proposed the following for conventional subsonic propellers entering into the drag divergence regime:

$$\eta_{pr} = \eta_{pr,nominal} - \frac{15}{100}\left(\frac{Ma_{tip} - Ma_{cr}}{0.1}\right) \qquad (3.61)$$

Thus, if one is cruising at flight conditions such that Ma_{tip} equals 0.9, propeller efficiency would go from a nominal value of say 0.83 down to a value of 0.76, an effective 9% decrease in cruise fuel consumption efficiency.

3.6 Activity Factor

Activity factor (AF) is a design parameter associated with the propeller blade's geometry [1, 3]. The more slender the blade (larger radius, smaller chord), the lower the AF value. The standard equation for AF, in terms of nondimensional radial distance x along the blade, is as follows:

$$AF = \frac{100000}{16} \int\limits_{x_h}^{1} \frac{c}{d_p} \cdot x^3 dx \tag{3.62}$$

Example: Consider a low-speed propeller of conventional design having a mean chord-to-diameter value of 0.06 (a relatively slender shape), a hub radius of 0.31 ft, and a propeller radius of 3.1 ft (0.95 m):

$$AF = 6250 \int\limits_{0.05}^{1.0} 0.06 \cdot x^3 dx = \frac{6250}{R^4} \int\limits_{0.31}^{3.1} 0.06 \cdot r^3 dr = \frac{6250(0.06)}{3.1^4} \left\{ \frac{r^4}{4} \right\}_{0.31}^{3.1}$$

$$= \frac{6250(0.06)}{3.1^4(4)} [3.1^4 - 0.31^4] = \frac{6250(0.06)}{4} \left[1 - \left(\frac{0.31}{3.1} \right)^4 \right] = 94$$

The AF value just below 100 is close to the quoted value for the specific propeller in question. Since the hub term to the fourth power turns out to be relatively small in practice, one notes that:

$$AF \approx 1563 \frac{\bar{c}}{d_p} \tag{3.63}$$

As a second example, consider a relatively thick blade as seen for the propellers used by the Lockheed C-130 Hercules transport aircraft's turboprop engines (see lower right of Fig. 3.1), whose mean chord-to-diameter value is closer to 0.11. Via Eq. 3.63, one has an AF of about 170, which is in fact close to the quoted value. Higher AF values are seen for propellers used on turboprop engines (as opposed to piston engines); they allow for a lower d_p at a higher n, to keep the blade tips within the compressibility tip loss limit.

3.7 Blade Number

One has the option of setting the number of blades, B, for a given application. While one has a minimum of two blades to choose from, one can presently go as high as around six blades on the high-performance end for an unducted propeller. With respect to avoiding excessive compressibility tip losses noted earlier, one can

Fig. 3.8 Photo of Fairey Gannett carrier-borne anti-submarine/AEW aircraft, employing two contra-rotating rows of four propeller blades each on a co-axial shaft setup, powered by a 3,000-hp Armstrong Siddeley Twin Mamba turboprop engine

choose to go with more blades, versus a fewer number of longer blades, to up the thrust delivery. If the engine power is sufficiently high, one can alternatively or in conjunction go to a wider blade (i.e., higher AF) for a given length to deliver a higher thrust level if so required. Beyond six blades, aerodynamic interference/wake effects will typically diminish the performance of an unducted conventional propeller. Ducting (shrouding) will mitigate this interference phenomenon to some degree, to allow for even more blades; propellers using many blades are referred to as fans. Research continues in regard to increased numbers of blades in the unducted case, and through such transonic-wing tactics as sweeping the leading edge and reducing the blade's thickness, one can see the potential use for 8 up to 12 blades. On occasion, one also sees the use of two contra-rotating rows of blades, to get more thrust delivery from one engine (Figs. 3.1, 3.8).

Due to some nonlinearity in adding or subtracting one or more blades for a given propeller on delivered performance (especially as one gets into higher B values), $C_T(J)$ and $C_P(J)$ charts are generally given for a set number of blades, e.g., Fig. 3.5 for a three-bladed propeller, and a separate chart for a four-bladed variant as given in Fig. 3.6.

3.8 Note on Helicopter Rotors

As one might expect, helicopter rotors (main and tail) share a number of similarities with airplane propellers. Some analysis done above for propellers can certainly be applied to rotors. The orientation of the rotor disk will of course be

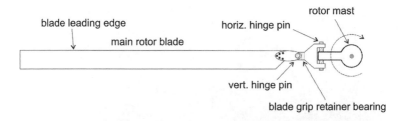

Fig. 3.9 Schematic diagram illustrating the various components of a conventional helicopter fully articulated main rotor system (plan view, showing just one of two or more blades that would comprise a main rotor). By convention, European rotors by design typically rotate clockwise (as shown), while American-built main rotors typically rotate in the counterclockwise direction, for single-rotor helicopters. Photo of Boeing-Vertol CH-47 Chinook with front rotor rotating counterclockwise, and rear rotor rotating clockwise, driven by two 3750-hp Textron Lycoming T55-L-712 turboshaft engines

somewhat different from that of the propeller, with respect to the resultant incoming air flow [1]. A rotor blade will advance into the air flow when in forward flight, and then retreat during the other half of the rotational cycle. Additionally, helicopter main rotors differ from propellers in that they must perform two principal functions. They must deliver forward thrust, and unlike the propeller, provide substantial lift as a good portion of their thrust delivery, as a substitute for airplane wing lift. Additionally, the main rotor (see Fig. 3.9), sometimes in combination with the smaller tail rotor, must provide the means for directional control of the vehicle, e.g., provide side thrust to steer the vehicle left or right, as opposed to the use of wing ailerons and the horizontal tail elevator in turning a conventional airplane.

The tail rotor primarily controls yaw forces and moments [primarily main-rotor-induced torque] on the helicopter, if only having one main rotor; a tandem-rotor helicopter, with two contra-rotating main rotors, would not need a tail rotor; see Fig. 3.10). One alternative to the use of a tail rotor for yaw stability and control is a boundary-layer/ejector approach. One well-known variant of this approach involves using an internal tailboom-positioned fan powered by the turboshaft engine producing pressurized air as the medium for exhausting from tangential-blowing slots, to exploit a side-force-inducing Coanda effect in the main rotor

Fig. 3.10 Photo at *left* illustrating a conventional two-bladed tail rotor for helicopter yaw control for Bell 205 employing a single main rotor, and at *right*, a tandem-rotor helicopter (Boeing Vertol CH-113 Labrador) not requiring a tail rotor

downwash, complemented by the usage of cold-gas side thruster nozzle(s). This system was developed and produced by McDonnell Douglas/Hughes Helicopters, and is referred to as NOTAR, i.e., "no tail rotor". Another alternative to the conventional tail rotor that has also been used in practice in various formats is the use of a shrouded or ducted fan in the tail area, as may be seen in Fig. 3.11.

To lift a relatively heavy vehicle off the ground into a hover or slow ascent manoeuvre, and then transfer to forward flight, requires a large main rotor (or two, as per Fig. 3.10, and in the case of needing a very impressive lifting capability, three main rotors have on occasion been employed). Typical main rotor blades have a greater length-to-chord ratio (aspect ratio), versus a conventional propeller, and thus can be observed to deflect substantially from root to tip (deflect downward [droop] at rest on the ground, or deflect upward [form an effective cone shape for the rotor disc] with the generation of a large lifting force). Typical rotation speeds for main rotors are understandably lower than those seen for propellers, on the order of 200–400 rpm, versus 1,000–1,500 rpm. A main rotor may have two blades as a common design choice for smaller helicopters, but the blade number can go as high as six, say for large-vehicle heavy-lift applications. Tail rotors typically have two or four blades, and a lower aspect ratio (larger chord, vs. length). Some tail rotors can be rotated, on command, several degrees clockwise and counterclockwise about the tailboom axis (from its reference angle relative to the horizontal), for better flight control.

The amount of lift generated by a main rotor is controlled by two means: (a) the engine throttle setting for desired level of main rotor rotational speed, and (b) *collective* pitch setting, which sets the angle of incidence of the main rotor blades collectively to produce the desired uniform lifting force on the vehicle (e.g., higher lift required, a higher blade incidence angle setting is needed, for the same rotor rotational speed). Rotation of the vehicle's body in pitch or roll or some combination thereof is largely via the *cyclic* pitch setting of the main rotor, whereby the individual main rotor blades will have their incidence vary as they complete a

Fig. 3.11 Example of a ducted tail fan used by Aerospatiale/Eurocopter HH-65 Dolphin. Photo courtesy of United States Coast Guard

given revolution about the vehicle, depending on the desired direction of the rotational moment. Allied to the use of cyclic pitch, in the helicopter main rotor mast and hub assembly (see Fig. 3.12), one commonly sees the use of a swashplate or moveable disk as a means of actuation, which via various alternative techniques causes the main rotor disk to effectively deflect or tilt in the direction indicated by the pilot, e.g., tilt the rotor disk forward to introduce a forward thrust component to move the vehicle forwards.

In-plane motion refers to lead-lag (drag) movement of the main rotor blade forward and backward during its rotation about the hub, while out-of-plane motion refers to the up-and-down flapping of the blade. Flapping of a main rotor blade up and down over one rotation cycle while in forward flight has the beneficial effect of leveling out the amount of lifting force being generated by the blade during the cycle. The structural design of the rotor blade root attachment to the hub influences the blade's command and innate dynamic behavior. Soft in-plane behavior refers to a low lag response (<1/rev, i.e., less than one oscillation cycle per main rotor revolution), while stiff in-plane behavior refers to a higher lag response (>1/rev). Given the various stresses acting on a main rotor blade, they are commonly hinged horizontally and vertically at the blade root where they attach to the hub (fully articulated hub attachment), in order to allow some limited motion and corresponding stress relief (see Fig. 3.9). In those cases where hinging is only done in one direction, it is referred to as a semi-rigid or teetering hub (rotor blades rigidly attached to hub, but hub can tilt [teeter] as required, giving a seesaw flapping motion [one up, one down] to the opposing rotor blades in a two-blade design). Higher performance requirements, and improvements in structural materials (largely, composites) and design (e.g., elastomeric bearings) with regard to allowed flexure and rotation, now results in the occasional more recent use of hingeless, and bearingless/hingeless, rotor/hub designs. Modern microprocessor-based active rotor control (ARC), sometimes proposed in conjunction with the use of individual blade control (IBC), allows for further new approaches to be considered [4]. ARC may also allow for the reduction of main rotor blade vibration and noise production, two problem areas commonly associated with helicopters. Like propellers, main and tail rotors can be constructed of aluminum, steel or lighter composite materials.

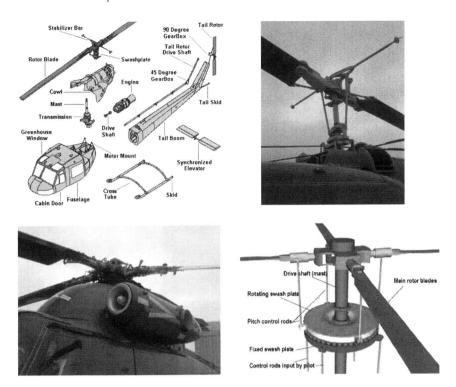

Fig. 3.12 Schematic diagram at *left* illustrating the various components making up a small conventional helicopter. The schematic diagram at *right* illustrates a conventional main rotor mast (rotorhead), with the hub above the mast connecting the rotor blades to the drive shaft in a fully articulated design (hinged); a swashplate approach is being used to control the effective main rotor disk deflection and tilt direction thereof. Right diagram courtesy of HowStuffWorks.com

As a final note on helicopter rotors, one can also encounter hybrid aircraft designs like the tiltwing V/STOL Canadair CL-84 shown in Fig. 3.13, where the rotor can play the role of propeller for part of the flight mission. Like the more modern tiltrotor Bell-Boeing V-22 Osprey of Fig. 3.14, the CL-84 exploits the advantages of a helicopter rotor for vertical takeoffs and landings, and by transitioning the position of the wing (in the case of the CL-84) or the engines (in the case of the Osprey), one can gain the cruise performance advantage of the propeller. Earlier in the history of aircraft development, before the mainstream use of helicopters, autogyros (a.k.a., autogiros, gyrocopters) were quite popular. Forward thrust was typically delivered by a piston-engine-powered conventional propeller, while lift was produced by a small fixed wing working in conjunction with an unpowered autorotating (windmilling) main rotor. This combination provided an impressive STOL (short takeoff/short landing) capability for its time. Example gyrocopters can be seen in Figs. 3.15 and 3.16.

Fig. 3.13 Photos and diagram of the Canadair CL-84 Dynavert tiltwing transport prototype aircraft (non-production; only four prototypes were ultimately built)

Fig. 3.14 Photos of the tiltrotor Bell-Boeing V-22 Osprey. Courtesy of United States Marine Corps

Fig. 3.15 Photo of McCulloch Super J-2 gyrocopter, utilizing a 3-bladed variable-pitch Hartzell propeller in a pusher configuration for forward thrust, powered by a 180-hp Lycoming O-360 4-cylinder horizontally-opposed spark-ignition piston engine. Aircraft displayed at Pima Air & Space Museum

Fig. 3.16 Photo of a Kellett K-2 autogiro, with its forward-thrust two-bladed propeller powered by a 165-hp Continental A-70 7-cylinder air-cooled radial piston engine. On display at the National Museum of the USAF

3.9   Example Problems

3.1. A radio-controlled (R/C) model aircraft has a 1-hp powerplant driving a 1-ft diameter propeller. Assuming the ideal actuator disk power is 75% of the above (i.e., this figure accounts for other factors like profile drag and vortex losses), estimate the static thrust produced by the propeller at standard sea level (SSL) conditions. In addition, ascertain the thrust produced at a flight speed of 60 ft/s, and at a flight speed of 120 ft/s.

3.2. A light utility aircraft employs a 180-hp normally-aspirated piston engine, with a 6-ft diameter propeller. Assuming the ideal actuator disk power is 75% of the above, ascertain the static thrust at SSL conditions, and also at flight speeds of 60 and 120 ft/s.

3.3. A twin-engine light cargo aircraft has a takeoff weight of 10,000 lb, and for a reasonable takeoff distance, requires a F/W at 50 mph of 0.25 at SSL conditions. The propeller diameter is 7 ft. Assuming the required engine shaft power to drive the propeller is 1.33 times the ideal actuator disk power, estimate the needed shaft power delivered by each engine.

3.4. As a first estimate of the thrust and power coefficient of a 3-bladed 6-ft-diameter low-speed propeller moving at an advance ratio of 0.6, simplify the standard momentum-blade element integration procedure by separating the propeller blade into 3 mean-property lengthwise sections. The hub ends at $0.2R$ (no thrust produced from $x = 0$ to $x = 0.2$), and the first section will apply from $x = 0.2$ to $x = 0.33$, with the mean blade chord being 0.3 ft and the mean blade pitch angle of 50°. The second section will apply from $x = 0.33$ to $x = 0.66$, with the mean blade chord being 0.42 ft and the mean blade pitch angle of 33°. The third section will apply from $x = 0.66$ to $x = 1.0$, with the mean blade chord being 0.33 ft and the mean blade pitch angle of 21°. You may assume the section lift-curve slope a_o of the propeller airfoil is 5.7 rad^{-1} (up to stall $C_{\ell\,max}$ of 1.2), and for section drag coefficient:

$$C_d = 0.01, C_\ell < 0.15$$

$$C_d = 0.01 + 0.02(C_\ell - 0.15)^2, C_\ell > 0.15$$

With this summation approach, find overall C_T, C_P and η_{pr} at $J = 0.6$.

3.5. Estimate the activity factor of the propeller of Problem 3.4 above, using the simplified integration/summation approach of that case.

3.6. Using the actuator disk approach, derive analytical expressions for the static and advancing thrust of a ducted fan/propeller, and the ideal induced power. The major difference from the unducted case is to assume that the propeller disk area and the wake cross-sectional area are the same (i.e., no slipstream area reduction). Correlate your resulting equation to that of the unducted case.

3.7. Refer to the propeller of Problem 3.4. As a design exercise, evaluate the effects of increasing the outer blade pitch angles, i.e., 36° for the middle blade segment (vs. 33°), and 26° for the outer blade segment (vs. 21°) in using the momentum/blade-element approach. For an advance ratio of 0.6, determine the new resulting values for C_T, C_P and η_{pr}.

3.10 Solution for Problems

3.1. General info:

Ideal actuator power $P = P_S \times$ correction factor $= 1.0 \times 0.75 = 0.75$ hp $=$ 412.5 ft lb/s, $A_1 = \pi d_p^2/4 = \pi(1)^2/4 = 0.79$ ft^2, $\rho_{S/L} = 0.002378$ slug/ft^3

Static sea level (SSL) thrust:
$F(0) = P^{2/3}(2\rho A_1)^{1/3} = 412.5^{0.667}(2 \cdot 0.002378 \cdot 0.79)^{0.333} = 8.65$ lbf, for a true power loading P_S/A_1 of $1.0/0.79 = 1.266$ hp/ft^2.
$V_o = 60$ ft/s (41 mph), sea level:
From textbook, for $V_o > 0$, ideal actuator power is:

$$P = F(V_o + w) = F\left(V_o - \frac{V_o}{2} + \frac{1}{2}\sqrt{V_o^2 + \frac{2F}{\rho A_1}}\right)$$

The above equation can be re-arranged into a cubic equation:

$$\frac{F^3}{2\rho A_1} + PV_o F - P^2 = 0$$

A cubic equation has three possible solutions (roots). In the context of the present problem, only one of the solutions will make sense. One notes that thrust drops moderately as airspeed increases.

$$266.2F^3 + 24750F - 170156 = 0$$

Solve iteratively to get: $F(V_o) = 5.28$ lbf
$V_o = 120$ ft/s (82 mph), sea level:

$$266.2F^3 + 49500F - 170156 = 0$$

Solve iteratively to get: $F(V_o) = 3.25$ lbf

3.2. General info:

Ideal actuator power $P = P_S \times$ correction factor $= 180.0 \times 0.75 = 135$ hp $=$ 74250 ft lb/s,
$A_1 = \pi d_p^2/4 = \pi(6)^2/4 = 28.3$ ft^2, $\rho_{S/L} = 0.002378$ slug/ft^3

Static sea level (SSL) thrust:
$F(0) = P^{2/3}(2\rho A_1)^{1/3} = 74250^{0.667}(2 \cdot 0.002378 \cdot 28.3)^{0.333} = 905$ lbf, for a true power loading P_S/A_1 of $180.0/28.3 = 6.36$ hp/ft^2 .

$V_o = 60$ ft/s (41 mph), sea level:
From Practice Problem 3.1, for $V_o > 0$,

$$\frac{F^3}{2\rho A_1} + PV_o F - P^2 = 0$$

$7.43F^3 + 4455000F - 5.513 \times 10^9 = 0$

Solve iteratively to get: $F(V_o) = 690$ lbf

$V_o = 120$ ft/s (82 mph), sea level:

$7.43F^3 + 8.91 \times 10^6 F - 5.513 \times 10^9 = 0$

Solve iteratively to get: $F(V_o) = 509$ lbf

3.3. General info:

Actual engine shaft power $P_S = P \times$ correction factor $= P \times 1.33$
$A_1 = \pi d_p^2/4 = \pi(7)^2/4 = 38.5$ ft^2, $\rho_{S/L} = 0.002378$ slug/ft^3
$F/W = 0.25$ at 50 mph (74 ft/s) for airplane using two engines/props, at sea level,
so for a single engine/propeller,
$F_{eng} = 0.5(0.25\ W) = 0.5\ (0.25 \cdot 10000) = 1250$ lbf

For ideal actuator power, for one engine:

$$P_{eng} = F_{eng}(V_o + w) = F_{eng}\left(\frac{V_o}{2} + \frac{1}{2}\sqrt{V_o^2 + \frac{2F_{eng}}{\rho A_1}}\right)$$

$$= 1250(74/2 + 0.5[74^2 + 2(1250)/(0.002378 \cdot 38.5)]) = 159412\,\text{ft lb/s}$$

So, for one engine, actual shaft power required is:

$P_{S\ eng} = 159412 \times 1.33 = 212017$ ft lb/s, or 385.5 hp

3.4. General info:

$B = 3$ blades, $C_{\ell max} = 1.2, a_0 = 5.7\,\text{rad}^{-1} = 0.1\ \text{deg}^{-1}$,
$C_{do} = 0.01\,(\text{up to } C_\ell = 0.15, C_d = C_{do})$

$\alpha_{max} = C_{\ell max}/a_0 = 12° = 0.21$ rad, beyond which the blade section will stall
(some drop off in C_ℓ, and substantial increase in C_d).

$$\phi = \tan^{-1}\left(\frac{V_\infty}{\omega r}\right) = \tan^{-1}\left(\frac{V_\infty}{\omega R x}\right) = \tan^{-1}\left(\frac{\lambda}{x}\right) = \tan^{-1}\left(\frac{J}{\pi x}\right),$$
where $x = r/R$, $J = V_\infty/(n \cdot 2R) = V_\infty/(\omega R/\pi)$,

and $\lambda = V_\infty/(\omega R) = J/\pi$.
Propeller blade tip speed, $V_T = \omega R = (2\pi n)d_p/2 = \pi n d_p$, $R = 3$ ft, $d_p = 6$ ft
Overall (ref.) propeller solidity, $\sigma_{ref} =$(blade area)/(disk area)$= B(c_{ref}R)/(\pi R^2) =$

$Bc_{ref}/(\pi R)$

Local propeller solidity (at radial location r, local chord c), $\sigma = Bc/(\pi R) = x \cdot Bc/(\pi r)$

$V_R \sin \alpha_i = w$ and $V_R \cos \phi = \omega r = V_T x$ from propeller blade velocity-vector diagram.

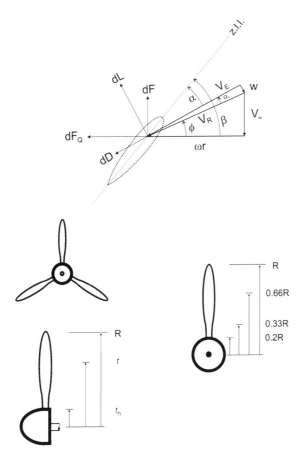

Employ the standard momentum/blade-element integration approach, with the propeller blade divided in this example into three segments, where mean properties will be assumed for each of the segments.

$\lambda = J/\pi = 0.191$

Blade segment #1, from $x = 0.2$ to $x = 0.33$:

$$\bar{x} = (0.2 + 0.33)/2 = 0.265, \bar{\phi} = \tan^{-1}\frac{J}{\pi x} = \tan^{-1}(0.6/(\pi \cdot 0.265)) = 35.8°,$$

$$\bar{\beta} = 50°$$

$$\lambda/\bar{x} = 0.191/0.265 = 0.72, \, V_R/V_T = \bar{x}/\cos\overline{\phi} = 0.265/\cos 35.8° = 0.33$$

$$\bar{\sigma} = \frac{B\bar{c}}{\pi R} = 3(0.3)/(\pi \cdot 3) = 0.096$$

$$\bar{\alpha}_i = \frac{1}{2}\left\{ -\left(\frac{\lambda}{x} + \frac{\sigma a_0 V_R}{8x^2 V_T}\right) + \left[\left(\frac{\lambda}{x} + \frac{\sigma a_0 V_R}{8x^2 V_T}\right)^2 + \frac{\sigma a_0 V_R}{2x^2 V_T}(\beta - \phi)\right]^{1/2} \right\}$$

$$= 0.5\{-(0.72 + 0.096(5.7)0.33/(8 \cdot 0.265^2)$$

$$+ [(0.72 + 0.096(5.7)0.33/(8 \cdot 0.265^2))^2$$

$$+ 0.096(5.7)0.33/(2 \cdot 0.265^2) \cdot (50 - 35.8)\pi/180]^{0.5}\}$$

$$= 0.5\{-1.041 + 1.184\} = 0.072 \text{ rad or } 4.1°$$

$C_\ell = a_o(\beta - \alpha_i - \phi) = 0.1(50 - 4.1 - 35.8) = 1.01$, so below $C_{\ell max}$ by a reasonable margin.

$$C_d = C_{do} + 0.02(C_\ell - 0.15)^2 = 0.01 + 0.02(1.01 - 0.15)^2 = 0.025$$

$$C_{T,1} = \frac{\pi}{8} \int_{x_{start}}^{x_{end}} \sigma(J^2 + \pi^2 x^2)[C_\ell \cos(\phi + \alpha_i) - C_d \sin(\phi + \alpha_i)]dx$$

$$= \frac{\pi}{8} \cdot \bar{\sigma}(J^2 + \pi^2\bar{x}^2)[C_\ell \cos(\phi + \alpha_i) - C_d \sin(\phi + \alpha_i)]\Delta x$$

$$= \pi/8 \cdot 0.096\,(0.6^2 + \pi^2(0.265^2))[1.01 \cos(35.8 + 4.1)$$

$$- 0.025 \sin(35.8 + 4.1)](0.33 - 0.2)$$

$$= 0.00392$$

$$C_{P,1} = \frac{\pi^2}{8} \int_{x_{start}}^{x_{end}} \sigma x(J^2 + \pi^2 x^2)[C_\ell \sin(\phi + \alpha_i) + C_d \cos(\phi + \alpha_i)]dx$$

$$= \frac{\pi^2}{8} \cdot \bar{\sigma} \cdot \bar{x}(J^2 + \pi^2\bar{x}^2)[C_\ell \sin(\phi + \alpha_i) + C_d \cos(\phi + \alpha_i)]\Delta x$$

$$= \pi^2/8 \cdot 0.096 \cdot 0.265 \times (0.6^2 + \pi^2(0.265^2))[1.01 \sin(35.8 + 4.1)$$

$$+ 0.025 \cos(35.8 + 4.1)](0.33 - 0.2)$$

$C_{P,I} = 0.00287$

Blade segment #2, from $x = 0.33$ to $x = 0.66$:

$$\bar{x} = (0.33 + 0.66)/2 = 0.5, \, \overline{\phi} = \tan^{-1}\frac{J}{\pi x} = \tan^{-1}(0.6/(\pi \cdot 0.5)) = 21°, \bar{\beta} = 33°$$

$\lambda/\bar{x} = 0.191/0.5 = 0.38, V_R/V_T = \bar{x}/\cos\overline{\phi} = 0.5/\cos 21° = 0.54$

$$\bar{\sigma} = \frac{B\bar{c}}{\pi R} = 3(0.42)/(\pi \cdot 3) = 0.13$$

$\alpha_i = 0.5\{-(0.38 + 0.13(5.7)0.54/(8 \cdot 0.5^2) + [(0.38 + 0.13(5.7)0.54/(8 \cdot 0.5^2))^2$

$\qquad + 0.13(5.7)0.54/(2 \cdot 0.5^2) \cdot (33 - 21)\pi/180]^{0.5}\}$

$\qquad = 0.5\{-0.58 + 0.71\} = 0.065$ rad or $3.7°$

$C_\ell = a_o(\beta - \alpha_i - \phi) = 0.1(33 - 3.7 - 21) = 0.83$, so below $C_{\ell max}$ by a healthy argin. If one wanted to maximize thrust at this advance ratio, one could increase blade pitch angle β a bit (bearing in mind that this is a fixed-pitch prop, so some compromise needed in moving from low J at takeoff and higher J at cruise).

$$C_d = C_{do} + 0.02(C_\ell - 0.15)^2 = 0.01 + 0.02(0.83 - 0.15)^2 = 0.02$$

$C_{T,2} = \pi/8 \cdot 0.13\,(0.6^2 + \pi^2(0.5^2))[0.83\cos(21 + 3.7) - 0.02\sin(21 + 3.7)]$

$\qquad \times (0.66 - 0.33) = 0.036$

$C_{P,2} = \pi^2/8 \cdot 0.13 \cdot 0.5\,(0.6^2 + \pi^2(0.5^2))[0.83\sin(21 + 3.7) + 0.02\cos(21 + 3.7)]$

$\qquad \times (0.66 - 0.33) = 0.027$

Blade segment #3, from $x = 0.66$ to $x = 1.00$:

$\bar{x} = (0.66 + 1.0)/2 = 0.83, \overline{\phi} = \tan^{-1}\dfrac{J}{\pi x} = \tan^{-1}(0.6/(\pi \cdot 0.83)) = 13°, \overline{\beta} = 21°$

$\lambda/\bar{x} = 0.191/0.83 = 0.23, V_R/V_T = \bar{x}/\cos\overline{\phi} = 0.83/\cos 13° = 0.85$

$$\bar{\sigma} = \frac{B\bar{c}}{\pi R} = 3(0.33)/(\pi \cdot 3) = 0.11$$

$\alpha_i = 0.5\{-(0.23 + 0.11(5.7)0.85/(8 \cdot 0.83^2)$

$\qquad + [(0.23 + 0.11(5.7)0.85/(8 \cdot 0.83^2))^2$

$\qquad + 0.11(5.7)0.85/(2 \cdot 0.83^2) \cdot (21 - 13)\pi/180]^{0.5}\}$

$\qquad = 0.5\{-0.33 + 0.4\} = 0.036$ rad or $2.1°$

$C_\ell = a_o(\beta - \alpha_i - \phi) = 0.1(21 - 2.1 - 13) = 0.59$; could bump that value up a lot, if needed

$$C_d = C_{do} + 0.02(C_\ell - 0.15)^2 = 0.01 + 0.02(0.59 - 0.15)^2 = 0.014$$

$$C_{T,3} = \pi/8 \cdot 0.11 \, (0.6^2 + \pi^2(0.83^2))[0.59\cos(13 + 2.1) - 0.014\sin(13 + 2.1)]$$
$$\times (1.0 - 0.66) = 0.06$$

$$C_{P,3} = \pi^2/8 \cdot 0.11 \cdot 0.83 \, (0.6^2 + \pi^2(0.83^2))[0.59\sin(13 + 2.1) + 0.014\cos(13 + 2.1)]$$
$$\times (1.0 - 0.66) = 0.046$$

Overall:
$C_T = \sum C_{Tj} = 0.00392 + 0.036 + 0.06 = 0.10$, on the low side for this propeller
$C_P = \sum C_{Pj} = 0.00287 + 0.027 + 0.046 = 0.076$, on the low side for this propeller
$\eta_{pr} = J\frac{C_T}{C_P} = 0.6(0.10)/0.076 = 0.79$, a reasonable figure if near cruise conditions, but a bit high if still wanting some early flight thrust production (more thrust = lower propulsive efficiency).

3.5. Refer to Practice Problem 3.4 for information on propeller blade segments. General info:

$d_p = 6$ ft, $R = 3$ ft, $x = r/R$; blade chord c varies in moving from hub to tip (0.3, 0.42, 0.33 ft)
Activity factor:

$$AF = \frac{100000}{16} \int\limits_{x_h}^{1} \frac{c}{d_p} \cdot x^3 dx$$

Separate propeller blade into three segments as per Problem 3.4, and assume mean properties over each segment, such that:

$$AF = 6250 \int\limits_{0.2}^{0.33} \frac{0.3}{6} \cdot x^3 dx + 6250 \int\limits_{0.33}^{0.66} \frac{0.42}{6} \cdot x^3 dx + 6250 \int\limits_{0.66}^{1.0} \frac{0.33}{6} \cdot x^3 dx$$

$$= 6250 \cdot \frac{0.3}{6} \left\{\frac{x^4}{4}\right\}_{0.2}^{0.33} + 6250 \cdot \frac{0.42}{6} \left\{\frac{x^4}{4}\right\}_{0.33}^{0.66} + 6250 \cdot \frac{0.33}{6} \left\{\frac{x^4}{4}\right\}_{0.66}^{1.0}$$

$$= 6250 \cdot \frac{0.3}{6} \left[\frac{0.33^4}{4} - \frac{0.2^4}{4}\right] + 6250 \cdot \frac{0.42}{6} \left[\frac{0.66^4}{4} - \frac{0.33^4}{4}\right] + 6250 \cdot \frac{0.33}{6} \left[\frac{1.0^4}{4} - \frac{0.66^4}{4}\right]$$

$$= 0.8 + 19.5 + 69.6 = 89.9$$

Since the activity factor for the Clark-Y low-speed propeller is about 100, this suggests that using only 3 lengthwise mean-property sections in Problem 3.4 might be leading to an underprediction of the propeller's actual expected performance. The use of more than 3 sections is certainly encouraged by this result on estimated AF, say, double the number to 6 sections in moving from the hub to the blade tip.

3.6. Refer to textbook for baseline case, for unducted (unshrouded) propeller. In the case of a ducted propeller or fan, the surrounding shroud or duct would extend rearward from the disk (of area A_1), such that there would be no reduction in the slipstream (of area A_3), giving:

$$A_3 = A_1$$

and

$$\dot{m} = \rho V_3 A_3 = \rho V_3 A_1$$

Static case ($V_o = 0$):
Ideal static power,

$$P_o = P_{i,o} = \frac{1}{2}\dot{m}V_3^2 = \frac{1}{2}\rho A_3 V_3 \cdot V_3^2 = \frac{1}{2}\rho A_1 V_3^3$$

Ideal static thrust (substitute from above results),

$$F_o = \dot{m}V_3 = \frac{2P_o}{V_3} = \frac{2P_o}{\left(\frac{2P_o}{\rho A_1}\right)^{1/3}} = 2^{2/3}P_o^{2/3}(\rho A_1)^{1/3} = 1.59P_o^{2/3}(\rho A_1)^{1/3}$$

which is 26% more thrust delivered, versus the unducted propeller from actuator disk theory, for the same power input.
Advancing ($V_o > 0$):
Let induced velocity w be estimated as downstream at station ③, so that for thrust,

$$F = \dot{m}V_3 - \dot{m}V_o = \rho A_1(V_o + w)^2 - \rho A_1(V_0 + w)V_o = \rho A_1(V_ow + w^2)$$

$$= \rho A_1 w(V_o + w)$$

From the quadratic equation solution to the above, for induced velocity,

$$w = -\frac{V_o}{2} + \frac{1}{2}\sqrt{V_o^2 + \frac{4F}{\rho A_1}}$$

For ideal power,

$$P = \frac{1}{2}\dot{m}V_3^2 - \frac{1}{2}\dot{m}V_o^2 = \frac{1}{2}\rho A_1(V_o + w)[(V_o + w)^2 - V_o^2]$$

$$= \frac{1}{2}\rho A_1 w[2V_o^2 + 3wV_o + w^2]$$

Thus for a given thrust requirement at flight speed V_o, one would first calculate the resulting w, and then calculate the required ideal power P as a function of w and V_o. The airfoil shape of the surrounding shroud may provide a small forward thrust component via the resultant overall lift vector. However, viscous drag over the shroud surface will certainly subtract from the delivered thrust. The weight of the shroud is also a penalty, from an overall aircraft weight viewpoint (more weight = more fuel to fly). So, if one still felt that one would get some benefit from shrouding a propeller or fan, it would be typical not to extend the shroud too far rearward from the prop/fan disk location.

3.7. Refer to Practice Problem 3.4 for original case. Outer blade pitch angles a bit higher here.
General info:

$$B = 3 \text{ blades}, C_{\ell max} = 1.2, a_o = 5.7 \text{ rad}^{-1} = 0.1 \text{ deg}^{-1},$$
$$C_{do} = 0.01 \text{ (up to } C_\ell = 0.15, C_d = C_{do})$$

$\alpha_{max} = C_{\ell max}/a_o = 12° = 0.21 \text{ rad}$, beyond which the blade section will stall (some drop off in C_ℓ, and substantial increase in C_d).
Propeller blade tip speed, $V_T = \omega R = (2\pi n)d_p/2 = \pi n d_p, R = 3 \text{ ft}, d_p = 6 \text{ ft}$
Employ the standard momentum/blade-element integration approach, with the propeller blade divided in this example into three segments, where mean properties will be assumed for each of the segments.

$$\lambda = J/\pi = 0.191$$

Blade segment #1, from $x = 0.2$ to $x = 0.33$:

$$\bar{x} = (0.2 + 0.33)/2 = 0.265, \bar{\phi} = \tan^{-1}\frac{J}{\pi x}$$
$$= \tan^{-1}(0.6/(\pi \cdot 0.265)) = 35.8°, \bar{\beta} = 50°$$

$$\lambda/\bar{x} = 0.191/0.265 = 0.72, V_R/V_T = \bar{x}/\cos\bar{\phi} = 0.265/\cos 35.8° = 0.33$$

$$\bar{\sigma} = \frac{B\bar{c}}{\pi R} = 3(0.3)/(\pi \cdot 3) = 0.096$$

$$\bar{\alpha}_i = \frac{1}{2}\left\{-\left(\frac{\lambda}{x}+\frac{\sigma a_0 V_R}{8x^2 V_T}\right) + \left[\left(\frac{\lambda}{x}+\frac{\sigma a_0 V_R}{8x^2 V_T}\right)^2 + \frac{\sigma a_0 V_R}{2x^2 V_T}(\beta - \phi)\right]^{1/2}\right\}$$

$$= 0.5\{-(0.72 + 0.096(5.7)0.33/(8 \cdot 0.265^2) + [(0.72 + 0.096(5.7)0.33/(8 \cdot 0.265^2))^2$$
$$+ 0.096(5.7)0.33/(2 \cdot 0.265^2) \cdot (50 - 35.8)\pi/180]^{0.5}\}$$
$$= 0.5\{-1.041 + 1.184\} = 0.072 \text{ rad or } 4.1°$$

$C_\ell = a_o(\beta - \alpha_i - \phi) = 0.1(50 - 4.1 - 35.8) = 1.01$, so below $C_{\ell max}$ by a reasonable margin.

$$C_d = C_{do} + 0.02(C_\ell - 0.15)^2 = 0.01 + 0.02(1.01 - 0.15)^2 = 0.025$$

$$C_{T,1} = \frac{\pi}{8} \int_{x_{\text{start}}}^{x_{\text{end}}} \sigma(J^2 + \pi^2 x^2)[C_\ell \cos(\phi + \alpha_i) - C_d \sin(\phi + \alpha_i)]dx$$

$$= \frac{\pi}{8} \cdot \bar{\sigma}(J^2 + \pi^2 \bar{x}^2)[C_\ell \cos(\phi + \alpha_i) - C_d \sin(\phi + \alpha_i)]\Delta x$$

$$= \pi/8 \cdot 0.096 \, (0.6^2 + \pi^2 (0.265^2))[1.01 \cos(35.8 + 4.1) - 0.025 \sin(35.8 + 4.1)]$$

$$\times (0.33 - 0.2) = 0.00392$$

$$C_{P,1} = \frac{\pi^2}{8} \int_{x_{\text{start}}}^{x_{\text{end}}} \sigma x(J^2 + \pi^2 x^2)[C_\ell \sin(\phi + \alpha_i) + C_d \cos(\phi + \alpha_i)]dx$$

$$= \frac{\pi^2}{8} \cdot \bar{\sigma} \cdot \bar{x}(J^2 + \pi^2 \bar{x}^2)[C_\ell \sin(\phi + \alpha_i) + C_d \cos(\phi + \alpha_i)]\Delta x$$

$$= \pi^2/8 \cdot 0.096 \cdot 0.265 \, (0.6^2 + \pi^2 (0.265^2))$$

$$\times [1.01 \sin(35.8 + 4.1) + 0.025 \cos(35.8 + 4.1)](0.33 - 0.2)$$

$$C_{P,1} = 0.00287$$

Blade segment #2, from $x = 0.33$ to $x = 0.66$:

$$\bar{x} = (0.33 + 0.66)/2 = 0.5, \bar{\phi} = \tan^{-1}\frac{J}{\pi x} = \tan^{-1}(0.6/(\pi \cdot 0.5)) = 21°, \bar{\beta} = 36°$$

$$\lambda/\bar{x} = 0.191/0.5 = 0.38, V_R/V_T = \bar{x}/\cos\bar{\phi} = 0.5/\cos 21° = 0.54$$

$$\bar{\sigma} = \frac{B\bar{c}}{\pi R} = 3(0.42)/(\pi \cdot 3) = 0.13$$

$$\alpha_i = 0.5\{-(0.38 + 0.13(5.7)0.54/(8 \cdot 0.5^2) + [(0.38 + 0.13(5.7)0.54/(8 \cdot 0.5^2))^2$$

$$+ 0.13(5.7)0.54/(2 \cdot 0.5^2) \cdot (36 - 21)\pi/180]^{0.5}\}$$

$$= 0.5\{-0.58 + 0.74\} = 0.08 \text{ rad or } 4.6°$$

$C_\ell = a_o(\beta - \alpha_i - \phi) = 0.1(36 - 4.6 - 21) = 1.04$, so closer to $C_{\ell\text{max}}$ by a reasonable margin.

$C_d = C_{\text{do}} + 0.02(C_\ell - 0.15)^2 = 0.01 + 0.02(1.04 - 0.15)^2 = 0.026$

$$C_{T,2} = \pi/8 \cdot 0.13 \, (0.6^2 + \pi^2 (0.5^2))[1.04 \cos(21 + 4.6) - 0.026 \sin(21 + 4.6)]$$

$$\times (0.66 - 0.33) = 0.044$$

$$C_{P,2} = \pi^2/8 \cdot 0.13 \cdot 0.5 \, (0.6^2 + \pi^2 (0.5^2))[1.04 \sin(21 + 4.6) + 0.026 \cos(21 + 4.6)]$$

$$\times (0.66 - 0.33) = 0.035$$

Blade segment #3, from $x = 0.66$ to $x = 1.00$:

$$\bar{x} = (0.66 + 1.0)/2 = 0.83, \overline{\phi} = \tan^{-1}\frac{J}{\pi x} = \tan^{-1}(0.6/(\pi \cdot 0.83)) = 13E, \overline{\beta} = 26°$$

$$\lambda/\bar{x} = 0.191/0.83 = 0.23, V_R/V_T = \bar{x}/\cos\overline{\phi} = 0.83/\cos 13° = 0.85$$

$$\bar{\sigma} = \frac{B\bar{c}}{\pi R} = 3(0.33)/(\pi \cdot 3) = 0.11$$

$$\alpha_i = 0.5\{-(0.23 + 0.11(5.7)0.85/(8 \cdot 0.83^2)$$
$$+ [(0.23 + 0.11(5.7)0.85/(8 \cdot 0.83^2))^2$$
$$+ 0.11(5.7)0.85/(2 \cdot 0.83^2) \cdot (26 - 13)\pi/180]^{0.5}\}$$
$$= 0.5\{-0.33 + 0.44\} = 0.055 \,\text{rad}\, or\, 3.2°$$

$C_\ell = a_o(\beta - \alpha_i - \phi) = 0.1(26 - 3.2 - 13) = 0.98$; reasonable margin below $C_{\ell\max}$

$$C_d = C_{do} + 0.02(C_\ell - 0.15)^2 = 0.01 + 0.02(0.98 - 0.15)^2 = 0.024$$

$$C_{T,3} = \pi/8 \cdot 0.11\,(0.6^2 + \pi^2(0.83^2))[0.98\cos(13 + 3.2) - 0.024\sin(13 + 3.2)]$$
$$\times (1.0 - 0.66) = 0.098$$

$$C_{P,3} = \pi^2/8 \cdot 0.11 \cdot 0.83\,(0.6^2 + \pi^2(0.83^2))$$
$$\times [0.98\sin(13 + 3.2) + 0.024\cos(13 + 3.2)](1.0 - 0.66) = 0.081$$

Overall:
$C_T = \sum C_{Tj} = 0.00392 + 0.044 + 0.098 = 0.146$, reasonable for this propeller
$C_P = \sum C_{Pj} = 0.00287 + 0.035 + 0.081 = 0.119$, reasonable for this propeller
$\eta_{pr} = J\frac{C_T}{C_P} = 0.6\,(0.146)/0.119 = 0.74$, a reasonable figure for an early flight requirement

References

1. McCormick BW (1995) Aerodynamics, aeronautics and flight mechanics, 2nd edn. Wiley, New York
2. Anonymous (2004) Airplane flying handbook. FAA-H-8083-3A, Airman Testing Standards Branch, Federal Aviation Administration, U.S. Department of Transportation, Oklahoma City
3. Dommasch DO (1953) Elements of propeller and helicopter aerodynamics. Pitman, New York
4. Padfield GD (1996) Helicopter flight dynamics: the theory and application of flying qualities and simulation modeling. AIAA, Reston (Virginia)

Chapter 4
Internal Combustion Engines

4.1 Introduction

Historically, one can say that the internal combustion engine [where fuel is burned in reacting with incoming air, within the engine, to produce the required heat] is the successor to the external combustion engine, best represented by the steam engine [where the working medium (water vapor) is produced by heating of water in an external diesel-, coal- or wood-fired furnace]. There are a number of different powerplants (spark-ignition, compression-ignition, rotary) under the internal combustion (IC) engine category that can be used for driving a propeller shaft. Each share the characteristic of intermittent combustion in their cyclic operation. Continuous combustion will be seen later for the gas turbine engine variants, including the turboprop engine that can also be used for driving a propeller shaft.

4.2 Spark-Ignition Engines

We are most likely familiar with this category of IC engine from the automotive world: spark ignition (SI), 4-stroke operational cycle, with reciprocating pistons in cylinders used to drive (rotate) the main driveshaft (crankshaft), as illustrated in Fig. 4.1 [1]. This kind of engine has been referred to as a positive displacement machine (say, in comparison to a turbomachine, which describes the gas turbine engines to be discussed later in the book). Some variants may have fuel injection, and perhaps turbosupercharging, to improve performance. A number of present day lighter aircraft are powered by some variant of the above, usually using two, four or six air-cooled cylinders in a horizontally opposed configuration for a given engine (see Figs. 4.2 and 4.3), with lower-cost cases likely using the older carburetion approach for fuel–air vapor delivery to the cylinders via a carburetor and intake manifold, versus fuel injection directly or almost directly into the cylinder. Carburetors are susceptible to icing up in their venturi as one gains

D. R. Greatrix, *Powered Flight*, DOI: 10.1007/978-1-4471-2485-6_4,
© Springer-Verlag London Limited 2012

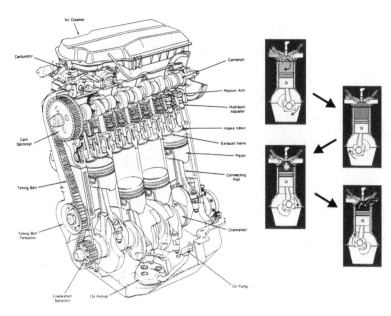

Fig. 4.1 Cutaway diagram of an in-line four-cylinder Chrysler 4-stroke spark-ignition engine for automotive applications [1]. A connecting rod connects the reciprocating piston (*up-and-down* motion) to the rotating crankshaft. The intake-compression-combustion-exhaust process is illustrated at the *right*. Diagram at *left* courtesy of Chrysler Group LLC

altitude and the outside air temperature drops [2]. As a result, carburetor heating is a standard requirement for preventing ice buildup, and when in doubt, pilots are advised to switch said heat on. Lower-performance piston engines may be unsupercharged, but higher performance ones are likely turbosupercharged. An example control setup for an SI engine is shown in Fig. 4.4 [3].

SI piston engines generally run on gasoline (automotive gasoline sometimes called mogas), and for aircraft piston engines, aviation gasoline (avgas), which has a higher octane level than mogas to prevent detonation (knocking). Density of liquid gasoline is around 720 kg/m^3 (6.1 lb/U.S. gallon), and gasoline has a lower heating value of 44.6 MJ/kg of fuel, when reacted with air (i.e., H_2O vapor in combustion products applies to LHV, as opposed to liquid H_2O, which applies to the higher heating value [HHV]). While gasoline is in fact a mixture of some 40 hydrocarbon molecules of varying sizes, it is amenable to assume a formula of octane, C_8H_{18}, for some analyses, or the pairing of octane and a lower percentage of heptane (C_7H_{16} has the characteristic of increasing the detonability of the fuel, as compared to octane). Fuels are characterized in part by their flash point. The flash point is the lowest temperature at which they will ignite in air or oxygen under standard conditions. The combustion process needs some fuel vapor present to proceed, i.e., some vapor pressure. Thus, the higher the volatility of the given fuel (i.e., its vapor pressure), the lower the flash point. The Reid vapor pressure (RVP) is a standard measurement of a fuel's volatility that is often quoted for a

Fig. 4.2 Photos of various piston engine configurations used over the years: **a** Pratt & Whitney R-1690 Hornet air-cooled 9-cylinder radial engine; **b** Allison/General Motors V-1710 liquid-cooled 12-cylinder supercharged V-12 engine; **c** Isotta-Fraschini Volo V.4B water-cooled 6-cylinder "straight-six" in-line engine; **d** Continental O-170 (A65) air-cooled 4-cylinder "flat-four" horizontally opposed piston engine

fuel's specifications. Gasoline is a fairly volatile fuel, and one needs to take some care in handling it. A distilled fuel like diesel fuel may also have a quoted end point (on the order of 300°C), which is the nominal highest temperature seen by the fuel during the distillation process. A fuel with a lower end point is observed to produce less smoke and other pollutants. In burning gasoline and diesel fuels in a combustion chamber, there are different techniques that have been developed over the years for improving the energy output and efficiency of the combustion process, including sequentially staging combustion through various means, rather than a lumped burning of a homogeneous fuel/air mixture. For example, a stratified (layered) charge fuel-injection approach involves only initially burning a richer portion of the available fuel/air mixture, and then burning the remainder of the

Fig. 4.3 Light utility aircraft (Cirrus SR22 Turbo) employing a horizontally opposed six-cylinder spark-ignition turbosupercharged air-cooled piston engine (fuel-injected 310-hp Teledyne Continental IO-550-N, with Tornado Alley turbonormalizing/intercooler kit or optional Vortech V-1 S-Trim Supercharger system kit for added power at altitude) to drive its 3-bladed Hartzell propeller

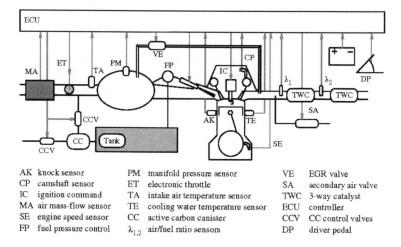

AK	knock sensor	PM	manifold pressure sensor	VE	EGR valve
CP	camshaft sensor	ET	electronic throttle	SA	secondary air valve
IC	ignition command	TA	intake air temperature sensor	TWC	3-way catalyst
MA	air mass-flow sensor	TE	cooling water temperature sensor	ECU	controller
SE	engine speed sensor	CC	active carbon canister	CCV	CC control valves
FP	fuel pressure control	$\lambda_{1,2}$	air/fuel ratio sensors	DP	driver pedal

Fig. 4.4 Example control setup for a modern automotive SI engine [3]. For the aircraft case, the analogy for the automobile driver pedal (accelerator) to control the engine's shaft rotation speed (and torque) would be the airplane pilot's manual throttle setting

leaner mixture in an extended process; this allows for higher effective compression ratios, and better fuel economy, without risking detonation.

Fuel is typically stored in the wing or in wing-tip tanks, for greatest safety in the event of a crash. Tanks can be constructed of aluminum, since that lighter metal

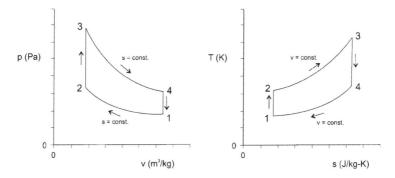

Fig. 4.5 Diagrams of p–v and T–s profiles for ideal Otto cycle

does not chemically react significantly with aviation fuels. Synthetic rubber or nylon fuel bladders may be used for containing fuel in some applications, e.g., for tanks constructed of composite materials having micro-pores that would leak fuel, otherwise. Bladders, while adding weight to the aircraft (not a good thing), do provide additional safety in the event of a crash. Fuel lines between the tank and the engine will typically be made of rigid aluminum metal alloy tubing and/or flexible synthetic rubber or Teflon hoses. Vapor lock is a problem associated with fuel lines, whereby higher atmospheric temperatures result in the fuel vaporizing and blocking the flow of liquid fuel in the line.

The piston engine's crankshaft, directly or through a reduction gear, rotationally drives the propeller. In a typical engine, the crankshaft will in addition drive a series of smaller gears, that in turn drive various items like engine lubricating oil pumps, cooling liquid pumps (if a liquid-cooled engine), cooling fans (to assist air cooling), fuel pumps, magnetos, generators, air compressors, etc. Lubricating oil contributes to engine cooling, provides an effective seal between the piston rings and cylinder walls (preventing gas "blow-by"), and as a lubricant, prevents wear by the presence of an oil film separating metal-to-metal contact. Oil also acts to clean and flush the engine of internal contaminants that have entered the engine, or been formed during the process of combustion, with the passage of the oil through a collecting filter. Oils are classified by their viscosity index (if too high for the present atmospheric conditions, one will observe a high oil pressure; conversely, if too low an index, one will observe a low oil pressure). Oil ideally should have a high flash point (i.e., not ignite at the highest temperature it might experience). Having said that, occasionally oil does enter into the combustion chamber; a high quality oil will leave little carbon (coking) or wax (lacquer) deposits should this happen. An oil ideally should also have a low pour point (i.e., point of solidification at a temperature substantially below that which might be anticipated in cold weather conditions). Oil dilution using the addition of avgas is sometimes employed for a temporary correction of overly viscous lubricating oil. The avgas can later be "boiled off" by running the engine hotter than usual, and venting the gasoline vapor out of the engine crankcase.

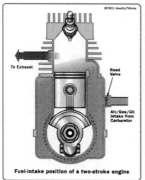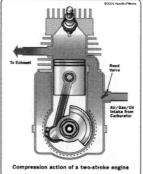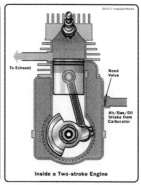

Fig. 4.6 Illustration of operational cycle of 2-stroke spark-ignition piston engine. Flapping reed or leaf valve commonly employed for intake of fresh fuel–air-oil mixture from carburetor into cylinder. The schematic shows the finning about the cylinder walls and head, as an efficient means for dissipating heat from the metal structure (more surface area) in an air-cooling approach. Diagram courtesy of HowStuffWorks.com

The ideal air-standard (air as working medium) Otto cycle approximates the open-cycle performance of a spark-ignition IC engine. As may be seen in Fig. 4.5, the characteristics of the operational cycle can be viewed through its p–v (pressure vs. specific volume) and T–s (temperature vs. entropy) diagrams. Recall that specific volume is the inverse of gas density. Process 1 → 2 is an isentropic compression of air as the piston moves from crank-end (bottom dead center, BDC) to head-end (top dead center, TDC, at cylinder head). Process 2 → 3 is representative of heat being added at a constant volume while the piston is momentarily near TDC. In the actual case, one has ignition of the fuel–air mixture by a spark plug, with subsequent burning and further resulting pressurization. Process 3 → 4 is an isentropic expansion, as the piston moves downward toward BDC. Process 4 → 1 is representative of heat being rejected from the air while the piston is in the vicinity of BDC.

For the actual 4-stroke cycle (see Fig. 4.1), beginning with outward stroke 1 from TDC to BDC, one has a fresh fuel–air mixture being taken into the cylinder. Inward stroke 2 from BDC to TDC is compression of this gaseous mixture, followed by ignition via a spark and subsequent burning. Outward power stroke 3 from TDC to BDC results in the expansion of the hot combustion products as the net cylinder volume increases. Inward stroke 4 from BDC to TDC results in the exhausting of these products out of the cylinder. The typically less efficient 2-stroke cycle combines the fuel–air intake and exhaust strokes from the 4-stroke case into a shorter transitional process near BDC. This process is called scavenging [4], and utilizes some variant of a pump/port mechanism (see example setup of Fig. 4.6) to implement the intake of fresh fuel–air mixture and exhaust of combusted gas. For this process to run smoothly (with low mechanical friction between various parts), some lubricating oil needs to be added to the fuel (presuming the lubricity of the fuel is insufficient otherwise). One advantage of

2-stroke engines is that they do have a higher power-to-weight ratio, given the fact that intermittent power delivery is every driveshaft revolution, as opposed to every two driveshaft revolutions for the 4-stroke case.

If we define Q_H as the heat input at constant volume, and Q_L as the heat loss at constant volume, then one can proceed to define the cycle thermal efficiency as follows:

$$\eta_{th} = \frac{Q_H - Q_L}{Q_H} = 1 - \frac{Q_L}{Q_H} = 1 - \frac{mC_v(T_4 - T_1)}{mC_v(T_3 - T_2)} \approx 1 - \frac{T_1\left(\frac{T_4}{T_1} - 1\right)}{T_2\left(\frac{T_3}{T_2} - 1\right)} \qquad (4.1)$$

In the above, we are assuming that C_v does not change appreciably in value over the range of temperatures in question, and also that the combusted gas has similar properties to unreacted fuel–air (reasonable, but not an overly accurate approximation). For the isentropic compression and expansion processes, given the quantity of mass is not changing too much during heating or cooling, and $V_1 = V_4$ & $V_2 = V_3$, we have:

$$\frac{T_2}{T_1} = \left(\frac{V_1}{V_2}\right)^{\gamma-1} = \left(\frac{V_4}{V_3}\right)^{\gamma-1} = \frac{T_3}{T_4} \qquad (4.2)$$

If the above is true, then

$$\frac{T_3}{T_2} = \frac{T_4}{T_1} \qquad (4.3)$$

Via substitution, one arrives at the following result for thermal efficiency:

$$\eta_{th} = 1 - \frac{T_1}{T_2} = 1 - r_v^{1-\gamma} = 1 - \frac{1}{r_v^{\gamma-1}} \qquad (4.4)$$

noting that volumetric compression $r_v = V_1/V_2 = V_4/V_3$.

Thus for the ideal Otto cycle, (thermal) efficiency increases with compression ratio. In practice, fuel detonation (rapid burning of fuel–air via detonation, versus slower burning via deflagration) occurs above a certain pressure threshold. Depending upon the anti-knock (-detonation) characteristics of the fuel in question, this threshold may vary from fuel to fuel. Aviation gasoline traditionally has a high anti-knock pressure threshold, allowing for compression ratios from 7:1 to as high as 9:1.

The peak temperature seen in the cycle, T_3, would ideally be close to the flame temperature of a stoichiometric fuel–air mixture undergoing complete combustion, i.e., no unburned reactants absorbing heat. Cylinder head walls typically should not have their surface temperature exceed some value closer to 540 K, substantially below the central gas temperature for T_3, which may have to be in the range of 2500 K plus for adequate performance. As a result, liquid or air cooling of the cylinder walls is required. Also, it is not uncommon to run the engine fuel-rich (i.e., higher fuel/air mixture than stoichiometric) for extended flight periods, since the excess fuel's evaporation within the combustion chamber acts to cool the combusted gas. The spark plugs (commonly two plugs per cylinder) used for intermittent cyclic ignition are commonly charged in a timed delivery from two magnetos (permanent magnets that produce a transient high voltage electric current from a low voltage source, say from an alternator or generator). The use of two magnetos (and two spark plugs per cylinder) is a classic example of a fail-safe flight operations approach, should one magneto fail at some point in a flight mission. An improperly timed magneto can result in overheated cylinders and damaged piston rings with mistimed sparking, so regular inspection and maintenance of this item's performance is typically required.

The parameter, mean effective pressure (mep or indicated mep, imep, i.e., before accounting for friction losses, etc.), is defined as the pressure which, if acting on the piston during the entire power stroke, would do the amount of useful work actually done on the piston, i.e., work output of power stroke minus work input of compression stroke:

$$w_{net} = mep(v_3 - v_1), \quad J/kg \text{ of air} \tag{4.5}$$

For an ideal cycle (no friction losses, etc.), cyclic work output is equal to cyclic heat input, so that:

$$w_{net} = q_H - q_L = \eta_{th} q_H \tag{4.6}$$

The ideal (before friction) power output for a 4-stroke reciprocating engine can in turn be determined via:

$$P_{S,i} = mep \cdot V_h \cdot \left(N \cdot \frac{1}{2} \right) \tag{4.7}$$

Here, V_h is the volumetric head displacement (total of all cylinders, m^3) and N is the engine drive shaft speed in revolutions per second (the ½ factor is evident due to there being 1 power stroke for every 2 revolutions of the shaft; this factor becomes unity for a 2-stroke engine, which suggests a potential avenue for a higher power-to-engine weight ratio). Note again that it is not uncommon to run the engine shaft at a higher rotation speed N than the propeller shaft speed at n, necessitating a speed reduction unit that can be a substantial addition in weight to the airplane. Note that the lower the mep, the higher the piston displacement to compensate, for a given power requirement; however, the greater the piston displacement, the larger the friction losses, so likely the mep will have to be increased in turn.

Brake specific fuel consumption is an important performance parameter, especially with respect to flight economics and attainable airplane range for propellered aircraft:

$$BSFC = \frac{\dot{m}_f}{P_S}, \quad kg(\text{of fuel})/(h\,W) \text{ or } lb/(h\,hp) \tag{4.8}$$

(a)

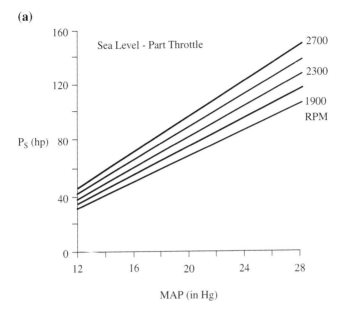

(b)

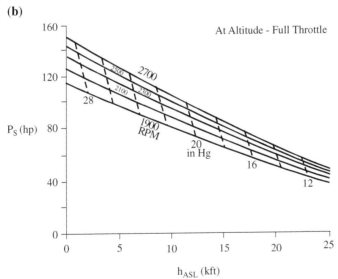

Fig. 4.7   a Shaft power as a function of intake manifold air pressure (MAP; in inches of mercury, where 1 atm = 29.9 in Hg = 760 mm Hg) and engine shaft speed (rpm) at full and part throttle, at sea level. The data is representative of a small 4-stroke normally aspirated carbureted spark-ignition piston engine, the air-cooled horizontally opposed four-cylinder Lycoming O-320-A2B **b** Shaft power as a function of altitude and engine shaft speed at full throttle, Lycoming O-320-A2B. Corresponding *dashed curves* for MAP are also shown

Typical values for airplane SI piston engines are around 0.5 lb/(h hp) (0.3 kg/(h kW)). Smaller piston engines may have a bit higher BSFC than that, and conversely, larger engines may have a bit lower value. Example performance charts for a small, normally aspirated (i.e., not supercharged) piston engine are provided in Fig. 4.7. It is common for a number of air-breathing engines to be more efficient as one goes to larger sizes, undoubtedly due in part to the higher development investment and care put into designing and building larger engines. Ease of manufacturing to high precision of larger scale components of the larger engines may also influence this trend. Turboprop and turboshaft engines have a comparable BSFC.

Also pertinent to aircraft design is the ratio of power to empty (dry, i.e., no fuel) engine weight, P_S/W_E. Obviously, the higher the ratio, the better, given the importance of keeping the airplane overall weight as low as possible. Small SI piston engines (say less than 100 hp or 75 kW peak power output) can have a ratio of around 0.4 hp/lb (0.66 kW/kg), medium engines (150–350 hp, 110–260 kW) can have a value closer to 0.7 hp/lb (1.15 kW/kg), while larger SI piston engines can have ratios exceeding 1 hp/lb (1.65 kW/kg). Turboprop and turboshaft engines may have ratios in the 2–4 hp/lb range, giving them a clear advantage for higher performance applications.

Please refer to the solutions for Example Problems 4.1 and 4.4 at the end of this chapter, for completed sample Otto cycle engine analyses. For students learning this material as part of a course, I would encourage them to attempt the given example problem first, and then check the solution to see if any mistakes were made.

4.3 Compression-Ignition Engines

A second class of internal combustion engines in the reciprocating piston category are the compression-ignition (CI) engines, also known as diesel engines. By the nature of their design, fuel injection (versus carburetion) directly into the cylinder is required, and for higher performance for aircraft applications, turbosupercharging is also likely required. Since largely only air is compressed during the compression stroke, diesels can operate at higher compression ratios of 15:1 up to 20:1, without too much concern about fuel detonation (knock). There can be 2- or 4-stroke variants (see Fig. 4.8 for a simple 2-stroke schematic) of this class of engine. The 2-stroke is actually more popular for aircraft, given the higher power-to-weight ratio discussed earlier. Diesel engines run on diesel fuel, with a lower heating value (LHV) of 44.2 MJ/kg of fuel, reacted with air. While like gasoline a mix of various hydrocarbon molecules, for some analyses, diesel fuel can be treated as dodecane, $C_{12}H_{26}$. An example control setup for a turbocharged CI engine is provided in Fig. 4.9 [3].

The ideal air-standard Diesel cycle [5] approximates the open-cycle performance of a CI piston engine. As may be seen in Fig. 4.10, the characteristics of the

Fig. 4.8 Illustration of internal workings of a 2-stroke compression-ignition piston engine. Diagram courtesy of HowStuffWorks.com

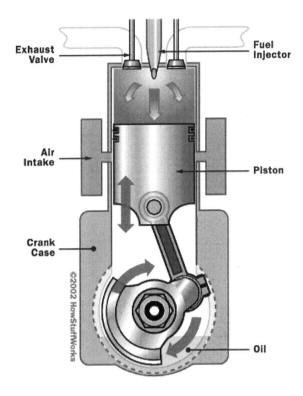

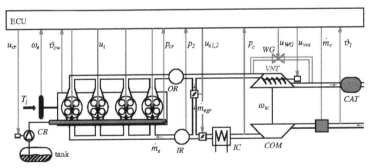

CAT	oxidation catalytic converter	$u_{e1,2}$	EGR valve(s) command	p_c	pressure after COM
COM	compressor	u_{cr}	CR pump command	p_2	intake pressure
CR	common-rail system	u_i	injection command	p_{cr}	CR injection pressure
IR	intake receiver	u_{vnt}	turbine nozzle command	$\dot{m}_c$	intake air mass flow
OR	outlet receiver	u_{WG}	WG command	ϑ_l	intake air temperature
IC	intercooler	T_l	load torque at the flywheel	ϑ_{cw}	cooling water temperature
VNT	variable nozzle turbine	ω_{tc}	turbocharger speed	ω_e	engine speed
WG	waste-gate (alternative to VNT)	$\dot{m}_e$	total engine-in mass flow	$\dot{m}_{egr}$	exhaust gas recirculation

Fig. 4.9 Example control setup for a turbocharged automotive CI engine [3]

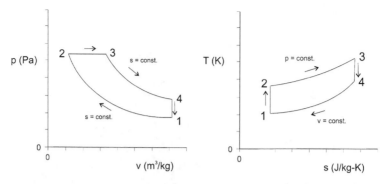

Fig. 4.10 Diagrams of p–v and T–s profiles for ideal Diesel cycle

operational cycle can be viewed through its p–v (pressure vs. specific volume) and T–s (temperature vs. entropy) diagrams. Process $1 \rightarrow 2$ represents an isentropic compression of air as the piston moves from BDC to TDC. Process $2 \rightarrow 3$ is representative of heat being added at a constant pressure while the piston is moving downward from TDC to some intermediate point in the power stroke, until combustion is completed. In the actual case, one has fuel injection with subsequent, relatively slower burning (note: a glow plug heat source may be required for starting the engine initially, but thereafter would typically not be required for sustained operation), and further resulting pressurization. Process $3 \rightarrow 4$ represents an isentropic expansion, as the piston moves further downward until reaching BDC. Process $4 \rightarrow 1$ is representative of heat being rejected from the air and fresh air in turn being take in, while the piston is in the vicinity of BDC.

The thermal efficiency of the Diesel cycle is given by:

$$\eta_{th} = 1 - \frac{Q_L}{Q_H} = 1 - \frac{mC_v(T_4 - T_1)}{mC_p(T_3 - T_2)} = 1 - \frac{T_1}{\gamma T_2} \frac{\left(\frac{T_4}{T_1} - 1\right)}{\left(\frac{T_3}{T_2} - 1\right)} \qquad (4.9)$$

One can substitute the compression ratio into the above:

$$\eta_{th} = 1 - \frac{1}{\gamma r_v^{\gamma-1}} \frac{\left(\frac{T_4}{T_1} - 1\right)}{\left(\frac{T_3}{T_2} - 1\right)} \qquad (4.10)$$

In practice, given that the isentropic expansion ratio from 3 to 4 is substantially less than compression from 1 to 2, and thus $T_4/T_1 > T_3/T_2$ by a substantial margin, the Diesel cycle has a lower efficiency for the same compression ratio versus the Otto cycle (even with the assist of the $1/\gamma$ for the Diesel). In order to compensate then, one must go to higher compression ratios. This approach is limited by the allowed stress on the engine structure, which forces increased engine dry weight. To date, typical power-to-weight ratios can be in the vicinity of 0.3 hp/lb

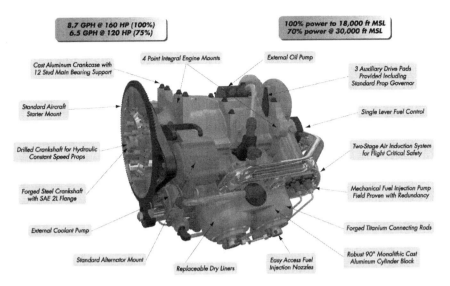

Fig. 4.11 Illustration of a turbosupercharged 160-hp two-stroke four-cylinder (V-4) DeltaHawk aircraft diesel engine (DH160A4/V4/R4). The engine can run on diesel or jet fuel. Diagram courtesy of DeltaHawk Engines, Inc

(0.5 kW/kg). The ideal mean effective pressure for a diesel engine may be correlated to work being output via:

$$w_{\text{net}} = \text{mep}(v_1 - v_2), \quad \text{J/kg(of air)} \tag{4.11}$$

Also, by substitution,

$$w_{\text{net}} = \eta_{\text{th}}q_H = q_H - q_L \tag{4.12}$$

In the past, diesel engines have not been selected for a given airplane very often. This was largely in part due to the higher engine weights needed at the higher compression ratios. The BSFC is potentially quite low, on the order of 0.35 lb/(h hp) (0.21 kg/(h kW)) or less for existing aircraft diesels, thus there is some incentive to continue research and development on such items as lighter weight engine materials. Refer to Fig. 4.11 for an example of a modern production diesel aircraft engine.

4.4 Rotary Engines

A third class of IC engines is the rotary engine. In the early twentieth century, one observed an old piston rotary approach being used, namely having the crankcase and cylinders in a radial configuration rotating in conjunction with the propeller, while the crankshaft was nominally fixed to the airplane (see example of

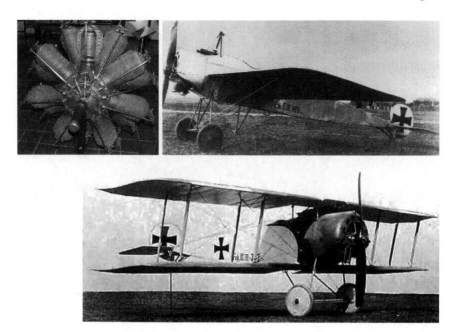

Fig. 4.12 Photo at *top left* of an Oberursel U.III 14-cylinder (twin cylinder banks) air-cooled 160-hp piston rotary engine on display at the Deutsches Museum, built circa early WWI. The twin cylinder banks rotated in the same direction (not contra-rotating), thus generating an uncomfortably large roll torque for pilots. Airplanes employing the engine included the Fokker E.IV (*upper right* photo) and the Fokker D.III (*lower right* photo from the San Diego Air & Space Museum)

Fig. 4.12). Since this approach is not particularly effective at higher performance levels (in addition to the aerodynamic drag losses with the rotating cylinders, above around 100 hp, the roll torque resulting on the airframe becomes excessive and progressively uncomfortable for flight control) and engine sizes relative to more modern SI piston engines, it effectively died out as an option. In the present day, the most common pistonless rotary design is the Wankel (see Fig. 4.13), which uses an offset cam-shaped rotor (vane) rotating about a center output driveshaft, which also rotates in conjunction. Because their volumetric displacement versus total engine size is quite high relative to the reciprocating piston engines, the power-to-weight ratio, P_S/W_E, can be quite high, on the order of 1–3 hp/lb (as compared to 0.5–1 hp/lb for SI piston engines). These engines typically use spark-ignition, with the intake, compression, ignition, expansion (power) and exhaust processes occurring smoothly in going around one cycle of the rotor.

The main concern with rotary engines is the longevity of the seals required for proper operation, noting the seals are required to cover a lot more surface area than their piston-ring counterparts in SI piston engines. These engines, for the SI version, can be evaluated using the Otto cycle as a basis. Gasoline is the principal fuel used with rotary engines. The BSFC is comparable to SI piston engines of a similar

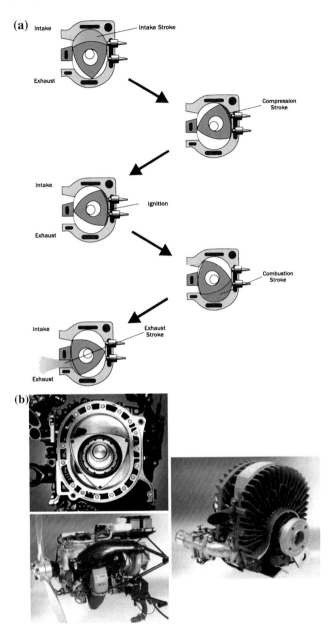

Fig. 4.13 a Illustration of internal operations of a Wankel rotary engine. Diagram courtesy of HowStuffWorks.com **b** Photo at *top left* of opened rotary engine, courtesy of Mazda Motor Corporation. Photo at *right* of AR731 air-cooled single-rotor 38-hp Wankel rotary engine for unmanned aircraft applications, courtesy of UAV Engines Ltd. Photo at *bottom left* of AR682 water-cooled twin-rotor 95-hp Wankel rotary engine for unmanned aircraft applications, courtesy of UAV Engines Ltd

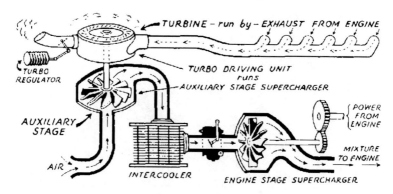

Fig. 4.14 Illustration of turbosupercharger in aircraft piston engine operation

power output, or occasionally a bit worse (with fuel leakage and the requirement for a fuel/lubricating oil mix), especially for smaller engines. The compression ratio is similar to SI piston engines (8:1 up to 10:1). Finally, these engines typically run at a higher driveshaft speed N, in order to produce the power comparable to a conventional IC engine. As a result, it is more likely that a propeller speed reduction unit will be required, to gear down to the propeller's shaft speed n.

4.5 Turbosupercharging

The performance of an IC engine (spark- or compression-ignition) can be enhanced by supercharging [4, 6], i.e., compressing the air entering the intake. This technique can be used to enhance sea-level performance by compressing the incoming air to a level substantially above the ambient air pressure, which is commonly the principal application for the automotive world. However, for airplanes, it can also be used for maintaining the engine's rated sea-level performance up to a higher threshold altitude, by compressing the thinner ambient outside air, at altitude, to a sea-level value upon entering the engine's combustion chambers.

Originally, superchargers were gear-driven off the main engine driveshaft, and as a result, their performance was somewhat limited. In order to largely avoid this issue of bleeding off the main shaft's power to run a front-end compressor, turbosuperchargers (TSCs) were introduced. TSCs employ a turbine in the exhaust stream, exploiting available energy that would otherwise be lost, as illustrated in Figs. 4.14 and 4.15. The back-end turbine in turn drives a front-end compressor for mechanical air pressurization, and thus bypass the extraction of power from the driveshaft. Example aircraft that employed a supercharging approach for their respective piston engines may be viewed in Figs. 4.16 and 4.17. Example performance charts for a small, turbosupercharged spark-ignition piston engine are provided in Fig. 4.18.

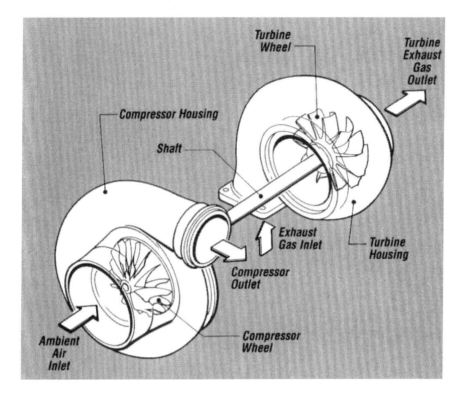

Fig. 4.15 Schematic diagram of turbosupercharger

There is a limit on the TSC's operational compression ratio, based on two primary factors. The first factor is the fuel–air detonation pressure limit in the downstream combustion chambers, which should not be exceeded. The second factor is the stress limit on the surrounding engine structure under the heightened pressure loading. An intercooling (aftercooling; charge air cooling) approach is sometimes used to enhance performance (via increased combustor intake air density), by using a heat exchanger to remove heat (cooling the air) that is leaving the TSC, prior to entering the combustion chamber.

Viewing a diagram of a turbosupercharger as in Fig. 4.15, one can see the role that TSCs, first applied to aircraft engines during World War I, played as a forerunner to the gas turbine engine, which was introduced for aircraft during World War II. The principal difference being, of course, the combustor placed in the middle of gas turbine engines.

Fig. 4.16 Front view of the Hawker Hurricane fighter of WWII Battle of Britain fame. The three-bladed variable-pitch metal Rotol propeller is being driven by a 1620-hp liquid-cooled Rolls-Royce Merlin XX spark-ignition piston engine (supercharged [2-speed gear]; V-12 cylinder arrangement). Note the ventral (belly) air intakes for air delivery to the engine's gear-driven supercharger and intake manifold, and for cooling the engine's liquid coolant and lubricant oil. The ejector exhaust pipes for the starboard cylinders may be viewed curving out of the starboard side of the front fuselage (the thrust component from this exhaust flow was utilized by this ejector approach). Turbosupercharging was tried at the time on the Merlin XX, but ultimately did not replace the well-proven gear-driven supercharging approach for that engine (one nominal advantage of a supercharger over a turbosupercharger is a quicker response, noting the "turbo lag" characteristic of existing turbosupercharged engines)

4.6 UAVs: A Realm for Innovation and Alternative Technologies

Unmanned (uninhabited) air vehicles (UAVs) provide an avenue for exploring the use of different technologies, without the safety and weight constraints associated with carrying passengers or crew. IC engines have been a common UAV powerplant choice for driving the propellers that deliver the required thrust for a number of lower-speed UAV applications.

In parallel with propellered UAVs, on the smaller scale, radio-controlled (R/C) model airplanes have a long heritage with various IC engines powering their propellers. Because one typically has a low wing loading (W/S) with low-speed, lightweight R/C airplanes, the thrust requirement is sometimes low enough to bring into consideration alternative powerplants. The use of wound elastic bands or compressed gas cartridges has some history, but usually for relatively short flights. Battery-powered brush direct-current [DC] motors have been utilized in the past for electric R/C airplanes; while having the advantage of being relatively

Fig. 4.17 Front view of the Boeing B-17 Flying Fortress bomber of WWII fame. The three-bladed variable-pitch propeller in view (one of four) is being driven by a 1200-hp Wright R-1820 Cyclone 9 air-cooled spark-ignition piston engine (turbosupercharged; radial 9-cylinder arrangement). The General Electric designed turbosupercharger allowed the engine to maintain sea-level power up to 25,000 ft in altitude. Radial engines always come in an odd number of cylinders (like 9) due to the alternate cylinder firing order sequence requirement (e.g., 1-3-5-7-9-2-4-6-8)

quiet, the performance of the brush motors with regard to power and wear was typically quite inferior to comparable IC engines for the given category of R/C craft. Batteries, e.g., nickel–cadmium, were also quite heavy and short-lived. With the more recent use of longer-life batteries (e.g., nickel metal hydride, lithium-ion and polymer lithium-ion), and more powerful brushless DC [BLDC] motors, electric R/C airplanes are presently finding a stronger niche in the market. In conjunction, with the miniaturization of such items as cameras, battery-powered electric motored UAVs are finding their place for various applications.

Battery-powered micro air vehicles [MAVs], UAVs on the scale of 10–20 cm in size, are being developed for civil and military applications like aerial photography and surveillance. Alternatives to the use of propellers are also being investigated for MAVs. Emulating flying insects and certain birds to generate both thrust and lift (like the hovering hummingbird [6]), MAVs employing a flapping wing are of interest. Flight vehicles in this category are referred to as ornithopters [7]).

For longer flight durations, at some point electric airplanes need to reduce the weight that comes with using batteries. Solar power is a popular solution. NASA's Helios (Fig. 4.19) used an array of solar panels positioned on the upper surface of

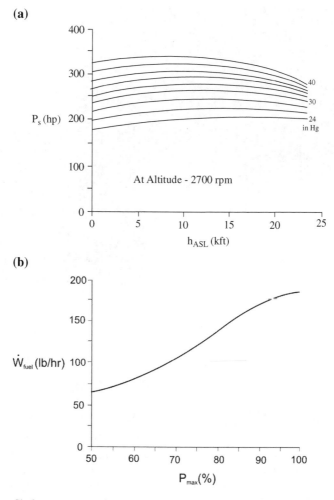

Fig. 4.18 a Shaft power as a function of altitude and MAP at an engine shaft speed of 2,700 rpm. The data is representative of a small 4-stroke turbosupercharged fuel-injected spark-ignition piston engine, the air-cooled horizontally opposed six-cylinder Teledyne Continental TSIO- 520-WB. **b** Fuel consumption rate as a function of percentage of maximum rated shaft power (max. rated MAP = 39 in Hg), turbosupercharged Teledyne Continental TSIO-520-WB

its long wing, to allow for recharging of its batteries during daytime, and running on batteries during the night, for continuously powering its fourteen DC motor-powered propellers for long endurance missions. An alternative to solar energy for powering an electric airplane is a beamed energy approach. The Canadian SHARP project (Stationary High Altitude Relay Platform) proposed the use of a focused microwave beam to power its long-duration high-altitude circling UAV (Fig. 4.20). A final working system to prove the concept was not completed in that case. The use of on-board fuel cells that chemically generate electricity for

Fig. 4.19 Photo of the unmanned solar-powered Helios prototype vehicle, built by AeroViron-ment for NASA's Environmental Research and Sensor Technology (ERAST) program. The vehicle reached an altitude record of 96,800 ft for level, fixed-wing flight in August 2001. In June 2003, the vehicle broke up in flight after encountering significant turbulence. Photo courtesy of NASA

Fig. 4.20 A photo of a sub-scale prototype of the SHARP vehicle, in this case with the propeller's DC motor being powered solely by nickel cadmium batteries, circa 1987. The full-scale version was to have a 40-m wingspan. Note the mock microwave-energy collecting antenna dish just ahead of the vertical tail. Work on this project for the most part ceased in 1988

powering DC motored UAVs has also been proposed, and successfully demon-strated (e.g., the 9-h flight by the prototype AeroVironment Puma vehicle in March 2008).

4.7 Example Problems

4.1. Perform a cycle analysis for a 4-stroke spark-ignition engine with a volumetric head displacement of 5 l, running at 4500 rpm. Assume a fuel/air intake mixture of 0.043, with a heat of reaction of 1.87×10^6 J/kg of air. Assume a volumetric compression ratio of 8, with ambient conditions being 101 kPa and 20°C. Gas properties of the working fluid may be presumed to be that of air. As you complete the cycle analysis, ascertain the various pressures and temperatures, thermal efficiency, mean effective pressure (mep) and the ideal shaft power output (neglecting friction, etc.).

4.2. Given that shaft power output of an aircraft piston engine is proportional to air massflow being taken in, show why actual power delivered is corrected for different ambient temperatures from ISA (International Standard Atmosphere) conditions via

$$P_{S,\text{act}} = P_{S,\text{ISA}}(T_{\text{ISA}}/T_{\text{act}})^{1/2}$$

Further hint: assume Mach number of intake flow remains constant at different temperatures, and in addition, remember that at a stated pressure altitude, the ISA pressure is the same as the actual ambient pressure.

4.3. Perform a cycle analysis for the engine of Problem 4.1, but for a lower volumetric compression ratio of 6. Evaluate the effects of this lower value on the various performance parameters.

4.4. Perform a cycle analysis for the engine of Problem 4.1, but for flight at 3,000 m above sea level (ASL). You may assume ISA properties (see Appendix I). Evaluate the effects of altitude on the various performance parameters.

4.5. Consider the performance charts for the normally aspirated Lycoming O-320-A2B spark-ignition piston engine of Fig. 4.7. (a) For an aircraft flying with that engine at full throttle at 10,000 ft (ISA) at an engine shaft speed of 2700 rpm, determine the fuel consumption rate (U.S. gallons/hour) and the percentage of maximum rated power (150 hp for this engine) that is presently being delivered by the engine. (b) With the aircraft now flying at 5,000 ft (ISA) at part throttle (engine shaft speed of 2100 rpm, intake manifold air pressure of 22 in Hg), determine the fuel consumption rate and percentage of maximum rated power that the engine is running at.

4.6. Perform a cycle analysis for a 2-stroke compression-ignition (diesel) engine having a volumetric heat displacement of 850 in³, running at 2300 rpm. Assume a heat of reaction of 800 Btu/lbm of air, and a volumetric compression ratio of 17. Gas properties may be assumed to be that of air. The ambient pressure is 14.7 psia, and the ambient temperature is 60°F. As you complete the cycle analysis, ascertain the various pressures and temperatures, thermal efficiency, mean effective pressure (mep), and the ideal shaft power output (neglecting friction, etc.).

4.8 Solutions to Example Problems

4.1. Evaluate 4-stroke spark-ignition engine using Otto cycle equations.

Head displacement $V_h = 5.0 \, 1 = 0.005 \, \text{m}^3$; shaft speed, $N = 4500/60 = 75$ revs./s; Fuel heat of combustion $q_R = 4.34 \times 10^7$ J/kg of fuel, which for fuel/air $f = 0.043$, gives $q_H = f \, q_R = 0.043(4.34 \times 10^7) = 1.87 \times 10^6$ J/kg of air; volumetric compression ratio $r_v = 8$

Assume gas properties of air: $\gamma = 1.4$, $R = 287$ J/kg K, $C_p = 1004$ J/kg·K, $C_v = 717$ J/(kg K)

Specific volume at state ① :

$v_1 = \frac{1}{\rho_1} = \frac{RT_1}{p_1} = 287(293)/101000 = 0.833 \, \text{m}^3/\text{kg}$. Moving to state ②,

$\frac{T_2}{T_1} = \left(\frac{V_1}{V_2}\right)^{\gamma-1} = (r_v)^{\gamma-1} = 8^{0.4} = 2.3$ so that $T_2 = 2.3(293) = 674$ K,

$\frac{p_2}{p_1} = \left(\frac{V_1}{V_2}\right)^{\gamma} = 8^{1.4} = 18.4$ so that $p_2 = 18.4 \, (101) = 1858$ kPa

$v_2 = v_1/r_v = 0.833/8 = 0.104 \, \text{m}^3/\text{kg}$. Moving to state ③, noting $v_3 = v_2$,

$q_{2 \to 3} = q_H = 1.87 \times 10^6 \text{J/kg of air} = C_v \, (T_3 - T_2)$;

$T_3 = T_2 + \frac{q_H}{C_v} = 674 + 1.87 \times 10^6/717 = 3282$ K; if in fact $C_{v,\text{gas}} > C_{v,\text{air}}$, T_3 will be lower

$\frac{p_3}{p_2} = \frac{T_3}{T_2} = 3282/674 = 4.87$ so that $p_3 = 4.87(1858) = 9049$ kPa. Moving to state ④,

$\frac{T_4}{T_3} = \left(\frac{V_3}{V_4}\right)^{\gamma-1} = \left(\frac{1}{r_v}\right)^{\gamma-1} = (1/8)^{0.4} = 0.435$ so that $T_4 = 0.435 \, (3282) = 1428$ K

$\frac{p_4}{p_3} = \left(\frac{V_3}{V_4}\right)^{\gamma} = (1/8)^{1.4} = 0.0544$ so that $p_4 = 0.0544 \, (9049) = 492$ kPa

$v_4 = r_v \, v_3 = 8(0.104) = 0.833 \, \text{m}^3/\text{kg}$

$\eta_{\text{th}} = 1 - \frac{1}{r_v^{\gamma-1}} = 1 - 1/8^{0.4} = 0.565$

$q_L = C_v(T_4 - T_1) = 717 \, (1428 - 293) = 0.814 \times 10^6$ J/kg of air

$w_{\text{net}} = q_H - q_L = 1.87 \times 10^6 - 0.814 \times 10^6 = 1.056 \times 10^6$ J/kg of air, net specific work

$\text{mep} = \frac{w_{net}}{v_1 - v_2} = 1.056 \times 10^6/(0.833 - 0.104) = 1449$ kPa, mean effective pressure

$P_{S,i} = \text{mep} \cdot V_h \cdot N/2 = 1449 \times 10^3 \, (0.005) \, (75/2) = 272{,}000$ W or 272 kW or 365 hp, ideal (indicated).

This is a high indicated estimate, to be sure. Actual brake shaft power delivered (i.e., after friction losses) closer to 240 hp, with a brake mean effective pressure *bmep* of 955 kPa.

4.2. Derive the correction factor on shaft power delivered by a piston engine, for the given atmospheric conditions. Consider mass flow of air, at a relatively constant Mach number, being taken into a piston engine:

$$\dot{m}_a = \rho V A_{\text{eff}} = \rho(a \cdot Ma)A_{\text{eff}}$$
$$= \left(\frac{p}{RT}\right)(\sqrt{\gamma RT} \cdot Ma)A_{\text{eff}}$$
$$= \frac{1}{\sqrt{T}} \cdot p \cdot \sqrt{\frac{\gamma}{R}} \cdot Ma \cdot A_{\text{eff}}$$

$\dot{m}_a = \frac{1}{\sqrt{T}} \cdot p \cdot K$, where K is in relative terms a constant-value coefficient.

Since at a given pressure altitude h_p the outside air pressure p is the same for both the ISA and actual atmosphere cases, this would leave only the temperature term as the variable factor acting on massflow, namely:

$\sqrt{T} \cdot \dot{m}_a = $ constant, so that

$\sqrt{T_{\text{ISA}}} \cdot \dot{m}_{a,\text{ISA}} = \sqrt{T_{\text{act}}} \cdot \dot{m}_{a,\text{act}}$, and thus:

$$\dot{m}_{a,\text{act}} = \dot{m}_{a,\text{ISA}} \cdot \sqrt{\frac{T_{\text{ISA}}}{T_{\text{act}}}}$$

Since for a piston engine, the delivered shaft power is directly proportional to the incoming massflow of air that feeds the combustion process, then the following is thus true:

$$\frac{P_{S,\text{act}}}{P_{S,\text{ISA}}} = \frac{\dot{m}_{a,\text{act}}}{\dot{m}_{a,\text{ISA}}} = \sqrt{\frac{T_{\text{ISA}}}{T_{\text{act}}}}$$

4.3. Evaluate 4-stroke spark-ignition engine using Otto cycle equations, at a lower r_v. Head displacement $V_h = 5.0 \text{ l} = 0.005 \text{ m}^3$; shaft speed, $N = 4500/60 = 75$ revs./s; Fuel heat of combustion $q_R = 4.34 \times 10^7$ J/kg of fuel, which for fuel/air $f = 0.043$, gives $q_H = f \, q_R = 0.043(4.34 \times 10^7) = 1.87 \times 10^6$ J/kg of air; volumetric compression ratio $r_v = 6$ Assume gas properties of air: $\gamma = 1.4$, $R = 287$ J/(kg K), $C_p = 1004$ J/(kg K), $C_v = 717$ J/(kg K)

Specific volume at state ① :

$v_1 = \frac{1}{\rho_1} = \frac{RT_1}{p_1} = 287(293)/101000 = 0.833 \text{ m}^3/\text{kg}$. Moving to state ②,

$\frac{T_2}{T_1} = \left(\frac{V_1}{V_2}\right)^{\gamma-1} = (r_v)^{\gamma-1} = 6^{0.4} = 2.05$ so that $T_2 = 2.05(293) = 601$ K,

$\frac{p_2}{p_1} = \left(\frac{V_1}{V_2}\right)^{\gamma} = 6^{1.4} = 12.3$ so that $p_2 = 12.3 \,(101) = 1242$ kPa

$v_2 = v_1/r_v = 0.833/6 = 0.139 \text{ m}^3/\text{kg}$. Moving to state ③, noting $v_3 = v_2$,

$q_{2\to3} = q_H = 1.87 \times 10^6$ J/kg of air $= C_v(T_3 - T_2)$;

$$T_3 = T_2 + \frac{q_H}{C_v} = 601 + 1.87 \times 10^6/717 = 3209 \text{ K}; \text{ if in fact } C_{v,\text{gas}} > C_{v,\text{air}},$$
T_3 will be lower

$$\frac{p_3}{p_2} = \frac{T_3}{T_2} = 3209/601 = 5.34 \text{ so that } p_3 = 5.34(1242) = 6632 \text{ kPa}.$$

Moving to state ④,

$$\frac{T_4}{T_3} = \left(\frac{\Psi_3}{\Psi_4}\right)^{\gamma-1} = \left(\frac{1}{r_v}\right)^{\gamma-1} = (1/6)^{0.4} = 0.488 \text{ so that}$$
$$T_4 = 0.488 \ (3209) = 1566 \text{ K}$$

$$\frac{p_4}{p_3} = \left(\frac{\Psi_3}{\Psi_4}\right)^{\gamma} = (1/6)^{1.4} = 0.0813 \text{ so that } p_4 = 0.0813(6632) = 539 \text{ kPa}$$

$v_4 = r_v \, v_3 = 6(0.139) = 0.833 \text{ m}^3/\text{kg}$

$\eta_{\text{th}} = 1 - \frac{1}{r_v^{\gamma-1}} = 1 - 1/6^{0.4} = 0.511 \text{ (vs. } 0.565, \text{ Problem 4.1)}$

$q_L = C_v(T_4 - T_1) = 717 \ (1566 - 293) = 0.913 \times 10^6 \text{ J/kg of air}$

$w_{\text{net}} = q_H - q_L = 1.87 \times 10^6 - 0.913 \times 10^6 = 0.957 \times 10^6 \text{ J/kg of air, net}$
specific work

$$mep = \frac{w_{\text{net}}}{v_1 - v_2} = 0.957 \times 10^6/(0.833 - 0.139) = 1379 \text{ kPa, mean effective}$$

pressure

$P_{S,i} = mep \cdot \Psi_h \cdot N/2 = 1379 \times 10^3 \ (0.005) \ (75/2) = 258{,}600 \text{ W or } 259 \text{ kW}$
or 348 hp, ideal

(indicated), versus 365 hp of Problem 4.1.

4.4. Evaluate 4-stroke spark-ignition engine using Otto cycle equations, at 3,000 m altitude.

From **Appendix I**, $p_\infty = 70121 \text{ Pa}$, $T_\infty = 268.7 \text{ K}$, $\rho_\infty = 0.909 \text{ kg/m}^3$.
Head displacement $\Psi_h = 5.01 = 0.005 \text{ m}^3$; shaft speed, $N = 4500/60 = 75 \text{ revs./s}$;
Fuel heat of combustion $q_R = 4.34 \times 10^7 \text{ J/kg of fuel}$, which for fuel/air
$f = 0.043$, gives

$q_H = f \ q_R = 0.043(4.34 \times 10^7) = 1.87 \times 10^6 \text{ J/kg of air; volumetric}$
compression ratio $r_v = 8$

Assume gas properties of air: $\gamma = 1.4$, $R = 287 \text{ J/(kg K)}$, $C_p = 1004 \text{ J/(kg K)}$,
$C_v = 717 \text{ J/(kg K)}$

Specific volume at state ① :

$$v_1 = \frac{1}{\rho_1} = \frac{RT_1}{p_1} = 287(268.7)/70121 = 1.10 \text{ m}^3/\text{kg. Moving to state ②,}$$

$$\frac{T_2}{T_1} = \left(\frac{\Psi_1}{\Psi_2}\right)^{\gamma-1} = (r_v)^{\gamma-1} = 8^{0.4} = 2.3 \text{ so that } T_2 = 2.3(268.7) = 618 \text{ K},$$

$$\frac{p_2}{p_1} = \left(\frac{\Psi_1}{\Psi_2}\right)^{\gamma} = 8^{1.4} = 18.4 \text{ so that } p_2 = 18.4 \ (70.1) = 1290 \text{ kPa}$$

$v_2 = v_1/r_v = 1.10/8 = 0.138 \text{ m}^3/\text{kg}$. Moving to state ③, noting $v_3 = v_2$,
$q_{2\to3} = q_H = 1.87 \times 10^6 \text{J/kg of air} = C_v(T_3 - T_2)$;

$$T_3 = T_2 + \frac{q_H}{C_v} = 618 + 1.87 \times 10^6/717 = 3226 \text{ K}; \quad \text{if in fact } C_{v,\text{gas}} > C_{v,\text{air}},$$

T_3 will be lower

$$\frac{p_3}{p_2} = \frac{T_3}{T_2} = 3226/618 = 5.22 \text{ so that } p_3 = 5.22(1290) = 6734 \text{ kPa}.$$

Moving to state ④,

$$\frac{T_4}{T_3} = \left(\frac{V_3}{V_4}\right)^{\gamma-1} = \left(\frac{1}{r_v}\right)^{\gamma-1} = (1/8)^{0.4} = 0.435 \text{ so that}$$

$T_4 = 0.435(3226) = 1403 \text{ K}$

$$\frac{p_4}{p_3} = \left(\frac{V_3}{V_4}\right)^{\gamma} = (1/8)^{1.4} = 0.0544 \text{ so that } p_4 = 0.0544(6734) = 366 \text{ kPa}$$

$v_4 = r_v v_3 = 8(0.138) = 1.10 \text{ m}^3/\text{kg}$

$$\eta_{th} = 1 - \frac{1}{r_v^{\gamma-1}} = 1 - 1/8^{0.4} = 0.565, \text{ same as Problem 4.1}$$

$q_L = C_v(T_4 - T_1) = 717\,(1403-293) = 0.796 \times 10^6 \text{ J/kg of air}$

$w_{net} = q_H - q_L = 1.87 \times 10^6 - 0.796 \times 10^6 = 1.074 \times 10^6 \text{ J/kg of air, net}$

specific work

$$mep - \frac{w_{net}}{v_1 - v_2} = 1.074 \times 10^6/(1.10 - 0.138) = 1116 \text{ kPa}, \quad \text{mean effective}$$

pressure

$P_{S,i} = \text{mep} \cdot V_h \cdot N/2 = 1116 \times 10^3\,(0.005)\,(75/2) = 209300 \text{ W or } 209 \text{ kW}$

or 280 hp, ideal

(indicated), versus 365 hp of Problem 4.1.

Side note: $P_{S,\text{alt}}/P_{S,S/L} \approx \sigma^m$, or $(280/365) = (0.909/1.225)^m$, or $m \approx \ell n\,(0.767)/\ell n\,(0.742) = 0.89$

4.5. (a) Flying at full throttle, at an ISA altitude of 10,000 ft, at 2700 rpm engine shaft speed (at altitude). Referring to Fig. 4.7b for full-throttle performance at altitude, this indicates we are running at around 112 hp, at an MAP of around 21 in Hg. Referring to Fig. 4.7a for sea-level performance at an MAP of 21 in Hg and 2700 rpm, this indicates a sea-level hp of around the same value (112 hp). Via Fig. 4.7c, at 21 MPa and 2700 rpm, one has a fuel consumption rate of a little over 9 US gallons/hour. The percentage of maximum rated power is around $112/150 \times 100\% = 75\%$ at that fuel consumption rate.

(b) Flying at part throttle, at an ISA altitude of 5,000 ft, at 2100 rpm shaft speed (at altitude) and an MAP of 22 in Hg. Referring to Fig. 4.7b for full-throttle performance at altitude, one has an MAP of about 25 in Hg at full throttle producing about 110 hp. At a MAP of 22 in Hg at 2100 rpm, one is producing around 90+ hp at full throttle at 9,500 ft ASL, according to 4.7(b). Referring to Fig. 4.7c, at 2100 rpm and an MAP at part throttle of 22 in Hg, fuel consumption rate is around 7.5 US gph. The percentage of maximum rated power is around $90/150 \times 100\% = 60\%$.

4.6. Evaluate 2-stroke compression-ignition engine using Diesel cycle equations. Head displacement $V_h = 850 \text{ in}^3 = 0.49 \text{ ft}^3$; shaft speed, $N = 2300/60 = 38.3$ revs./sec; $q_H = 800 \text{ Btu/lbm of air} = 2.0 \times 10^7 \text{ ft lb/slug of air}$; volumetric compression ratio $r_v = 17$ Assume gas properties of air: $\gamma = 1.4$, $R = 1715$ ft lb/slug·°R, $C_p = 6002$ ft lb/slug·°R,

$C_v = 4287$ ft lb/slug·°R

$T_1 = 60°F = 519°R$, $p_1 = 14.7$ psia $= 2{,}117$ lb/ft^2

Specific volume at state ① :

$v_1 = \dfrac{1}{\rho_1} = \dfrac{RT_1}{p_1} = 1715(519)/2117 = 420\,\text{ft}^3/\text{slug}$. Moving to state ②,

$v_2 = v_1/r_v = 420/17 = 24.7\,\text{ft}^3/\text{slug}$.

$\dfrac{T_2}{T_1} = \left(\dfrac{V_1}{V_2}\right)^{\gamma-1} = (r_v)^{\gamma-1} = 17^{0.4} = 3.11$ so that $T_2 = 3.11(519) = 1614°R$,

$\dfrac{p_2}{p_1} = \left(\dfrac{V_1}{V_2}\right)^{\gamma} = 17^{1.4} = 52.8$ so that $p_2 = 52.8\ (2117) = 111{,}776$ lb/ft^2 or 776 psia

Moving to state ③, noting $v_3 > v_2$, constant–pressure heating ($p_3 = p_2$) :

$q_{2\to3} = q_H = 2.0 \times 10^7\,\text{ft} \cdot \text{lb/slug of air} = C_p(T_3 - T_2)$;

$T_3 = T_2 + \dfrac{q_H}{C_p} = 1614 + 2.0 \times 10^7/6002 = 4953°R$; if in fact $C_{p,\text{gas}} > C_{p,\text{air}}$, T_3 will be lower

$\dfrac{V_3}{V_2} = \dfrac{T_3}{T_2} = 4953/1614 = 3.07$ so that $v_3 = 3.07(24.7) = 75.8\ \text{ft}^3/\text{slug}$. Moving to state ④,

$\dfrac{T_4}{T_3} = \left(\dfrac{V_3}{V_4}\right)^{\gamma-1} = (75.8/420)^{0.4} = 0.504$ so that $T_4 = 0.504(4953) = 2496°R$;

$p_4 = 71$ psia.

$q_L = q_{4\to1} = C_v(T_4 - T_1) = 4287\ (2496 - 519) = 8.48 \times 10^6$ ft lb/slug of air

$w_{\text{net}} = q_H - q_L = 2.0 \times 10^7 - 8.48 \times 10^6 = 1.152 \times 10^7$ ft lb/slug of air, net specific work

$\eta_{\text{th}} = \dfrac{w_{\text{net}}}{q_H} = 1.152 \times 10^7/2.0 \times 10^7 = 0.576$

$mep = \dfrac{w_{\text{net}}}{v_1 - v_2} = 1.152 \times 10^7/(420 - 24.7) = 29142\,\text{lb}/\text{ft}^2$, mean effective pressure (203 psia)

$P_{S,i} = mep \cdot V_h \cdot N = 29142\ (0.49)\ (38.3) = 546{,}908$ ft lb/s or 995 hp, ideal (indicated).

This is a high indicated estimate, to be sure. Actual brake shaft power delivered (i.e., after friction losses) closer to 600 hp, with a brake mean effective pressure bmep of 120 psia.

4.7. (a) An airplane utilizing a turbosupercharged Teledyne Continental TSIO-520-WB spark-ignition piston engine is running at 2700 rpm, 26 in Hg MAP, at 15000 ft (ISA). Referring to Fig. 4.18a, the engine output is about 225 hp. Referring to Fig. 4.18(b), at $225/320 \times 100\% = 70\%$ of maximum rated power, the fuel consumption rate is 105 lb/h, or around $105/6.1 = 17.2$ US gph. Current BSFC is $105/225 = 0.47$ lb/(h hp).

(b) Later in the flight, at a different altitude (4000 m ISA) and flight speed (105 m/s), the airplane's 2.1-m-diameter, constant-speed 4-bladed propeller's current reference blade pitch angle β is at $40°$, rotating at 1600 rpm through a propeller speed reduction unit from the engine shaft speed of 2700 rpm. Refer to the propeller charts of Fig. 3.6.

$n = 1600/60 = 26.7$ revs/sec; $\rho_\infty = 0.8194$ kg/m^3; $J = V_\infty/(nd) = 105/(26.7 \cdot 2.1) = 1.87$

$C_P = 0.32$ from Fig. 3.6, and $C_T = 0.135$; $\eta_{pr} = J\,C_T/C_P = 0.82$

$P_S = C_P\,\rho n^3 d^5 = 0.32(0.8194)26.7^3 2.1^5 = 203834$ W or 203.8 kW, or 274 hp

$h_{ASL} = 4000(3.28) = 13{,}120$ ft; from Fig. 4.18a, MAP around 32 in Hg.

Percentage of maximum rated power is around $274/320 \times 100\% = 86\%$.

From Fig. 4.18b, fuel consumption is around 170 lb/h, so BSFC = $170/274 = 0.62$ lb/(h hp).

References

1. Heywood JB (1988) Internal combustion engine fundamentals. McGraw-Hill, New York
2. Anonymous (1996) From the ground up, 27th edn. Aviation Publishers, Ottawa
3. Guzzella L, Onder CH (2010) Introduction to modeling and control of internal combustion engine systems. Springer, Berlin
4. Stone R (1992) Introduction to internal combustion engines, 2nd edn. SAE, Warrendale
5. Van Wylen GJ, Sonntag RE (1973) Fundamentals of classical thermodynamics, 2nd edn. Wiley, New York
6. Archer RD, Saarlas M (1996) An introduction to aerospace propulsion. Prentice-Hall, New Jersey
7. De Laurier JD (1999) The development and testing of a full-scale piloted ornithopter. Can Aeronaut Space J 45:72–82

Chapter 5
Pulsejet Engines

5.1 Introduction

The pulsejet engine can be considered a transitional development between the older reciprocating piston engines and the newer gas turbine engines. PJs and piston engines are similar in their intermittent fuel–air burning cycle, as compared to the continuous burning seen with gas turbines. On the other hand, PJs are similar to turbojets (TJs) in exhausting a hot jet to provide thrust (as opposed to using a propeller). Also, conventional PJs have a relatively linear-path left-to-right throughput of air/fuel from intake to exhaust, somewhat comparable to TJs (as opposed to the more circuitous path that air/fuel takes in internal combustion engines, which at higher vehicle speeds, would be an issue with respect to flow momentum losses), as may be seen in Fig. 5.1. Now, unlike the TJ, the PJ has no mechanical compressor or turbine. A conventional PJ has one moving part, the combustor intake valve (a reed valve design [1] is shown in Fig. 5.1) that opens and closes as the pressure drops and rises cyclically in the combustor, as per Fig. 5.2. A valveless (aerovalved) PJ has no moving parts. Since a high-speed, continuous-burn ramjet (RJ) engine in a similar fashion has no mechanical compressor or turbine (see Chap. 13), or moving parts, the PJ is sometimes referred to as an "intermittent ramjet". The V-1 cruise missile developed by Germany, and used in World War II, is the most infamous example of a pulsejet-powered air vehicle (see Fig. 5.3).

Pulsejets can run on gasoline, diesel fuel, and kerosene. Gasoline is in practice a better fuel choice for PJs, because of its relatively narrow range of flammability, i.e., will flame out (blow out) if the fuel–air mixture is too lean or too rich in the combustor. This characteristic better enables intermittent burning (as opposed to continuous burning). Hydrogen, H_2, is a poor PJ fuel in general, because of its wide flammability range that tend to cause prolonged or even continuous burning in the PJ combustor as it moves through its required operational cycle.

D. R. Greatrix, *Powered Flight*, DOI: 10.1007/978-1-4471-2485-6_5,

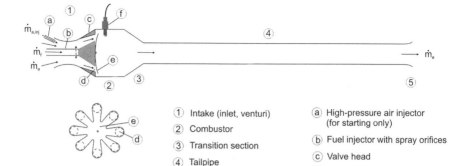

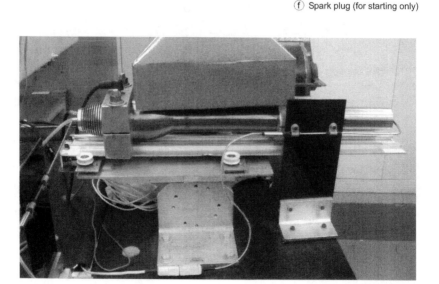

Fig. 5.1 Schematic diagram of conventional valved pulsejet, and corresponding photo of a customized Dyna-Jet RedHead valved pulsejet engine on a test stand at Ryerson University. A cooling fan/blower is in place above the forward portion of the engine's combustion tube

5.2 Performance Considerations

For practical flight applications of pulsejet engines, the supply of incoming air (and fuel) through the intake to the combustor should be such that the peak over pressure in the combustor in the operational cycle would be on the order of 100% over the stagnation intake air pressure (i.e., $p_4/p_1 \approx 2$ using the flow state numbering scheme adopted later) [1, 2]. Figure 5.4 provides an idealized sinusoidal representation of the combustor pressure profile with time. It shows the upper and lower absolute pressure limits about the idealized baseline pressure being

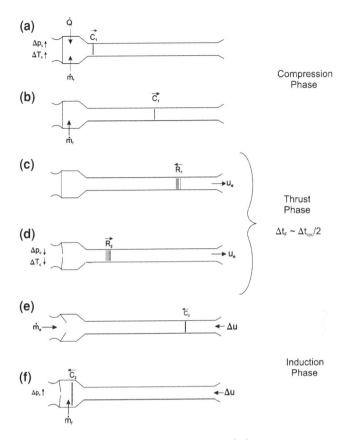

Fig. 5.2 Illustration of operational cycle of conventional pulsejet

Fig. 5.3 Photo of an American "copy" of a German pulsejet-powered V-1 "*buzz-bomb*" cruise missile, reproduced in large measure from reverse-engineering those seized by American forces towards the end of WWII, for post-war flight testing and technology demonstration by the USAF. Eventually, a lot of these copies were used, and destroyed, as target drones

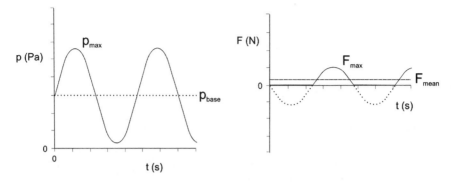

Fig. 5.4 Illustration of idealized (sinusoidal) combustor pressure, and corresponding thrust, as a function of time

approximately equal in magnitude, and indeed limited to a peak Δp of $\pm$ 100% for a symmetrical cycle (vacuum limit, $p \approx 0$, as the most negative trough point possible in the idealized sinusoidal cycle acting to enforce this limitation on performance). Referring to Fig. 5.1, the compressed high-pressure air source at the intake, and the spark plug igniter at the combustor, are only needed for the initial start-up.

Once started, the engine will continue to operate on its own, with hot residual combustion gases subsequently in a cyclic manner igniting incoming fresh fuel–air mixtures in a low–high–low passive pressure cycle. The base temperature in the combustor will continue to rise with the cyclic heat energy input, until the countering cooling effect of cyclic fresh air intake brings the system to a quasi-equilibrium state. At steady state, the baseline gas temperature can get quite high, T_1 approaching 1000 K, thus producing a peak transient combustor temperature (T_4) as high as 2000 K. For flight vehicle applications, convective air cooling of the intake valve head, combustor and tailpipe is a common approach for keeping the surrounding engine structure below the threshold for permanent heat damage. Complementary liquid cooling may also be used, as required (channeling of heat-absorbing liquid through the structure).

Valveless (aerovalved) PJ designs remove the need for a front-end combustor intake valve [1]. These designs can vary (as may be seen by Figs. 5.5 and 5.6), but typically depend on accurate tuning of the pressure wave passage in the applicable duct (combustion tube) system in order to operate efficiently. These designs allow for backflow of air out of the intake, which can be a problem if excessive, whereas the valved designs prevent backflow altogether with a mechanical valve. One incentive for pursuing valveless designs is the typical rapid wear of mechanical valves for valved PJs, necessitating frequent replacement (a cost consideration).

As illustrated by Fig. 5.2, a conventional PJ, in acoustic terminology, operates on a 4L wave system [3]. That is to say, 4 pressure waves (compression [left-ward]–rarefaction [rightward]–rarefaction [leftward]–compression [rightward],

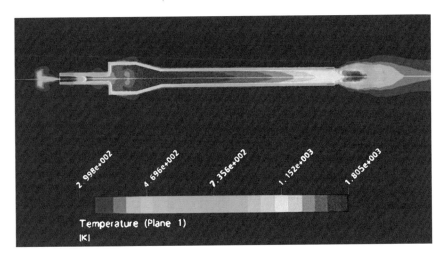

Fig. 5.5 Temperature distribution in one type of valveless pulsejet, $V_\infty = 50$ m/s. Courtesy of Applied Energy Research Laboratory, North Carolina State University (Dr. William Roberts)

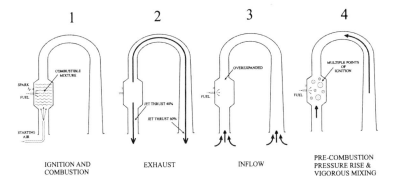

Fig. 5.6 Illustration of operational cycle of a second type of valveless pulsejet. A spark plug and fuel injector are shown at the left side of the primary combustion chamber, which itself is positioned on the left side of the apparatus

$\vec{C} - \vec{R} - \vec{R} - C$) moving at or slightly above the local gas sound speed (superimposed on the local gas velocity) sequentially traverse the duct system (combustor + tailpipe = combustion tube of length $\ell_{c/p}$) in one operational cycle. Thus the operational period for one cycle in seconds is approximately:

$$\Delta t_{cyc} \approx \frac{4\ell_{c/p}}{a_1} \approx \frac{1}{f_{cyc}} \tag{5.1}$$

where f_{cyc} is the operational frequency in cycles/sec or Hz. One may note that traveling compression wavelets tend to converge with time into one sharp front (a shock front, if strong enough), while rarefaction (expansion) wavelets tend to diverge with time, i.e., spread out. A compression wave induces a positive flow velocity behind its front, while a rarefaction wave induces a negative flow component relative to the direction of its traveling front. A weak pressure wave reflecting off a solid wall boundary will produce a reflection approximately double the impinging wave strength (pressure up for a compression, pressure down for a rarefaction). Conversely, a weak pressure wave reflecting off an open air boundary (e.g., channel opening up into a large reservoir) produces a reversal in identity (compression becomes a rarefaction, and visa versa) and strength (same magnitude of Δp, but reversed sign).

Referring to Fig. 5.2, one can examine the various components that make up a conventional PJ's operational cycle. Beginning with the compression phase at Fig. 5.2a, one has the ignition and subsequent burning of the fuel–air mixture in the combustor, noting the fuel injection mass flow rate is defined as $\dot{m}_f$. The front-end intake valve is closed when under overpressure (pressure above atmospheric). Fuel may be injected during part of the cycle (intermittently), or alternatively may be delivered continuously (for simplicity if not overly concerned with fuel economy, or, to cool the gas mixture sufficiently to help keep the surrounding structure below its temperature threshold for permanent heating damage). Autoignition of the fuel–air mixture f in the presence of hot residual gases will occur within a certain range of equivalence ratio around unity (stoichiometric, where $\phi = f_{act}/f_{stoich}$), where the mixture is fuel-lean while $\phi<1$, and fuel-rich when $\phi>1$. With PJs, there is some advantage with respect to smoothness of operation in having a fuel that does not have too wide a range of ϕ for burning. As noted, during this Fig. 5.2a segment, pressure and temperature will rise in the combustor, reaching a nominal peak of p_4 (or p_{max}) and T_4 respectively. A rightward-moving compression wave will arise from this building pressure, and commence moving into the lower-pressure zone of the tailpipe.

Analogous to assuming the idealized sinusoidal combustor pressure–time profile of Fig. 5.4, one can assume a corresponding thrust–time profile, as also shown in Fig. 5.4, that would result from such a profile. With a further assumption that one does not have any appreciable negative thrust resulting from the above mentioned backflow during the induction phase of the operational cycle, the remaining positive components of the cyclic thrust profile would produce a net mean thrust of:

$$\overline{F} \approx \frac{F_{max}}{\pi} \tag{5.2}$$

Mean thrust, $\overline{F}$ or F_{mean}, is useful to know for flight performance calculations, and is the basis for such parameters as thrust-specific fuel consumption (TSFC) quoted for pulsejets.

For pulsejet performance evaluation, let us first look at estimating the correlation between combustor pressure and temperature. Consider an idealized conservation of energy representation of the combustion process in the combustor:

$$\Delta Q + mC_pT_1 = mC_pT_4 \tag{5.3}$$

Here, ΔQ is the heat input from the complete combustion process (for one cycle), m is the mass of the fuel–air mixture in the combustion chamber, T_1 is the base temperature, and T_4 is the peak combustor temperature. Further from this result,

$$T_4 = T_1 + \frac{\Delta Q}{mC_p} \approx T_1 + \frac{\dot{m}_f \Delta t_{c,i} q_R \eta_b}{\rho_C \mathbb{V}_C C_p} \tag{5.4}$$

Here, $\dot{m}_f$ is the instantaneous (i.e., not average) mass flow rate of fuel being injected, $\Delta t_{c,i}$ is the ideal combustion time period (to complete combustion for one cycle; note, this value can be substantially less than the actual fuel injection time for a given cycle, for reasons noted above), and η_b is the burning efficiency (default assumption, $\eta_b \approx 1$). The heat of reaction (LHV) of the fuel–air mixture, q_R , would have a value in the vicinity of 40 MJ/kg (of fuel). If one assumes that mass in the combustor of volume $\mathbb{V}_C$ is relatively constant during the combustion segment, then the gas density therein should be:

$$\rho_C = \frac{m}{\mathbb{V}_C} \approx \text{constant} \tag{5.5}$$

From the equation of state,

$$p = \rho R T \tag{5.6}$$

where the specific gas constant R for the gas through the combustion process still remains close to the nominal value for air (R_{air}), one can as a result establish that:

$$\frac{p_4}{p_1} \approx \frac{T_4}{T_1} \approx 1 + \frac{\dot{m}_f \Delta t_{c,i} q_R \eta_b}{\rho_C \mathbb{V}_C C_p T_1} \tag{5.7}$$

Please refer to the solution for example Problem 5.1 at the end of this chapter, for a completed sample pulsejet combustor performance analysis. For students learning this material as part of a course, I would encourage them to attempt the given example problem first, and then check the solution to see if any mistakes were made.

5.3 Wave Analysis

From gasdynamics, one can undertake some basic analysis of the wave motion within the pulsejet combustion tube, in order to gain further information on the performance capabilities of the given engine. Referring to Fig. 5.7(a), for a constant-area combustion tube, one can analyze a simplified representation of the wave activity arising from a pressurized combustor. Flow state 4 is upstream in the combustor, assumed to initially be at its peak pressure p_4 and T_4, while flow state 1 is the baseline flow condition in the tailpipe (near atmospheric pressure, but depending on the time of analysis, T_1 might be quite hot if later into a firing). The following estimate for sound speed can be made, based on previous information above:

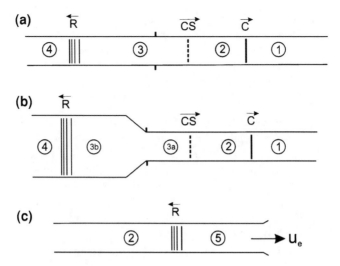

Fig. 5.7 Illustration of representative wave activity in various locations of the pulsejet

$$\frac{a_4}{a_1} \approx \sqrt{\frac{T_4}{T_1}} \approx \sqrt{\frac{p_4}{p_1}} \tag{5.8}$$

assuming γ and R do not change too much through the potentially large difference in temperature. An upstream-moving rarefaction wave will move into the combustor in acting to lower the pressure in that region, and accelerating the flow rightward down the tailpipe in the process. In the case of Fig. 5.7a, one will have flow state 3 downstream of the rarefaction, while in the case of Fig. 5.7b for a PJ having a transition section (thus a combustor with a larger cross-sectional area), one will have flow state 3b upstream of the transition and state 3a downstream. Referring to the 5.7(b) case, from gasdynamics (characteristic wave theory) one has the following correlation through the rarefaction:

$$u_4 + \frac{2a_4}{\gamma - 1} = u_{3b} + \frac{2a_{3b}}{\gamma - 1} \tag{5.9}$$

noting that u_4 is approximately zero. From isentropic flow theory:

$$\frac{p_{3b}}{p_4} = \left(\frac{a_{3b}}{a_4}\right)^{\frac{2\gamma}{\gamma - 1}} \tag{5.10}$$

and

$$\frac{\rho_{3b}}{\rho_4} = \left(\frac{p_{3b}}{p_4}\right)^{\frac{1}{\gamma}} \tag{5.11}$$

If an area transition exists such as for 5.7(b), one observes the following area-Mach number relation for upstream and downstream area values:

$$\frac{A_u}{A_d} = \frac{Ma_{3a}}{Ma_{3b}} \left[\frac{2 + (\gamma - 1)Ma_{3b}^2}{2 + (\gamma - 1)Ma_{3a}^2}\right]^{\frac{\gamma + 1}{2(\gamma - 1)}} \tag{5.12}$$

Across the rightward-moving contact surface (CS) moving at the local gas velocity, the following is true:

$$\begin{aligned} u_{3a} = u_2, \quad p_{3a} = p_2 \\ a_{3a} \neq a_2, \quad \rho_{3a} \neq \rho_2 \end{aligned} \tag{5.13}$$

Assuming the rightward-moving compression is still isentropic (not yet formed into a sharp-fronted non-isentropic shock), characteristic wave theory gives:

$$u_2 - \frac{2a_2}{\gamma - 1} = u_1 - \frac{2a_1}{\gamma - 1} \tag{5.14}$$

One may assume that u_1 is approximately zero. From isentropic flow theory:

$$\frac{p_2}{p_1} = \left(\frac{a_2}{a_1}\right)^{\frac{2\gamma}{\gamma - 1}} \tag{5.15}$$

and

$$\frac{\rho_2}{\rho_1} = \left(\frac{p_2}{p_1}\right)^{\frac{1}{\gamma}} \tag{5.16}$$

In the constant-area combustion tube case, Fig. 5.7(a), one can solve the above equations in a direct fashion. One does not need Eq. 5.12, and flow states 3a and 3b are the same, namely 3. Via Eqs. 5.9 and 5.13,

$$\frac{2a_4}{\gamma - 1} = u_2 + \frac{2a_3}{\gamma - 1} \tag{5.17}$$

Via Eqs. 5.14 and 5.17:

$$\frac{2a_1}{\gamma - 1} = \frac{2a_2}{\gamma - 1} - u_2 = \frac{2(a_2 - a_4 + a_3)}{\gamma - 1} \tag{5.18}$$

From Eqs. 5.10, 5.13 and 5.15:

$$\frac{a_3}{a_4} = \left(\frac{p_3}{p_4}\right)^{\frac{\gamma - 1}{2\gamma}} = \left(\frac{p_2}{p_1} \cdot \frac{p_1}{p_4}\right)^{\frac{\gamma - 1}{2\gamma}} = \frac{a_2}{a_1}\left(\frac{p_1}{p_4}\right)^{\frac{\gamma - 1}{2\gamma}} \tag{5.19}$$

Substituting the above into Eq. 5.18, one obtains:

$$\frac{2a_1}{\gamma - 1} = \frac{2}{\gamma - 1}\left(a_2 - a_4 + a_2\frac{a_4}{a_1}\left(\frac{p_1}{p_4}\right)^{\frac{\gamma - 1}{2\gamma}}\right) \qquad (5.20)$$

Re-arranging the above, one arrives at a direct solution for a_2:

$$a_2 = \frac{a_1 + a_4}{1 + \dfrac{a_4}{a_1}\left(\dfrac{p_1}{p_4}\right)^{\frac{\gamma - 1}{2\gamma}}} \qquad (5.21)$$

From there, one can progress to solving directly the remaining properties of interest:

$$p_2 = p_1\left(\frac{a_2}{a_1}\right)^{\frac{2\gamma}{\gamma - 1}} \qquad (5.22)$$

$$u_2 = \frac{2(a_2 - a_1)}{\gamma - 1} \qquad (5.23)$$

$$\rho_2 = \rho_1\left(\frac{p_2}{p_1}\right)^{\frac{1}{\gamma}} \qquad (5.24)$$

Let us now lay the groundwork for analyzing a PJ with an area transition between the combustor and the tailpipe (Fig. 5.7(b)), where we incorporate the influence of Eq. 5.12 at some point. From Eq. 5.9,

$$u_{3b} = a_{3b}Ma_{3b} = \frac{2}{\gamma - 1}(a_4 - a_{3b}) \qquad (5.25)$$

so that

$$a_{3b} = \frac{a_4}{1 + \frac{\gamma - 1}{2}Ma_{3b}} \qquad (5.26)$$

From Eqs. 5.13 and 5.14,

$$a_2 = \frac{\gamma - 1}{2}\left[a_{3a}Ma_{3a} + \frac{2}{\gamma - 1}a_1\right] \qquad (5.27)$$

From Eqs. 5.13, 5.15 and 5.27:

$$a_2 = a_1\left(\frac{p_{3a}}{p_4}\cdot\frac{p_4}{p_1}\right)^{\frac{\gamma - 1}{2\gamma}} = a_1\frac{a_{3a}}{a_4}\left(\frac{p_4}{p_1}\right)^{\frac{\gamma - 1}{2\gamma}} = \frac{\gamma - 1}{2}a_{3a}Ma_{3a} + a_1 \qquad (5.28)$$

From Eq. 5.28, we establish one route (A) to finding a_{3a}:

$$a_{3a}^A = a_1 / \left[\frac{a_1}{a_4} \left(\frac{p_4}{p_1} \right)^{\frac{\gamma-1}{2\gamma}} - \frac{\gamma-1}{2} Ma_{3a} \right] \tag{5.29}$$

A second route (B) to finding a_{3a} is conservation of energy through the area transition:

$$a_{3a}^B = a_{3b} \left(\frac{1 + \frac{\gamma-1}{2} Ma_{3b}^2}{1 + \frac{\gamma-1}{2} Ma_{3a}^2} \right)^{1/2} \tag{5.30}$$

An iterative solution is required to establish the value for a_{3a} that agrees with both Eqs. 5.29 and 5.30. One could begin the calculation process by guessing a value for Ma_{3b}, and proceed to solve Ma_{3a}, a_{3b} and a_{3a}^A, and after calculating a_{3a}^B, check for closeness to a_{3a}^A. If not close enough, repeat the process with a better guess of Ma_{3b}, etc., until confident that the value for a_{3a} is converged. At that juncture, one can proceed to directly solving for a_2, p_2 and u_2 via equations provided above.

Now, let us move to estimate thrust, by examining the wave activity in the vicinity of the tailpipe exit. Here, as shown by Fig. 5.7c, the thrust phase has begun with the rarefaction wave moving upstream toward the combustor. The tailpipe in practice may be as shown (a small expansion lip, to accommodate the possibility of flow choking such that there will be a supersonic exit flow at the edge of the lip, with sonic flow at the entry to the expansion lip; however, for a lower performance PJ, the flow will most likely be for the most part subsonic throughout, with the exhaust jet area comparable to the upstream tailpipe area). As a result then, for this case, assume u_5 downstream of the rarefaction is approximately u_e in this idealized representation. Similarly, one can in this case assume p_5 and p_e are close to ambient atmospheric, so in turn, approximately p_1. From characteristic wave theory,

$$u_2 + \frac{2a_2}{\gamma - 1} = u_5 + \frac{2a_5}{\gamma - 1} \tag{5.31}$$

From isentropic flow theory:

$$\frac{a_5}{a_2} = \left(\frac{p_5}{p_2} \right)^{\frac{\gamma-1}{2\gamma}} \approx \left(\frac{p_1}{p_2} \right)^{\frac{\gamma-1}{2\gamma}} \approx \frac{a_1}{a_2} \tag{5.32}$$

which suggests $a_5 \approx a_1$ and therefore $T_5 \approx T_1$, and via the equation of state, $\rho_5 \approx \rho_1$. Returning to Eq. 5.31,

$$u_5 \approx u_2 + \frac{2(a_2 - a_1)}{\gamma - 1} \tag{5.33}$$

Collecting the above information, for ideal maximum (peak) thrust, one can use the following thrust equation:

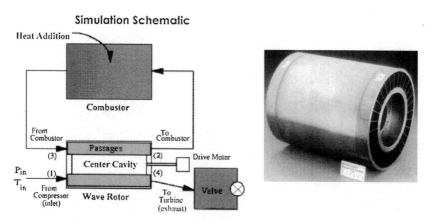

Fig. 5.8 Schematic block diagram of a wave rotor placed in parallel with a gas turbine engine combustor [4]. Photo of 4-port wave rotor unit developed for engine testing by NASA Glenn Research Center. Courtesy of NASA

$$F_{\text{ideal,max}} = \dot{m}_e(u_e - V_\infty) = \rho_5 u_5 A_p(u_5 - V_\infty) \tag{5.34}$$

where A_p is the tailpipe area upstream of the nozzle expansion. Mean thrust can be estimated as a function of this ideal maximum thrust value, assuming the sinusoidal pressure–time profile about ambient pressure of Fig. 5.3, via Eq. 5.2 above.

Please refer to the solution for example Problem 5.3 at the end of this chapter, for a completed sample pulsejet wave and thrust estimation analysis.

5.4 Related Propulsion Technology

5.4.1 Wave Rotor

While the pulsejet itself does not presently see wide usage with respect to air vehicle applications (largely due to fuel inefficiency, cooling and structural vibration issues), elements of its technology are being considered for other propulsion systems. Pressure exchangers of various kinds operate on an acoustic cycle basis [1]. One type, the dynamic pressure-exchanger, is commonly referred to as a wave rotor. Wave rotor "topping-cycle" (meaning performance-augmenting) pressure-gain units for gas turbine engines, placed in parallel with a conventional combustor (see Fig. 5.8), can potentially raise the core gas pressure and temperature at that engine location such that it improves specific fuel consumption (BSFC or TSFC) by as much as 15%, for the same power or thrust output by the engine [4].

Like the PJ, the passage of traveling compression and rarefaction waves within the wave rotor is fundamental to its operation (the transient wave motion produces the net increase in pressure and temperature in the core flow, but allows for the

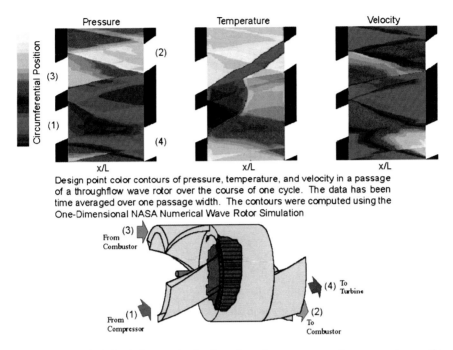

Design point color contours of pressure, temperature, and velocity in a passage of a throughflow wave rotor over the course of one cycle. The data has been time averaged over one passage width. The contours were computed using the One-Dimensional NASA Numerical Wave Rotor Simulation

Fig. 5.9 Numerical simulation results for flow through one passage of a wave rotor, as the passage rotates past the various ports [4]. Courtesy of NASA

surrounding walls of the given wave rotor passage to remain cooler than one would see for a quasi-steady continuous approach, e.g., afterburning, which will be discussed later for TJs). The effective combustor temperature could be raised up to 25%, while still retaining the original nominal compressor and turbine gas temperatures upstream and downstream respectively. The effective pressure entering the turbine section, unlike the temperature, would be substantially higher than the original nominal level.

A wave rotor consists of a series of constant cross-sectional area passages, with each passage rotating about a central axis. Via this rotation, the ends of the passages (upstream and downstream) are periodically exposed to various circumferentially arranged ports connecting the wave rotor to the upstream compressor, the central main combustor, or the downstream turbine (see Figs. 5.8 and 5.9). From gasdynamics, one knows that the sudden appearance of an opening at one end of a pressurized channel will produce a traveling pressure wave (compression or expansion [rarefaction], depending on the relative conditions inside and outside the given passage). Each passage of the wave rotor will experience hot and cold flow over a given cycle, and this helps to keep the passage wall temperatures well below are that seen for the peak gas temperature being discharged from the main combustor. In a typical wave rotor operational cycle, air from the compressor would be passed through the wave rotor for additional compression, with that compressed air then sent on to the combustor for heating via combustion of the fuel–air mixture. That reacted gas mixture would then be returned to the wave rotor for effective expansion, and then ultimately released to the turbine. It turns out that the temperature of the gas entering the turbine will be the same as the original non-wave-rotor level, but the pressure will now be substantially higher, a principal beneficial contribution of a wave rotor.

At this juncture, there has been one derivative wave rotor unit that has gone beyond the research/testing phase, and entered into production, namely the Comprex pressure-wave supercharger produced by ABB (Asea Brown Boveri), for improving the performance of automotive diesel engines [1, 5].

5.4.2 Pulse Detonation Engine

Pulse detonation engines (PDEs; [6]), presently being researched for high-speed flight applications as a potential competitor to RJs noted above, also share some similarities to pulsejets. As a significant difference, their operation depends more on higher-speed, higher-heat-input, higher-strength supersonic *detonation* waves moving downstream (rightward) in the intake/combustion/exhaust (detonation/combustion tube) duct system (flame front moves with the compression shock front), as opposed to weaker slightly supersonic compression and sonic rarefaction waves moving significantly faster than the subsonic *deflagration* front of the flame within the PJ combustor and tailpipe. Depending on the fuel-oxidizer combination being used in the detonation tube, the Chapman-Jouguet equations may be used for predicting detonation wave strengths, in a manner analogous to the Rankine-Hugoniot equations for normal shock waves. PDEs can optionally function (depending on the system's approach) on a nominal 2L wave system, $\vec{D} - \overleftarrow{R}$ (fast downstream-moving compressive detonation wave followed by a slower upstream-moving rarefaction wave, as per Fig. 5.10; [7]), or more like a 4L approach $(\vec{D} - \overleftarrow{R} - \vec{S} - \overleftarrow{R}$, where the first leftward-moving rarefaction in the cycle, upon reflection with the open intake, eventually transforms into a rightward-moving compressive shock wave, given the intake on the left side of the detonation tube, originally closed, is now opened to allow for air admittance). PDEs can produce higher effective chamber pressures and correspondingly significantly better values for specific impulse (I_{sp}) than PJs, which makes them a competitive candidate for high-speed flight relative to RJs. As one might expect, the vibrational nature of the propulsion system's operation is a potential drawback for manned flight vehicle applications (or flight vehicles with sensitive electronics!), unless a suitable damping approach can be employed effectively between the engine and vehicle.

One can use a Humphrey cycle representation as a basis for thermodynamic cycle analysis of a PDE, as represented by the pressure–volume and temperature-entropy diagrams of Fig. 5.11. In turn, one can estimate the thermal efficiency as follows:

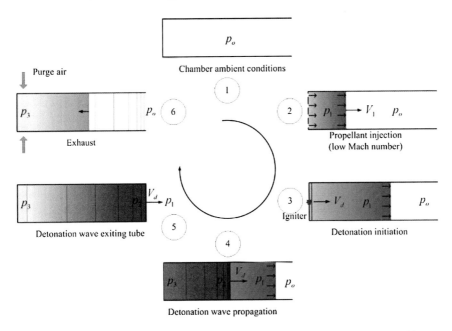

Fig. 5.10 Illustration of operational cycle of conventional pulse detonation engine [7]. Diagram courtesy of Aerodynamics Research Center, University of Texas at Arlington (Dr. Frank K. Lu)

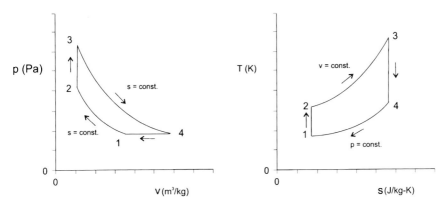

Fig. 5.11 Diagrams of $p-v$ and $T-s$ profiles for ideal Humphrey cycle

$$\eta_{th} = 1 - \gamma \frac{T_1}{T_2} \cdot \left\{ \frac{\left(\frac{T_3}{T_2}\right)^{1/\gamma} - 1}{\frac{T_3}{T_2} - 1} \right\} \tag{5.35}$$

One can compare the efficiency of this process, to that of a jet engine's Brayton cycle, which will be introduced in the next chapter.

5.5 Example Problems

5.1. A pulsejet has a combustor with a diameter of 5 cm and length of 10 cm, and an overall effective combustor/tailpipe length of 40 cm. Assume the baseline (prior to burning pulse) combustor temperature is 300 K (i.e., at or near start-up), baseline gas density is 1.225 kg/m³, gas molecular weight is 29 amu, and gas ratio of specific heats is 1.33. Assume the fuel injection time is half the operational cycle time period, and neglect counterflow velocity Δu in your calculations. Assume the effective combustion time in the combustor is one-quarter the operational cycle time period. The operational compression ratio of this PJ is 2. The heat of reaction of the fuel/air mixture is 40 MJ/kg of fuel, and you may assume 100% combustion efficiency. You may assume the tailpipe ambient gas temperature is comparable to the baseline combustor temperature. With this information, estimate the fuel consumption rate at this juncture (kg/hr).

5.2. Estimate the static thrust produced by the pulsejet of Problem 5.1 above, using the start-up wave analysis approach. The tailpipe diameter is 5 cm, the tailpipe exhaust is unchoked in operation, and the ambient tailpipe gas temperature is similar to the baseline combustor temperature.

5.3. As part of a start-up wave analysis of a pulsejet with a combustor/tailpipe area convergence of 2, estimate the strength of the initial downstream-moving wave (find p_2/p_1, u_2/a_1) and compare the values to the constant-area PJ of Problem 5.2. The operational compression ratio of the combustor is 2, the gas specific heat ratio is 1.33, the molecular weight of the gas is 29 amu, and one can assume that the gas in the tailpipe is at a temperature that is similar to the combustor baseline temperature.

5.4. Repeat Problem 5.1, but for a more practical compression ratio of 1.5 for a small pulsejet engine.

5.5. Derive Eq. 5.35 with the information provided by Fig. 5.11.

5.6 Solutions to Problems

5.1. Evaluate fuel consumption $\dot{m}_f$ of conventional valved pulsejet.

$$R = \frac{R}{M} = 8312/29 = 287 \text{ J/kg} \cdot \text{K}, \, a_1 = \sqrt{\gamma R T_1} = (1.33 \cdot 287 \cdot 300)^{0.5} = 338 \text{ m/s}$$

$$C_P = \frac{\gamma R}{\gamma - 1} = 1.33 \, (287)/0.33 = 1157 \text{ J/kg} \cdot \text{K}$$

Time period for one operational cycle, $\Delta t_{cyc} \approx \dfrac{4\ell_{c/p}}{a_1} \approx 4(0.4)/338 = 0.00473 \text{ s}$

Combustor volume, $V_C \approx \dfrac{\pi d_C^2}{4} \cdot \ell_C = \pi(0.05)^2/4 \cdot 0.10 = 0.0002 \text{ m}^3$

Effective time period for ideal combustion for one cycle, $\Delta t_{c,i} \approx \dfrac{\Delta t_{cyc}}{4} \approx 0.0012$ s.

Compression ratio of 2:

$$\frac{p_4}{p_1} \approx \frac{T_4}{T_1} \approx 1 + \frac{\dot{m}_f \Delta t_{c,i} q_R \eta_b}{\rho_c \Psi_C C_P T_1} = 2, \text{ or}$$

$$\dot{m}_f \approx \frac{\rho_c \Psi_C C_P T_1}{\Delta t_{c,i} q_R \eta_b} \left(\frac{p_4}{p_1} - 1 \right)$$

$$= 1.225(0.0002)1157(300)/(00012 \cdot 40 \times 10^6 \cdot 1.0)(2 - 1) = 0.0018 \, \text{kg/s}$$

which is the instantaneous fuel injection rate. In this case, it is stated that the fuel injection occurs only for half the operational cycle, so that

$$\dot{m}_{f,\text{overall}} = \dot{m}_f \cdot \frac{\text{inj}}{\Delta t_{cyc}} = 0.0018 \, (1/2) = 9 \times 10^{-4} \, \text{kg/s} = 3.24 \, \text{kg/hr}$$

If one found it beneficial to have continuous fuel injection for the entire operational cycle, e.g., for keeping the engine structure cool enough (or simply for easier operation), then one would have a fuel consumption rate of around 6.5 kg/hr for this example. Similarly then, if the PJ is known to produce a static sea-level *mean* thrust of about 18 N, then the thrust-specific fuel consumption (TSFC), for continuous fuel injection, would be: TSFC = 6.5 kg/hr/18 N = 0.36 kg/hr·N. A conventional small subsonic turbojet would have a static sea-level TSFC of around 0.07 kg/hr·N, or about five times better fuel economy.

5.2. Estimate startup static thrust of conventional valved pulsejet of Problem 5.1. Tailpipe diameter is 5 cm (same as the combustor diameter), pipe cross-sectional area thus about 0.002 m².

Given $\dfrac{T_4}{T_1} \approx \dfrac{p_4}{p_1} = 2$ in this example, then $\dfrac{a_4}{a_1} = \sqrt{\dfrac{T_4}{T_1}} = 2^{0.5} = 1.414$; for

$a_1 = \sqrt{\gamma R T_1} = (1.33 \cdot 287 \cdot 300)^{0.5} = 338$ m/s, $\quad a_4 = 1.414(338) = 478$ m/s.

Also, $T_4 = 2(300) = 600$ K.

From textbook:

$$a_2 = \frac{a_1 + a_4}{1 + \dfrac{a_4}{a_1} \left(\dfrac{p_1}{p_4} \right)^{\frac{\gamma-1}{2\gamma}}} = (338 + 478)/(1 + 1.414(0.5)^{0.124}) = 355.2 \, \text{m/s},$$

$$p_2 = p_1 \left(\frac{a_2}{a_1} \right)^{\frac{2\gamma}{\gamma-1}} = 101000(355.2/338)^{8.06} = 150683 \, \text{Pa}$$

$$u_2 = \frac{2(a_2 - a_1)}{\gamma - 1} = 2(355.2 - 338)/0.33 = 104.2 \, \text{m/s}$$

$$\rho_2 = \rho_1 \left(\frac{p_2}{p_1}\right)^{\frac{1}{\gamma}} = 1.225(150.7/101)^{0.75} = 1.654 \, \text{kg/m}^3$$

Exhaust flow unchoked, so assume $p_5 \approx p_\infty = 101000$ Pa $\approx p_1$.

$$a_5 = a_2 \left(\frac{p_5}{p_2}\right)^{\frac{\gamma-1}{2\gamma}} = 355.2(101/150.7)^{0.124} = 338 \, \text{m/s} = a_1 \text{ in this case, so}$$
$T_5 = T_1$ as a result.

$$u_5 = u_2 + \frac{2(a_2 - a_5)}{\gamma - 1} = u_2 + \frac{2a_2}{\gamma - 1} - \frac{2a_2}{\gamma - 1}\left(\frac{p_5}{p_2}\right)^{\frac{\gamma-1}{2\gamma}}$$
$$= 104.2 + 2(335.2)/0.33 - 2(355.2)/0.33 \cdot (101/150.7)^{0.124} = 208.5 \, \text{m/s}$$

$$\rho_5 = \rho_2 \left(\frac{p_5}{p_2}\right)^{\frac{1}{\gamma}} = 1.654(101/150.7)^{0.752} = 1.225 \, \text{kg/m}^3 = \rho_1 \text{ in this case. Ideal}$$
max. thrust at startup,

$$F_{\text{ideal,max}} = \dot{m}_e(u_e - V_\infty) = \rho_5 u_5 A_p(u_5 - V_\infty) = 1.225(208.5)(0.002)(208.5 - 0)$$
$$= 106.5 \, \text{N}$$

Mean thrust estimate, assuming sinusoidal pressure–time profile about ambient pressure:

$$\bar{F} \approx \frac{F_{\text{max}}}{\pi} = 106.5/\pi = 33.9 \, \text{N}$$

$\bar{F}/A_p = 33.9/0.002 = 17000 \, \text{N/m}^2 = 17 \, \text{kN/m}^2$, which is a low from a performance viewpoint for a conventional pulsejet (35 kN/m^2 a typical mean static-thrust performance value for a PJ).

5.3. Estimate startup flow properties of the conventional valved pulsejet of Problem 5.1, Tailpipe diameter is 3.54 cm (smaller than the combustor diameter of 5 cm), for a pipe cross-sectional area of about 0.001 m^2.

Given $\dfrac{T_4}{T_1} \approx \dfrac{p_4}{p_1} = 2$ in this example, then $\dfrac{a_4}{a_1} = \sqrt{\dfrac{T_4}{T_1}} = 2^{0.5} = 1.414$; for

$$a_1 = \sqrt{\gamma R T_1} = (1.33 \cdot 287 \cdot 300)^{0.5} = 338 \text{ m/s}, \quad a_4 = 1.414(338) = 478 \text{ m/s}.$$

Also, $T_4 = 2(300) = 600$ K.

From textbook, for an area convergence of area ratio 2, moving from station 3b upstream to station 3b downstream of area transition:

$$\frac{A_u}{A_d} = \frac{Ma_{3a}}{Ma_{3b}} \left[\frac{1 + \frac{\gamma-1}{2}Ma_{3b}^2}{1 + \frac{\gamma-1}{2}Ma_{3a}^2} \right]^{\frac{\gamma+1}{2(\gamma-1)}} = 2 = \frac{Ma_{3a}}{Ma_{3b}} \left[\frac{1 + 0.165Ma_{3b}^2}{1 + 0.165Ma_{3a}^2} \right]^{3.53}$$

First guess: let $Ma_{3b} = 0.1$, which via iteration gives $Ma_{3a} = 0.204$ downstream via the area-Mach no. relation above. Then, following solution route A in the textbook, we have:

$$\frac{a_{3b}}{a_1} = \frac{a_4/a_1}{1 + \frac{\gamma-1}{2}Ma_{3b}} = \frac{\sqrt{2}}{1 + 0.165Ma_{3b}} = 1.391$$

$$\frac{a_{3a}^A}{a_1} = \left[\frac{a_1}{a_4}\left(\frac{p_4}{p_1}\right)^{\frac{\gamma-1}{2\gamma}} - \frac{\gamma-1}{2}Ma_{3a} \right]^{-1} = \left[\frac{1}{\sqrt{2}}(2)^{0.124} - 0.165\,Ma_{3a} \right]^{-1} = 1.357$$

$$\frac{a_{3a}^A}{a_{3b}} = \frac{a_{3a}^A/a_1}{a_{3b}/a_1} = 1.357/1.391 = 0.9756. \text{ Compare this value to that from}$$

solution route B in the textbook:

$$\frac{a_{3a}^B}{a_{3b}} = \left(\frac{1 + \frac{\gamma-1}{2}Ma_{3b}^2}{1 + \frac{\gamma-1}{2}Ma_{3a}^2} \right)^{0..5} = \left(\frac{1 + 0.165Ma_{3b}^2}{1 + 0.165Ma_{3a}^2} \right)^{0.5} = 0.9982. \text{ Not bad, but let}$$

us get a little closer.

Second guess: let $Ma_{3b} = 0.15$, which via iteration gives $Ma_{3a} = 0.313$ from the area-Mach no. relation above. Then,

$$\frac{a_{3b}}{a_1} = 1.38, \quad \frac{a_{3a}^A}{a_1} = 1.391, \quad \frac{a_{3a}^A}{a_{3b}} = \frac{a_{3a}^A/a_1}{a_{3b}/a_1} = 1.008, \text{ which one compares to:}$$

$$\frac{a_{3a}^B}{a_{3b}} = 0.9938, \text{ which tells us we have gone too high on } Ma_{3b}.$$

Third guess: let $Ma_{3b} = 0.13$, which via iteration gives $Ma_{3a} = 0.27$ from the area-Mach no. relation above. Then,

$$\frac{a_{3b}}{a_1} = 1.385, \quad \frac{a_{3a}^A}{a_1} = 1.377, \quad \frac{a_{3a}^A}{a_{3b}} = \frac{a_{3a}^A/a_1}{a_{3b}/a_1} = 0.995, \text{ which one compares to:}$$

$$\frac{a_{3a}^B}{a_{3b}} = 0.995. \text{ Close enough! Thus, from the textbook,}$$

$$\frac{a_2}{a_1} = \frac{\gamma - 1}{2}\left[\frac{a_{3a}}{a_1}Ma_{3a} + \frac{2}{\gamma - 1}\right] = 1.061, \ a_2 = 1.061(338) = 359 \ \text{m/s}$$

$$\frac{p_2}{p_1} = \left(\frac{a_2}{a_1}\right)^{\frac{2\gamma}{\gamma - 1}} = 1.611 \ \text{(vs. 1.491 for the constant-area combustor-pipe case),}$$
$$p_2 = 1.611(101000) = 162711 \ \text{Pa}$$
$$\frac{u_2}{a_1} = \frac{2}{\gamma - 1}\left(\frac{a_2}{a_1} - 1\right) = 0.37 \ \text{(vs. 0.3079 for the constant-area combustor-pipe}$$
case)
$$u_2 = 0.37(338) = 125 \ \text{m/s}$$

$$\rho_2 = \rho_1\left(\frac{p_2}{p_1}\right)^{\frac{1}{\gamma}} = 1.225(162.7/101)^{0.75} = 1.752 \ \text{kg/m}^3$$

Exhaust flow assumed to be unchoked, so assume $p_5 \approx p_\infty = 101000$ Pa $\approx p_1$.

$$a_5 = a_2\left(\frac{p_5}{p_2}\right)^{\frac{\gamma - 1}{2\gamma}} = 359(101/162.7)^{0.124} = 338 \ \text{m/s} \ = a_1 \quad \text{in} \quad \text{this} \quad \text{case}, \quad \text{so}$$
$T_5 = T_1$ as a result.

$$u_5 = u_2 + \frac{2(a_2 - a_5)}{\gamma - 1} = u_2 + \frac{2a_2}{\gamma - 1} - \frac{2a_2}{\gamma - 1}\left(\frac{p_5}{p_2}\right)^{\frac{\gamma - 1}{2\gamma}} = 125 + 2(359)/0.33 -$$
$$2(359)/0.33 \cdot (101/162.7)^{0.124} = 249.9 \ \text{m/s}$$

$$\rho_5 = \rho_2\left(\frac{p_5}{p_2}\right)^{\frac{1}{\gamma}} = 1.752(101/162.7)^{0.752} = 1.225 \ \text{kg/m}^3 = \rho_1 \ \text{in this case.}$$

Ideal max. thrust at startup,

$$F_{\text{ideal,max}} = \dot{m}_e(u_e - V_\infty) = \rho_5 u_5 A_p(u_5 - V_\infty) = 1.225(249.9)(0.001)(249.9 - 0)$$
$$= 76.5 \ \text{N},$$

is in fact less than the 106.5 N predicted for the constant-area case (note: double the exit nozzle area for the original case). However, if the baseline comparison between the two PJs had been for the same tailpipe diameter, the second case with the area convergence would be superior in performance.

Mean thrust estimate, assuming sinusoidal pressure–time profile about ambient pressure:

$\bar{F} \approx \frac{F_{\text{max}}}{\pi} = 76.5/\pi = 24.4$ N, compared to 33.9 N for the previous case

$\bar{F}/A_p = 24.4/0.001 = 24400 \ \text{N/m}^2 = 24.4 \ \text{kN/m}^2$, which is a bit better from a performance viewpoint than the previous case (17 kN/m^2), for a conventional pulsejet (35 kN/m^2 a typical mean static-thrust performance value for a PJ).

5.4. Evaluate fuel consumption $\dot{m}_f$ of conventional valved pulsejet, for a lower compression ratio.

$$R = \frac{R}{M} = 8312/29 = 287 \text{ J/kg} \cdot \text{K}, a_1 = \sqrt{\gamma R T_1} = (1.33 \cdot 287 \cdot 300)^{0.5} = 338 \text{ m/s}$$

$$C_P = \frac{\gamma R}{\gamma - 1} = 1.33\,(287)/0.33 = 1157 \text{ J/kg} \cdot \text{K}$$

Time period for one operational cycle, $\Delta t_{cyc} \approx \dfrac{4\ell_{c/p}}{a_1} \approx 4(0.4)/338 = 0.00473$ s

Combustor volume, $V_C \approx \dfrac{\pi d_C^2}{4} \cdot \ell_C = \pi(0.05)^2/4 \cdot 0.10 = 0.0002 \text{ m}^3$

Effective time period for ideal combustion for one cycle, $\Delta t_{c,i} \approx \dfrac{\Delta t_{cyc}}{4} \approx 0.0012$ s.

Compression ratio of 1.5:

$$\frac{p_4}{p_1} \approx \frac{T_4}{T_1} \approx 1 + \frac{\dot{m}_f \Delta t_{c,i} q_R \eta_b}{\rho_C V_C C_P T_1} = 1.5, \text{ or}$$

$$
\begin{aligned}
\dot{m}_f &\approx \frac{\rho_C V_C C_P T_1}{\Delta t_{c,i} q_R \eta_b}\left(\frac{p_4}{p_1} - 1\right) \\
&= 1.225(0.0002)1157(300)/(0.0012 \cdot 40 \times 10^6 \cdot 1.0)(1.5 - 1) \\
&= 0.0009 \text{ kg/s},
\end{aligned}
$$

which is the instantaneous fuel injection rate. In this case, it is stated that the fuel injection occurs only for half the operational cycle, so that overall fuel consumption rate is:

$$\dot{m}_{f,\text{overall}} = \dot{m}_f \cdot \frac{\text{inj}}{\Delta t_{cyc}} = 0.0009\,(1/2) = 4.5 \times 10^{-4} \text{ kg/s} = 1.62 \text{ kg/hr}$$

5.5. Referring to the thermodynamic cycles analyses of Chaps. 4 and 6, one defines thermal efficiency for the Humphrey cycle as:

$$\eta_{th} = 1 - \frac{Q_L}{Q_H} = 1 - \frac{mC_p(T_4 - T_1)}{mC_v(T_3 - T_2)} = 1 - \gamma\frac{(T_4 - T_1)}{(T_3 - T_2)} - 1 - \frac{\gamma T_1}{T_2}\frac{(T_4/T_1 - 1)}{(T_3/T_2 - 1)}$$

One notes for the isentropic processes that:

$$\frac{p_2}{p_1} = \left(\frac{T_2}{T_1}\right)^{\frac{\gamma}{\gamma-1}} \quad \text{and} \quad \frac{p_3}{p_4} = \left(\frac{T_3}{T_4}\right)^{\frac{\gamma}{\gamma-1}}$$

For the constant-volume process:

$v_3 = v_2$, $\rho_3 = \rho_2$, so $p_3/(RT_3) = p_2/(RT_2)$ and therefore $p_3/p_2 = T_3/T_2$

For the constant-pressure process:

$p_4 = p_1$; substituting from earlier, $p_4 = p_3 \left(\frac{T_4}{T_3}\right)^{\frac{\gamma}{\gamma-1}} = p_1 = p_2 \left(\frac{T_1}{T_2}\right)^{\frac{\gamma}{\gamma-1}}$ so that

$$\frac{p_3}{p_2} = \frac{\left(\frac{T_1}{T_2}\right)^{\frac{\gamma}{\gamma-1}}}{\left(\frac{T_4}{T_3}\right)^{\frac{\gamma}{\gamma-1}}} = \left(\frac{T_3}{T_2} \cdot \frac{T_1}{T_4}\right)^{\frac{\gamma}{\gamma-1}} = \frac{T_3}{T_2}$$ via further substitution; continuing,

$$\frac{T_1}{T_4} = \left(\frac{T_3}{T_2}\right)^{\frac{\gamma-1}{\gamma}-1} = \left(\frac{T_3}{T_2}\right)^{-\frac{1}{\gamma}}$$ which gives $$\frac{T_4}{T_1} = \left(\frac{T_3}{T_2}\right)^{\frac{1}{\gamma}}$$

Substitute this result back into the earlier expression for η_{th}:

$$\eta_{th} = 1 - \frac{Q_L}{Q_H} = 1 - \frac{\gamma T_1}{T_2} \frac{([T_3/T_2]^{1/\gamma} - 1)}{(T_3/T_2 - 1)}$$

References

1. Kentfield JAC (1993) Nonsteady, one-dimensional, internal, compressible flows–theory and applications. Oxford University Press, New York
2. Barr PK, Dwyer HA, Bramlette TT (1988) A one-dimensional model of a pulse combustor. Combust Sci Technol 58:315–336
3. Geng T, Kiker A, Ordon R, Kuznetsov AV, Roberts WL (2007) Combined numerical and experimental investigation of a hobby-scale pulsejet. J Propul Power 23:186–193
4. Paxson DE, Nalim MR (1999) A modified through-flow wave rotor cycle with combustor bypass ducts. J Propul Power 15:462–467
5. Stone R (1992) Introduction to internal combustion engines, 2nd edn. SAE, Warrendale
6. El-Sayed AF (2008) Aircraft propulsion and gas turbine engine. CRC Press, Boca Raton (Florida)
7. Lu FK (2009) Prospects for detonations in propulsion. In: Proceedings of 9th international symposium on experimental and computational aerothermodynamics of internal flows, Gyeongju (Korea), pp 8–11 Sept 2009

Chapter 6
Gas Turbine Engines: Fundamentals

6.1 Introduction

The big difference between gas turbine engines and internal combustion engines for powering aircraft is the ability of gas turbines to throughput large quantities of air for a given engine size. Gas turbine engines also employ internal combustion, but historically, they are slotted in a category separate from the predominantly older IC engines discussed in Chap. 4. The original development of the gas turbine engine was motivated by the desire to produce thrust via a high-speed, high-mass-flow exhaust jet (Whittle, von Ohain concepts for turbojet [TJ]; [1]) which cannot be produced practically with an IC approach. Later variants of the gas turbine (turboshaft [TS] for driving helicopter rotors, turboprop [TP] for driving airplane propellers, and turbofan [TF] for driving ducted many-bladed fans) have largely replaced the IC engine in systems that can still use both approaches. This is largely due to the high power-to-weight (commonly quoted as static sea-level shaft power over dry engine weight, P_{So}/W_{eng}) that accompanies the high air intake capability.

The primary core of any gas turbine engine is the gas generator, which is comprised of the compressor, combustion chamber, and turbine. Depending on the variant, different components must in turn be added to complete the given engine, e.g., the intake at the front end and the exhaust nozzle at the aft end of the engine. The air-standard Brayton cycle approximates the ideal open-cycle performance of a simple gas turbine engine (see the corresponding p–v and T–s charts of Fig. 6.1).

Referring to Fig. 6.1, process $1 \rightarrow 2$ is an isentropic compression of air (in practice, through the intake/diffuser and mechanical compressor); process $2 \rightarrow 3$ represents heat addition to the air at constant pressure (in practice, fuel injected and burning in combustor); process $3 \rightarrow 4$ represents isentropic expansion of air (in practice, through the turbine and exhaust nozzle); process $4 \rightarrow 1$ represents rejection of heat of the working medium (air, or exhaust gas in the actual case) at constant pressure (to the outside ambient environment).

D. R. Greatrix, *Powered Flight*, DOI: 10.1007/978-1-4471-2485-6_6,
© Springer-Verlag London Limited 2012

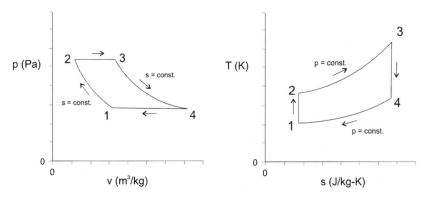

Fig. 6.1 Diagrams of p–v and T–s profiles for ideal Brayton cycle

The thermal efficiency of the Brayton cycle [2] is given by

$$\eta_{th} = 1 - \frac{Q_L}{Q_H} = 1 - \frac{mC_p(T_4 - T_1)}{mC_p(T_3 - T_2)} = 1 - \frac{T_1}{T_2}\frac{(T_4/T_1 - 1)}{(T_3/T_2 - 1)} \tag{6.1}$$

Here, Q_L is the heat lost from the system, and Q_H is the system heat input (from the combustion process). Defining r here as the compression ratio p_2/p_1, we note that

$$\frac{p_2}{p_1} = \frac{p_3}{p_4} = r \tag{6.2}$$

From isentropic flow theory,

$$\frac{p_2}{p_1} = \left(\frac{T_2}{T_1}\right)^{\frac{\gamma}{\gamma-1}} = \frac{p_3}{p_4} = \left(\frac{T_3}{T_4}\right)^{\frac{\gamma}{\gamma-1}} \tag{6.3}$$

As a result,

$$\frac{T_3}{T_4} = \frac{T_2}{T_1} \tag{6.4}$$

and therefore

$$\frac{T_3}{T_2} = \frac{T_4}{T_1} \tag{6.5}$$

Via substitution then, one arrives at the following for the cycle thermal efficiency:

$$\eta_{th} = 1 - \frac{T_1}{T_2} = 1 - \frac{1}{r^{\frac{\gamma-1}{\gamma}}} \tag{6.6}$$

Thus the thermal efficiency is shown to be a function of the isentropic compression process, which in turn is a function of such parameters as ambient air pressure and density, free stream flight Mach number Ma_4, Ma of the flow at the

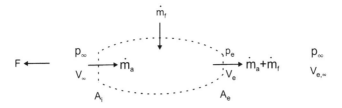

Fig. 6.2 Control volume diagram illustrating the fundamental operational processes associated with a turbojet engine, towards delivering thrust

compressor inlet, and the mechanical compressor's capability. The compressor's compression delivery level is a function of the compressor rotor shaft speed N (for a single shaft [spool] engine; for a two-spool engine, the majority of the compression would be done by the high-pressure spool running at speed N_2 [a.k.a., N_{HP}], while there would be less compression work being done by the lower pressure spool running at a lower rotation speed N_1 [a.k.a., N_{LP}]), which of course is the same rotation speed of the turbine attached to said shaft downstream.

6.2 An Introduction to Turbojet Engines

Referring to Fig. 6.2, consider the fundamental thrust equation for jet propulsion:

$$F = (\dot{m}_a + \dot{m}_f)V_e - \dot{m}_a V_\infty + (p_e - p_\infty)A_e \qquad (6.7)$$

The above equation applies for a conventional non-afterburning turbojet. For air-breathing jet engines like the turbojet [3], one can define the fuel–air ratio as:

$$f = \frac{\dot{m}_f}{\dot{m}_a} \qquad (6.8)$$

such that

$$F = \dot{m}_a[(1 + f)V_e - V_\infty] + (p_e - p_\infty)A_e \qquad (6.9)$$

One can note the momentum drag term associated with the intake of air at true flight airspeed V_∞ on the *net* thrust generated (excluding that term, one would have a higher *gross* thrust value). Noting that the thrust component produced by the pressure differential at the nozzle exit is not particularly efficient (although potentially substantial), the primary function of the jet engine is to exhaust the gas at a speed greater than that coming in, in order to produce the principal thrust. The more airflow $\dot{m}_a$ that can be drawn into the engine, the higher will be the thrust.

At this point, let us define jet propulsive efficiency η_p (a parameter that is different from propeller propulsive efficiency η_{pr}) as the ratio of the useful (thrust) power to the rate of kinetic energy production by the propulsion system:

$$\eta_p = \frac{FV_\infty}{\dot{m}_a\left[(1+f)\frac{V_{e,\infty}^2}{2} - \frac{V_\infty^2}{2}\right]} \tag{6.10}$$

Given a typical value for f is in the range of 0.025 for TJ engines (i.e., well below one in value), and assuming that in the ideal case that the exhaust jet can fully expand down to the ambient outside air pressure such that V_e goes to $V_{e,\infty}$, Eq. 6.10 can be simplified to the following well-known expression, sometimes referred to as Froude's equation for jet efficiency:

$$\eta_p \approx \frac{2 \cdot V_\infty/V_{e,\infty}}{1 + V_\infty/V_{e,\infty}} \tag{6.11}$$

While one needs $V_{e,\infty} > V_\infty$ for finite thrust delivery, it is interesting that Eq. 6.11 suggests that peak propulsive efficiency occurs ($\eta_p \to 1$) when $V_{e,\infty} \to V_\infty$, i.e., at a nominally very low thrust delivery. While this might seem like an impractical and contradictory result, comparable to that seen for propeller thrust versus efficiency, it does suggest going to a higher air mass flow intake $\dot{m}_a$ (as compensation for bringing down the value of $V_{e,\infty}$ closer to V_∞), which does to some degree make practical sense. In any case, it is not realistic to pursue a high efficiency value in isolation of other flight mission factors, e.g., the need for high thrust for a short period of time during takeoff and early climbout supersedes efficiency concerns, as compared to the much longer period of time, at a lower thrust value but significantly higher efficiency, at cruise conditions.

The thermal efficiency η_{th} for jet engines is defined as the ratio of kinetic energy addition to the gas flow relative to the total energy consumption rate $\dot{m}_f q_R$, where q_R is the heat of reaction [LHV] of the fuel (in reacting with air, typical jet fuel $q_R \approx 4.34 \times 10^7$ J/kg of fuel):

$$\eta_{th} = \frac{\dot{m}_a\left[(1+f)\frac{V_{e,\infty}^2}{2} - \frac{V_\infty^2}{2}\right]}{\dot{m}_f q_R} = \frac{(1+f)\frac{V_{e,\infty}^2}{2} - \frac{V_\infty^2}{2}}{f \cdot q_R} \tag{6.12}$$

Note that for a turboprop or turboshaft engine, the thermal efficiency is defined in terms of power (as opposed to thrust for TJs):

$$\eta_{th} = \frac{P_S}{\dot{m}_f q_R} = \frac{1}{\text{BSFC} \cdot q_R} \tag{6.13}$$

Please refer to the solution for Example Problem 6.1 at the end of this chapter, for a completed sample turboprop performance efficiency analysis that requires propeller performance calculations as covered in Chap. 3. For students learning this material as part of a course, I would encourage them to attempt the given example problem first, and then check the solution to see if any mistakes were made.

Overall efficiency η_o is the ratio of useful (thrust) power to energy consumption rate:

$$\eta_o = \frac{FV_\infty}{\dot{m}_f q_R} = \eta_p \eta_{th} \tag{6.14}$$

For the case that $f \ll 1$,

$$\eta_o \approx 2\eta_{th} \cdot \frac{V_\infty/V_{e,\infty}}{1 + V_\infty/V_{e,\infty}} \tag{6.15}$$

One can see that if one pursued in isolation a high propulsive efficiency such that $V_{e,\infty} \to V_\infty$, thermal efficiency would on the other hand go to zero, and thus overall one would have a low overall efficiency, which as a result better aligns with practical expectations.

Considering the static thrust case in isolation ($V_\infty \to 0$), one notes the following from Eq. 6.9:

$$F_o \approx \dot{m}_a V_{e,\infty} \tag{6.16}$$

while from Eq. 6.12,

$$F_o \approx \frac{2\eta_{th} \dot{m}_f q_R}{V_{e,\infty}} \tag{6.17}$$

The above suggests that for a given thermal efficiency and fuel flow, takeoff thrust may be increased by accelerating a large amount of air (as per Eq. 6.16) to a smaller exhaust velocity (as per Eq. 6.17). Again, one needs to be wary in interpreting such results in isolation, without considering other factors that might actually come into play for a given component of a flight mission. It is true that the turbofan engine does exploit the use of a higher $\dot{m}_a$ at a somewhat lower $V_{e,\infty}$ in order to improve its efficiency (flight economy), within a certain flight speed/altitude range.

Thrust specific fuel consumption (TSFC; typically provided in units of kg of fuel consumed per hour per newton of thrust delivered, or lb of fuel per hour per lb of thrust) is an important performance parameter for a given jet engine's flight economy:

$$\text{TSFC} = \frac{\dot{m}_f}{F} \approx \frac{\dot{m}_f}{\dot{m}_a[(1+f)V_{e,\infty} - V_\infty]} = \frac{f}{[(1+f)V_{e,\infty} - V_\infty]} \tag{6.18}$$

Equation 6.18 suggests that for flight economy, one should try to fly at as low a fuel–air ratio as possible (again, one would not be wise to consider this in isolation; e.g., flames have a lower extinguishment limit for lean mixtures, so f must be greater than that threshold value). Equation 6.18 in addition suggests that if one

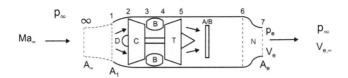

Fig. 6.3 Schematic diagram illustrating the various afterburning turbojet engine components and associated station numbering

were flying with a fixed $V_{e,\infty}$ (say with a choked jet exhaust nozzle), one could potentially see a worsening (increase) in TSFC as flight speed V_∞ is increased. This is more likely to be observed in practice (see later).

Another parameter of interest in evaluating engine performance is specific thrust, commonly referenced to the incoming massflow of air:

$$\frac{F}{\dot{m}_a} \approx (1+f)V_{e,\infty} - V_\infty \qquad (6.19)$$

For sizing a jet engine and estimating the actual delivered thrust, one needs to establish the expected value for air mass flow. This parameter, $\dot{m}_a$, will depend on such factors as intake cross-sectional entry area (A_1), flight speed and altitude, and further downstream through the engine, the air-swallowing capacity of the various engine components all the way to the nozzle exit of area A_e.

6.3 Cycle Analysis of Turbojet Engine

Before going into detail on the various engine components of a turbojet or other gas turbine variant (e.g., intake, centrifugal and axial compressors, combustor, axial turbine, afterburner, nozzle), let us undertake a cycle analysis of a TJ, in order to understand some of the factors affecting its performance. We will follow the engine station numbering scheme as presented in Fig. 6.3.

In order to account for inherent losses in real systems, we will want to define the adiabatic component efficiency. For the diffuser (inlet, intake), in moving from ambient outside atmospheric conditions to the inlet of the compressor, η_d will be defined as the ratio of the ideal to actual stagnation enthalpy change through the diffusion process:

$$\eta_d = \frac{h_{02,\text{ideal}} - h_\infty}{h_{02} - h_\infty} \qquad (6.20)$$

where at the inlet of the mechanical compressor (station 2), one utilizes stagnation temperature T_{02} as follows:

$$h_{02} = C_p T_{02} = C_p T_2 + \frac{V_2^2}{2} \qquad (6.21)$$

Note that the static enthalpy is used at the ambient air boundary, i.e.,

$$h_\infty = C_p T_\infty \tag{6.22}$$

Typical values for η_d can range from 0.7 to 0.9 or more, depending in part on the given flight conditions (i.e., flight Ma_∞) and how far one is away from the design point (i.e., off-design versus on-design), e.g., the nominal cruise speed and altitude. For adiabatic flow (no heat exiting or entering the system), conservation of energy gives

$$T_{02} = T_\infty \left(1 + \frac{\gamma_d - 1}{2} Ma_\infty^2\right) \tag{6.23}$$

From isentropic flow theory,

$$\left.\frac{T_{02}}{T_\infty}\right|_{ideal} = \left(\frac{p_{02}}{p_\infty}\right)^{\frac{\gamma_d-1}{\gamma_d}} \tag{6.24}$$

so that via Eq. 6.20:

$$T_{02} - T_\infty = \frac{1}{\eta_d}(T_{02,ideal} - T_\infty) \tag{6.25}$$

For non-isentropic flow then,

$$p_{02} = p_\infty \left[1 + \eta_d \left(\frac{T_{02}}{T_\infty} - 1\right)\right]^{\frac{\gamma_d}{\gamma_d-1}} \tag{6.26}$$

As noted by the diffuser subscript designation, one can allow for varying values of the ratio of specific heats γ through the various engine stations. With respect to air at the compressor inlet, unless moving into the supersonic flight regime where the aerodynamic ram compression (and associated temperature rise) is at a substantial level, one can assume that γ_d will still be close to 1.4 (value for a diatomic gas [e.g., N_2, O_2]).

For the mechanical compressor (axial, centrifugal or some combination thereof; see examples of Figs. 6.4 and 6.5), define the adiabatic component efficiency η_c as the ratio of the work required in an isentropic process to that for the actual process:

$$\eta_c = \frac{h_{03,ideal} - h_{02}}{h_{03} - h_{02}} \tag{6.27}$$

which says

$$T_{03} - T_{02} = \frac{1}{\eta_c}(T_{03,ideal} - T_{02}) \tag{6.28}$$

Typical values for η_c are from 0.85 to 0.9. One can specify a compressor stagnation pressure ratio π_c such that at the compressor outlet (station 3):

$$p_{03} = \pi_c p_{02} \tag{6.29}$$

Fig. 6.4 Photo of cutaway museum display of a General Electric J85-GE-17A non-afterburning turbojet engine employing a multi-staged axial compressor (note the engine's tail cone at right extends to the exhaust nozzle exit in this example, providing a smoothly expanding flow cross-sectional area downstream of the turbine section)

Fig. 6.5 Schematic diagram illustrating a cutaway display of a small Turbine Technologies Ltd. SR-30 non-afterburning turbojet engine employing a single-stage centrifugal compressor (impeller + diffuser). The front convergent cone (seen at the *left* of the schematic diagram, and from the front in the corresponding Ryerson University photo) facilitates the entry of air into the compressor. Note that the compact design of the engine requires that the air flow exiting the compressor must reverse its direction through the combustor, and then return to its original direction when passing into the single-stage axial turbine (stator + rotor) section and the convergent exhaust nozzle further downstream

The value for π_c is in practice a function of the compressor rotor shaft speed (N if a single spool engine, N_1 or N_2 if a two-spool engine, etc.) and air flow parameters. Shaft speeds can be as low as 10,000 rpm for the low-pressure spool on a large engine, and up to 100,000 rpm for the engine shaft of a small engine.

At the compressor outlet, noting that

$$\left.\frac{T_{03}}{T_{02}}\right|_{\text{ideal}} = \left(\frac{p_{03}}{p_{02}}\right)^{\frac{\gamma_c-1}{\gamma_c}} \tag{6.30}$$

then

$$T_{03} = T_{02}\left[1 + \frac{1}{\eta_c}\left[\pi_c^{\frac{\gamma_c-1}{\gamma_c}} - 1\right]\right] \tag{6.31}$$

Here, γ_c is the average value through the compression process, and for conventional subsonic engines it will still be around 1.4. At higher compressions, as the compressed air goes to higher temperatures as seen for supersonic engines, the value for γ_c will drop below 1.4.

In the burner (main combustor), one can define the adiabatic component efficiency η_b as the measure of completeness of combustion, as illustrated within the following conservation of energy equation:

$$(\dot{m}_a + \dot{m}_f)C_{p,\text{be}}T_{04} - \dot{m}_a C_{p,c}T_{03} = \eta_b \dot{m}_f q_R \tag{6.32}$$

The symbols $C_{p,c}$ and $C_{p,\text{be}}$ respectively refer to the constant-pressure specific heat of air in the compressor (average value) and the constant-pressure specific heat of the hot gas at the burner exit (combustor outlet). Given the relatively low value for f, and the fact that without dissociation the molecular mass of a gas does not change and therefore its corresponding specific gas constant (R) value will also remain unchanged, one can estimate $C_{p,\text{be}}$ and C_p values for the hot gas further downstream in the engine as follows, allowing for a variable γ:

$$C_p \approx \frac{\gamma}{\gamma - 1} \cdot R_{\text{air}} \tag{6.33}$$

Returning to Eq. 6.32, one can reduce it to:

$$(1+f)C_{p,\text{be}}T_{04} - C_{p,c}T_{03} = \eta_b f \cdot q_R \tag{6.34}$$

Re-arranging the above, one can produce the following expression for the combustor exit stagnation temperature, sometimes called T_{max} because it is the nominal maximum gas temperature possible in the engine (unless that engine has an afterburner, in which case T_{06} may be greater than T_{04}):

$$T_{04} = \frac{\eta_b f \cdot q_R + C_{p,c}T_{03}}{(1+f)C_{p,\text{be}}} \tag{6.35}$$

If T_{04} is known earlier, say as a design target, then via Eq. 6.35, one can solve for f as follows:

$$f = \frac{\dfrac{T_{04}}{T_{03}} - \dfrac{C_{p,c}}{C_{p,\text{be}}}}{\dfrac{\eta_b q_R}{C_{p,\text{be}}T_{03}} - \dfrac{T_{04}}{T_{03}}} \tag{6.36}$$

Typical values for η_b can range from 0.95 to 0.99 for good to excellent combustor designs operating near their design conditions. One should likely account for a stagnation pressure loss through the combustor section of the engine, as stipulated by:

$$p_{04} = \pi_b \, p_{03} \tag{6.37}$$

Typical values for π_b can range from 0.9 to 0.95. At the combustor outlet, one has a high-temperature gas mixture comprised of the combustion products and unreacted air. As a result, the value for γ at this exit location will now be substantially below that for the cooler air upstream, perhaps having a value as low as around 1.33.

In the turbine, we can define the adiabatic component efficiency as the ratio of the actual work done by the gas on the turbine to that work corresponding to isentropic expansion:

$$\eta_t = \frac{h_{04} - h_{05}}{h_{04} - h_{05,\text{ideal}}} \tag{6.38}$$

which says

$$T_{04} - T_{05} = \eta_t(T_{04} - T_{05,\text{ideal}}) \tag{6.39}$$

Typical values for η_t can range from 0.9 to 0.95. For steady-state operation, the turbine supplies the power required by the compressor (in a two-spool case, the LP turbine powers the LP compressor, and the HP turbine powers the HP compressor), so that for adiabatic flow (and accounting for turbomachinery bearing friction through a mechanical efficiency η_m, around 0.99 in value),

$$\eta_m \dot{m}_t C_{p,t}(T_{04} - T_{05}) = \dot{m}_c C_{p,c}(T_{03} - T_{02}) \tag{6.40}$$

In practice, $\dot{m}_t$ and $\dot{m}_c$ are not quite the same in value, given the fuel input $\dot{m}_f$ by the burner to augment $\dot{m}_t$, and the typical bleeding off of some air from the compressor for cooling, etc., to effectively reduce $\dot{m}_c$. However, for preliminary analysis where one tends to approximate various values in any case, let us assume the following:

$$\dot{m}_t C_{p,t} \approx \dot{m}_c C_{p,c} \tag{6.41}$$

so that

$$T_{05} \approx T_{04} - \frac{1}{\eta_m}(T_{03} - T_{02}) \tag{6.42}$$

Given

$$\left.\frac{T_{05}}{T_{04}}\right|_{\text{ideal}} = \left(\frac{p_{05}}{p_{04}}\right)^{\frac{\gamma_t - 1}{\gamma_t}} \tag{6.43}$$

then one arrives at

$$p_{05} = p_{04} \left[1 - \frac{1}{\eta_t} \left(1 - \frac{T_{05}}{T_{04}} \right) \right]^{\frac{\gamma_t}{\gamma_t - 1}} \tag{6.44}$$

The average value for γ in the turbine will remain similar to that seen at its entrance, in the vicinity of 1.33 or a bit higher. Engine pressure ratio (EPR), typically defined as p_{05}/p_{02} (i.e., low-pressure turbine outlet stagnation pressure divided by the low-pressure compressor inlet stagnation pressure), is commonly measured by sensors in-flight, and monitored by the pilot as a measure of thrust output effectiveness (noting that there presently is no directly explicit means for monitoring actual thrust delivery values in-flight). Typical EPR values range from 1.5 to 2, substantially lower than overall pressure ratio [OPR] values that can range from 10 to 40.

Since the issue of using an afterburner (A/B) was raised earlier, let us include it in our discussion here. If the gas exiting the turbine contains a substantial amount of unreacted air (fuel-lean), then one can potentially use that condition for running an afterburner (thrust augmentor, reheater). In the afterburner, one instigates further fuel input and burning to raise the temperature of the exhaust gas entering the nozzle downstream of the A/B, and thus augments the overall thrust delivery of the engine. The cycle analysis equations for the A/B will be similar in form to that seen for the main combustor upstream. For example, at the A/B exit (station 6, as per Fig. 6.3),

$$T_{06} \approx \frac{\eta_{AB} f_{AB} q_R + C_{p,t} T_{05}}{(1 + f_{AB}) C_{p,ABe}} \tag{6.45}$$

The adiabatic component efficiency η_{AB} can range from 0.85 to 0.95. The fuel-gas ratio for the A/B may be found via:

$$f_{AB} = \frac{\dot{m}_{f,AB}}{\dot{m}_a + \dot{m}_f} = \frac{\dot{m}_{f,AB}}{\dot{m}_a(1 + f)} \approx \frac{\dfrac{T_{06}}{T_{05}} - \dfrac{C_{p,t}}{C_{p,ABe}}}{\dfrac{\eta_{AB} q_R}{C_{p,ABe} T_{05}} - \dfrac{T_{06}}{T_{05}}} \tag{6.46}$$

One can also specify the stagnation pressure loss through the afterburner section via the following:

$$p_{06} = \pi_{AB} p_{05} \tag{6.47}$$

The value for π_{AB} can range from 0.9 to 0.95. In a conventional jet engine with no afterburner, then one can indicate the following:

$$T_{06} = T_{05}, \quad p_{06} = p_{05}$$

In the case that the engine has an afterburner but it is not being used at the mission time of interest (running "dry" with A/B off, running "wet" with A/B on), one may need to allow for a small drop in value of both T_{06} and p_{06}, due to viscous flow losses through the "cold" A/B section of the engine.

As per Figs. 6.3, 6.4 and 6.5, let us assume we have a simple convergent nozzle (no divergence section), a common choice for a subsonic TJ engine. Let us define the adiabatic component efficiency η_n as the ratio of the actual change in enthalpy through the flow acceleration process, to the ideal case:

$$\eta_n = \frac{h_{06} - h_7}{h_{06} - h_{7,\text{ideal}}} \qquad (6.48)$$

Note that for steady adiabatic flow through the nozzle, $h_{07} = h_{06}$. Conservation of energy dictates that

$$h_{07} = h_7 + \frac{V_e^2}{2} = C_{p,n}T_7 + \frac{V_e^2}{2} \qquad (6.49)$$

From earlier then,

$$h_{07} - h_7 = h_{06} - h_7 = \eta_n(h_{06} - h_{7,\text{ideal}}) \qquad (6.50)$$

so that

$$V_e = \sqrt{2\eta_n(h_{06} - h_{7,\text{ideal}})} = \sqrt{2\eta_n\left(C_{p,n}T_{06} - C_{p,n}T_{06}\left(\frac{T_7}{T_{06}}\right)\Big|_{\text{ideal}}\right)} \qquad (6.51)$$

Realizing that

$$\frac{T_7}{T_{06}}\Big|_{\text{ideal}} = \left(\frac{p_7}{p_{06}}\right)^{\frac{\gamma_n - 1}{\gamma_n}} \qquad (6.52)$$

then the following is applicable:

$$V_e = \sqrt{2\eta_n C_{p,n}T_{06}\left[1 - \left(\frac{p_7}{p_{06}}\right)^{\frac{\gamma_n - 1}{\gamma_n}}\right]} \qquad (6.53)$$

Typical values for η_n can range from 0.9 to 0.98, depending on the nozzle design and how close it is to its design condition. Values for γ_n will be for a cooler gas on average than that just upstream, so in the vicinity of around 1.36.

The value for nozzle exit pressure p_7 will depend on whether the convergent exit (throat) is choked or not. If the nozzle exit flow is unchoked ($Ma_e < 1$), one would apply the following:

$$p_7 \to p_\infty$$

$$V_e \to V_{e,\infty}$$

$$V_{e,\infty} = \sqrt{2\eta_n C_{p,n}T_{06}\left[1 - \left(\frac{p_\infty}{p_{06}}\right)^{\frac{\gamma_n - 1}{\gamma_n}}\right]} \qquad (6.54)$$

$$A_{e,\infty} \approx A_e$$

$$T_{e,\infty} \approx T_{06} \left(\frac{p_\infty}{p_{06}}\right)^{\frac{\gamma_n - 1}{\gamma_n}}$$

$$\rho_{e,\infty} = \frac{p_\infty}{RT_{e,\infty}}$$

Except at low shaft rotation speeds N at lower altitudes (say in landing approach, engine(s) near idle), a turbojet engine would typically run in a choked-nozzle condition. One can check for choking following the criterion as noted below:

$$\frac{p_{06}}{p_\infty} > \left(\frac{\gamma_n + 1}{2}\right)^{\frac{\gamma_n}{\gamma_n - 1}}, \quad \text{nozzle throat is choked} \tag{6.55}$$

In the case of choked flow, the following is applicable:

$$\left.\frac{T_7}{T_{06}}\right|_{\text{ideal}} = \frac{2}{\gamma_n + 1} \tag{6.56}$$

$$V_e = \sqrt{2\eta_n C_{p,n} T_{06} \left[\frac{\gamma_n - 1}{\gamma_n + 1}\right]} \tag{6.57}$$

$$p_e = p_7 = p_{06} \left[\frac{T_7}{T_{06}}\bigg|_{\text{ideal}}\right]^{\frac{\gamma_n}{\gamma_n - 1}} = p_{06} \left[\frac{2}{\gamma_n + 1}\right]^{\frac{\gamma_n}{\gamma_n - 1}} \tag{6.58}$$

$$\rho_7 = \rho_{06} \left(\frac{2}{\gamma_n + 1}\right)^{\frac{1}{\gamma_n - 1}} = \frac{p_{06}}{RT_{06}} \left(\frac{2}{\gamma_n + 1}\right)^{\frac{1}{\gamma_n - 1}} \tag{6.59}$$

$$a_7 = \sqrt{\gamma_n RT_7} = \sqrt{\frac{2\gamma_n RT_{06}}{\gamma_n + 1}} \tag{6.60}$$

In the choked flow case, the pressure-area term, $(p_e - p_\infty)A_e$, has potentially a finite positive value (unlike the unchoked case, where $p_e \to p_\infty$). As a result, when calculating specific thrust $F/\dot{m}_a$, which for a non-afterburning TJ via Eq. 6.9 is given by

$$\frac{F}{\dot{m}_a} = (1+f)V_e - V_\infty + \frac{A_e}{\dot{m}_a}(p_e - p_\infty) \tag{6.61}$$

one will need the following:

$$\frac{A_e}{\dot{m}_a} = \frac{A_e}{\dot{m}}(1+f) = \frac{1+f}{\rho_7 a_7} = (1+f)\frac{RT_{06}}{p_{06}} \left[\frac{\gamma_n + 1}{2}\right]^{\frac{1}{\gamma_n - 1}} \left[\frac{\gamma_n + 1}{2\gamma_n RT_{06}}\right]^{1/2} \tag{6.62}$$

Throughout the above cycle analysis, we have been using various adiabatic component efficiencies. This is the most common approach for analyzing jet engines. However, there is an alternative approach that can be taken, where one uses polytropic efficiencies instead of adiabatic efficiencies. The polytropic efficiencies act on the isentropic exponents, such that one would be modelling a polytropic process, rather than a corrected isentropic process. For example, consider the non-isentropic flow through a diffuser. Earlier, we modelled this process as follows, for finding stagnation pressure at the exit of the intake:

$$p_{02} = p_\infty \left[1 + \eta_d \left(\frac{T_{02}}{T_\infty} - 1 \right) \right]^{\frac{\gamma_d}{\gamma_d - 1}} \tag{6.63}$$

Treating the decelerated flow as a polytropic process, one has the alternative equation for the exit stagnation pressure,

$$p_{02} = p_\infty \left[1 + \left(\frac{T_{02}}{T_\infty} - 1 \right) \right]^{\frac{e_d \gamma_d}{\gamma_d - 1}} = p_\infty \left[\frac{T_{02}}{T_\infty} \right]^{\frac{e_d \gamma_d}{\gamma_d - 1}} \tag{6.64}$$

where e_d is the component polytropic efficiency (in this case, for the diffuser).

In the preliminary evaluation of a new engine design, possibly before setting the engine's size (e.g., before setting the expected value for nozzle exit area, A_e), one would initially be looking at parameters like the aforementioned specific thrust and TSFC. It is typically observed that an increase in a turbojet's turbine inlet temperature (combustor exit temperature) T_{04} improves the level of the specific thrust. However, at lower compressor pressure ratios (π_c) and lower flight speeds (V_∞), one commonly observes that raising T_{04} causes an increase in TSFC (not desirable, in general); a higher temperature produces a higher jet exhaust speed (V_e), which lowers propulsive efficiency η_p. At higher flight speeds, higher altitudes, and higher compressor pressure ratios, the trend on TSFC can reverse, which can be inferred to result from more favourable aerodynamic ram effects on the air intake and resulting value for η_p.

In Fig. 6.6, one can observe the typical trend of specific thrust dropping with increasing airspeed (momentum drag effect), while mass flow will tend to increase at higher airspeeds with increased ram compression, for flight at a given altitude. The resulting product of the two parameters produces a net thrust profile in Fig. 6.6 that is fairly characteristic of turbojet engines at most altitudes, as discussed in the next paragraph.

Further into the preliminary design process, one will at some point have pinned down the desired thrust level to be produced by the new engine. At that point, one will establish the air massflow needed ($\dot{m}_a$), which in turn more or less sets the expected nozzle exit area A_e and engine intake area A_1, which effectively establishes the size of the engine. An example thrust chart may be found in Fig. 6.7 for a larger turbojet engine, where maximum continuous (normal rated) thrust (similar to maximum climb thrust, i.e., higher than

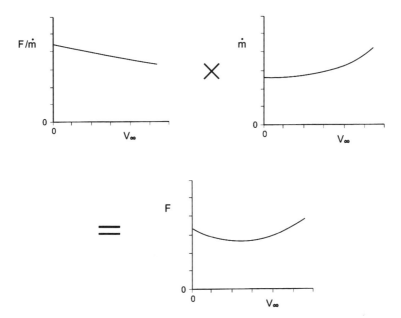

Fig. 6.6   At a given altitude, looking at the product of specific thrust and mass flow in chart form (as function of airspeed), to produce the resulting chart for net thrust of an example turbojet engine

maximum cruise throttle setting, but lower than takeoff thrust setting, to save on engine wear) is given as a function of flight airspeed and altitude. The chart also includes TSFC curves. One can observe from the chart that net thrust in general decreases as one goes up in altitude (corresponding to dropping values for ambient air pressure and density). It should be noted that specific thrust, $F/\dot{m}_a$, on the other hand increases with increasing altitude, due to the decreasing outside air temperature (air temperature, however, levels off at 217 K beyond the tropopause, h_{ASL} around 11 km, as the troposphere gives way to the lower stratosphere). TSFC can also show some improvement with altitude, given the similar dependence on ambient air temperature. For example, referring to Fig. 6.7, flying at 120 m/s at sea level, TSFC is around 0.9 kN/h-kN, while at the same speed at 9 km altitude, TSFC is lower at about 0.84 kN/h-kN. One also sees the trend of TSFC worsening with increasing airspeed in Fig. 6.7, which corresponds to the earlier discussion surrounding Eq. 6.18.

At a given altitude, one observes from Fig. 6.7 that thrust tends to decrease to varying degrees as airspeed increases (see earlier discussion on momentum drag term, $\dot{m}_a V_\infty$). However, it also displays a trend reversal at higher airspeeds, where one can attribute the recovery and increase in thrust to the favorable effects of aerodynamic ram compression at the intake, and increased $\dot{m}_a$.

One can compare the performance of a larger turbojet engine like that of Fig. 6.7 with a significantly smaller turbojet, whose thrust and TSFC performance can be observed in Figs. 6.8 and 6.9. An engine of this size might be used for an

Fig. 6.7 Thrust chart for
example turbojet engine
(maximum continuous thrust
throttle setting) at various
altitudes. Dashed lines are
TSFC curves, in kN/h-kN.
Engine performance
comparable to Pratt &
Whitney JT4A, with
compression ratio (π_c) of 12:1

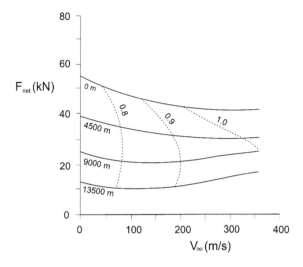

Fig. 6.8 Thrust chart for
example small turbojet
engine (maximum continuous
thrust throttle setting) at
various altitudes. Engine
performance comparable to
Teledyne-Continental J402-
CA-702, with compression
ratio (π_c) of 8.4:1

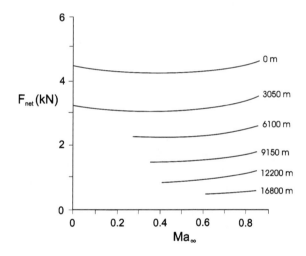

unmanned drone. The favourable effects of ram compression are evident at higher flight Mach numbers on the thrust delivery of this engine (Fig. 6.8). The worsening of TSFC (i.e., increasing in value) is quite evident as one picks up airspeed, at any altitude (Fig. 6.9). For a given airspeed, the TSFC graph for this engine shows that TSFC may improve (become lower) as one gains altitude from sea level, but then experience several back-and-forth trend reversals at some higher altitude ranges. The benefits of a cooler outside air temperature is undoubtedly influencing this behavior.

Please refer to the solution for Example Problem 6.2 at the end of this chapter, for a completed sample turbojet engine cycle analysis.

Fig. 6.9 TSFC chart for
example small turbojet
engine (maximum continuous
thrust throttle setting) at
various altitudes. Engine
performance comparable to
Teledyne-Continental J402-
CA-702

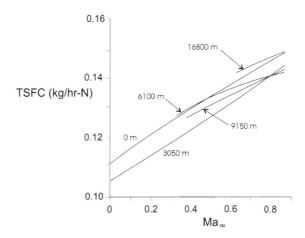

6.4 Intake

Engine components whose operation involves a rise in static pressure (such as
through the forward air intake, or mechanical compressor section) as opposed to a
fall in pressure (such as through the turbine section, or the exhaust nozzle) must
deal with the effects of an adverse pressure gradient on viscous boundary layer
growth and as a result, possible flow separation (boundary layer detachment from
the internal or external airframe surface). For intakes (a.k.a. inlets, diffusers), this
is true both for the internal duct flow from the inlet lip to the compressor fan inlet
face, as well as the external flow about the forward portion of the engine nacelle
(cowl) and airframe in proximity to the intake. Supersonic intakes must in addition
deal with flow momentum losses through various shock wave patterns in and about
the inlet, as well as aerodynamic drag increases due to shockwave interaction with
the internal and external boundary layers.

6.4.1 Subsonic Intake

The main function of the intake is to bring into the engine a suitable air mass
flow $\dot{m}_a$, at different flight speeds (V_∞), altitudes (h_{ASL}) and attitudes (α, β), and to
decelerate (diffuse) this flow to a flow Mach number suitable for mechanical
compressor operation (Ma_2 from 0.2 to 0.6, typically). If designed well and near
design-point flight conditions, the diffusion process of flow deceleration will
experience relatively low stagnation pressure losses. External views of a few
subsonic engine intakes are provided in Fig. 6.10.

Under static or low-V_∞ high-thrust operation (say, during takeoff), the engine
will be demanding a large quantity of air, such that external acceleration of air near
the inlet will occur, and the effective undisturbed upstream capture area A_∞ of the

Fig. 6.10 *Photo at upper left* of Boeing 737-700 right engine nacelle, giving external view of engine intake region. *Photo at right* of Canadair CL-600 Challenger rear fuselage mounted left engine nacelle, giving view of intake area. *Lower photo* of nose intake of North American F-86 Sabre, to feed air to a single central embedded turbojet engine

incoming streamtube of air will be substantially bigger than the inlet entrance plane area A_1. The area ratio A_∞/A_1 is sometimes referred to as the capture ratio. In the ideal case where there is no bleed-off of engine air (say, from the compressor section), for a turbojet engine, steady-state mass conservation would suggest the following correlation between entering and exiting engine gas flow:

$$\dot{m}_a = \rho_\infty V_\infty A_\infty \approx \rho_e V_e A_e/(1+f) \tag{6.65}$$

Re-arranging, this allows for an estimate of the upstream capture area:

$$A_\infty \approx \frac{\rho_e V_e A_e}{\rho_\infty V_\infty (1+f)} \tag{6.66}$$

For static tests, A_∞ effectively goes to infinity, and in practice, a bell-mouth entry apparatus (as may be seen in Fig. 6.11) is attached to the engine intake entrance to avoid lip flow boundary layer separation. As the airplane's flight speed increases, A_∞ tends to decrease, and typically at cruise speed and altitude, A_∞ is somewhat lower in value than A_1, thus some external flow deceleration occurs as the inlet entrance is approached. Typical capture ratios at subsonic cruise conditions range

Fig. 6.11 Schematic of incoming airflow into engine, without and with Borda bell-mouth entry, static test conditions

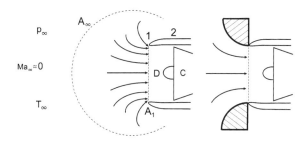

Fig. 6.12 Schematic of incoming airflow into engine, subsonic cruise conditions

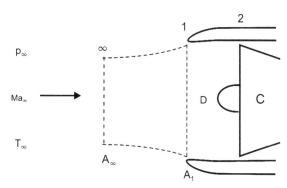

from 0.5 to 0.8 (see Fig. 6.12). Consider the following example gas dynamic calculation for estimating the inlet flow Mach number Ma_1:

$$\frac{A_\infty}{A_1} \approx 0.8 = \frac{Ma_1}{Ma_\infty}\left[\frac{2+(\gamma-1)Ma_\infty^2}{2+(\gamma-1)Ma_1^2}\right]^{\frac{\gamma+1}{2(\gamma-1)}} = \frac{Ma_1}{0.82}\left[\frac{2+(0.4)0.82^2}{2+(0.4)Ma_1^2}\right]^3$$

The above example scenario with a flight Mach number of 0.82 provides an estimate of 0.53 for Ma_1, which seems reasonable, relative to the expected equivalent or somewhat lower flow Mach number downstream in the compressor ($Ma_2 < Ma_1$, $A_2 > A_1$).

It is important to consider the flow external to the captured stream tube, since it will tend to accelerate over the nacelle (cowl) lip, with the associated risk of downstream boundary layer separation. Internally, flow separation is a possibility along the internal intake wall with diffusion (flow deceleration, static pressure rise), and with acceleration followed by further downstream deceleration moving over the center-body nose upstream of the compressor inlet face. On the positive side, a forward thrust component (of a resultant lift force) can be generated by the effective camber (curved airfoil) shape of the forward inlet shroud (duct), to in essence decrease the overall aerodynamic drag acting on the aircraft. Finally, one can note that a fixed geometry intake will typically give acceptable performance over the subsonic speed regime (from takeoff to cruise). On occasion, for enhanced performance, flexibility or flight economics, one will see the use of variable geometry for subsonic-flight intakes.

6.4.2 Supersonic Intake

Current conventional supersonic intakes, designed for high transonic or supersonic flight, must deliver a lower subsonic flow (as noted above) to the compressor inlet, for adequate performance by the mechanical compressor. Deceleration of the flow ahead of the compressor face will largely be through one or a series of normal and/ or oblique shock waves ahead of and within the intake. For lower stagnation pressure losses over a wider flight speed range, variable geometry of the intake will likely be required. For operation of the airplane over subsonic and supersonic flight speeds, a variable-geometry intake would employ mechanically actuated ramps, doors, valves and diverters to effectively modify the flow geometry to allow for a more efficient ingestion of flow from the outside atmosphere at a given flight condition. One may also note that above a certain supersonic flight Mach number, say 2.5 to 3 (depending on the flight vehicle and application), one might obtain sufficient aerodynamic ram compression in the diffuser for functional engine operation, without the use of a mechanical compressor (ramjet engine approach).

6.4.3 Supersonic Intake, Pitot Type

Ideally, one would like a shock-free isentropic diffusion process through the intake to the compressor face. In an ideal situation, one might consider a Kantrowitz-Donaldson inlet, also known as a convergent-divergent pitot-type design, as illustrated in Fig. 6.13. The upper diagram of Fig. 6.13 shows the ideal case for this type of nozzle (no standing shocks present). In practice, it can be somewhat difficult for the pilot to fly the aircraft efficiently into that ideal shockless state, say at the design-point supersonic cruise flight condition. Below the cruise Ma_∞, at subsonic flight conditions, the flow is subsonic throughout the K-D intake, so no problem there. However, entering into the transonic flight regime (high subsonic/ low supersonic) the intake throat will at some point choke (reach a sonic condition). Above that airspeed, there will be air spillage around the intake (the ideal mass flow can not be accommodated due to the choked condition), and above a flight Mach number of 1, there will be a standing normal or bow shock upstream of the inlet face (see Fig. 6.14). Only at slightly above the design cruise Ma_∞ does the upstream shock become swallowed by the intake, spillage more or less ceases, and a weaker standing normal shock appears in the intake's divergent section downstream (this process, from supersonic wind tunnel terminology, is called "starting" the intake; the enabling process of bringing the aircraft to a speed above the design condition is called "overspeeding"). Again, in the ideal case, the pilot would decelerate the aircraft a little bit at this point, to bring that standing shock very close to the intake throat, where the shock will be very weak (approaching isentropic flow conditions). In practice, as shown by the lower diagram of Fig. 6.13,

Fig. 6.13 Schematic
diagram of convergent-
divergent pitot-type
supersonic intake, ideal
(*upper*) and practical (*lower*)
flight states

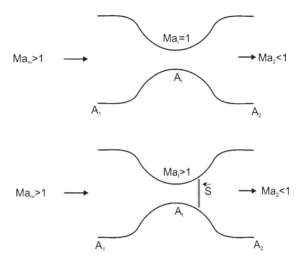

some transient-flow margin is allowed for, such that a stronger standing shock on average is tolerated in the divergent section at the cruise design point. This setting will avoid the risk of unstarting the intake or encountering overly intense oscillatory inflow conditions.

An alternative to overspeeding the airplane as a starting method is to use a variable-area diffuser throat. With that geometric modification capability, one would open up the diffuser throat to swallow the upstream normal shock, and then in turn reduce the diffuser throat area to bring forward the divergence section's standing shock to a position close to the throat (and thus weakening the shock). Venting or opening bleed valves in the intake region are other mechanical means for drawing in the forward shock.

In practice, the conventional K-D intake is inefficient and unpredictable when flying at off-design conditions. As a result, one in practice tended to compromise by choosing a simple divergence shape for the intake, as shown in Fig. 6.14. This shape essentially avoids the issue of starting and unstarting the diffusion process at supersonic flight speeds, but on the penalty side, retains the normal upstream shock. As such, this approach is typically only used up to a flight Mach number of about 1.6. Above this Mach number, the flow momentum losses through the normal shock are too substantial for efficient supersonic flight. An example of a simple divergent pitot design is given in Fig. 6.15.

6.4.4 Supersonic Intake, Oblique Shock Type

The use of a center-body forward nose or spike (three-dimensional applications), or ramp (two-dimensional; see Fig. 6.16), induces a standing external oblique shock wave in front of the inlet entrance, allowing for more efficient

Fig. 6.14 Schematic
diagram of simple divergent
supersonic pitot intake

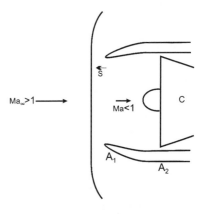

Fig. 6.15 Photo giving
external view of divergent
supersonic pitot intake, North
American F-100 Super Sabre

compression than that is seen with single normal upstream shocks discussed earlier. Referring to Fig. 6.17a, this is the optimal state ("critical" mode) for the best pressure recovery, with the smaller internal normal shock positioned very close to the inlet of the cowl (outer forward portion of nacelle). One practical concern with the shock positioned there is the possibility of local boundary layer separation as the normal shock interacts with the boundary layer. At a lower supersonic flight speed, or if there is increased flow resistance being encountered downstream in the engine, spillover around the cowl is a likely possibility, as illustrated in Fig. 6.17b, for "subcritical" operation of the intake. The small normal shock has moved out of the intake and become attached to the external main oblique shock. In this case then, with external air spillage, there is less than the maximum mass flow being accommodated. Conversely, at a higher supersonic flight speed, or decreased flow resistance downstream in the engine, the normal shock moves further aft down the intake annular channel, as shown in Fig. 6.17c. In this "supercritical" mode, the intake is accommodating the maximum air mass flow

Fig. 6.16 Left side air intake for McDonnell F-101B Voodoo supersonic interceptor, an aircraft employing two side-by-side Pratt & Whitney J57-P-55 turbojet engines. The leading forward plate at the front of the air intake has a gap between it and the main fuselage to allow for diversion of fuselage-created boundary-layer flow away from the intake; the plate's leading edge will also act to initiate the desired oblique shock ahead of the intake during supersonic flight. Lower photo courtesy of USAF

possible for the intake, but at sub-optimal pressure recovery (more entropy and flow momentum losses). To allow for some margin of transient flow (i.e., avoid overly oscillatory engine operation) as well as ensuring the maximum mass flow to the engine, one in practice would likely fly in supercritical mode when at supersonic cruise conditions.

As a trade-off between efficient compression versus the prevention of boundary layer separation, the single-shock intake design referred to above would be a common choice for flight Mach numbers between say 1.5 and about 2. Above Ma_∞ values of 2, multiple-shock intakes would provide more efficient compression. These designs are commonly credited to Oswatitsch, based on his studies. As an example, see the double-shock design of Fig. 6.18, where two oblique shocks will result in less entropy losses, versus a single oblique shock, for the same compression. As a 2D variant of this approach, the single-engined General Dynamics F-16 fighter jet aircraft employs a fixed double-ramp entry for the upper wall of the ventral (belly) intake positioned beneath the main fuselage (see Fig. 6.19). The Concorde intake design is a second 2D example (Fig. 6.20). As noted earlier for the normal shock in the diffuser passage, there is some risk involved with the interaction of the second oblique shock with the boundary layer causing a subsequent local separation of the layer.

As noted earlier, moving to flight Mach numbers of about three, one is approaching the flight regime where a mechanical compressor is no longer necessary for the gas turbine engine to operate effectively. From gas dynamics,

Fig. 6.17 Schematic of
simple center-body single-
shock supersonic intake:
a critical mode, **b** subcritical
mode (some air spillage),
c supercritical mode

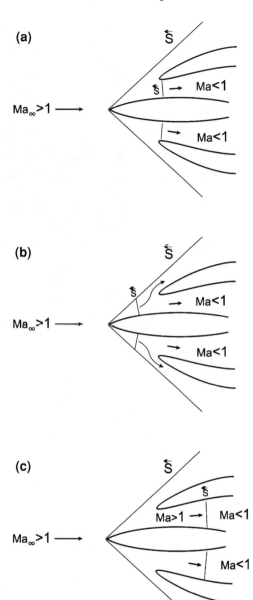

we know that the external front oblique shocks will be getting steeper and
steeper at higher values of Ma_∞, as one pertinent design consideration. From
flight Mach numbers of five and higher, one is in the hypersonic flight regime.
The scramjet engine (supersonic-combustion ramjet engine) is one of a few
current candidates for air-breathing propulsion applications within the hypersonic

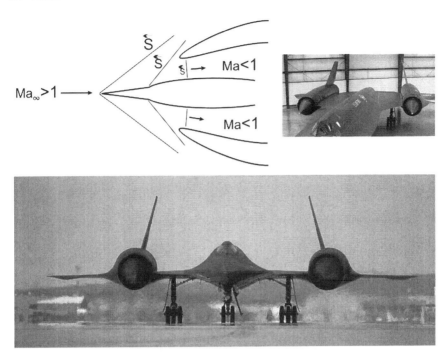

Fig. 6.18 Schematic diagram at upper left of center-body double-shock supersonic Oswatitsch intake. Photos of Lockheed SR-71 Blackbird supersonic reconnaissance aircraft, displaying the two wing-mounted afterburning turbojet engines' front intakes (engine: Pratt & Whitney J58)

Fig. 6.19 Photos providing external view of General Dynamics/Lockheed Martin F-16 Fighting Falcon ventral intake. Courtesy of USAF

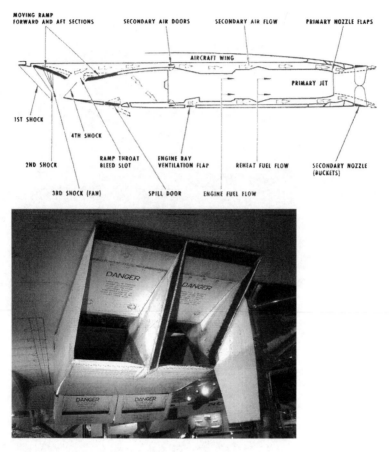

Fig. 6.20 Schematic diagram of an engine intake for the Aérospatiale/British Aircraft Corporation Concorde airliner, with the internal variable geometry (doors, valves, etc.) set up for supersonic cruise flight

Fig. 6.21 Schematic of hypersonic Busemann intake, for supersonic-flow combustor downstream

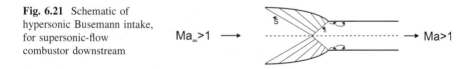

regime. One potential scramjet engine intake candidate, illustrated in Fig. 6.21, is the design credited to Busemann. The Busemann intake to a large degree acts as a reverse nozzle, and thus is somewhat analogous to the lower-speed K-D pitot intake. As shown by Fig. 6.21, the compression process is made efficient by a series of strengthening oblique shocks, rather than one or two oblique shocks as noted in earlier cases. A conical shock that reflects from the centre-line completes the oblique shock wave pattern in the diffuser, with a slowed-down but still supersonic air flow passing downstream to the combustor. A noted drawback of the Busemann design is the tendency of the boundary layer to separate at or just downstream of the diffuser throat, as illustrated by the swirling flow in Fig. 6.21.

6.5 Centrifugal Compressor

The centrifugal or radial compressor was first used in some early (World War II-era) British (see Fig. 6.22) and American turbojet engines, largely due to predictable, robust performance (during that period, the Germans opted, for the most part, axial compressors, which will be discussed in more detail later). A lot of knowledge had been gained earlier with piston-engine superchargers that would commonly employ a centrifugal compressor. With its larger cross-section and lower efficiency with multi-staging (typically, a maximum of two stages was the practical limit, as compared to numerous stages being possible for the axial variant), the centrifugal compressor is presently more often used for smaller thrust or power applications, where durability and simplicity of operation are as important as efficiency of mechanical compression. These latter positive characteristics, and the shorter resulting engine length, makes the centrifugal compressor a viable choice for helicopter powerplants.

On occasion, a low-pressure set of axial compressor stages (each axial stage = one moving rotor + one fixed stator) is placed in front of a high-pressure centrifugal compressor stage, to get the combined advantages of both designs. A single centrifugal compressor stage would typically give a compression ratio of 4 or 5:1 ($\pi_{c,\text{st}}$), as compared to an axial stage's typical top-end of 2:1. Centrifugal compressors tend to function better at lower air mass flows (say, less than 10 kg/s) given the nature of the impeller design, as compared to axial compressors with short blade heights in the aft portion of the compressor being hampered by boundary layer problems. With stronger materials that can withstand high centrifugal stresses, and modern rotor tip speeds reaching as high as 650 m/s (clearly supersonic) with tolerable shock flow losses, it is possible for a centrifugal stage to potentially reach a 10:1 compression.

In a conventional (single-entry) centrifugal compressor (see Figs. 6.22, 6.23, 6.24, 6.25; [3]), the incoming air is drawn through the static (non-rotating) inlet guide vanes (IGVs) into the compressor rotor (impeller) by the forward rotating inducer vanes (blades) near the centre-line of the impeller (location referred to as the impeller eye). In a less conventional centrifugal compressor, one can have air entering on both the forward and aft portions of the rotor (double-entry, via use of a plenum chamber for evenly distributed air delivery; refer to example of Boeing T60 turboshaft engine in [3]), rather than just the forward face, so as to permit a lower rotor speed for the same air mass flow. Returning to the conventional case, the air passing through the centre location is then centrifuged to the outer rotor

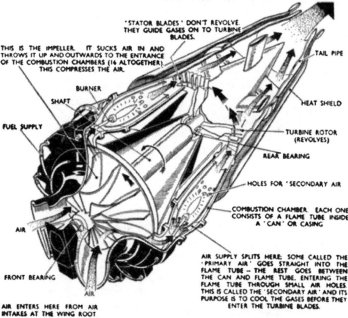

Fig. 6.22 Photo of cutaway museum display of a de Havilland Goblin II turbojet engine, at the de Havilland Aircraft Heritage Centre. The single-stage centrifugal compressor's impeller is clearly evident at the left of the photo. The overall pressure ratio (OPR) is rated at 3.3:1, producing a peak static thrust of approximately 3000 lbf (13.3 kN). This engine was first used in the de Havilland Vampire Mk. 1 fighter, circa 1946

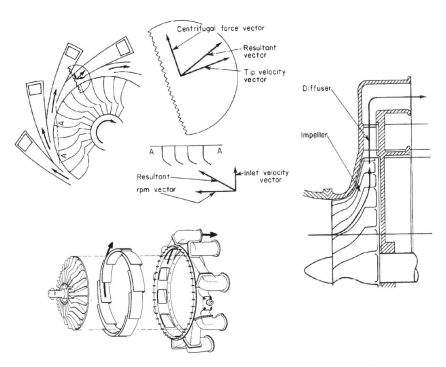

Fig. 6.23 Schematic diagram of one centrifugal compressor stage (impeller + diffuser), showing the various air flow directions and corresponding flow vector components [3]. Reprinted with permission of The McGraw-Hill Companies, Inc

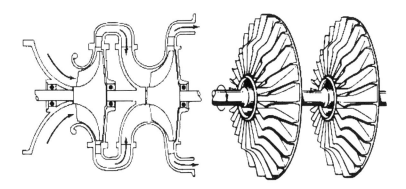

Fig. 6.24 Schematic diagrams of a two-stage centrifugal compressor

radius through the main impeller section. The air moves out of the rotor into a radial diffuser passage (referred to as the collector scroll). The passage may contain static diffuser vanes that act to remove the air's swirl (tangential motion) as well as slowing or diffusing it. Beyond the diffuser, the air may enter a second

Fig. 6.25 Example views of two different impeller designs for a centrifugal compressor. The upper left design is a simple conventional impeller. The second design shown by the remaining photos of a quarter section of the impeller reveals an alternating full impeller blade + inducer vane sequence. Photos courtesy of John Karpynczyk

centrifugal compressor stage (as noted above, likely a maximum of two centrifugal stages in a typical aircraft engine), or, enter the downstream combustor.

One can do a preliminary evaluation of a centrifugal compressor stage's performance. The rotor's tip speed U_t is given by:

$$U_t = \omega R = \frac{2\pi N}{60} \cdot R, \ \text{m/s} \tag{6.67}$$

where N is the shaft rotation speed in revolutions per minute (rpm), R is the outer radius of the impeller (in meters), and ω is the shaft rotation in radians per second (rad/s). Let us identify a slip factor ε such that the air exiting the impeller has a swirl (tangential) velocity Δw such that:

$$\Delta w = \varepsilon U_t, \ \text{m/s} \tag{6.68}$$

The following correlation exists between ε and number of impeller blades n_b:

$$\varepsilon \approx 1 - \frac{2}{n_b} \qquad (6.69)$$

A typical value for ε is 0.9, which corresponds to an n_b of 20. One recalls that work is only done by rotating components in the gas turbine engine. The work W_c done by one rotor from inlet to diffuser exit may be established via the following energy balance (stagnation enthalpy versus kinetic energy, of a unit mass m of the working fluid):

$$\frac{W_c}{m} = C_{p,c}T_{o,e} - C_{p,c}T_{o,i} = \Delta w \cdot U_t = \varepsilon U_t \cdot U_t = \varepsilon U_t^2 \qquad (6.70)$$

Because one is ultimately distributing all the air out from the blade tip (rather than in a distributed centre-line-to-blade-tip fashion, say as done in moving air left-to-right through a propeller) there is not a 1/2 factor appearing in Eq. 6.70. Using the station numbering used earlier in the course, for a single compressor stage,

$$T_{03} - T_{02} = \frac{\varepsilon U_t^2}{C_{p,c}} \qquad (6.71)$$

Applying an adiabatic component efficiency, η_c, for non-isentropic compression, one finds the following compression ratio:

$$\pi_c = \frac{p_{03}}{p_{02}} = \left[1 + \eta_c \cdot \frac{\varepsilon U_t^2}{C_{p,c}T_{02}}\right]^{\frac{\gamma_c}{\gamma_c-1}} \qquad (6.72)$$

Consider the following example, for a rotor tip speed of 450 m/s:

$$\pi_c = \left[1 + 0.85 \cdot \frac{0.9(450^2)}{1063(248)}\right]^{\frac{1.37}{1.37-1}} = 5.53$$

The compression prediction is typical of a conventional centrifugal rotor stage.

In presenting performance charts for a compressor stage or a complete compressor section, one must allow for differing ambient conditions from the standard reference sea-level case. One has the option of producing a large book of charts for differing altitudes, or, provide one chart that has correction factors acting on the pertinent parameters, like local air temperature and pressure on shaft speed and air mass flow in predicting the expected compression. Assuming the latter approach is most efficient, let's begin with a correction for shaft speed N. From the analysis above, one observes the following proportional relationships:

$$\pi_c \propto \frac{U_t^2}{T_{02}} \propto \frac{N^2}{T_{02}} \qquad (6.73)$$

Re-arranging,

$$N \propto \sqrt{T_{02}} \cdot \sqrt{\pi_c} \tag{6.74}$$

Let us define the following non-dimensional air temperature ratio:

$$\theta_2 = \frac{T_{02}}{T_{02,\text{REF}}} \tag{6.75}$$

Given the above information, one can then define the temperature-corrected shaft rotation speed N_{corr}:

$$N_{corr} = \frac{N}{\sqrt{\theta_2}} \tag{6.76}$$

Essentially, Eq. 6.76 says that one will get the same compression delivered at the actual N and θ_2, as one would get at reference (say, standard sea level) conditions at N_{corr}. As a result, this in turn says that we need only generate engine test data points at one known atmospheric condition and re-adjust said data to standard sea level conditions for placement on the one published chart, and this will suffice for predicting compressor performance over the entire flight regime from sea level to, say, 14 km in altitude.

Similarly, for air mass flow, given

$$\delta_2 = \frac{p_{02}}{p_{02,\text{REF}}} \tag{6.77}$$

and observing from conservation of mass,

$$\dot{m}_a = \rho_2 a_2 \cdot Ma_2 \cdot A_{\text{eff}} = \frac{p_2}{RT_2} \sqrt{\gamma RT_2} \cdot Ma_2 \cdot A_{\text{eff}} = \frac{p_2}{\sqrt{T_2}} \cdot K_1 \approx \frac{\delta_2}{\sqrt{\theta_2}} \cdot K_2 \tag{6.78}$$

then one can define the corrected mass flow as follows:

$$\dot{m}_{a,\text{corr}} = \dot{m}_a \cdot \frac{\sqrt{\theta_2}}{\delta_2} \tag{6.79}$$

Note that $T_{02,\text{REF}}$ and $p_{02,\text{REF}}$ are commonly set as T_{SL} and p_{SL} (standard sea level values; this is also true for other engine locations).

One can refer to Fig. 6.26 for a simplified illustrative version of a centrifugal compressor performance map, where compression ratio π_c is provided as a function of corrected mass flow, with various equilibrium curves at increasing corrected shaft speeds. An actual performance map can be viewed in Fig. 6.27 [4, 5]. The influence of local rotor blade angle-of-attack (α) on the internal aerodynamics of rotor operation, as illustrated by Fig. 6.28, can be quite important, and certainly plays a role in the positioning of the surge line on the compressor map of Fig. 6.26. Surge by definition refers to violent oscillations in the shaft rotational speed N,

Fig. 6.26 Centrifugal compressor map, showing compression ratio π_c as a function of corrected airflow, for various corrected shaft rotation speeds. The steady-state operating line is shown

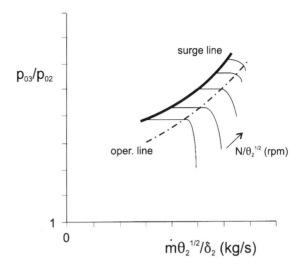

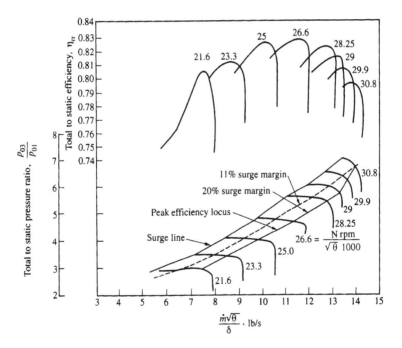

Fig. 6.27 Example centrifugal compressor map, showing total compression ratio (lower part of graph) and compressor adiabatic efficiency (upper part of graph) as a function of corrected airflow, for various corrected shaft rotation speeds [4, 5]. Reprinted with permission of the American Institute of Aeronautics and Astronautics [4]

Fig. 6.28 Representative
flow velocity triangle,
compressor rotor blade
section at a given radial
position

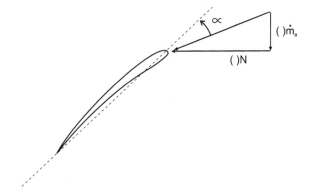

which will in turn result in a transient drop and rise in delivery pressure downstream to the combustor and turbine. The occurrence of surge corresponds to
compressor stall, where at some point on the rotor blade the local α is high enough
(in a positive or negative inclination to the local flow) to induce aerodynamic
stalling (substantial boundary layer separation with large wake development, and
loss of lift force). Referring to Figs. 6.26 and 6.28, realizing that an increasing
mass flow lowers that airfoil section's positive α, as does a decreasing N, then this
should correspond to the general trend of the surge line (it does). In accelerating up
to equilibrium rotation speed as one starts the aircraft engine, one moves from a
low N to a high N. The transient acceleration operating line will to some degree be
above the steady-state equilibrium operating line illustrated in Fig. 6.26, and thus
for a short period will bring the engine closer to the surge line (i.e., the surge
margin will be reduced), for a given $\pi_c(t)$. In the event that $\pi_c(t)$ becomes too close
to entering into the surge domain, blowoff valves may be opened or the nozzle exit
area opened up (if a variable-geometry nozzle) to reduce the engine pressure, in
combination with a reduction in fuel flow $\dot{m}_f$. If the inlet guide vanes (IGVs) at the
front of the compressor are adjustable, then these too may be temporarily deflected
to a new incidence angle, to help bring down $\pi_c(t)$ below the surge line. Modern
engines employ an on-board microcomputer for transient implementation of the
above strategies, as part of a FADEC (full authority digital electronic/engine
control) system approach for effective and safe automated engine control working
in conjunction with manual control inputs (e.g., throttle settings) by the pilot.

6.6 Axial Compressor

The advantages of the axial compressor over the centrifugal compressor include
higher efficiency and compression (overall, with the capability of many stages in
series), using a substantially lower cross-section. As may be seen in Figs. 6.29 and
6.30, the flow moves rearward in a more direct axial manner than the circuitous
path of a one- or two-stage centrifugal compressor. A single axial compressor

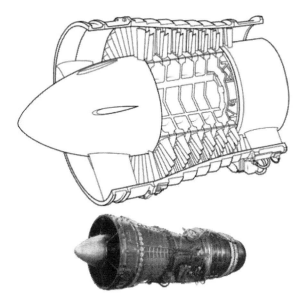

Fig. 6.29 Schematic diagram of a six-stage low-pressure axial compressor section for a two-spool Bristol/Rolls-Royce Olympus turbojet engine (stationary inlet and outlet guide vanes shown, in addition to the rotor blades of the six stages; for clarity, stator vanes for the six stages not included in diagram). This engine has a seven-stage high-pressure compressor section downstream of the above LP section. Photo shows the complete engine. A later afterburning version of the Olympus engine (the 593 by Rolls-Royce/SNECMA) was employed for powering the Aérospatiale/BAC Concorde supersonic airliner. Courtesy of Rolls-Royce plc

stage (rotor + stator) would typically give a compression ratio of up to 2:1, and with many stages possible (see [3]), one can presently deliver up to 40:1 overall (e.g., General Electric GE90 at takeoff throttle setting). Rotor blade tip Mach numbers in the forward transonic stages can reach up to 1.7, in order to produce the required compression.

One can see similarities in evaluating the aerodynamic performance of individual rotor blades, as was done earlier for propeller blades (see Fig. 6.31 for views of a single blade). This is more explicitly demonstrated when one represents a mean radius section as an unwrapped row of cambered airfoil sections (cascade field representation; see Fig. 6.32). With the axial flow passing through, each rotor blade and stator vane will have a low-pressure ("suction") zone on the convex surface, and a high-pressure ("pressure") zone on the concave surface. As depicted in Fig. 6.33, a representative velocity triangle can be drawn in passing through each station as one moves downstream. These triangles help to establish the resultant flow direction and effective angle of attack α of a given airfoil section. From an analytical viewpoint, one can apply a momentum-blade element approach for estimating the lift, drag and resulting torque acting on a given blade, as done earlier for a propeller blade in Chap. 3. The local rotor blade velocity at a given radial position r,

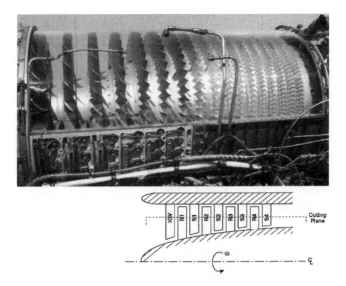

Fig. 6.30 Photo of the axial compressor for a single-shaft General Electric J79 turbojet engine. The 17 rotor rows are visible (the stators between the rotors have been removed, for clarity) in this museum display. This 17-stage compressor was capable of producing a 13:1 compression of the incoming air. *Lower schematic diagram* illustrates the axial compressor's basic setup of a forward inlet guide vane followed by various rotor + stator stages, with the flow cross-sectional area decreasing as one moves left-to-right through the compressor

Fig. 6.31 Differing views of a conventional axial compressor rotor blade. Note the I-shaped root at the bottom of the blade, for attaching the blade to the disk/shaft. Photos courtesy of John Karpynczyk

$$U = \omega r \qquad (6.80)$$

imparts work on the air (note: with no motion, the stator vanes do not) and increases the net tangential (swirl) velocity component to Δw, such that for one stage (inlet to exit):

$$\left.\frac{p_{o,e}}{p_{o,i}}\right|_{\text{stage}} = \left[1 + \eta_{\text{stage}} \frac{U \cdot \Delta w}{C_{p,c} T_{o,i}}\right]^{\frac{\gamma_c}{\gamma_c - 1}} \qquad (6.81)$$

Fig. 6.32 Diagram of adjoining compressor rotor blade sections at a given radial position (cascade field representation)

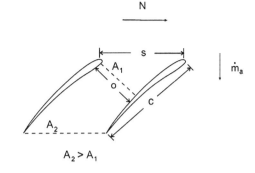

$A_2 > A_1$

Fig. 6.33 Velocity triangles for compressor stage rotor and stator blade sections at a given radial position (cascade field representation). The above airfoil combination corresponds to an approximate 50% degree of reaction ($v_{1,rel}$ similar in magnitude to v_2, v_3 similar in magnitude to $v_{2,rel}$)

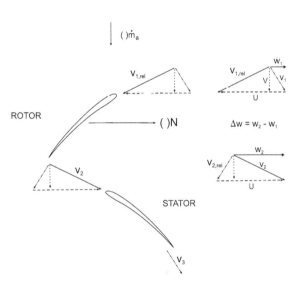

Note that η_{stage} is the component adiabatic efficiency for one stage ($\eta_{\text{stage}} < 1$). In practice, boundary layer development moving downstream will tend to degrade η_{stage} in stages further aft, as blade heights become shorter with the shrinking flow passage (see Fig. 6.30). From mass conservation, one reduces the flow area moving downstream as air pressure and density go up, in order to keep (more or less) the same axial velocity (for a conventional engine, say 60–90 m/s, or a flow Mach number around 0.4 to 0.6). While the stator vanes do not impart work, they do act to diffuse (decelerate) the flow, and re-direct the flow for efficient entry into the next station. A pertinent parameter in stage design is the degree of reaction (for a compressor, one uses the symbol $°R_c$), defined by:

$$°R_c = \frac{\text{rotor static enthalpy rise}}{\text{stage static enthalpy rise}} = \frac{h_{\text{mid}} - h_i}{h_e - h_i} \approx \frac{T_{\text{mid}} - T_i}{T_e - T_i} \qquad (6.82)$$

for one stage. The static temperature rise indicated by Eq. 6.82 also to a large degree will correspond to the static pressure rise through the stage. In order to share the burden of blade and vane loading with this static pressure increase, and as a result best avoid boundary layer separation in an adverse pressure gradient situation, a typical value for $°R_c$ would be in the neighborhood of 0.5. Referring to Fig. 6.33, one can make the following substitutions for enthalpy with equivalent energy and work (remember, the stationary stator does no work) into Eq. 6.82:

$$°R_c = \frac{h_2 - h_1}{h_3 - h_1} = \frac{\left(v_{1,rel}^2/2 - v_{2,rel}^2/2\right)}{U(w_2 - w_1)} \tag{6.83}$$

In the conventional situation where axial velocity V does not change much while moving downstream through the compressor, the trigonometry of Fig. 6.33 provides for the following on Eq. 6.83:

$$v_{1,rel}^2 - v_{2,rel}^2 = V^2 + (U - w_1)^2 - V^2 - (U - w_2)^2 = w_1^2 - w_2^2 - 2U(w_1 - w_2),$$

$$°R_c = \frac{\left(v_{1,rel}^2 - v_{2,rel}^2\right)}{2U(w_2 - w_1)} = \frac{w_1^2 - w_2^2}{2U(w_2 - w_1)} + 1 = 1 - \frac{(w_2 + w_1)}{2U} \tag{6.84}$$

Referring to Fig. 6.32, design factors such as rotor blade chord length c, separation (spacing) distance s (which effectively stipulates the number of blades for a given stage), and opening distance o are determined from various empirical rules, CFD and experimental test data. These and other design parameters for the blades and vanes can vary from the norm, depending on axial location in the compressor, and the influence of boundary layer behavior on aerodynamic performance from blade root to tip at that location.

Early in the design process, one would commonly do an initial "mean line" design study, evaluating the performance needs along the mean streamline through the engine component, in this case the compressor (see Fig. 6.30). Radial and circumferential variations of flow-related parameters would not be considered at this juncture. Overall geometric parameters (axial distance for a given stage, stator and rotor blade chords, inner and outer radii for the effective flow passage at a given location, clearances between blade tips and shroud, etc.) and mean velocity triangles would be established, with peak efficiency being a general objective where possible. Parameters of interest for the compressor would include inlet stagnation pressure and temperature, mass flow, stage and overall pressure ratios, shaft rotational speed, torque and power requirements, and adiabatic efficiencies at on- and off-design operational conditions. A software program like NPSS (Numerical Propulsion System Simulation, developed by NASA) may be used as the framework for evaluating the expected aerothermodynamic behavior of the various engine components (like the compressor in this case) acting together in a complete engine system. A program like NPSS can act as a "virtual test cell" [6], or a more rapidly developed specifications deck (collection of specifications files outlining on- and off-design performance, etc.).

Fig. 6.34 Axial compressor map, showing compression ratio π_c as a function of corrected airflow, for various corrected shaft rotation speeds. Adiabatic compressor efficiency curves also illustrated. The steady-state operating line is shown

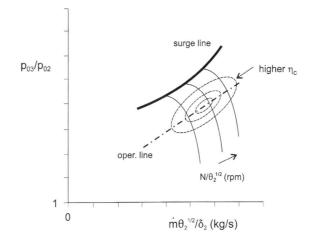

Moving from the initial mean-line study as a point of reference, one would commence a through-flow design study for evaluation and establishment of parameters and performance from hub to shroud (i.e., vane or blade radial location, root to tip), allowing for radial variation of flow-related properties in conjunction with axial variations. Airfoil aerodynamic performance from leading to trailing edge at a given radial location (e.g., chord-wise pressure distribution in conjunction with the applicable local velocity triangles, flow losses) moving from root to tip would be established, in moving left-to-right through the compressor. The compressor exit flow profile (velocity [axial, swirl components], temperature, pressure) would be defined, in order to accommodate the design of the combustor downstream of the compressor.

Moving from the through-flow design study, one would progress to detailed stator and rotor blade airfoil design from root-to-tip (i.e., radially "stacking" up a complete set of 2D airfoils so as to complete the given 3D vane or blade along its so-called stacking axis), blade-to-blade, and finally stage-to-stage. Detailed three-dimensional CFD and structural (finite-element) analysis (and experimentation) may be done at this juncture, to ascertain whether the results of the previous, largely 1D and 2D analysis has proven adequate for proceeding to prototype hardware development. If not adequate, further detailed study via CFD, etc., may be necessary. A similar design approach, as described above, would apply to the turbine section downstream.

A compressor performance map can be produced for a single stage, or for the overall axial compressor (see a representative illustrative example in Fig. 6.34, and an actual example stage map in Fig. 6.35 [7]). In comparing to the centrifugal case, the N-dependent curves for the axial compressor appear steeper, with less of a step prior to approaching the surge line when reducing the air flow. The problems of surge and compressor blade stall are present in the axial case. One also has the possibility of a phenomenon called "rotating stall", where the aerodynamic stall

Fig. 6.35 Example compressor map for one transonic axial stage, showing compression ratio $\pi_{c,stage}$ (*lower part* of graph) and stage efficiency $\eta_{c,stage}$ (*upper part* of graph) as a function of corrected airflow, for various corrected shaft rotation speeds (in terms of percentage of design speed) [7]. Reprinted with permission of the American Institute of Aeronautics and Astronautics

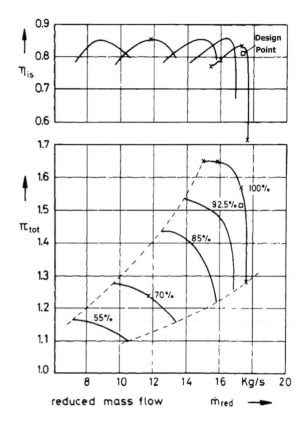

pattern progresses around a given blade row (the wake from one stalled blade will adversely affect the blade next to it, etc.), and then possibly propagates downstream to the next stage to worsen things. Variable-incidence stator vanes help ease operational difficulties like the above over the range of operational conditions expected for the engine (i.e., at on- and off-design conditions), as a FADEC application. Other FADEC strategies noted earlier for centrifugal compressors would also potentially be of use for the axial case.

For larger gas turbine engines, such as the example of Fig. 6.29, one may have a low-pressure (LP) compressor on one shaft, and a high-pressure (HP) compressor on a second shaft. While the two shafts would be mechanically independent, where the LP shaft rotates slower, the two would to some degree be aerodynamically coupled. For even greater performance and flexibility in big engines, one may see the use of three shafts, where one would have an intermediate-pressure (IP) shaft in the middle.

Since the temperatures in conventional gas turbine compressors for subsonic-flight engines do not become too high, the material for centrifugal and axial compressor components (blades, vanes, disks, etc.; see [3]) can be less expensive

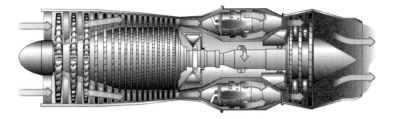

Fig. 6.36 Schematic diagram of a single-shaft General Electric J79 turbojet engine, illustrating the combustor cans positioned between the upstream axial compressor and the downstream turbine (10 combustor cans in total) [8]

high-strength long-life metal alloys of aluminum or steel. For lighter weight, corrosion resistance and high strength at higher temperatures, but at additional cost, titanium alloys are an alternative choice for some compressor components, perhaps being desirable when getting closer to the combustor.

6.7 Combustor

As illustrated by the jet engine schematic of Fig. 6.36 [8], the pressurized air flowing rightward from the compressor enters into the combustor. The air, essentially the working fluid of the engine, is heated by the injection, vaporization, and subsequent burning of the jet fuel (upon reacting with the incoming air). The above process is done continuously (i.e., not intermittent), once the engine has been started (initial fuel–air ignition accomplished, and the engine's shaft is spinning up toward equilibrium operation). Common jet fuels, like Jet A or Jet A-1, are kerosene-based (as shown in Table 6.1 [9], density around 810 kg/m^3), with good overall combustion properties for civil aircraft usage. For example, a fuel that has a large aromatics content (heavy hydrocarbon molecule presence) is undesirable, given the characteristic of producing highly luminous radiative flames that can lead to overheated surrounding combustor wall structures, and producing soot that can lead to excessive carbon buildup on combustor liners (coking). The process of "cracking," i.e., the breaking up of larger hydrocarbon molecules, may be applied in the distillation process, to mitigate this characteristic. Cracking generally causes an increase in volatility. As a second example, a fuel with a low flash point (like Jet B, a so-called "wide-cut" fuel having a wide boiling range, as opposed to a "narrow-cut" fuel with a narrow, higher temperature range of boiling), while easier to ignite at lower temperatures and pressures and thus is cleaner, is not in fact desirable for widespread civil usage, given the major concerns of fire safety in handling and in the event of a crash.

Table 6.1 Mass properties of various liquids at room conditions [9]

Liquid	Specific gravity	Specific weight, N/ℓ	Density, kg/m^3
Alcohol	0.81	7.95	810.0
Benzene	0.899	8.82	899.0
Biodiesel	0.88	8.63	880.0
Carbon tetrachloride	1.595	15.64	1595.0
Castor oil	0.969	9.50	969.0
Diesel	0.85	8.34	850.0
Ethanol (ethyl alcohol)	0.79	7.75	790.0
Ethylene glycol	1.12	10.98	1120.0
Gasoline	0.72	7.06	720.0
Glycerine	1.261	12.36	1261.0
Jet A fuel	0.81	7.94	810.0
Jet A-1 (JP-8) fuel	0.802	7.87	802.0
JP-1 fuel	0.80	7.84	800.0
JP-3 fuel	0.775	7.60	775.0
JP-4 (Jet B) fuel	0.785	7.70	785.0
JP-5 fuel	0.817	8.01	817.0
JP-6 fuel	0.81	7.94	810.0
Kerosene	0.82	8.04	820.0
Mercury	13.546	132.80	13546.0
Motor oil (lubricant)	0.89	8.73	890.0
Octane	0.702	6.88	702.0
RP-1 rocket fuel	0.797	7.82	797.0
Sea water	1.025	10.05	1025.0
Synthetic oil	0.928	9.10	928.0
Water	1.00	9.81	1000.0

Reprinted with permission of the American Insititute of Aeronautics and Astronautics

Given a typical heat of reaction q_R of around 4.34×10^7 J/kg (of fuel) for a jet fuel–air reaction, one must deliver a fuel–air ratio f to the combustor exit (turbine inlet) such that the peak stagnation gas temperature T_{04} does not exceed the allowed level for the turbine blades. For an engine using uncooled blades made of a nickel-based steel, the maximum allowed temperature would be around 1300 K, and with cooled blades, the maximum might be 1700 K. For a typical jet fuel–air mixture, the stoichiometric mass-based fuel–air ratio f_{st} is around 0.067, which corresponds to an equivalence ratio of 1. Equivalence ratio is defined as:

$$\phi = \frac{f_{actual}}{f_{st}} \tag{6.85}$$

such that ϕ is less than one for a lean mixture (excess of air), and greater than one for a rich mixture (excess of fuel). Typical exhaust products for a hydrocarbon fuel reacted with air will be oxides of nitrogen (e.g., NO_2) and carbon (e.g., CO_2; possibly some carbon monoxide [CO]), water vapor (H_2O), and various hydrocarbons of differing size smaller than the reactant hydrocarbon molecules (i.e.,

Fig. 6.37 Example combustion system stability loop diagram, for an aircraft engine combustor, at a constant combustor entry pressure (p_3)

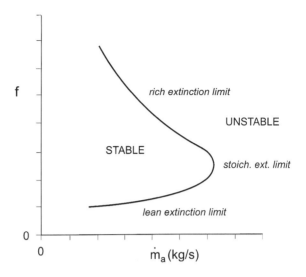

cracked hydrocarbons being in the exhaust), depending on ϕ. The appearance of soot or smoke is evidence of unburnt fuel in the exhaust. A typical overall f for the gas exiting the combustor will be well below stoichiometric, in part to keep T_{04} low enough to avoid creep (temperature-induced fatigue strain) of the turbine blades (if stoichiometric, T_{04} would approach 2500 K in value). Having said this, one should realize that a fuel–air combustion reaction will have a lower lean extinction (extinguishment) limit, given an f below which stable burning can not be sustained at that or a higher mass flow rate. As illustrated by the example combustion system stability loop diagram of Fig. 6.37, one also has a rich extinction limit, given an f above which burning cannot be sustained at that or a higher mass flow. It is possible to have an overall typical operational f of 0.02–0.03 that in fact lies below the lean limit value for that air mass flow. If such is the case, locally in the combustion chamber a restricted flow of air can be used for primary ignition in the primary combustor zone and subsequent continuous burning (thus raising the local value for f, possibly in combination with a lower local mass flow, to ensure stable burning) into the secondary or middle combustor zone downstream, while the remaining unreacted air can be re-directed upstream, or further downstream to dilute and cool the combustion products in the aft (tertiary) combustor zone before entry to the turbine section. In place of air mass flow, one would sometimes use a dimensional scaling factor called the combustor loading parameter [CLP] for performance charts, say for assessing the flame's stability as a function of equivalence ratio ϕ :

$$\text{CLP} = \frac{\dot{m}_a}{p_{03}^n \cdot \mathcal{V}_c}, \quad \text{kg}/(\text{s} \cdot \text{atm}^n \cdot \text{m}^3) \tag{6.86}$$

Fig. 6.38 Photo of combustor section of the GE J79 engine, showing several of the combustor cans and internal perforated liners in this museum display; axial flow moving left to right in above view

where the exponent n acting on combustor entry stagnation pressure p_{03} (in this example, in atmospheres [atm]) is in the neighborhood of 1.8–1.9 for a jet-fuel/air reaction, and V_c is the effective combustor volume.

Combustor walls or liners (see Fig. 6.38) would typically be made of high-temperature materials (sometimes several in combination, depending on the local peak temperature seen at that location in the combustor) like nickel-based high-performance (high-stress, high-temperature) super alloys or chromium-based (stainless) steels, or in more modern engines, possibly a ceramic composite material, or some use of thermal barrier coatings overlaid on the base metal alloy or composite structure. Film cooling is commonly applied for inner liner surface cooling, using bled-off compressor air. One would want to avoid excessive carbon buildup on the structure near the fuel injection region (primary combustor zone), by avoiding rich fuel pockets, and by avoiding the quenching (chilling) of intermediate transitionary combustion products by mixing jets. Gas turbine combustors are expected to function for several thousands of flight hours before any major maintenance or overhaul, i.e., in line with the common 3000-hour hot-end engine repair-free requirement (see Chap. 7).

A number of factors will affect the quality and corresponding intensity (W/m^3) of the combustion process in the chamber. The so-called diffusion flame (more pertinently, a mixing turbulent diffusion flame, versus a premixed laminar flame) will behave as a function of mixture turbulence and chemical kinetics (rates of reaction of various molecular compounds when they are brought together), where flow speed, swirling, and re-circulation (assisted by obstructions in the flow) are factors that potentially aid or at least influence diffusion flame strength. Premixed flames are more influenced by the local pressure (higher pressure, higher flame temperature). A typical jet engine combustor flame has elements of both flame types. With a lower pressure p_3 in the combustor at higher flight altitudes, combustion efficiency η_b will drop off as a result. As reflected by Fig. 6.37, loss of

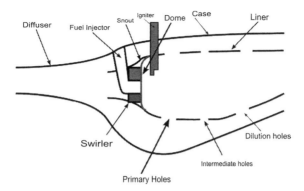

Fig. 6.39 Schematic cutaway diagram of a jet engine axial-flow cannular (can-annular) combustor (also applicable to annular design), showing the various components around one fuel injector, with axial flow moving left-to-right [10–12]. The diffuser slows the air coming in from the compressor. The snout collects and distributes compressor air for passage over the liner to help in its convective cooling, and delivering air for primary and secondary combustion and dilution. The dome produces a region of high turbulence and flow shear, and in conjunction the swirler enhances swirl in the diffusion core flame, to help mixing of fuel and air which in turn increases the intensity of the flame. Some modern annular *staged* combustors may employ a *main* fuel injector in conjunction with a lower- power *pilot* fuel injector (secondary injector useful for starts, idle, low power/low pollution situations) [13]

flame stability or ultimately flame extinction (extinguishment), also known as blowout, blowoff or flameout, is a function of p_3, f (and thus ϕ), and $\dot{m}_a$ (and thus, CLP). The effective concentration of O_2 being made available to the local combustion process may in practice vary from the nominal expectation; vitiated (degraded or corrupted) combustion refers to the case where the oxygen content is less than desirable, with larger concentrations of CO_2, H_2O, etc., being present, the net result being to act to lower the effective flame temperature. As one might expect, vitiation of the air that reaches an afterburner downstream of the main burner and turbine section can be quite pronounced, and of significant concern for adequate afterburner performance.

With respect to the combustion chamber geometry, there are three general categories: (1) can or tubular (see the J79 example, Figs. 6.36 and 6.38), (2) can-annular or *cannular* (see Fig. 6.39), and the most modern, (3) annular (Fig. 6.40) [10–13]. The original turbojet engines used the individual tubular burner cans which, because of the small working chamber volumes, provided easier control and stability of the flame, but at the expense of lower burning efficiency η_b and higher combustor stagnation pressure losses ($\pi_b \approx 0.93$). In the newer cannular designs, the individual tubular perforated burner liners (but not the surrounding can casings) are retained for better flame stability (a *can* advantage), but with higher burning efficiency with liner-to-liner air exchange around the annulus and a more uniform outlet flow (an *annular* advantage) to the turbine inlet ($\pi_b \approx 0.94$). In the most modern full annular designs, the individual tubular cans and tubular liners are eliminated altogether, with the result being the most uniform outlet flow

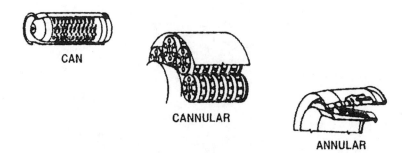

Fig. 6.40 Schematic diagrams of the three main combustor geometry categories: (1) can; (2) cannular; (3) annular [10, 12]

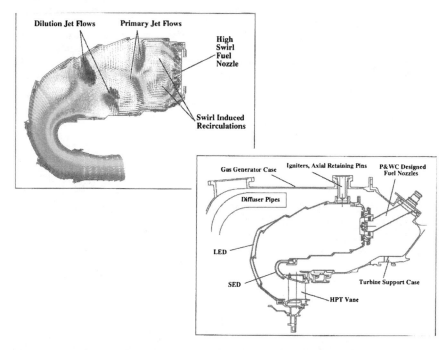

Fig. 6.41 Schematic diagram of CFD-predicted combustor flow and temperature distribution in the annular combustor of a Pratt & Whitney Canada PW150 turboprop engine [14]. Diagrams used with the permission of the Canadian Aeronautics and Space Institute

of the three approaches, and the least stagnation pressure losses ($\pi_b \approx 0.95$). However, this approach requires a more sophisticated flame control with the one large chamber volume. As a result, one typically only sees an annular combustor approach with larger modern jet engines. More detailed analyses, including the substantial use of CFD simulations as well as experimentation, allows for this progress to higher efficiency and performance (see example of Fig. 6.41; [14]).

Table 6.2 Spontaneous ignition temperatures of various fuels reacting with air [15]	Fuel	SIT, K
	Propane	767
	Butane	678
	Pentane	558
	Hexane	534
	Heptane	496
	Octane	491
	Nonane	479
	Decane	481
	Hexadecane	478
	Iso-octane	691
	Kerosene, JP-8, Jet A	501
	JP-3	511
	JP-4	515
	JP-5	506

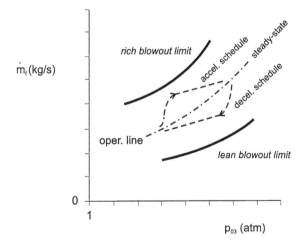

Fig. 6.42 Fuel-flow/combustor pressure chart, for steady-state and transient operation

The spontaneous ignition lower-limit temperature (SIT ≈ 500 K for jet-fuel/air; see Table 6.2 [10, 15]) is well below T_{04}, so a high-temperature heat source is only needed for initial ignition in starting or re-starting the engine. After ignition (usually from a relatively high-energy short-term spark, flame [torch], glow plug, or comparable transient heat source), the combustion process is spontaneous and continuous, as long as the proper air and fuel flow is passing through the combustor (see Fig. 6.42 as an example chart for required fuel delivery as a function of combustor pressure, including allowing for slam [rapid full-throttle-up] acceleration and slam [rapid full-throttle-down] deceleration commands by the pilot). Note that the starting process (bringing the rotor shaft speed up to the point that turbine power output can sustain acceleration) may be accomplished by a variety of

Fig. 6.43 Photo showing the positioning of an APU in the lower tail cone section of a commercial airliner, viewed from the rear of the airplane. As the name suggests, in addition to helping start one or more of the airplane's main engines, an APU can be used for powering a number of airplane accessories and systems, e.g., the environmental control/air-conditioning system for maintaining a desired cabin temperature, especially while on the ground. The exhaust port of the APU may be viewed at the far rear of the airplane (*top right of above photo*)

means. Small engines can use an electric starter motor for this purpose. Air (pneumatic) or pyrotechnic-cartridge starts involve passing high-pressure gas over the rotor of an attached or portable starter turbine (no compressor), which in turn will deliver rotation to the main engine shaft via a clutch/gear connection (see [3]). An air impingement start (used for smaller engines) involves passing high-pressure air through a valve directly onto the engine's turbine wheel, or alternatively, onto the compressor rotor. The on-board auxiliary power unit (APU; this is in fact a small gas turbine [turboshaft] engine, as per the example of Fig. 6.43) or another main engine (if a multi-engined aircraft) can be valved to deliver high-pressure air to an engine's starter turbine. For a re-start ("re-light"; [10]) in flight, the engine, under wind milling conditions, should be able to reaccelerate up to the required minimum rotation speed for the turbine to take over.

On occasion, gas turbine engines experience transient pressure wave motion associated with the combustor's operation. Depending on the perceived frequency and character of the resulting noise, the combustion instability phenomenon is commonly referred to as "rumble" or "hooting". Hooting (producing a humming sound) is commonly associated with the resonant frequency of the fuel fluid flow in the fuel injector geometry upstream of the spray injector heads. A common resolution for resonant flow in chambers is the use of a small side chamber at the quarter wave length of the offending symptom's perceived wave length. Rumble is commonly associated with the fuel injector head fuel delivery, and the downstream

Fig. 6.44 Photo of an axial turbine section for a single-shaft General Electric J79 turbojet engine. At the far left, the stationary nozzle guide vanes for entry into the turbine section are visible (as are the cooling channels within each vane). Downstream of the NGVs, the three turbine rotor blade rows at different axial locations are visible (the stators between the rotors have been removed, for clarity, in this museum display). A smaller schematic diagram [8] shows the complete engine (this engine has 17 axial compressor stages to produce a 13:1 compression, and three axial turbine stages to drive them, on one shaft), without an afterburner; again, for clarity of viewing, the stationary inlet guide vanes and stator rows used through the engine have been removed from the picture. The engine's tail cone provides a smoothly expanding flow area aft of the turbine section

spray pattern's interaction with the combustion chamber geometry. Altering the spray pattern sometimes reduces or eliminates the rumble noise.

6.8 Axial Turbine

Hans von Ohain's first turbojet prototype used both a centrifugal compressor and centripetal turbine, as part of Germany's early efforts to develop that engine. While the centrifugal (radial) compressor has some advantages, the centripetal turbine is relatively inefficient at higher mass flows. As a result, von Ohain ultimately switched to axial compressors and axial turbines in later models, to obtain better and higher performance. Today, one would possibly see the use of a centripetal turbine in very small gas turbine engines, but it is relatively rare. The axial turbine is typically made up of several stages, where each stage is comprised of a set of stator vanes ("nozzles") and rotor blades ("buckets"), with entry into the turbine via the stationary nozzle guide vanes [NGVs], as may be observed in Fig. 6.44. The axial turbine is very efficient in extracting work from the high-temperature gas flow as the pressure drops through the various turbine stages.

Fig. 6.45 Photo of a turbine rotor blade, with film cooling ports visible on the blade surface. Note the distinctive "fir tree" root (lower portion below the blade's aerodynamic working area), for affixing the blade to the rotor disk, which in turn is affixed to the HP or LP rotor shaft. Photo courtesy of NASA

Since the pressure is dropping as one moves downstream, boundary layer development is better managed and much less of a design issue, as compared to the intake and compressor upstream. So for the same pressure increment, in going up (upstream) and then down (downstream), one is typically forced to use more compressor stages, to avoid boundary layer separation problems in an adverse pressure gradient environment. Turbine blade lengths can be substantially longer than the upstream counterpart in the compressor, and thus for the same shaft rotation speed N, the mean tangential velocity U will commonly be higher for the turbine rotors. For these reasons, one will commonly see a lower number of turbine stages, versus compressor stages. In addition to the above, one additional reason for more compressor stages, for a thrust-producing turbojet, is that the turbine is there to drive the compressor, while the compressor is there to pressurize the air for the combustor and for contributing to thrust delivery via the exhaust nozzle. Note however, centrifugal stresses at the higher radius values are thus higher for turbine blades; this can be a significant structural concern.

A principal design issue with an axial turbine is the high temperature of the gas moving through this section. Depending on the materials being used and the operating temperature expected, elaborate cooling procedures applied both internally and externally to the turbine blades and vanes may be a necessity. Internal cooling, with coolant (air bled off from the compressor section, or less frequently, a liquid

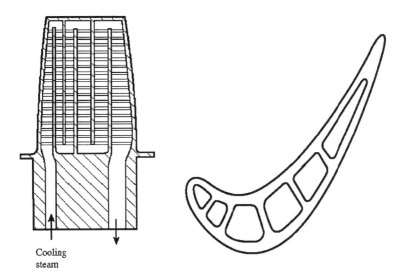

Cooling
steam

Fig. 6.46 Differing cross-sectional views of a steam-cooled turbine rotor blade, with internal cooling channels and sets of triangular-shaped ribs placed within the blade's internal structure [16]

such as the on-board fuel) absorbing heat as it passes through channels within the blade or vane, is predominantly via a convective heat transfer approach, and thus referred to as convective cooling. Impingement cooling, where a jet of coolant impinges on the inner surface of the heated structure to take away heat, may be applied as an additional internal means of cooling. Film (transpiration) cooling is the predominant means of external cooling, where a thin film of cool bleed air is blown along the blade or vane outer surface (see Fig. 6.45 for view of cooling ports positioned on the turbine blade surface). One may see the use of very small cylindrical protuberances placed into the internal cooling channel flow, called pin fins or pedestals, that effectively act as individual micro-heat-exchangers in taking heat away from the vane or blade main structure. Other usages of protuberances can be for turbulence promotion (commonly rib-shaped [15], sometimes called turbulators), to enhance the heat transfer process between the structure and the internal channel flow (see Fig. 6.46). Lubricating oil may also complement the cooling process where applicable, in a dual role of surface-to-surface friction reduction (primarily for the shaft bearing assemblies), and removing some heat from the structure. The heated oil may subsequently need to be air- or liquid-cooled in the cycle of operation, e.g., fuel is commonly used as an oil coolant.

In order to maintain a relatively constant subsonic axial velocity through the turbine as the gas is expanding and cooling as it applies work to the rotor shaft, the cross-sectional area of the flow passage must get bigger in moving downstream, with the result of having fewer but larger (and wider) vanes and blades per stage. As depicted by Fig. 6.47 for two adjoining rotor blade airfoil sections, a cascade field representation can be useful for stator and rotor design. As noted earlier for axial compressor blades and vanes, through various studies, one can establish

Fig. 6.47 Diagram of adjoining turbine rotor blade sections at a given radial position (cascade field representation)

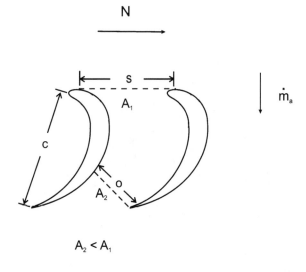

$$A_2 < A_1$$

important design parameters such as chord length c, separation distance s, and opening distance o for each turbine stage. As illustrated by Fig. 6.48, a representative velocity triangle can be drawn as one enters a stator vane's local aerodynamic region, and moving downstream, a second velocity triangle can be shown as one enters a rotor blade's local aerodynamic region. One can observe the reversal of function relative to that seen earlier for a compressor rotor/stator combination. The stator nozzles are used to accelerate and re-direct the flow to give an acceptable angle-of-attack for the rotor blades moving at a tangential velocity of $U = \omega r$ at the given radial location r. The work done by the gas on the rotor (per unit mass) in moving from a stage's inlet to exit is:

$$\frac{W_t}{m} = C_{p,t} T_{o,i} - C_{p,t} T_{o,e} = \Delta w \cdot U \tag{6.87}$$

where Δw is the effective positive change in the gas tangential velocity through the rotor. Assuming U and Δw are determined at an appropriate mean radial location, the pressure drop through the stage would thus be

$$\left. \frac{p_{o,e}}{p_{o,i}} \right|_{stage} = \left[1 - \eta_{stage} \frac{U \cdot \Delta w}{C_{p,t} T_{o,i}} \right]^{\frac{\gamma_t}{\gamma_t - 1}} \tag{6.88}$$

The degree of reaction $^\circ R_t$ for a turbine stage is an important performance parameter in setting the design of the stator + rotor combination:

$$^\circ R_t = \frac{\text{rotor static enthalpy drop}}{\text{stage static enthalpy drop}} = \frac{h_{mid} - h_e}{h_i - h_e} \approx \frac{T_{mid} - T_e}{T_i - T_e} \tag{6.89}$$

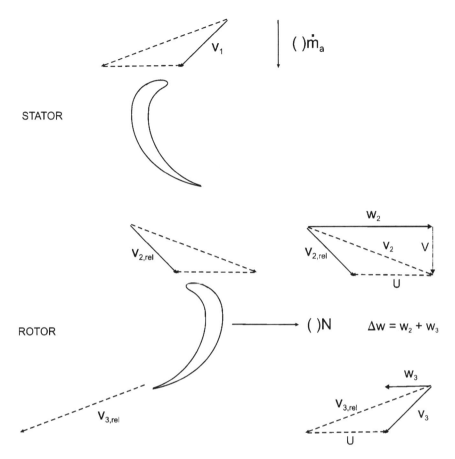

Fig. 6.48 Velocity triangles for axial stage stator and rotor blade sections at a given radial position (cascade field representation). The above airfoil combination corresponds to an approximate 50% degree of reaction (v_3 similar in magnitude to $v_{2,\text{rel}}$, v_2 similar in magnitude to $v_{3,\text{rel}}$)

The static temperature drop indicated by Eq. 6.89 also to a large degree will correspond to the static pressure drop through the stage. A 50% reaction design, similar to that represented by Fig. 6.48, would have the gas expansion shared equally between the stator and rotor rows. Referring to Fig. 6.48, one can make the following substitutions for enthalpy with equivalent energy and work into Eq. 6.89:

$$°R_t = \frac{h_2 - h_3}{h_1 - h_3} = \frac{\left(v_{3,\text{rel}}^2/2 - v_{2,\text{rel}}^2/2\right)}{U(w_3 + w_2)} \tag{6.90}$$

In the conventional situation where axial velocity V does not change much while moving downstream through the turbine, the trigonometry of Fig. 6.48 provides for the following on Eq. 6.90:

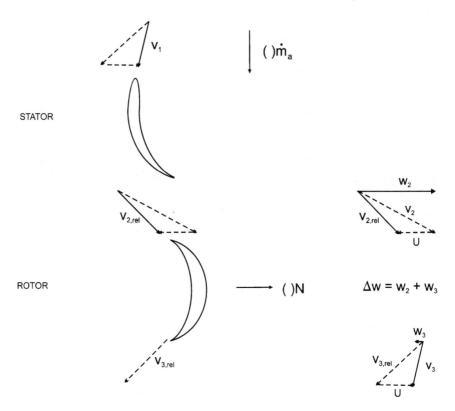

Fig. 6.49 Velocity triangles for axial stage stator and rotor blade sections at a given radial position (cascade field representation). The above airfoil combination corresponds to an approximate 0% degree of reaction (relative velocity of gas entering, $v_{2,rel}$, and exiting rotor, $v_{3,rel}$, similar in magnitude)

$$v_{3,rel}^2 - v_{2,rel}^2 = V^2 + (U + w_3)^2 - V^2 - (U - w_2)^2 = w_3^2 - w_2^2 + 2U(w_3 + w_2),$$

$$^\circ R_t = \frac{(v_{3,rel}^2 - v_{2,rel}^2)}{2U(w_3 + w_2)} = 1 - \frac{(w_2 - w_3)}{2U} \tag{6.91}$$

One might expect most designs to operate in this 50% region at the mean stage diameter (radial location), but in practice, one does see a significant range for mean $^\circ R_t$ in conventional subsonic engines, from 30–50%. Figure 6.49 illustrates a 0% reaction design, at one end of the spectrum (rotor just acts to redirect the flow). Figure 6.50 illustrates a 100% reaction design, at the other end of the spectrum (stator just acts to re-direct the flow).

Other performance parameters may come into play, to influence the effective value seen above for $^\circ R_t$. The stage loading coefficient ψ (also identified as the temperature-drop coefficient or the turbine work ratio) is typically defined by:

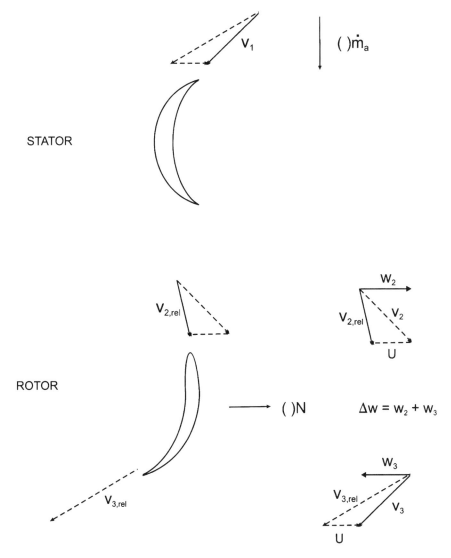

Fig. 6.50 Velocity triangles for axial stage stator and rotor blade sections at a given radial position (cascade field representation). The above airfoil combination corresponds to an approximate 100% degree of reaction (velocity of gas entering, v_1, and exiting stator, v_2, similar in magnitude, and absolute velocity of gas exiting rotor, v_3, similar in magnitude to v_2)

$$\psi = \frac{W_t/m}{U^2} = \frac{\Delta h_o}{U^2} = \frac{C_p \Delta T_o}{U^2} = \frac{\Delta w}{U} \tag{6.92}$$

In practice, values for ψ can range from 1 to 2.5. Above 2.5, this indicates that exit swirl is becoming large. All things being equal, a lower exit swirl (tangential

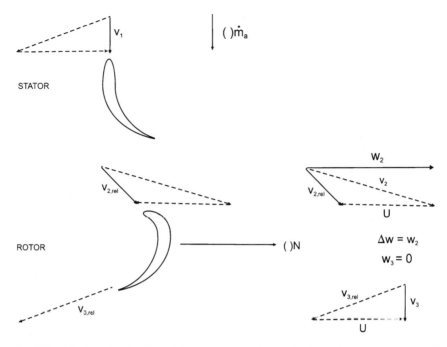

Fig. 6.51 Velocity triangles for axial stage stator and rotor blade sections at a given radial position (cascade field representation). The above airfoil combination corresponds to a zero exit swirl requirement, as represented by stator entry velocity v_1 and rotor exit velocity v_3 being aligned with the engine longitudinal axis (in essence, v_3 equal to V)

velocity Δw) reduces viscous flow losses and delivers more work for a given pressure drop through a turbine stage. Figure 6.51 illustrates a stator + rotor combination that delivers a zero exit swirl. This example shows that one can have a zero exit swirl at a given radial location, with a $°R_t$ of around 40% (referring as well to Eq. 6.91). From an efficiency viewpoint, it would ideally be desirable to have a lower value of ψ, but this would mean a larger number of stages for a desired overall expansion. Thus, conversely, a higher value of ψ lowers the engine weight, but at the expense of a lower η_{stage}. In addition, a higher ψ acts to lower the effective $°R_t$. Given this information, one can see why some engine designers have chosen, on those occasions where engine weight was a critical design factor, to set the mean $°R_t$ below 50%.

Another parameter of interest is the flow coefficient ϕ (also identified as the axial velocity ratio) typically defined by

$$\phi = \frac{V}{U} \tag{6.93}$$

which typically ranges in practice from 0.4 to 0.6. Here, as noted, V is the local axial velocity, and U is the local rotor tangential velocity. From an efficiency viewpoint, the lower the value for ϕ, the better.

Fig. 6.52 Axial turbine map, showing expansion ratio p_{04}/p_{05} as a function of corrected mass flow, for various corrected shaft rotation speeds. Adiabatic turbine efficiency curves also illustrated. The steady-state operating line is shown

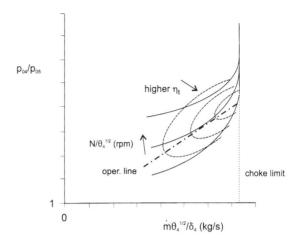

In the preliminary design phase, it is common to make a free-vortex (meaning zero vorticity, i.e., irrotational flow, about a given radial vortex filament) flow assumption [5, 17] in the early stages of compressor and turbine blade design, which in essence means neglecting the radial velocity components that in practice can exist (because of a spanwise gradient in pressure along a given blade) in having the gas flow moving axially through a contracting (compressor) or expanding (turbine) annular flow passage left-to-right (upstream to downstream through the engine section). Referring in part to Eq. 6.91 as well as the various velocity-triangle diagrams for the work-related swirl velocity components in question, the free-vortex assumption stipulates that along a given turbine blade (from root to tip, radial distance r increasing):

$$r \cdot w_2 = \text{constant}, \quad r \cdot w_3 = \text{constant} \tag{6.94}$$

For example, as one moves outward radially along a given turbine blade, w_2 (and w_3) would need to drop in value for the given local 2D airfoil section. Similarly, for an axial compressor blade, referring to Eq. 6.84, a free-vortex assumption would imply:

$$r \cdot w_1 = \text{constant}, \quad r \cdot w_2 = \text{constant} \tag{6.95}$$

An ideal free-vortex design would produce a uniform flow at the exit of either the compressor or turbine, which in general is desirable. However, a free-vortex stipulation generally requires that one use a substantial blade twist that is not necessarily practical or entirely effective. That being said, there are flow energy losses associated with the actual secondary radial flow noted above for a non-free-vortex [NFV] compressor or turbine. For more detailed calculations, an NFV design for the compressor and turbine rotor blades is commonly accounted for, to more accurately predict expected performance.

Limits on performance can also be enforced by such factors as high-tempera-ture creep of the turbine blades (induced elongation over a long period of time) and

Fig. 6.53 Example single-stage axial turbine map, showing corrected non-dimensionalized mass flow as a function of expansion ratio p_{04}/p_{05}, for various corrected shaft rotation speeds (in percentage of corrected design speed) [18]. The older format for presentation of the data as seen here is the reverse of the format used for Fig. 6.52

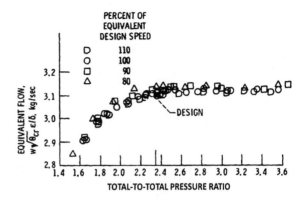

centrifugal stresses produced by 10,000 to 1,00,000 g acceleration loading levels acting on the rotor blades, and the disks they are attached to. In order to attain the benefits of operating with a higher temperature gas flow (toward thrust or power delivery), we noted earlier the various cooling methods for keeping the structure of the blades and vanes at hundreds of degrees below the local gas stagnation temperature. As a turbomachine, a gas turbine engine will require an effective lubricating oil system for reduction of friction at shaft bearing load locations. In parallel, the oil lubricant will act to cool hot engine components, in complementing the cooling provided from incoming outside air (ram air intake) or air bled off from the compressor section. Blades and vanes are commonly cast or machined from nickel-based steel alloys, which provide high strength at the high temperatures faced by the structure. For higher strength, single-crystal blades may be produced (continuous crystal lattice, without any grain boundaries, right to the edges of the blade). The blades may be welded or integrally cast to the rotor disk (the so-called *blisk*), an attachment that couples the rotor blades to the shaft (see Fig. 6.45).

An example illustrative turbine performance map is provided in Fig. 6.52 (an example actual map is provided in Fig. 6.53) [18], showing the pressure drop p_{04}/p_{05} and adiabatic efficiency η_t as a function of the corrected mass flow:

$$\dot{m}_{\text{corr}} = \dot{m} \cdot \frac{\sqrt{\theta_4}}{\delta_4} \tag{6.96}$$

where

$$\theta_4 = \frac{T_{04}}{T_{\text{SL}}}, \quad \delta_4 = \frac{p_{04}}{p_{\text{SL}}} \tag{6.97}$$

A number of equilibrium curves at increasing corrected rotation speed are given in Fig. 6.52, as well as the steady-state operating line. Note that a maximum value for corrected mass flow is reached when choking occurs (local *Ma* of flow goes to unity) in either the stator vane (nozzle) throats or at the turbine annular outlet, at and above some critical value of p_{04}/p_{05}. One can see that prior to and after choking, there is the expected larger stagnation pressure drop with increasing shaft rotation speed N. There is no aerodynamic stall concern here, but the use of variable-incidence stator vanes in the turbine can control outflow such that it can help control compressor surge problems upstream (FADEC application).

We noted earlier in our cycle analysis for a single-shaft turbojet engine that the turbine provides that amount of work required to operate the upstream compressor at some design value for π_c, such that:

$$\frac{W_t}{m} = C_{p,t}(T_{04} - T_{05}) \approx \frac{1}{\eta_m} C_{p,c}(T_{03} - T_{02}) \qquad (6.98)$$

We recall that η_m is the mechanical efficiency, usually just less than 1, due to bearing friction, etc. Over a range of operating conditions (typically optimal at the on-design point, less than optimal at off-design conditions away from the design point), the turbine should be designed to match up well to that work which will be required by the compressor.

As noted earlier for axial compressors, one is not restricted to a single shaft or spool for the compressor/turbine combination. Larger aircraft engines commonly employ a twin-spool setup, with an LP compressor and turbine on one shaft, and a higher-speed ($N_2 > N_1$) HP compressor and turbine on a second concentric shaft. While mechanically independent, their corresponding speeds are tied together aerodynamically. For engines requiring high pressure ratios and efficient operation, a triple-spool arrangement (LP, IP, HP concentric shafts) might be warranted.

6.9 Afterburner for Jet Engines

The afterburner (A/B) is a compromise solution for delivering large thrust output for short periods of time (e.g., for takeoff, manoeuvring). An augmented turbojet engine weighs much less than a standard TJ delivering the same maximum thrust, but it will deliver that thrust at a lower thermal efficiency and higher TSFC in this reheat approach. Afterburning is made possible by the very lean fuel–air mixture f passing through the main burner producing an oxygen-rich gas flow exiting the turbine outlet. This outlet flow can have additional fuel added to it and burned in a secondary combustion chamber formed by the jet pipe (see Figs. 6.54, 6.55; [3, 19, 20]), where high temperature constraints are relaxed. With increased total temperature at the nozzle inlet with this heat addition (T_{06} as high as 2000 K or more), an increased exit velocity for the exhaust jet will in turn result, and therefore produce greater thrust (up to 100% or more above the standard thrust level). Running the engine with the A/B active and higher corresponding thrust delivery is called wet or "maximum thrust" engine operation, and running without the A/B is called dry or "military" engine operation. In the

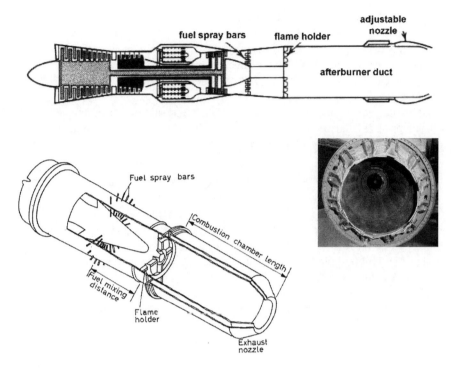

Fig. 6.54 *Upper schematic diagram* [19] outlines a turbojet engine, with an afterburner jet pipe incorporated; reprinted with permission of John Wiley & Sons, Inc. *Lower diagram* further illustrates positioning of various conventional afterburner components; reprinted with permission of Elsevier [20]. *Right photo* showing view looking upstream into a General Electric J85 turbojet engine's afterburner

Fig. 6.55 Photo of cutaway view of GE J79 turbojet engine's afterburner section, clearly showing flame holding ring gutters upstream, and downstream, the slots of the jet pipe's surface noise/ wave-suppression acoustic liner are evident

Fig. 6.56 View of Rockwell B-1B supersonic bomber in takeoff, with turbofan engine afterburners being employed (engine: General Electric F101-102). Courtesy of USAF

Fig. 6.57 Photo of Pratt & Whitney F119 engine being operated at full afterburner on a static test stand in a propulsion test facility (Tyndall AFB; USAF photo: Albert Bosco)

past, water injection or alcohol/water injection was often used for thrust augmentation over short periods of time (e.g., for takeoff), hence the "wet" reference. One clear benefit of adding water to the flow is to cool the air (as the water vaporizes, it absorbs heat); this is certainly useful for augmenting compressor performance. See Fig. 6.56 for an example of an aircraft employing afterburners in takeoff, to assist in shortening the takeoff distance, and Fig. 6.57, for a jet engine being tested in a test cell at full afterburner.

In order to maintain a stable flame in the A/B jet pipe, V-shaped gutters may be used as flameholders. Referring to Figs. 6.55 and 6.58 [21–23], fuel is injected via a spray ring or bar and ignited by a starting high-temperature spark, and impingement of this initial flame on the bluff body of the flame holders downstream produces a larger, main stationary flame downstream of the flame holders. There is a compromise between jet-pipe length and number of flame holder gutters for adequate combustion, since both to excess will contribute to greater stagnation pressure losses which will detract from thrust output. The liner of the surrounding

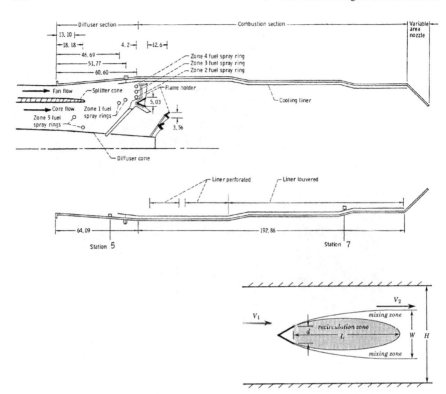

Fig. 6.58 Schematic diagram of the afterburner section of a low-bypass Pratt & Whitney TF30-P3 turbofan engine [21, 22]. Note the variable-area exhaust capability to allow for different temperatures of the mass flow, to keep the flow within the duct subsonic. *Lower diagram* illustrates framework for evaluating a V-gutter flame holder/stabilizer [11, 23]

A/B casing must be used for both cooling and as an acoustic absorber (so-called screech liner). The perforations in the screech liner act to damp out pressure waves (transverse: radial and/or tangential; axial) that can develop in a large, flexible duct. Note that these waves are intermediate (axial) to high (transverse) frequency, and thus produce a screech (axial) or howl (transverse) sound. Other transient-flow problems associated with afterburners include so-called hard starts, where unburned fuel accumulates in the jet pipe during an unusually prolonged starting process, resulting in subsequent strong transient axial pressure waves that are fed as that excess fuel ignites.

Given that an A/B jet pipe is typically of a relatively constant cross-sectional area with a substantial length-to-diameter (L/D) ratio (2.5 up to 4:1), one has to be concerned about flow Mach number increasing with friction (Fanno flow; [24]) and heat addition (Rayleigh flow; [24]) towards the sonic limit. For duct lengths beyond the choking limit length L_{max}, equilibrium dynamics will force a reduction in $\dot{m}$ to compensate, and thus produce a lower thrust. Note as well that both friction and heat addition cause a reduction in stagnation pressure (friction causes a reduction in stagnation temperature, while clearly heat addition increases stagnation temperature). For example, given an A/B inlet flow Ma_5 of 0.2, purely Fanno flow would produce a reduction in stagnation pressure to the choke limit (assuming γ of 1.4) of $p_{05}/p_{06}^* = 2.96$ (i.e., $\pi_{AB}^* = 1/2.96 = 0.34$) when axial distance x along the A/B jet pipe reaches L_{max}, noting $fL_{max}/D = 14.5$ (Darcy-Weisbach friction factor $f \approx 0.02$ for a smooth pipe surface, and substantially higher at 0.08 for a rough pipe surface). Alternatively, purely Rayleigh flow would produce a reduction in stagnation pressure to the choke limit of $p_{05}/p_{06}^* = 1.23$ (i.e., $\pi_{AB}^* = 1/1.23 = 0.81$) when $T_{06}^*/T_{05} = 5.76$. Note again that the abovementioned V-gutters will contribute to reducing the stagnation pressure in the jet pipe.

The following relations for an afterburning TJ may be useful. For specific thrust,

$$\frac{F}{\dot{m}_a} = (1 + f + [1 + f]f_{AB})V_e - V_\infty + \frac{A_e}{\dot{m}}(1 + f + [1 + f]f_{AB})(p_e - p_\infty) \quad (6.99)$$

where the A/B fuel-gas ratio is given by

$$f_{AB} = \frac{\dot{m}_{f.AB}}{\dot{m}_a + \dot{m}_f} = \frac{\dot{m}_{f.AB}}{\dot{m}_a(1 + f)} \quad (6.100)$$

Finally, for overall engine thrust specific fuel consumption,

$$\text{TSFC} = \frac{\dot{m}_f + \dot{m}_{f.AB}}{F} = \frac{f + (1 + f)f_{AB}}{F/\dot{m}_a} \quad (6.101)$$

An example fuel control system setup for running an older afterburning J47 turbojet engine under normal and augmented-thrust conditions may be found in [3].

6.10 Exhaust Nozzle for Gas Turbine Engines

The simple convergent exhaust nozzle (as opposed to the convergent/divergent category), utilizing a central tail cone that allows for a smooth expansion in flow area aft of the turbine section (and maintain subsonic flow upstream of the nozzle exit), is a common choice for subsonic aircraft employing turbojet or turbofan engines (see Fig. 6.36). In general it will weigh less than a C-D nozzle, and while the diverging exhaust jet will not be guided optimally, the lack of skin friction (and allowance for a potentially larger cross-section for the downstream exhaust jet) with no area divergence section may in some cases produce a net thrust advantage for subsonic-flight applications. If the convergent nozzle's geometry is fixed, then typically at low-pressure ratios (nozzle entry, station ⑥ to nozzle throat/exit, station ⑦) which correspond to low engine shaft rotational speeds (N), the convergent nozzle exit will be unchoked ($Ma_e < 1.0$).

Fig. 6.59 Variable-geometry nozzle at end of afterburner of General Electric J79 turbojet engine on display in Pima Air & Space Museum (*upper photo*). *Lower photo* shows engine with nozzle fully opened for peak mass flow under full afterburner operation

At higher N values corresponding to higher thrust delivery, the nozzle exit flow will likely be choked (and expand supersonically downstream). In this particular case, the choked mass flow rate can be estimated via

$$\dot{m} = p_{06}A_e\sqrt{\frac{\gamma_n}{RT_{06}}}\left[\frac{2}{\gamma_n+1}\right]^{\frac{\gamma_n+1}{2(\gamma_n-1)}} \qquad (6.102)$$

Fig. 6.60 View of the two exhaust nozzles employed for the two afterburning Pratt & Whitney J57-P-55 turbojet engines powering the McDonnell F-101B Voodoo supersonic fighter-interceptor. Note the generous clearance provided for the empennage (tail structure) in avoiding exhaust plume damage

The optimal setting for nozzle exit area A_e is when the flow just chokes, i.e., at a greater A_e the flow would not choke, or possibly choke downstream of the exit. If on the other hand, A_e is smaller than optimal, the mass flow must drop (and therefore thrust) in order to maintain system equilibrium.

For supersonic flight, the use of a C-D nozzle to expand the exhaust flow to a supersonic value, and to provide a more uniform flow profile at the nozzle expansion exit, is a more appropriate choice for effective thrust delivery. However, structural weight constraints may shorten the nozzle divergence section somewhat, to lower thrust delivery a little bit. One criterion for selecting a C-D nozzle over a convergent nozzle is that for most of the flight mission, $p_{06}/p_\infty > 6$. Optimal thrust is obtained when nozzle exit pressure p_e is close to outside air pressure p_∞, so one

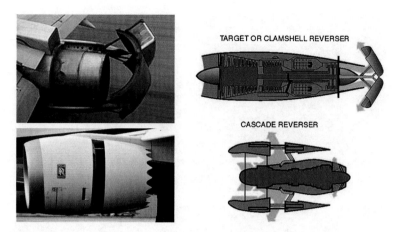

Fig. 6.61 Two common reverse thrust approaches (core exhaust nozzle clamshell/blocker, *upper*; cascade vent opening from turbofan engine's fan duct, *lower*) [8]

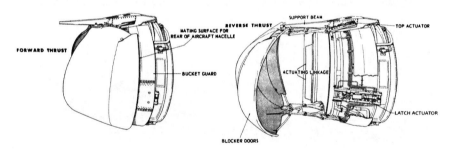

Fig. 6.62 Schematic diagram of a Pratt & Whitney JT8D low-bypass (0.96:1 BPR) turbofan engine, illustrating the positioning of two complementary thrust reverser mechanisms (cascade vanes + clamshell doors) at the rear of the engine

must likely factor in altitude, as it affects p_{06} and p_{∞}, on the given nozzle's performance.

Nozzles with variable geometry (i.e., variable throat and exit area) allow for better flow management, by keeping $\dot{m}$ as high as possible under different flight conditions, i.e., the operating conditions upstream in the engine will not be adversely affected by a mass flow constriction problem, and the thrust output via the nozzle exit plane setting will be optimal. The variable-geometry nozzle can help avoid compressor surge problems during the starting run-up (open up nozzle throat to lower pressure temporarily). See Fig. 6.59 for a photo of a variable-geometry exhaust nozzle positioned at the end of a jet engine's afterburner jet pipe (mechanical iris type, with hydraulically-actuated overlapping petals/leaves, not necessarily requiring use of bypass air or gas flow). Typically

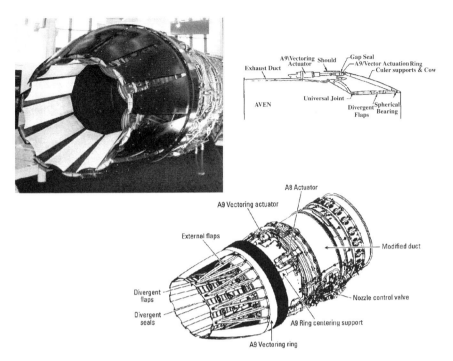

Fig. 6.63 Vectorable (3D), variable-geometry (iris design: hydraulically adjustable, overlapping petals) nozzle for modern fighter aircraft engine on static display. Right schematic diagrams are of the Axisymmetric Vectoring Exhaust Nozzle (AVEN) developed and flight tested by General Electric on a modified F110 low-bypass turbofan engine powering a USAF F-16 fighter aircraft. Diagrams courtesy of NASA [25]

older supersonic jet engines, such as used by the fighter aircraft in Fig. 6.60, use an ejector nozzle approach, where bypass air (may be pulled in from outside the engine, or from the compressor) and/or gas is variably channeled through small secondary/tertiary nozzles and ports (doors) in such a manner as to modify the effective main nozzle flow area, where needed at that juncture in the flight.

Other nozzle design considerations might include thrust reversal (for slowing the aircraft upon landing) and thrust vectoring (for augmented aircraft manoeuvrability). While a turbofan engine might use the fan duct's high-pressure air flow re-directed with a forward vector component to produce a reverse thrust capability, one might use a comparable approach (using cascade vents or clamshell-like blockers; [6]) downstream, near or beyond the main nozzle exit, for re-directing the exhaust flow forward to some degree (see Figs. 6.61 and 6.62). Thrust reversal can be as high as 50% of the nominal maximum forward thrust capability.

With respect to thrust vector control (TVC), a few fighter aircraft, some with vertical takeoff and landing (VTOL) capability, have exhaust nozzles with some ability to swivel 20° or more (see Figs. 6.63 [25] and 6.64 [26]). The Hawker

Fig. 6.64 Pratt & Whitney F119 afterburning turbofan with 2D vectoring capability in upper diagram (reprinted with permission of the American Institute of Aeronautics and Astronautics) [26]; used by the twin-engined Lockheed Martin F-22 Raptor (lower photos; courtesy of USAF)

Siddeley AV-8A Harrier, and its later naval variant, the British Aerospace Sea Harrier, are classic examples of a successful production VTOL jet aircraft employing a directed exhaust jet and ejector thrust approach [13]. These aircraft employ the Rolls-Royce Pegasus low-bypass turbofan engine, illustrated in Fig. 6.65.

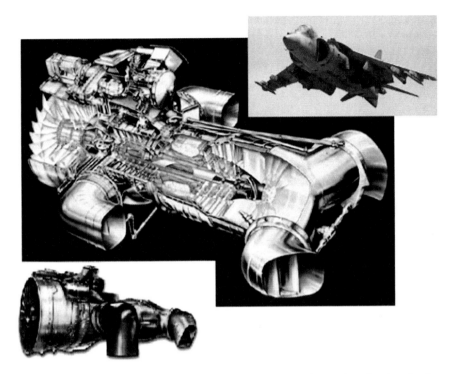

Fig. 6.65 Schematic diagram of the Rolls-Royce Pegasus turbofan engine, showing the four principal exhaust nozzles that can be rotated about a 98° arc for directed thrust in takeoff, landing and forward flight. In the main diagram, the nozzles are shown in the fully aft position, for conventional forward flight. The lower diagram shows the nozzles in the fully downward position, for vertical takeoff or landing of the Harrier. Engine diagrams courtesy of Rolls-Royce plc

6.11 Example Problems

6.1. A four-engined turboprop regional airliner flies at a Mach number of 0.65, at a cruise altitude of 4000 m (outside air temp. = 262.2 K, outside air pressure = 61.66 kPa, outside air density 0.819 kg/m^3). Each engine's propeller is rotating at 1100 rpm, with the four-bladed props having a diameter of 3.05 m. Each prop is producing a thrust coefficient of 0.5, at a power coefficient of 2.3. At this flight condition, each engine's BSFC is presently 0.3 kg/h kW. You may assume that the fuel in reaction with air will produce a heat of reaction of 42 MJ per kg of fuel. Find the thermal efficiency, and then the overall efficiency, of each turboprop engine. Be careful on the units being used for your calculations to arrive at the proper answers, especially with regard to propeller rotation speed and BSFC units, e.g., take note that 1 W = 1 J/s.

6.2. Perform a design-point cycle analysis for a non-afterburning turbojet engine at cruise flight conditions. Take note of the following performance parameters:

diffuser, $\eta_d = 0.97$; $\gamma_d = 1.4$; compressor, $\eta_c = 0.95$, $\gamma_c = 1.37$, $\pi_c = 20$;
burner, $\eta_b = 1.0$, $\gamma_b = 1.35$, $\pi_b = 0.95$, $T_{max} = 1500$ K
$q_R = 45$ MJ/kg of fuel; turbine, $\gamma_t = 1.33$; $\eta_t = 0.90$; $\eta_m = 0.99$
core nozzle, $\eta_n = 0.98$; $\gamma_n = 1.36$
current flight conditions at 12200 m cruise altitude,
cruise flight Mach number = 0.85

Given the above flight and engine information, determine the specific thrust, TSFC, thermal efficiency, propulsive efficiency, and overall efficiency.

6.3. Perform a design-point cycle analysis for an afterburning version of the turbojet engine of Problem 6.2 above, at its cruise Mach number of 0.85 and altitude of 12200 m. Take note of the following design parameters associated with the afterburner:

$\eta_{AB} = 0.9$, $\gamma_{AB} = 1.33$, $\pi_{AB} = 0.94$, $q_R = 45$ MJ/kg of fuel, $T_{max,AB} = 2000$ K

Ascertain the specific thrust and TSFC, and compare these results to the values for the non-afterburning version of the engine.

6.4. Perform a design-point cycle analysis for the non-afterburning turbojet engine of Problem 6.2 at its cruise Mach number of 0.85 and altitude of 12200 m, but now at a cooler combustor peak temperature of 1300 K (rather than 1500 K). Ascertain the specific thrust and TSFC, and note the trends relative to the higher peak combustor temperature of Problem 6.2.

6.5. Perform a design-point cycle analysis for the non-afterburning turbojet engine of Problem 6.2 at its cruise Mach number of 0.85 and altitude of 12200 m, but now at a higher compressor pressure ratio (25, rather than 20). Ascertain the specific thrust and TSFC, and note the trends relative to the lower compression ratio of Problem 6.2.

6.6. Perform a design-point cycle analysis for the non-afterburning turbojet engine of Problem 6.2 at its cruise Mach number of 0.85 and altitude of 12200 m, but now at a lower compressor pressure ratio (15, rather than 20). Ascertain the specific thrust and TSFC, and note the trends relative to the higher compression ratio of Problem 6.2.

6.7. Identify the type of intake being used by the airplane in Fig. 1.14 (F-22 Raptor), in Chap. 1.

6.8. An unmanned drone is presently flying at 6100 m altitude (ISA), at a flight Mach number of 0.7. The UAV employs a Teledyne-Continental J402-CA-702 turbojet engine. Assuming the engine performance charts of Figs. 6.8 and 6.9 apply (maximum continuous throttle setting), establish the current fuel consumption rate (kg/h) by the drone.

6.9. *Computer-Based Project*: Perform an off-design cycle analysis for a turbojet engine at its cruise throttle setting, at altitudes from 0 to 13,200 m ISA, at flight Mach numbers from zero to 0.9. Take note of the following design parameters:

diffuser, $\eta_d = 0.97$; $\gamma_d = 1.4$; compressor, $\eta_c = 0.93$, $\gamma_c = 1.37$, $\pi_c = 10.2$;
burner, $\eta_b = 0.98$, $\gamma_b = 1.35$, $\pi_b = 0.95$, $T_{max} = 1100$ K
$q_R = 43$ MJ/kg of fuel; turbine, $\gamma_t = 1.33$; $\eta_t = 0.90$; $\eta_m = 0.98$
core nozzle, $\eta_n = 0.96$; $\gamma_n = 1.36$; nozzle fixed exit diameter
$d_e = 0.55$ m

Plot net thrust F versus flight Mach number Ma_∞ at the four principal altitudes of 0, 4400, 8800 and 13,200 m ISA (generate enough data points so that each curve appears smooth in appearance). Ascertain the TSFC data points when values of 0.09, 0.10 and 0.11 kg/h-N are attained, and graph these cross-hatch TSFC curves onto the main $F(h_{ASL})$ versus Ma_∞ graph (the TSFC curves will have four data points each, and thus may not be particularly smooth in appearance as a result). Comment on the characteristics of this engine's thrust profile at different flight conditions.

6.12 Solution to Example Problems

6.1. General info:
$a_\infty = \sqrt{\gamma R T_\infty} = [1.4(287)262.2]^{1/2} = 324.6$ m/s, $V_\infty = a_\infty Ma_\infty = 324.6$
$(0.65) = 211$ m/s
BSFC $= 0.3$ kg/(h kW) $= 0.3/(3600$ s/h)/1000 W/kW) $= 8.33 \times 10^{-8}$ kg/(s W)

Thrust per engine, via propeller thrust coefficient:
$F_{eng} = C_T \rho n^2 d^4 = 0.5\,(0.819)\,(1100/60)^2(3.05)^4 = 11,910$ N or 11.9 kN

Total thrust for four-engined airplane:
$F_{tot} = 4\,F_{eng} = 47643$ N $= D$, aerodynamic drag acting on the airplane in SLF (steady level flight)

Shaft power per engine, via propeller power coefficient:
$P_{S,eng} = C_P \rho n^3 d^5 = 2.3(0.819)(1100/60)^3(3.05)^5 = 3.064 \times 10^6$ W $= 3.064$ MW, or 4109 hp

Propeller's propulsive efficiency:

$$\eta_{pr} = \frac{F_{eng} V_\infty}{P_{S,eng}} = 11910(211)/(3.064 H 10^6) = 0.82$$

Alternatively, using advance ratio J:

$$\eta_{pr} = \frac{C_T}{C_P} \cdot J = \frac{C_T}{C_P} \cdot \frac{V_\infty}{nd} = 0.5/2.3 \cdot (211)/((1100/60) \cdot 3.05) = 0.82$$

Engine thermal efficiency:

$$\eta_{th} = \frac{P_{S,eng}}{\dot{m}_{f,eng} \cdot q_R} = \frac{1}{\text{BSFC} \cdot q_R} = 1/(8.33 \times 10^{-8} \cdot 42 \times 10^6) = 0.286, \text{ noting fuel}$$

flow per engine is

$$\dot{m}_{f,eng} = \text{BSFC} \cdot P_{S,eng} = 0.255 \text{ kg/s}.$$

Engine overall efficiency:

$$\eta_o = \frac{F_{eng} V_\infty}{\dot{m}_{f,eng} \cdot q_R} = \eta_{pr} \cdot \eta_{th} = 0.82 (0.286) = 0.235$$

6.2. Looking at a non-afterburning turbojet engine, at a flight Mach no. Ma_∞ of 0.85, at 12200 m altitude ($p_\infty = 18.75$ kPa, $T_\infty = 216.7$ K, $\rho_\infty = 0.30$ kg/m^3). Starting from intake (diffuser):

$$T_{02} = T_\infty [1 + \frac{\gamma_d - 1}{2} Ma_\infty^2] = 216.7[1 + 0.2(0.85^2)] = 248 \text{ K, at diffuser outlet}$$

$$p_{02} = p_\infty [1 + \eta_d (\frac{T_{02}}{T_\infty} - 1)]^{\frac{\gamma_d}{\gamma_d-1}} = 18.75[1 + 0.97(1.144 - 1)]^{3.5} = 29.63 \text{ kPa,} \quad \text{at}$$

diffuser outlet
Moving through compressor:

$$T_{03} = T_{02}\left[1 + \frac{1}{\eta_c}[\pi_c^{\frac{\gamma_c-1}{\gamma_c}} - 1]\right] = 248[1 + 1/0.95 \cdot [20^{0.27} - 1]] = 573 \text{ K,} \quad \text{at}$$

compressor outlet

$p_{03} = \pi_c p_{02} = 20(29.63) = 592.6$ kPa, at compressor outlet; $C_{p,c} = \frac{\gamma_c R_{air}}{\gamma_c - 1} =$ 1.37(287)/0.37 = 1063 J/kg $\cdot$ K.
Moving through combustor (burner):

$$\bar{C}_{p,b} \approx \frac{C_{p,c} + C_{p,b,exit}}{2} = \frac{\gamma_b R_{air}}{\gamma_b - 1} = 1.35(287)/(1.35 - 1) = 1107 \text{ J/(kg K), so that}$$

at burner exit:

$$C_{p,b,exit} \approx 2\bar{C}_{p,b} - C_{p,c} = 2(1107) - 1063 = 1151 \text{ J/(kg K)}.$$

$T_{04} = T_{max} = 1500$ K, $p_{04} = \pi_b p_{03} = 0.95(592.6) = 563$ kPa, at burner exit.

$$f = \frac{\frac{T_{04}}{T_{03}} - \frac{C_{p,c}}{C_{p,b,exit}}}{\frac{\eta_b q_R}{C_{p,b,exit} T_{03}} - \frac{T_{04}}{T_{03}}} = \frac{\frac{1500}{573} - \frac{1063}{1151}}{\frac{1.0(45 \times 10^6)}{1151(573)} - \frac{1500}{573}} = 0.0258, \text{ fuel–air ratio (a reason-}$$

able value)

Moving through the turbine section of engine:

$T_{05} \approx T_{04} - \dfrac{1}{\eta_m}(T_{03} - T_{02}) = 1500 - 1/0.99 \cdot (573 - 248) = 1172$ K, at outlet of turbine

$p_{05} = p_{04}\left[1 - \dfrac{1}{\eta_t}\left(1 - \dfrac{T_{05}}{T_{04}}\right)\right]^{\frac{\gamma_t}{\gamma_t - 1}} = 563[1 - 1/0.9 \cdot (1 - 0.781)]^{4.03} = 183$ kPa,   at turbine outlet.

No afterburner, so $T_{06} = T_{05} = 1172$ K, and $p_{06} = p_{05} = 183$ kPa. Now, approaching subsonic converging nozzle, need to check for choking:

$\dfrac{p_{06}}{p_\infty} = \dfrac{183}{18.75} = 9.76$ and choking criterion $\left(\dfrac{\gamma_n + 1}{2}\right)^{\frac{\gamma_n}{\gamma_n - 1}} = 1.87 < \dfrac{p_{06}}{p_\infty}$,   so   nozzle flow is *choked*.

$$C_{p,n} = \dfrac{\gamma_n R_{air}}{\gamma_n - 1} = 1084 \text{ J/(kg K)}.$$

$$V_e = \sqrt{2\eta_n C_{p,n} T_{06}\left[\dfrac{\gamma_n - 1}{\gamma_n + 1}\right]} = \sqrt{2(0.98)1084(1172)[0.36/2.36]} = 616 \text{ m/s}$$

(1062 m/s if unchoked, $p_7 \to p_\infty$, $V_e \to V_{e,\infty} \cdots$ will use later for efficiency estimation)

$T_7 = \dfrac{2}{\gamma_n + 1}T_{06} = 2/2.36 \cdot 1172 = 993$ K, exit nozzle static temperature at sonic flow condition.

$p_7 = p_{06}\left[\dfrac{T_7}{T_{06}}\right]^{\frac{\gamma_n}{\gamma_n - 1}} = p_{06}\left[\frac{2}{\gamma_n + 1}\right]^{\frac{\gamma_n}{\gamma_n - 1}} = 183[2/2.36]^{3.778} = 97.9$ kPa $= p_e > p_\infty$.   For choked flow,

$\dfrac{A_e}{\dot{m}} = \left[\dfrac{\gamma_n + 1}{2}\right]^{\frac{1}{\gamma_n - 1}}\left[\dfrac{\gamma_n + 1}{2\gamma_n R T_{06}}\right]^{0.5}\dfrac{R T_{06}}{p_{06}} = \left[\dfrac{2.36}{2}\right]^{2.778}\left[\dfrac{2.36}{2(1.36)287(1172)}\right]^{0.5}\dfrac{287(1172)}{183000}$

$= 0.00468 \text{ m}^2\text{s/kg}.$

$V_\infty = a_\infty Ma_\infty = \sqrt{\gamma_{air} R_{air} T_\infty} \cdot Ma_\infty = 295(0.85) = 251$ m/s. For specific thrust,

$\dfrac{F}{\dot{m}_a} = (1 + f)V_e - V_\infty + \dfrac{A_e}{\dot{m}}(p_e - p_\infty)(1 + f)$

$= (1 + 0.0258)616 - 251 + 0.00468(97900 - 18750)(1.0258) = 380.9 + 380$

$= 760.9$ N s/kg or 0.761 kN s/kg(if had been unchoked, 0.836 kN s/kg), reasonable for TJ

$$\text{TSFC} = \frac{\dot{m}_f}{F} = \frac{f}{\frac{F}{\dot{m}_a}} = 0.0258/760.9 = 3.39 \times 10^{-5} \text{kg}/(\text{s N}) \text{ or } 0.1220 \text{ kg}/(\text{h N}) \text{ or}$$

1.2 lb/(h lb) ... if fully expanded in ideal case, 1.09 lb/(h lb).

$$\eta_o = \frac{F\, V_\infty}{\dot{m}_f\, q_R} = \frac{1}{\text{TSFC}} \frac{V_\infty}{q_R} = (1/3.39 \times 10^{-5})(251/45 \times 10^6) = 0.165,$$

overall efficiency.

$$\text{Given } V_{e,\infty} = \sqrt{2\eta_n C_{p,n} T_{06}\left[1 - \left(\frac{p_\infty}{p_{06}}\right)^{\frac{\gamma_n-1}{\gamma_n}}\right]} =$$

$$\sqrt{2(0.98)(1084)1172\left[1 - \left(\frac{18.75}{183}\right)^{0.265}\right]} = 1062 \text{ m/s},$$

$$\eta_{\text{th}} = \frac{(1+f)V_{e,\infty}^2/2 - V_\infty^2/2}{f \cdot q_R} = \frac{(1+0.0258)1062^2/2 - 251^2/2}{0.0258 \cdot 45 \times 10^6} = 0.48, \text{ thermal}$$

efficiency, and

$$\eta_p = \frac{\frac{F}{\dot{m}_a}V_\infty}{(1+f)\frac{V_{e,\infty}^2}{2} - \frac{V_\infty^2}{2}} = \frac{760.9(251)}{1.0258(1062^2)/2 - 251^2/2} = 0.349,$$

propulsive efficiency.

6.3. Looking at an afterburning version of the turbojet engine of Problem 6.2, at a flight Mach no. Ma_∞ of 0.85, at 12200 m altitude ($p_\infty = 18.75$ kPa, $T_\infty = 216.7$ K, $\rho_\infty = 0.30$ kg/m^3).

Same flight condition as Problem 6.2, so can skip a few steps (numbers already available in Problem 6.2 solution), given it is the same engine in front of the afterburner section.

At exit of the after burner, $T_{06} = T_{\text{max,A/B}} = 2000$ K, and $p_{06} = \pi_{\text{AB}} \cdot p_{05} = 0.94(193) = 172$ kPa.

Given $\gamma_t = 1.33$ and $\gamma_{\text{AB}} = 1.33$, $C_{p,\text{A/B,exit}} = C_{p,t} = 1.33(287)/0.33 = 1157$ J/kg K, so that:

$$f_{\text{AB}} = \frac{\dfrac{T_{06}}{T_{05}} - \dfrac{C_{p,t}}{C_{p,\text{AB,exit}}}}{\dfrac{\eta_{\text{AB}}q_R}{C_{p,\text{AB,exit}}T_{05}} - \dfrac{T_{06}}{T_{05}}} = \frac{\dfrac{2000}{1172} - \dfrac{1157}{1157}}{\dfrac{0.9(45 \times 10^6)}{1157(1172)} - \dfrac{2000}{1172}} = 0.025 = \frac{\dot{m}_{f,\text{AB}}}{\dot{m}_a + \dot{m}_f}$$

One checks that $f + f_{AB} = 0.0258 + 0.025 = 0.051 < f_{\text{stoich}}$ (0.067 for this air/fuel mixture), so can sustain the afterburner combustion process.

Now, approaching subsonic converging nozzle, need to check for choking:

$$\frac{p_{06}}{p_\infty} = \frac{172}{18.75} = 9.17 \text{ and choking criterion } \left(\frac{\gamma_n+1}{2}\right)^{\frac{\gamma_n}{\gamma_n-1}} = 1.87 < \frac{p_{06}}{p_\infty}, \text{ so nozzle}$$

flow is *choked*.

$$C_{p,n} = \frac{\gamma_n R_{air}}{\gamma_n - 1} = 1084 \text{ J/(kg K)}$$

$$V_e = \sqrt{2\eta_n C_{p,n} T_{06}\left[\frac{\gamma_n - 1}{\gamma_n + 1}\right]} = \sqrt{2(0.98)1084(2000)[0.36/2.36]} = 805.1 \text{ m/s}$$

$T_7 = \dfrac{2}{\gamma_n + 1} T_{06} = 2/2.36 \cdot 2000 = 1695$ K, exit nozzle static temperature at sonic flow condition.

$$p_7 = p_{06}\left[\frac{T_7}{T_{06}}\right]^{\frac{\gamma_n}{\gamma_n - 1}} = p_{06}\left[\frac{2}{\gamma_n + 1}\right]^{\frac{\gamma_n}{\gamma_n - 1}} = 172[2/2.36]^{3.778} = 92 \text{ } kPa = p_e > p_\infty. \qquad \text{For}$$

choked flow,

$$\frac{A_e}{\dot{m}} = \left[\frac{\gamma_n + 1}{2}\right]^{\frac{1}{\gamma_n - 1}}\left[\frac{\gamma_n + 1}{2\gamma_n R T_{06}}\right]^{0.5}\frac{R T_{06}}{p_{06}} = \left[\frac{2.36}{2}\right]^{2.778}\left[\frac{2.36}{2(1.36)287(2000)}\right]^{0.5}\frac{287(2000)}{172000}$$
$$= 0.0065 \text{m}^2\text{s/kg}.$$

For specific thrust,

$$\frac{F}{\dot{m}_a} = (1 + f + f_{AB})V_e - V_\infty + \frac{A_e}{\dot{m}}(p_e - p_\infty)(1 + f + f_{AB})$$
$$= (1 + 0.0258 + 0.025)805.1 - 251 + 0.0065(92000 - 18750)(1.051)$$
$$= 1095 \text{ N s/kg or } 1.095 \text{ kN s/kg, reasonable increase for TJ in A/B operation,}$$
$$\text{about } 45\% \text{ higher versus the non-afterburning TJ of Problem 6.2.}$$

$$\text{TSFC} = \frac{\dot{m}_f + \dot{m}_{f,AB}}{F} \approx \frac{\frac{f + f_{AB}}{\frac{F}{\dot{m}_a}}}{} = (0.0258 + 0.025)/1095 = 4.64 \times 10^{-5}\text{kg/(s N)} \quad \text{or}$$

0.167 kg/(h N) or 1.64 lb/(h lb) (reasonable, given higher fuel flow to feed a relatively inefficient A/B), vs. 1.2 lb/(h lb) for the non-A/B case.

Note: $T_{06}/T_{05} = 2000/1172 = 1.7$, and must be sure in the practical engine case, the flow Mach number at the A/B inlet, Ma_5, is low enough so that after the heat input of the A/B combustion process, Ma_6 at the A/B outlet is still sufficiently below 1 to avoid excessive stagnation pressure losses (from gasdynamics, recall that heating and friction will act to increase Ma along the A/B duct).

6.4. Looking at the non-afterburning turbojet engine from Problem 6.2, at a flight Mach no. Ma_∞ of 0.85, at 12,200 m altitude ($p_\infty = 18.75$ kPa, $T_\infty = 216.7$ K, $\rho_\infty = 0.30$ kg/m^3), running at a cooler combustor temperature (T_{max} of 1300 K, vs. 1500 K).

Same flight condition as Problem 6.2, so can skip a few steps (numbers already available in Problem 6.2 solution), given it is the same engine in front of the combustor section.

Moving through combustor (burner):
$$\bar{C}_{p,b} \approx \frac{C_{p,c} + C_{p,b,\text{exit}}}{2} = \frac{\gamma_b R_{\text{air}}}{\gamma_b - 1} = 1.35(287)/(1.35 - 1) = 1107 \text{ J}/(\text{kg K}), \text{ so that}$$
at burner exit:

$$C_{p,b,\text{exit}} \approx 2\bar{C}_{p,b} - C_{p,c} = 2(1107) - 1063 = 1151 \text{ J}/(\text{kg K}).$$

$T_{04} = T_{\max} = 1300 \text{ K}, \quad p_{04} = \pi_b p_{03} = 0.95(592.6) = 563 \text{ kPa, at burner exit.}$

$$f = \frac{\dfrac{T_{04}}{T_{03}} - \dfrac{C_{p,c}}{C_{p,b,\text{exit}}}}{\dfrac{\eta_b q_R}{C_{p,b,\text{exit}} T_{03}} - \dfrac{T_{04}}{T_{03}}} = \frac{\dfrac{1300}{573} - \dfrac{1063}{1151}}{\dfrac{1.0(45 \times 10^6)}{1151(573)} - \dfrac{1300}{573}} = 0.0204, \text{ fuel–air ratio (a reason-}$$

able value)

Moving through the turbine section of engine:
$$T_{05} \approx T_{04} - \frac{1}{\eta_m}(T_{03} - T_{02}) = 1300 - 1/0.99 \cdot (573 - 248) = 972 \text{ K, at outlet of}$$
turbine

$$p_{05} = p_{04}[1 - \tfrac{1}{\eta_t}(1 - \tfrac{T_{05}}{T_{04}})]^{\frac{\gamma_t}{\gamma_t - 1}} = 563[1 - 1/0.9 \cdot (1 - 972/1300)]^{4.03} = 150 \text{ kPa} \text{ at}$$
turbine outlet

No afterburner, so $T_{06} = T_{05} = 972$ K, and $p_{06} = p_{05} = 150$ kPa. Now, approaching subsonic converging nozzle, need to check for choking:

$$\frac{p_{06}}{p_\infty} = \frac{150}{18.75} = 8.0 \text{ and choking criterion } \left(\frac{\gamma_n + 1}{2}\right)^{\frac{\gamma_n}{\gamma_n - 1}} = 1.87 < \frac{p_{06}}{p_\infty}, \text{ so nozzle flow}$$
is *choked*.

$$C_{p,n} = \frac{\gamma_n R_{\text{air}}}{\gamma_n - 1} = 1084 \text{ J}/(\text{kg K})$$

$$V_e = \sqrt{2\eta_n C_{p,n} T_{06}\left[\frac{\gamma_n - 1}{\gamma_n + 1}\right]} = \sqrt{2(0.98)1084(972)[0.36/2.36]} = 561 \text{ m/s}$$

$$T_7 = \frac{2}{\gamma_n + 1} T_{06} = 2/2.36 \cdot 972 = 824 \text{ K, exit nozzle static temperature at sonic}$$
flow condition.

$$p_7 = p_{06}\left[\frac{T_7}{T_{06}}\right]^{\frac{\gamma_n}{\gamma_n - 1}} = p_{06}\left[\frac{2}{\gamma_n + 1}\right]^{\frac{\gamma_n}{\gamma_n - 1}} = 150[2/2.36]^{3.778} = 80.3 \text{ kPa} = p_e > p_\infty. \text{ For choked}$$
flow,

$$\frac{A_e}{\dot{m}} = \left[\frac{\gamma_n + 1}{2}\right]^{\frac{1}{\gamma_n - 1}}\left[\frac{\gamma_n + 1}{2\gamma_n RT_{06}}\right]^{0.5}\frac{RT_{06}}{p_{06}} = \left[\frac{2.36}{2}\right]^{2.778}\left[\frac{2.36}{2(1.36)287(972)}\right]^{0.5}\frac{287(972)}{150000}$$
$$= 0.0052 \text{m}^2\text{s/kg}.$$

For specific thrust,

$$\frac{F}{\dot{m}_a} = (1 + f)V_e - V_\infty + \frac{A_e}{\dot{m}}(p_e - p_\infty)(1 + f)$$
$$= (1 + 0.0204)561 - 251 + 0.0052(80300 - 18750)(1.0204) = 321.5 + 326.6$$
$$= 648 \text{ N s/kg or } 0.648 \text{ kN s/kg} < 0.761 \text{ of Problem 6.2}$$

$$\text{TSFC} = \frac{\dot{m}_f}{F} = \frac{f}{\frac{F}{\dot{m}_a}} = 0.0204/648 = 3.15 \times 10^{-5} \text{kg/(s N) or } 0.1133 \text{ kg/(h N) or}$$

1.112 lb/(h lb) < 1.2 lb/(h lb) of Prob. 6.2, so better here, but at a
lower relative thrust delivery as the price, for a given air
flow rate.

make note.

6.5. Looking at a non-afterburning turbojet engine, at a flight Mach no. Ma_∞ of
0.85, at 12200 m altitude ($p_\infty = 18.75$ kPa, $T_\infty = 216.7$ K, $\rho_\infty = 0.30$ kg/m³).

Starting from intake (diffuser):

$$T_{02} = T_\infty\left[1 + \frac{\gamma_d - 1}{2}Ma_\infty^2\right] = 216.7[1 + 0.2(0.85^2)] = 248 \text{ K, at diffuser outlet}$$

$$p_{02} = p_\infty\left[1 + \eta_d\left(\frac{T_{02}}{T_\infty} - 1\right)\right]^{\frac{\gamma_d}{\gamma_d - 1}} = 18.75[1 + 0.97(1.144 - 1)]^{3.5} = 29.63 \text{ kPa, at}$$

diffuser outlet.

Moving through compressor:

$$T_{03} = T_{02}\left[1 + \frac{1}{\eta_c}\left[\pi_c^{\frac{\gamma_c - 1}{\gamma_c}} - 1\right]\right] = 248[1 + 1/0.95 \cdot [25^{0.27} - 1]] = 609.5 \text{ K},$$

at compressor outlet

$$p_{03} = \pi_c p_{02} = 25(29.63) = 741 \text{ kPa, at compressor outlet; } C_{p,c} = \frac{\gamma_c R_{air}}{\gamma_c - 1}$$
$$= 1.37(287)/0.37 = 1063 \text{ J/(kg K)}.$$

Moving through combustor (burner):

$$\bar{C}_{p,b} \approx \frac{C_{p,c} + C_{p,b,exit}}{2} = \frac{\gamma_b R_{air}}{\gamma_b - 1} = 1.35(287)/(1.35 - 1) = 1107 \text{ J/(kg K), so that}$$

at burner exit:

$$C_{p,b,exit} \approx 2\bar{C}_{p,b} - C_{p,c} = 2(1107) - 1063 = 1151 \text{ J/(kg K)}.$$

$$T_{04} = T_{max} = 1500 \text{ K}, \quad p_{04} = \pi_b p_{03} = 0.95(741) = 704 \text{ kPa, at burner exit.}$$

$$f = \frac{\dfrac{T_{04}}{T_{03}} - \dfrac{C_{p,c}}{C_{p,b,\text{exit}}}}{\dfrac{\eta_b q_R}{C_{p,b,\text{exit}} T_{03}} - \dfrac{T_{04}}{T_{03}}} = \frac{\dfrac{1500}{609.5} - \dfrac{1063}{1151}}{\dfrac{1.0(45 \times 10^6)}{1151(609.5)} - \dfrac{1500}{609.5}} = 0.0249, \text{ fuel–air ratio}$$

Moving through the turbine section of engine:

$$T_{05} \approx T_{04} - \frac{1}{\eta_m}(T_{03} - T_{02}) = 1500 - 1/0.99 \cdot (609.5 - 248) = 1135 \text{ K}, \qquad \text{at}$$

outlet of turbine

$$p_{05} = p_{04}\left[1 - \frac{1}{\eta_t}\left(1 - \frac{T_{05}}{T_{04}}\right)\right]^{\frac{\gamma_t}{\gamma_t-1}} = 704[1 - 1/0.9 \cdot (1 - 0.757)]^{4.03} = 198 \text{ kPa}, \qquad \text{at}$$

turbine outlet

No afterburner, so $T_{06} = T_{05} = 1135$ K, and $p_{06} = p_{05} = 198$ kPa. Now, approaching subsonic converging nozzle, need to check for choking:

$\dfrac{p_{06}}{p_\infty} = \dfrac{198}{18.75} = 10.56$ and choking criterion $\left(\dfrac{\gamma_n + 1}{2}\right)^{\frac{\gamma_n}{\gamma_n-1}} = 1.87 < \frac{p_{06}}{p_\infty}$, so nozzle

flow is *choked*.

$$C_{p,n} = \frac{\gamma_n R_{\text{air}}}{\gamma_n - 1} = 1084 \text{ J/(kg K)}.$$

$$V_e = \sqrt{2\eta_n C_{p,n} T_{06}\left[\frac{\gamma_n-1}{\gamma_n+1}\right]} = \sqrt{2(0.98)1084(1135)[0.36/2.36]} = 606.5 \text{ m/s},$$

choked value.

If fully expanded (ideal case):

$$V_{e,\infty} = \sqrt{2\eta_n C_{p,n} T_{06}\left[1 - \left(\frac{p_\infty}{p_{06}}\right)^{\frac{\gamma_n-1}{\gamma_n}}\right]}$$

$$= \sqrt{2(0.98)1084(1135)\left[1 - \left(\frac{18.75}{198}\right)^{0.265}\right]} = 1058 \text{ m/s}$$

$$T_7 = \frac{2}{\gamma_n + 1}T_{06} = 2/2.36 \cdot 1135 = 962 \text{ K}, \text{ exit nozzle static temperature at sonic}$$

flow condition.

$$p_7 = p_{06}\left[\frac{T_7}{T_{06}}\right]^{\frac{\gamma_n}{\gamma_n-1}} = p_{06}\left[\frac{2}{\gamma_n+1}\right]^{\frac{\gamma_n}{\gamma_n-1}} = 198[.2/2.36]^{3.778} = 106 \text{ kPa} = p_e > p_\infty. \quad \text{For choked}$$

flow,

$$\frac{A_e}{\dot{m}} = \left[\frac{\gamma_n + 1}{2}\right]^{\frac{1}{\gamma_n-1}}\left[\frac{\gamma_n + 1}{2\gamma_n R T_{06}}\right]^{0.5}\frac{R T_{06}}{p_{06}}$$

$$= \left[\frac{2.36}{2}\right]^{2.778}\left[\frac{2.36}{2(1.36)287(1135)}\right]^{0.5}\frac{287(1135)}{198000} = 0.00425 \text{ m}^2\text{s/kg}.$$

$V_\infty = a_\infty Ma_\infty = \sqrt{\gamma_{air} R_{air} T_\infty} \cdot Ma_\infty = 295(0.85) = 251$ m/s. For specific thrust,

$$\frac{F}{\dot{m}_a} = (1+f)V_e - V_\infty + \frac{A_e}{\dot{m}}(p_e - p_\infty)(1+f)$$
$$= (1+0.0249)606.5 - 251 + 0.00425(106000 - 18750)(1.0249) = 370.6 + 380$$
$$= 750.6 \, \text{N s/kg or } 0.751 \, \text{kN s/kg} < 0.761 \text{ of Problem 6.2, but close in value.}$$

Ideal fully expanded case, $\frac{F}{\dot{m}_a} = (1+f)V_{e,\infty} - V_\infty = 1.0249(1058) - 251 = $ 833.4 kN s/kg or 0.833 kN s/kg $<$ ideal 0.836 of Problem 6.2, although very close in value.

$$\text{TSFC} = \frac{\dot{m}_f}{F} = \frac{f}{\frac{F}{\dot{m}_a}} = 0.0249/750.6 = 3.32 \times 10^{-5} \text{kg}/(\text{s N})$$

or 0.120 kg/(h N) or 1.17 lb/(h lb) < 1.2lb/(h lb) of Problem 6.2.

Ideal fully expanded case, TSFC = 1.055 lb/(h lb), vs. ideal 1.09 for Problem 6.2, a noticeable improvement.

made note

6.6. Looking at a non-afterburning turbojet engine, at a flight Mach no. Ma_∞ of 0.85, at 12200 m altitude ($p_\infty = 18.75$ kPa, $T_\infty = 216.7$ K, $\rho_\infty = 0.30$ kg/m^3).

Starting from intake (diffuser):

$$T_{02} = T_\infty \left[1 + \frac{\gamma_d - 1}{2} Ma_\infty^2\right] = 216.7[1 + 0.2(0.85^2)] = 248 \text{ K, at diffuser outlet}$$

$$p_{02} = p_\infty \left[1 + \eta_d \left(\frac{T_{02}}{T_\infty} - 1\right)\right]^{\frac{\gamma_d}{\gamma_d - 1}} = 18.75[1 + 0.97(1.144 - 1)]^{3.5} = 29.63 \text{ kPa}, \quad \text{at}$$

diffuser outlet

Moving through compressor:

$$T_{03} = T_{02} \left[1 + \frac{1}{\eta_c}\left[\pi_c^{\frac{\gamma_c - 1}{\gamma_c}} - 1\right]\right] = 248[1 + 1/0.95 \cdot [15^{0.27} - 1]] = 529.3 \text{ K}, \quad \text{at}$$

compressor outlet

$p_{03} = \pi_c p_{02} = 15(29.63) = 444.5$ kPa, at compressor outlet; $C_{p,c} = \frac{\gamma_c R_{air}}{\gamma_c - 1} =$ 1.37(287)/0.37 = 1063 J/(kg K).

Moving through combustor (burner):

$$\bar{C}_{p,b} \approx \frac{C_{p,c} + C_{p,b,exit}}{2} = \frac{\gamma_b R_{air}}{\gamma_b - 1} = 1.35(287)/(1.35 - 1) = 1107 \text{ J/(kg K), so that}$$

at burner exit:

$$C_{p,b,exit} \approx 2\bar{C}_{p,b} - C_{p,c} = 2(1107) - 1063 = 1151 \text{ J/(kg K)}.$$

$T_{04} = T_{max} = 1500$ K, $\quad p_{04} = \pi_b p_{03} = 0.95(444.5) = 422.3$ kPa, at burner exit.

$$f = \frac{\dfrac{T_{04}}{T_{03}} - \dfrac{C_{p,c}}{C_{p,b,\text{exit}}}}{\dfrac{\eta_b q_R}{C_{p,b,\text{exit}} T_{03}} - \dfrac{T_{04}}{T_{03}}} = \frac{\dfrac{1500}{529.3} - \dfrac{1063}{1151}}{\dfrac{1.0(45 \times 10^6)}{1151(529.3)} - \dfrac{1500}{529.3}} = 0.0269, \text{ fuel–air ratio}$$

Moving through the turbine section of engine:

$$T_{05} \approx T_{04} - \frac{1}{\eta_m}(T_{03} - T_{02}) = 1500 - 1/0.99 \cdot (529.3 - 248) = 1216 \text{ K, at outlet}$$

of turbine

$$p_{05} = p_{04}\left[1 - \frac{1}{\eta_t}\left(1 - \frac{T_{05}}{T_{04}}\right)\right]^{\frac{\gamma_t}{\gamma_t - 1}} = 422.3[1 - 1/0.9 \cdot (1 - 0.811)]^{4.03} = 163.3 \text{ kPa,}$$

at turbine outlet

No afterburner, so $T_{06} = T_{05} = 1216$ K, and $p_{06} = p_{05} = 163.3$ kPa. Now, approaching subsonic converging nozzle, need to check for choking:

$$\frac{p_{06}}{p_\infty} = \frac{163.3}{18.75} = 8.71 \text{ and choking criterion } \left(\frac{\gamma_n + 1}{2}\right)^{\frac{\gamma_n}{\gamma_n - 1}} = 1.87 < \frac{p_{06}}{p_\infty}, \text{ so nozzle}$$

flow is *choked*.

$$C_{p,n} = \frac{\gamma_n R_{\text{air}}}{\gamma_n - 1} = 1084 \text{ J/(kg K)}.$$

$$V_e = \sqrt{2\eta_n C_{p,n} T_{06}\left[\tfrac{\gamma_n - 1}{\gamma_n + 1}\right]} = \sqrt{2(0.98)1084(1216)[0.36/2.36]} = 627.8 \text{ m/s,}$$

choked value

If fully expanded (ideal case):

$$V_{e,\infty} = \sqrt{2\eta_n C_{p,n} T_{06}\left[1 - \left(\frac{p_\infty}{p_{06}}\right)^{\frac{\gamma_n - 1}{\gamma_n}}\right]}$$

$$= \sqrt{2(0.98)1084(1216)\left[1 - \left(\frac{18.75}{163.3}\right)^{0.265}\right]} = 1062 \text{ m/s}$$

$$T_7 = \frac{2}{\gamma_n + 1}T_{06} = 2/2.36 \cdot 1216 = 1031 \text{ K, exit nozzle static temperature at sonic}$$

flow condition.

$$p_7 = p_{06}\left[\frac{T_7}{T_{06}}\right]^{\frac{\gamma_n}{\gamma_n - 1}} = p_{06}\left[\frac{2}{\gamma_n + 1}\right]^{\frac{\gamma_n}{\gamma_n - 1}} = 163.3[2/2.36]^{3.778} =$$

87.4 kPa $= p_e > \; = 163.3[2/2.36]^{3.778} = 87.4$ kPa $= p_e > p_{4\infty}$. For choked flow,

$$\frac{A_e}{\dot{m}} = \left[\frac{\gamma_n + 1}{2}\right]^{\frac{1}{\gamma_n - 1}} \left[\frac{\gamma_n + 1}{2\gamma_n RT_{06}}\right]^{0.5} \frac{RT_{06}}{p_{06}}$$

$$= \left[\frac{2.36}{2}\right]^{2.778} \left[\frac{2.36}{2(1.36)287(1216)}\right]^{0.5} \frac{287(1216)}{163300} = 0.00534 \text{ m}^2\text{s/kg}.$$

$V_\infty = a_\infty Ma_\infty = \sqrt{\gamma_{air} R_{air} T_\infty} \cdot Ma_\infty = 295(0.85) = 251$ m/s. For specific thrust,

$$\frac{F}{\dot{m}_a} = (1+f)V_e - V_\infty + \frac{A_e}{\dot{m}}(p_e - p_\infty)(1+f)$$

$$= (1+0.0269)627.8 - 251 + 0.00534(87400 - 18750)(1.0269) = 393.7 + 376.5$$

$$= 770.2 \text{ Ns/kg or } 0.77 \text{ kNs/kg} > 0.761 \text{ of Problem } 6.2, \text{ but close in value.}$$

Ideal fully expanded case, $\frac{F}{\dot{m}_a} = (1+f)V_{e,\infty} - V_\infty = 1.0269(1062) - 251 = $ 839.6 N s/kg or 0.84 kN s/kg $>$ ideal 0.836 of Problem 6.2, although very close in value.

$$\text{TSFC} = \frac{\dot{m}_f}{F} = \frac{f}{\frac{F}{\dot{m}_a}} = 0.0269/770.2$$

$$= 3.493 \times 10^{-5} \text{ kg/(s N)} \quad \text{or} \quad 0.126 \text{ kg/(h N)} \quad \text{or} \quad 1.234 \text{ lb/(h lb)}$$

$$> 1.2 \text{ lb/(h lb) of Problem } 6.2.$$

Ideal fully expanded case, TSFC = 1.131 lb/(h lb), vs. ideal 1.09 for Problem 6.2, a noticeable decrease in efficiency with the lower compression ratio.

6.7. Oswatitsch multiple-shock inlet for efficient supersonic flight of the F-22.

6.8. An unmanned drone is presently flying at 6100 m altitude (ISA), at a flight Mach number of 0.7. The UAV employs a Teledyne-Continental J402-CA-702 turbojet engine. Assuming the engine performance charts of Figs. 6.8 and 6.9 apply (max. continuous throttle setting), establish the current fuel consumption rate (kg/h) by the drone.
Referring to Fig. 6.8, the thrust being delivered by the turbojet at the specified flight condition is around 2.5 kN (560 lbf). The corresponding TSFC for that flight condition, from Fig. 6.9, is 0.14 kg/(h N).
The resulting fuel consumption rate: $\dot{m}_f = F \cdot \text{TSFC} = 2500 \cdot 0.14 = 350$ kg/hr.

6.9. Perform an off-design cycle analysis for a turbojet engine at its cruise throttle setting, at altitudes from 0 to 13,200 m ISA, at flight Mach numbers from zero to 0.9. Take note of the following design parameters:

diffuser, $\eta_d = 0.97$; $\gamma_d = 1.4$; compressor, $\eta_c = 0.93$, $\gamma_c = 1.37$, $\pi_c = 10.2$; burner, $\eta_b = 0.98$, $\gamma_b = 1.35$, $\pi_b = 0.95$, $T_{max} = 1100$ K

$q_R = 43$ MJ/kg of fuel; turbine, $\gamma_t = 1.33$; $\eta_t = 0.90$; $\eta_m = 0.98$
core nozzle, $\eta_n = 0.96$; $\gamma_n = 1.36$; nozzle fixed exit diameter $d_e = 0.55$ m

Plot net thrust F versus flight Mach number Ma_∞ at the four principal altitudes of 0, 4400, 8800 and 13,200 m ISA (generate enough data points so that each curve appears smooth in appearance). Ascertain the TSFC data points when values of 0.09, 0.10 and 0.11 kg/h N are attained.

As a check, sample set of calculations, $Ma_\infty = 0.7$ at $h_{ASL} = 8800$ m:
$p_\infty = 32$ kPa, $T_\infty = 231$ K, $\rho_\infty = 0.48$ kg/m^3, $a_\infty = 305$ m/s
$A_e = \pi d_e^2/4 = 0.238$ m^2

Starting from intake (diffuser):

$$T_{02} = T_\infty \left[1 + \frac{\gamma_d - 1}{2} Ma_\infty^2\right] = 231[1 + 0.2(0.7^2)] = 253.6 \text{ K, at diffuser outlet.}$$

$$p_{02} = p_\infty[1 + \eta_d(\tfrac{T_{02}}{T_\infty} - 1)]^{\frac{\gamma_d}{\gamma_d - 1}} = 32[1 + 0.97(1.098 - 1)]^{3.5} = 44 \text{ kPa, at   diffuser}$$
outlet

Moving through compressor:

$$T_{03} = T_{02}\left[1 + \frac{1}{\eta_c}\left[\pi_c^{\frac{\gamma_c - 1}{\gamma_c}} - 1\right]\right] = 253.6[1 + 1/0.93 \cdot [10.2^{0.27} - 1]] = 491.4 \text{ K, \quad at}$$
compressor outlet

$$p_{03} = \pi_c p_{02} = 10.2(44) = 449 \text{ kPa, \quad at \quad compressor \quad outlet;} \quad C_{p,c} = \frac{\gamma_c R_{air}}{\gamma_c - 1} =$$
$1.37(287)/0.37 = 1063$ J/(kg K).

Moving through combustor (burner):

$$\bar{C}_{p,b} \approx \frac{C_{p,c} + C_{p,b,exit}}{2} = \frac{\gamma_b R_{air}}{\gamma_b - 1} = 1.35(287)/(1.35 - 1) = 1107 \text{ J/(kg K), so that}$$
at burner exit:

$$C_{p,b,exit} \approx 2\bar{C}_{p,b} - C_{p,c} = 2(1107) - 1063 = 1151 \text{ J/(kg K).}$$

$T_{04} = T_{max} = 1100$ K, $p_{04} = \pi_b p_{03} = 0.95(449) = 426.6$ kPa, at burner exit.

$$f = \frac{\dfrac{T_{04}}{T_{03}} - \dfrac{C_{p,c}}{C_{p,b,exit}}}{\dfrac{\eta_b q_R}{C_{p,b,exit} T_{03}} - \dfrac{T_{04}}{T_{03}}} = \frac{\dfrac{1100}{491.4} - \dfrac{1063}{1151}}{\dfrac{0.98(43 \times 10^6)}{1151(491.4)} - \dfrac{1100}{491.4}} = 0.0182, \text{ fuel–air ratio}$$

Moving through the turbine section of engine:

$$T_{05} \approx T_{04} - \frac{1}{\eta_m}(T_{03} - T_{02}) = 1100 - 1/0.98 \cdot (491.4 - 253.6) = 857.4 \text{ K, \quad at}$$
outlet of turbine

$$p_{05} = p_{04}\left[1 - \frac{1}{\eta_t}\left(1 - \frac{T_{05}}{T_{04}}\right)\right]^{\frac{\gamma_t}{\gamma_t - 1}} = 426.6[1 - 1/0.9 \cdot (1 - 0.78)]^{4.03} = 137.9 \text{ kPa, at tur-}$$
bine outlet

No afterburner, so $T_{06} = T_{05} = 857.4$ K, and $p_{06} = p_{05} = 137.9$ kPa. Now, approaching subsonic converging nozzle, need to check for choking:

$\frac{p_{06}}{p_\infty} = \frac{137.9}{32} = 4.31$ and choking criterion $\left(\frac{\gamma_n+1}{2}\right)^{\frac{\gamma_n}{\gamma_n-1}} = 1.87 < \frac{p_{06}}{p_\infty}$, so nozzle flow is *choked.*

$$C_{p,n} = \frac{\gamma_n R_{air}}{\gamma_n - 1} = 1084 \text{ J}/(\text{kg K}).$$

$$V_e = \sqrt{2\eta_n C_{p,n} T_{06} \left[\frac{\gamma_n - 1}{\gamma_n + 1}\right]} = \sqrt{2(0.98)1084(857.4)[0.36/2.36]} = 527.2 \text{ m/s}$$

$T_7 = \frac{2}{\gamma_n + 1} T_{06} = 2/2.36 \cdot 857.4 = 726.6$ K, exit nozzle static temperature at sonic flow condition.

$p_7 = p_{06} \left[\frac{T_7}{T_{06}}\right]^{\frac{\gamma_n}{\gamma_n-1}} = p_{06} \left[\frac{2}{\gamma_n+1}\right]^{\frac{\gamma_n}{\gamma_n-1}} = 137.9[2/2.36]^{3.778} = 73.8$ kPa $= p_e > p_\infty$. For choked flow,

$$\frac{A_e}{\dot{m}} = \left[\frac{\gamma_n + 1}{2}\right]^{\frac{1}{\gamma_n-1}} \left[\frac{\gamma_n + 1}{2\gamma_n R T_{06}}\right]^{0.5} \frac{R T_{06}}{p_{06}} = \left[\frac{2.36}{2}\right]^{2.778} \left[\frac{2.36}{2(1.36)287(857.4)}\right]^{0.5}$$
$$= 0.00531 \text{ m}^2\text{s/kg}.$$

$V_\infty = a_\infty Ma_\infty = \sqrt{\gamma_{air} R_{air} T_\infty} \cdot Ma_\infty = 305(0.7) = 214$ m/s. For specific thrust,

$$\frac{F}{\dot{m}_a} = (1+f)V_e - V_\infty + \frac{A_e}{\dot{m}}(p_e - p_\infty)(1+f)$$
$$= (1 + 0.0182)527.2 - 214 + 0.00531(73800 - 32000)(1.0182)$$
$$= 322.8 + 226 = 548.8 \text{ N s/kg or } 0.549 \text{ kN s/kg}$$

$$\text{TSFC} = \frac{\dot{m}_f}{F} = \frac{f}{\frac{F}{\dot{m}_a}} = 0.0182/548.8$$
$$= 3.32 \times 10^{-5} \text{kg}/(\text{s N}) \text{ or } 0.1194 \text{ kg}/(\text{h N}) \text{ or } 1.17 \text{ lb}/(\text{h lb})$$

$\rho_7 = p_7/(R T_7) = 73800/(287 \cdot 726.6) = 0.354 \text{ kg/m}^3 = \rho_e, = \rho_e V_e A_e$
$= 0.354(527.2)0.238 = 44.4$ kg/s.

$$\dot{m}_a = \dot{m}_e/(1 + f) = 44.4/(1.0182) = 43.6 \text{ kg/s (96.1 lbm/s)}$$

$$F = \dot{m}_a \cdot \frac{F}{\dot{m}_a} = 43.6 \cdot 548.8 = 23928 \text{ N (5380 lbf)}$$

References

1. Mattingly JD (1996) Elements of gas turbine propulsion. McGraw-Hill, New York
2. Van Wylen GJ, Sonntag RE (1973) Fundamentals of classical thermodynamics, 2nd edn. Wiley, New York
3. Treager IE (1970) Aircraft gas turbine engine technology. McGraw-Hill, New York
4. Kenny DP (1979) A novel correlation of centrifugal compressor performance for off-design prediction. In: Proceedings of fifteenth AIAA/SAE/ASME joint propulsion conference, Las Vegas, June 18–20
5. Hill PG, Peterson CR (1992) Mechanics and thermodynamics of propulsion, 2nd edn. Addison-Wesley, New York
6. Sampath R, Irani R, Balasubramaniam M et al (2004) High fidelity system simulation of aerospace vehicles using NPSS. In: Proceedings of 42nd AIAA aerospace sciences meeting, Reno, January 5–8
7. Dunker RJ, Hungenberg HG (1980) Transonic axial compressor using laser anemometry and unsteady pressure measurements. AIAA J 18:973–979
8. Anonymous (2004) Airplane flying handbook. FAA-H-8083-3A, Airman Testing Standards Branch, Federal Aviation Administration, U.S. Department of Transportation, Oklahoma City
9. Denno RR, Smith H, Hammer R (1987) AIAA aerospace design engineers guide (revised and enlarged). AIAA, New York
10. Mattingly JD, Heiser WH, Daley DH (1987) Aircraft engine design. AIAA, Reston (Virginia)
11. Henderson RE, Blazowski WS (1989) Turbopropulsion combustion technology. In: Oates GC (ed) Aircraft propulsion systems technology and design. AIAA, Washington
12. Henderson RE, Blazowski WS (1978) Turbopropulsion combustion technology. In: Oates GC (ed.) The aerothermodynamics of aircraft gas turbine engines. AFAPL-TR-78-52, USAF Aero Propulsion Laboratory
13. Archer RD, Saarlas M (1996) An introduction to aerospace propulsion. Prentice-Hall, Upper Saddle River
14. Hu TCJ, Sze RML, Sampath P (1998) Design and development of advanced combustion system for PW150 turboprop engine. In: Proceedings of 45th CASI annual conference, Calgary, May 12
15. Blazowski WS (1978) Fundamentals of combustion. In: Oates GC (ed.) The aerothermodynamics of aircraft gas turbine engines. AFAPL-TR-78-52, USAF Aero Propulsion Laboratory
16. Takeishi K, Tsuyoshi K, Matsuura M, Shimizu K (2003) Heat transfer characteristic of a triangular channel with turbulence promoter. In: Proceedings of international gas turbine congress, Tokyo, November 2–7
17. Farokhi S (2009) Aircraft propulsion. Wiley, New York
18. Stabe RG, Whitney WJ, Moffitt TP (1984) Performance of a high-work low aspect ratio turbine tested with a realistic inlet radial temperature profile. NASA TM 83655
19. McCormick BW (1995) Aerodynamics, aeronautics and flight mechanics, 2nd edn. Wiley, New York
20. Useller JW (1959) Effect of combustor length on afterburner combustion. Combust Flame 3:339–346
21. Zukoski EE (1985) Afterburners. In: Oates GC (ed) Aerothermodynamics of aircraft engine components. AIAA, Washington
22. McAulay JE, Abdelwahab M (1972) Experimental evaluation of a TF30-P3 turbofan engine in an altitude facility: afterburner performance and engine-afterburner operating limits. NASA TN D-6839
23. Zukoski EE (1978) Afterburners. In: Oates GC (ed.) The aerothermodynamics of aircraft gas turbine engines. AFAPL-TR-78-52, USAF Aero Propulsion Laboratory

24. John JEA (1984) Gas dynamics, 2nd edn. Prentice-Hall, Upper Saddle River
25. Vickers J (1994) Propulsion analysis of the F-16 multi-axis thrust vectoring aircraft. NASA CP-10143
26. Deskin WJ, Yankel JJ (2002) Development of the F-22 propulsion system. In: Proceedings of 38th AIAA/ASME/SAE/ASEE joint propulsion conference, Indianapolis, July 7–10

Chapter 7
Turbofan Engines

7.1 Introduction

The turbofan engine (a.k.a., fanjet; [1]) is a compromise between the propeller (high $\dot{m}_a$, relatively low V_e) and the turbojet (relatively low $\dot{m}_a$, high V_e), for peak efficiency over certain higher subsonic and lower supersonic flight conditions. The primary modification in going from a TJ is the introduction of a ducted or shrouded, many-bladed subsonic/transonic fan, which would typically be included as a front stage on the low-pressure compressor (see Figs. 7.1, 7.2, 7.3, 7.4, 7.5, 7.6; [2, 3]), or less commonly, as a stage on the LP turbine. A given fan blade can be solid or hollow, and commonly comprised of steel or titanium metal alloys, or more recently, of lighter composite construction (carbon fiber/epoxy matrix) with metal leading edges. Earlier, slimmer (smaller chord) fan blades that were more flexible would commonly employ a snubber (a.k.a. clapper), a mid-span shroud, to help instill some structural stiffness. More modern fan blades may be wide chord, and as a result, reducing the number of blades required, and also not requiring a snubber. An important parameter associated with TF performance is the bypass ratio $\mathcal{B}$ (alternative symbols/acronyms used in the literature include BPR, BR, α and β):

$$\mathcal{B} = \frac{\text{by pass mass flow (through fan duct)}}{\text{core mass flow (through gas generator)}} = \frac{\dot{m}_{by}}{\dot{m}_a} \qquad (7.1)$$

The value for this parameter can range from as low as close to zero (approaching a TJ in performance and characteristic behaviour) for higher supersonic flight applications, to around one for transonic and low supersonic flight, to as high as nine for high subsonic (low transonic) flight (e.g., the General Electric GE90 of Fig. 7.2 used by the Boeing 777 civil airliner has a bypass ratio of 8.4:1). The bypass airflow beyond the fan can exit as a separate exhaust stream out of a short (see Fig. 7.1) or long fan duct (fan duct exit at or a little forward of the core nozzle exit), a common choice for higher bypass engines. One often sees the use of outlet guide vanes (OGVs) before or at the fan duct exit to straighten the

D. R. Greatrix, *Powered Flight*, DOI: 10.1007/978-1-4471-2485-6_7,
© Springer-Verlag London Limited 2012

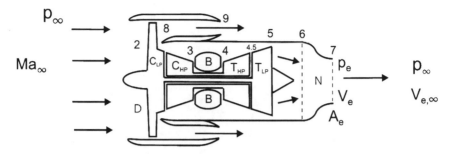

Fig. 7.1 Schematic diagram illustrating the various turbofan engine components and associated station numbering, for a forward fan position on a short-ducted, non-mixing (separate-stream) TF engine. Note that it is common to observe in turbofan engines that the central core engine tail cone extends substantially beyond the nominal exit plane of the core nozzle (unlike the above diagram), to provide the appropriate flow cross-sectional area for the hot core gas as it meets up with the surrounding airflow produced from the fan duct upstream (see Fig. 7.21)

Fig. 7.2 Photo of forward fan of General Electric GE90 turbofan engine, installed in the wing-mounted engine nacelle of a Boeing 777. One can note the curved or sickle-shaped leading-edge sweep angle of the 22 fan blades, spanwise, for more efficient high-speed aerodynamics and better distributed structural loading on the relatively wide blades. Note as well the absence of a mid-span snubber, with the wide-chord fan blades

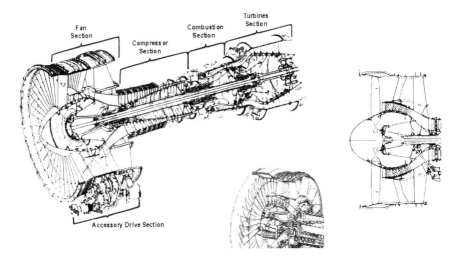

Fig. 7.3 Schematic cutaway diagrams illustrating the various turbofan engine components for a forward fan position on a short-ducted, non-mixing TF engine (General Electric CF6-6). One can see a snubber (clapper) being used at mid-span for stiffening the front fan blades on this older engine design. Courtesy of General Electric

Fig. 7.4 Photo of short-ducted, forward-fan, non-mixing TF engine on display stand (Pratt & Whitney PW2043)

flow out before exiting. In the passive case, the outer fan (cold) and inner core (hot) co-annular flows mix across the shear interface between them, at or downstream of the core exhaust nozzle. This co-axial mixing process acts to reduce the net exhaust flow temperature and speed, the latter effect being beneficial for noise reduction. A so-called "forced exhaust mixer" [4, 5], using lobes or corrugations on a peripheral shroud with intertwining hot and cold gas exit chutes in the vicinity or just upstream of the nozzle exit plane, allows for a more aggressive air/gas mixing process via axial vortices, if that is perceived as being desirable (one may

Fig. 7.5 Schematic diagram
of Rolls-Royce Trent 1000
high-bypass turbofan engine.
Courtesy of Rolls-Royce plc

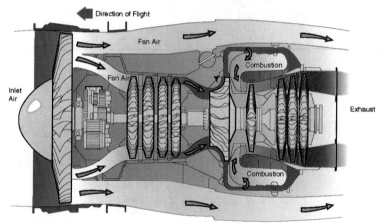

Fig. 7.6 Schematic diagram of a turbofan engine employing a short forward fan duct [3]

see a small gain in thrust delivery, depending on the viscous flow losses associated
with such an exit geometry; noise reduction is also attainable with this approach;
see schematic example of Fig. 7.7). The weight penalty associated with structural
exhaust mixers, or ejector shrouds downstream of the mixer (both being effective
for noise suppression), sometimes precludes their use. For lower bypass engines,
one may observe the remixing of the bypass air with the core gas stream beyond
the core turbine outlet (this is potentially useful for afterburning engines). Let's
consider the more common separate-stream case first. The thrust equation for such
an arrangement is as follows:

$$F = (\dot{m}_a + \dot{m}_f)V_e + \dot{m}_{by}V_{ef} - (\dot{m}_a + \dot{m}_{by})V_\infty + (p_e - p_\infty)A_e + (p_{ef} - p_\infty)A_{ef}$$

$$(7.2)$$

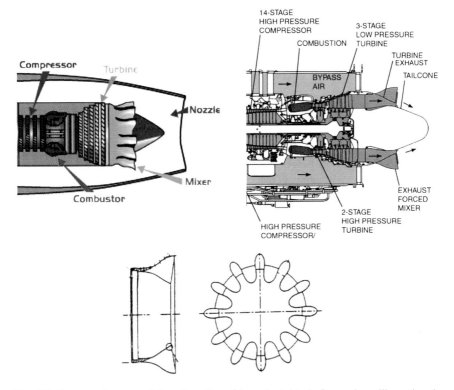

Fig. 7.7 Cutaway diagrams of the aft portion of long-ducted turbofan engines, illustrating the usage of a lobed forced exhaust mixer at the rear of the respective engines. Lower schematic diagram shows a 12-lobed mixer [5]. Courtesy of NASA

Via substitution, for specific thrust one arrives at:

$$\frac{F}{\dot{m}_a} = (1+f)V_e + \mathcal{B}V_{ef} - (1+\mathcal{B})V_\infty + \frac{A_e}{\dot{m}_a}(p_e - p_\infty) + \frac{A_{ef}}{\dot{m}_a}(p_{ef} - p_\infty) \quad (7.3)$$

For fuel consumption by TF engines, one would work in terms of thrust, as for TJs:

$$\text{TSFC} = \frac{f}{F/\dot{m}_a} \quad (7.4)$$

Letting for the moment $V_e \rightarrow V_{e,\infty}$ and $V_{ef} \rightarrow V_{ef,\infty}$ with exit pressure assumed to expand to ambient, then propulsive efficiency for TFs becomes:

$$\eta_p = \frac{FV_\infty}{\dot{m}_a\left[(1+f)\frac{V_{e,\infty}^2}{2} + \mathcal{B}\frac{V_{ef,\infty}^2}{2} - (1+\mathcal{B})\frac{V_\infty^2}{2}\right]} \quad (7.5)$$

Overall efficiency remains as before for TJs:

$$\eta_o = \frac{FV_\infty}{\dot{m}_f q_R} = \frac{1}{\text{TSFC}} \frac{V_\infty}{q_R} \tag{7.6}$$

7.2 Cycle Analysis of Separate-Stream Turbofan Engine

A lot of the elements of the TJ cycle analysis can be carried over to the TF. For the core stream, one can simply refer to the TJ equations already developed, with the important exception of the turbine section. Assuming a separate bypass stream with no fan duct burning, one can look at the fan inlet as a starting point. Here, one can assume the same conditions as the compressor inlet (station 2), given the same diffuser (thus giving T_{02} and p_{02} for an associated η_d and γ_d).

The fan's stagnation pressure ratio, as described by:

$$p_{08} = \pi_f p_{02} \tag{7.7}$$

will be a function of the LP compressor rotor (shaft) speed N_1 and other parameters. If a geared turbofan, π_f will be a function of a lower fan shaft speed n, reduced from N_1. At nominal cruise conditions, π_f can range from 1.4 to 2.6 for existing TF engines. Aft of the fan, the stagnation temperature can be estimated from:

$$T_{08} = T_{02}\left[1 + \frac{1}{\eta_f}[\pi_f^{\frac{\gamma_f - 1}{\gamma_f}} - 1]\right] \tag{7.8}$$

The fan's adiabatic component efficiency η_f may range from 0.8 to 0.85, while π_f remains close to 1.4 for most applications.

Beyond the fan, one moves past a row of stator vanes (that act to remove the swirl from the flow, i.e., straighten out the flow for improved thrust delivery) down the fan duct to the fan nozzle exit plane, which may or may not be choked, depending upon flight conditions. If the fan duct is relatively short, one can neglect the loss in stagnation pressure up to that point due to flow friction (if a long fan duct, one might want to include this). One can check for choking of the fan nozzle exit following the criterion as noted below:

$$\frac{p_{08}}{p_\infty} > \left(\frac{\gamma_{fn} + 1}{2}\right)^{\frac{\gamma_{fn}}{\gamma_{fn} - 1}}, \quad \text{fan nozzle throat is choked} \tag{7.9}$$

For the unchoked case, which may occur in practice at lower throttle settings and lower altitudes, one can note the following:

$$p_9 \to p_\infty$$

$$V_{ef} \rightarrow V_{ef,\infty}$$

$$V_{ef,\infty} = \sqrt{2\eta_{fn} C_{p,fn} T_{08} \left[1 - \left(\frac{p_\infty}{p_{08}} \right)^{\frac{\gamma_{fn}-1}{\gamma_{fn}}} \right]} \tag{7.10}$$

$$A_{ef,\infty} \approx A_{ef}$$

$$T_{ef,\infty} \approx T_{08} \left(\frac{p_\infty}{p_{08}} \right)^{\frac{\gamma_{fn}-1}{\gamma_{fn}}}$$

$$\rho_{ef,\infty} = \frac{p_\infty}{RT_{ef,\infty}}$$

The fan nozzle adiabatic efficiency can be in the vicinity of 0.97 in value, while γ_{fn} will be around 1.4, with specific gas constant R being close to the nominal sea-level air value.

If on the other hand the fan nozzle is choked, then the following might be useful:

$$\left. \frac{T_9}{T_{08}} \right|_{ideal} = \frac{2}{\gamma_{fn} + 1} \tag{7.11}$$

$$V_{ef} = \sqrt{2\eta_{fn} C_{p,fn} T_{08} \left[\frac{\gamma_{fn} - 1}{\gamma_{fn} + 1} \right]} \tag{7.12}$$

$$p_{ef} = p_9 = p_{08} \left[\frac{T_9}{T_{08}} \bigg|_{ideal} \right]^{\frac{\gamma_{fn}}{\gamma_{fn}-1}} = p_{08} \left[\frac{2}{\gamma_{fn} + 1} \right]^{\frac{\gamma_{fn}}{\gamma_{fn}-1}} \tag{7.13}$$

$$\rho_9 = \rho_{08} \left(\frac{2}{\gamma_{fn} + 1} \right)^{\frac{1}{\gamma_{fn}-1}} = \frac{p_{08}}{RT_{08}} \left(\frac{2}{\gamma_{fn} + 1} \right)^{\frac{1}{\gamma_{fn}-1}} \tag{7.14}$$

$$a_9 = \sqrt{\gamma_{fn} RT_9} = \sqrt{\frac{2\gamma_{fn} RT_{08}}{\gamma_{fn} + 1}} \tag{7.15}$$

In the choked flow case, the pressure-area term $(p_{ef} - p_\infty)A_{ef}$ has potentially a finite positive value (unlike the unchoked case, where $p_{ef} \rightarrow p_\infty$). For specific thrust (Eq. 7.3), one may need the following:

$$\frac{A_{ef}}{\dot{m}_a} = B \frac{A_{ef}}{\dot{m}_{by}} = B \frac{RT_{08}}{p_{08}} \left[\frac{\gamma_{fn} + 1}{2} \right]^{\frac{1}{\gamma_{fn}-1}} \left[\frac{\gamma_{fn} + 1}{2\gamma_{fn} RT_{08}} \right]^{1/2} \tag{7.16}$$

Given that the core engine (gas generator) is doing work on the fan stream, one will need to modify the expression for the stagnation temperature at the turbine outlet. Consider the following energy balance:

$$\eta_m \dot{m}_t C_{p,t}(T_{04} - T_{05}) = \dot{m}_a C_{p,c}(T_{03} - T_{02}) + \mathcal{B}\dot{m}_a C_{p,f}(T_{08} - T_{02}) \qquad (7.17)$$

Making the approximation that

$$\dot{m}_t C_{p,t} \approx \dot{m}_a C_{p,c} \approx \dot{m}_a C_{p,f} \qquad (7.18)$$

gives the following for the stagnation temperature:

$$T_{05} \approx T_{04} - \frac{1}{\eta_m}(T_{03} - T_{02}) - \frac{\mathcal{B}}{\eta_m}(T_{08} - T_{02}) \qquad (7.19)$$

In conjunction with the aforementioned equations from the TJ cycle analysis done earlier, one can now solve for various performance parameters, and produce graphs for specific thrust, TSFC and η_p as a function of a given π_f, combustor T_{04}, and flight conditions (h_{asl}, Ma_∞, for an example TF engine. In the case of a geared turbofan, one is taking power off of the LP turbine, to rotate a fan shaft, through a reduction gear (say, e.g., from 12,000 rpm down to 4,000 rpm), at a lower speed n:

$$\frac{\mathcal{B}}{\eta_m}(T_{08} - T_{02}) \approx \frac{P_S}{\eta_m \dot{m}_t C_{p,t}} \approx \frac{C_P \rho n^3 d^5}{\eta_m \dot{m}_t C_{p,t}} \qquad (7.20)$$

where the shaft power required to rotate the many-bladed fan will be governed by the shrouded fan's power coefficient C_P (like a propeller, a function of advance ratio J and blade pitch angle β), etc. The fan pressure ratio π_f can be correlated to Eq. 7.20 via:

$$\frac{T_{08}}{T_{02}} = \left[1 + \frac{1}{\eta_f}\left[\pi_f^{\frac{\gamma_f - 1}{\gamma_f}} - 1\right]\right] \qquad (7.21)$$

Refer to Fig. 7.8 for an example high-bypass TF thrust chart showing net thrust (at a maximum continuous thrust throttle setting) as a function of flight Mach number and altitude. The characteristic larger drop in thrust of a high-bypass TF (relative to a TJ) with forward airspeed at lower altitudes is evident with this example. This behavior at higher bypass ratios is indicative of the behavior seen with propellered aircraft. At higher altitudes, the respective behavior of TFs and TJs becomes more similar. Curves for TSFC are also shown in Fig. 7.8, and clearly reveal a substantially lower range of values than that seen for the TJ example (at cruise, say around 0.56 kN/h-kN for the TF, versus around 0.95 kN/h-kN for the TJ). For a different older TF engine, Fig. 7.9 reveals the effects of the throttle setting on the engine's thrust behavior at high altitude (10 km). As noted above, the throttle setting will control the effective value for π_f as well as π_c. Figure 7.10 provides an example of takeoff thrust performance for a smaller turbofan engine, at different air temperatures and altitudes. At sea level, beyond an air temperature of 27°C, the engine's static thrust begins to drop off, while at 1,500 m altitude, this decrease in performance begins at about 5°C. Please refer to the solution for example Problem 7.1 at the end of this chapter, for a completed sample separate-stream turbofan engine performance analysis. For

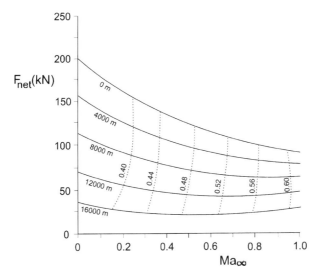

Fig. 7.8 Thrust chart for example turbofan engine (maximum continuous thrust throttle setting). Dashed lines are TSFC curves, in kN/h-kN. Engine performance comparable to General Electric TF39, with core compression ratio (π_c) of 22:1, and bypass ratio of 8

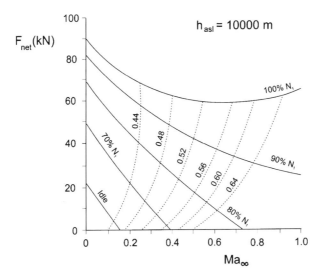

Fig. 7.9 Thrust chart for example turbofan engine at various throttle settings (from maximum continuous down to flight idle), at an altitude of 10 km. Dashed lines are TSFC curves, in kN/h-kN. Engine performance comparable to Pratt & Whitney JT9D-59A, with core compression ratio (π_c) of 22.5:1, and bypass ratio of 4.9

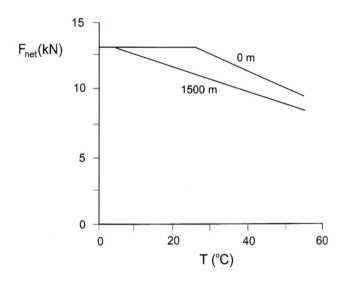

Fig. 7.10 Static thrust chart for example turbofan engine at takeoff throttle setting, at various outside air temperatures and airfield altitudes. Engine performance comparable to Pratt & Whitney JT15D-5B, with core compression ratio (π_c) of 12.3:1, and bypass ratio of 2.1:1

students learning this material as part of a course, I would encourage them to attempt the given example problem first, and then check the solution to see if any mistakes were made.

Further examples of useful results produced from cycle analysis, in this case at the nominal design cruise point, can be viewed in Figs. 7.11, 7.12 and 7.13. With the best fuel economy as the objective, one can observe the trough point for TSFC in the various TSFC curves, which for Fig. 7.11 happens to correspond to the peak thrust delivery at the optimal fan pressure ratio of around 1.64. In the optimal case of bypass ratio and turbine inlet temperature for best fuel economy, thrust delivery is a bit lower than theoretically attainable. For example, as reflected by Fig. 7.12, one could choose a lower bypass ratio than 8.4 to obtain a higher thrust, but it would be at the expense of poorer fuel economy. Similarly, as reflected by Fig. 7.13, one could choose to run at a higher combustor peak temperature than 1,350 K to get a higher thrust delivery, but it will cause fuel economy to suffer.

In the development of turbofan engines over the years, subsonic fans (fan blade tip Mach numbers <0.85) have given way to transonic fans, where fan blade tip Mach numbers can now reach or exceed 1.7. At these Mach numbers, shock waves are present, but are managed effectively through the use of thin airfoil sections and many blades, to avoid overloading individual blades and preventing (or limiting) boundary layer separation in the presence of rising pressure (i.e., adverse pressure gradients). Shrouding (ducting) assists in the above, as well as limiting tip leakage losses (with tiny separation distance between blade tip and shroud wall). Having said all this, if the optimal LP shaft

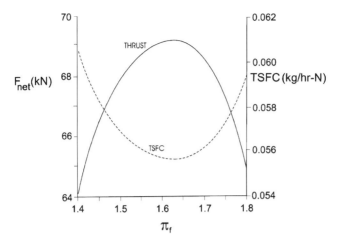

Fig. 7.11 Thrust and TSFC as a function of fan pressure ratio ($_f$) for example turbofan engine at cruise throttle setting, at a nominal cruise altitude of 10.7 km and flight Mach number of 0.85. Engine performance comparable to General Electric GE90, with overall compression ratio (including ram, at Ma_∞ of 0.85) of 40:1 and bypass ratio of 8.4 at cruise

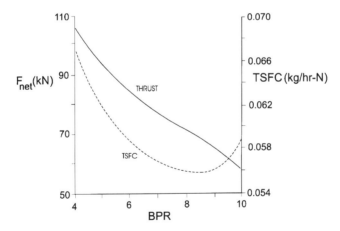

Fig. 7.12 Thrust and TSFC as a function of bypass ratio (B) for example turbofan engine at cruise throttle setting, at a nominal cruise altitude of 10.7 km and flight Mach number of 0.85. Engine performance comparable to General Electric GE90, with overall compression ratio (including ram, at Ma_∞ of 0.85) of 40:1 and bypass ratio of 8.4 at cruise

speed N_1 is too high for the fan blades to operate at, it is not uncommon to employ a speed reduction gear to rotate the fan at a lower rotation speed n, as noted earlier (albeit at the expense of extra engine weight; referred to as a geared turbofan). As fan blade diameters continue to grow along with bypass ratios

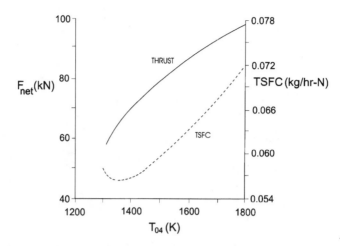

Fig. 7.13 Thrust and TSFC as a function of turbine inlet temperature (T_{04}) for example turbofan engine at cruise throttle setting, at a nominal cruise altitude of 10.7 km and flight Mach number of 0.85. Engine performance comparable to General Electric GE90, with overall compression ratio (including ram, at Ma_{∞} of 0.85) of 40:1 and bypass ratio of 8.4 at cruise

Fig. 7.14 Schematic cutaway diagram of a high-bypass turbofan engine produced by GE Aviation (GEnx series), with BPR approaching 10:1. Diagram courtesy of General Electric

(from high bypass, as in Fig. 7.14 approaching 10:1, towards ultra-high bypass ratios like 20:1), gearing will almost surely be required. By way of comparison, the effective mass-flow bypass ratio B for a turboprop engine/propeller combination is on the order of 20 to 50:1, and up to 250:1 for a turboshaft/main rotor combination for a helicopter, noting that the number of propeller or main rotor blades is significantly less than that for a fan. Shrouding (surrounding the periphery of the fan and fan duct) will likely continue to be employed at these higher turbofan engine bypass values. Shroudless unducted fan (UDF; propfan; see Fig. 7.15) designs have been investigated over the years by various research groups, but the high noise production from such a design continues to hamper its mainstream introduction.

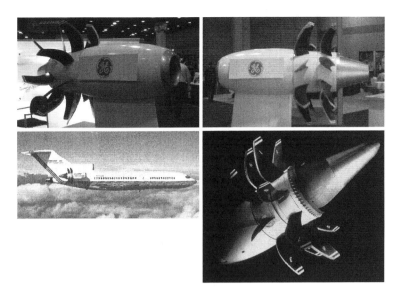

Fig. 7.15 Subscale mockup of an unducted fan engine, using two contra-rotating rows of blades as two aft stages on an LP turbine shaft (effective bypass ratio of 35:1). Illustration of Boeing 7J7 aircraft at lower left, and UDF engine at lower right, courtesy of General Electric

7.3 Cycle Analysis of Mixed-Stream Turbofan Engine

As may be viewed by the schematic diagram of Fig. 7.16, a turbofan engine can be designed to bring the fan air stream back into the core flow of the engine, after the turbine section. This is commonly done for low-bypass turbofan engines for supersonic fighters, with the common positioning of an afterburner jet pipe to take advantage of this additional air for reheat energy (see Fig. 7.17; [6]). Following the convention of earlier analyses, the thrust equation in the absence of after-burning is given by:

$$F = (\dot{m}_a + \dot{m}_f + \dot{m}_{by})V_e - (\dot{m}_a + \dot{m}_{by})V_\infty + (p_e - p_\infty)A_e \qquad (7.22)$$

Via substitution, for specific thrust one arrives at:

$$\frac{F}{\dot{m}_a} = (1 + f + \mathcal{B})V_e - (1 + \mathcal{B})V_\infty + \frac{A_e}{\dot{m}_a}(p_e - p_\infty) \qquad (7.23)$$

In the case where afterburning is employed, the thrust equation for a mixed-stream engine becomes:

$$F = (\dot{m}_a + \dot{m}_f + \dot{m}_{by} + \dot{m}_{f,AB})V_e - (\dot{m}_a + \dot{m}_{by})V_\infty + (p_e - p_\infty)A_e \qquad (7.24)$$

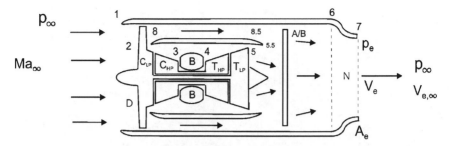

Fig. 7.16 Schematic diagram illustrating the various turbofan engine components and associated station numbering, for a forward fan position on a mixed-stream afterburning TF engine

Fig. 7.17 Schematic cutaway diagram of a GE Rolls-Royce F136 low-bypass turbofan engine, designed to power the single-engined Lockheed Martin F-35 Lightning II supersonic fighter aircraft. Diagram of engine courtesy of General Electric and Rolls-Royce plc. Photo of F-35 courtesy of USAF

The corresponding equation for specific thrust is given by:

$$\frac{F}{\dot{m}_a} = (1 + f + \mathcal{B} + [1 + f + \mathcal{B}]f_{AB})V_e - (1 + \mathcal{B})V_\infty + \frac{A_e}{\dot{m}_a}(p_e - p_\infty) \quad (7.25)$$

where in this case, the afterburner fuel-to-gas ratio would be

$$f_{AB} = \frac{\dot{m}_{f,AB}}{\dot{m}_a + \dot{m}_f + \dot{m}_{by}} = \frac{\dot{m}_{f,AB}}{\dot{m}_a(1 + f + \mathcal{B})} \quad (7.26)$$

Moving to the pertinent new information for a mixed-stream flow downstream of the turbine and fan duct exits, one can make some quasi-steady ideal mixture assumptions in passing through the mixing zone to station 5.5 at its nominal exit. Conservation of mass gives:

$$\dot{m}_{5.5} = \dot{m}_5 + \dot{m}_{8.5} = (1 + f + \mathcal{B})\dot{m}_a \quad (7.27)$$

Conservation of energy gives:

$$\dot{m}_5 C_{p,t} T_{05} + \dot{m}_{8.5} C_{p,f} T_{08.5} = \dot{m}_{5.5} C_{p,5.5} T_{05.5} \tag{7.28}$$

so that

$$T_{05.5} = \frac{\dot{m}_5 C_{p,t} T_{05} + \dot{m}_{8.5} C_{p,f} T_{08.5}}{\dot{m}_{5.5} C_{p,5.5}} \approx \frac{(1+f)}{(1+f+\mathcal{B})} T_{05} + \frac{\mathcal{B}}{(1+f+\mathcal{B})} T_{08.5} \tag{7.29}$$

In a non-isentropic mixing process, with a presumed adiabatic mixing efficiency of η_{mixer} passing through the nominal mixer volume, the following result would approximately apply:

$$p_{05.5} = p_{05} \left[1 - \frac{1}{\eta_{\text{mixer}}} \left(1 - \frac{T_{05.5}}{T_{05}} \right) \right]^{\frac{\gamma_t}{\gamma_t - 1}} \tag{7.30}$$

For a more accurate assessment of the mixer exit stagnation pressure, one can incorporate the conservation of linear momentum and applicable compressible gas relations into the problem [6]. The applicable equation for estimating the afterburner fuel-gas mixture is as follows:

$$f_{\text{AB}} = \frac{\dfrac{T_{06}}{T_{05.5}} - \dfrac{C_{p,t}}{C_{p,\text{AB,exit}}}}{\dfrac{\eta_{\text{AB}} q_R}{C_{p,\text{AB,exit}} T_{05.5}} - \dfrac{T_{06}}{T_{05.5}}} \tag{7.31}$$

One can estimate TSFC via the following:

$$TSFC = \frac{\dot{m}_f + \dot{m}_{f,\text{AB}}}{F} = \frac{f + [1 + f + \mathcal{B}]f_{\text{AB}}}{\dfrac{F}{\dot{m}_a}} \tag{7.32}$$

For the remainder of the analysis, without or with an afterburner in place, one can refer to the previous coverage in Chap. 6 and above. Please see the solution for example Problem 7.4 at the end of this chapter for a sample completed cycle analysis of a mixed-stream afterburning turbofan engine.

7.4 On-Design Versus Off-Design Cycle Analysis

In the propulsion literature, you will see references to "on-design" or "design-point" or "parametric" cycle analysis of a gas turbine engine. This is generally done in preliminary design with a "rubber" or "paper" engine, i.e., prior to building a prototype engine. The anticipated engine performance at a given value for a design parameter (e.g., the desired or maximum allowed compressor pressure ratio or combustor exit temperature) at a set of different flight conditions may be assessed during this process. Design point examples would be for expected cruise flight conditions, and takeoff at different altitudes and air temperatures.

"Off-design cycle" or "engine performance" analysis refers to assessment of the performance of an engine at all flight conditions and throttle settings (engine shaft speed N) expected. For example, the change in the engine's performance as one goes up in altitude might be classified as an off-design issue, or as a second example, flight at substantially reduced throttle settings. Off-design analysis would be for both steady-state and transient operation, typically. Additionally, this would more typically be done for an existing engine (prototype, or production unit) that requires a thorough evaluation, and to provide a database for related analysis like flight simulation, etc. Component efficiencies would typically be better established for the specific engine, in doing these calculations. The engine may readily meet its nominal design requirements, but off-design performance for startup, idle, and adequate acceleration may prove to be an additional and new challenge.

7.5 Engine Health Monitoring

It has long been recognized that frequent monitoring of an aircraft engine's health is an effective means for prevention of a catastrophic engine failure or a forced engine shutdown in flight, especially for those cases where the flaws or associated symptoms start small but progressively grow with time. By "frequent", it is often meant to mean that there is more monitoring of the engine's operational health, above and beyond that done pre- and post-flight (typically only minor inspections, checks and maintenance done at the airport in question, by aircraft maintenance engineers [AMEs], e.g., using a boroscope [flexible pipe-like viewing instrument] inserted into various internal viewing ports) and certainly more frequently than comprehensive scheduled major engine maintenance or overhaul periods at a separate ground facility. Pulling an engine off an aircraft to conduct prolonged major engine maintenance or overhaul in a commercial transport context is expensive, so there is some incentive to look for alternative means to expeditiously detect the development of a possibly serious engine problem. Piston engines historically are brought in around every 3,000 flight hours for major inspection and overhaul. Gas turbine engines to date generally have the hotter aft end of the engine overhauled every 3,000 flight hours, while the colder front end of the engine may not need major maintenance for as long as 20,000 flight hours. Note, however, that the above hours primarily apply to various parts of the engine wearing out in a consistent fashion over time. Failures due to poorly manufactured parts generally occur much sooner than the above (ideally, during the early shakedown period of a new engine, when AMEs are keeping a close eye on the engine), and engine failures due to some random occurrence or unusual confluence of events, or more likely series of events over a period of time, tend to happen on average a bit later in the life of the engine. As noted in Chap. 1, the result of failures can be mitigated by various design approaches (safe-life, fault-tolerant, etc.). Of course, more rigorous quality control in the first place, before the aircraft's first flight, is a recognized means for reducing the number of subsequent failure events.

Various means for engine health monitoring (EHM) have been proposed over the years. The baseline for more frequent monitoring (and less scheduled maintenance/checkups) is having the pilot, or flight engineer (if applicable), observe engine operational behaviour during each flight, and having the AMEs make observations pre- and post-flight. Collecting these observations over the course of many flights, it would be hypothetically possible to detect minor changes in the engine's capabilities or characteristics, and then correlate to existing data that would indicate whether an unusual symptom, suggestive of an engine flaw, is growing at a rate such that maintenance or repair would be necessitated by a certain time. With the introduction of modern, miniature sensor technology, one can now place a number of sensors on the engine for measuring gas temperature, gas pressure, gas flow rate, lubricating oil temperature, lubricating oil contamination, engine structural vibration, position translation, etc., at various locations, and in a more explicit, quantitative fashion, detect developing flaws earlier than that possible by less frequent human observation. For example, a marked change in gas flow rate is a solid indicator that something is likely amiss with the compressor. While not necessarily easy to measure accurately, a drop in thrust or power for a given fuel flow rate is another strong indicator of some engine degradation or growing problem. In order to capture the above sensor data in a comprehensive fashion, one can store the data by an on-board processor and corresponding data storage media during a given flight, and collect it after the flight for later processing. An alternative to this approach is to have the sensors transmit their information to a ground-based receiver for almost immediate processing (wireless approach) by the fault detection software, thus allowing for a more expeditious correction than would otherwise be possible.

With respect to turbofan engines, mechanisms that can lead to failure of a fan blade include cyclic-loading fatigue, bird strikes, erosion (FOD [foreign object damage], primarily during takeoff and landings; impact of intake-ingested airborne particles like hail, smoke or volcanic ash), or in the case of metal structural components, corrosion (through oxidation). With respect to FOD or its counterpart, DOD [domestic object damage, resulting from loose internal engine items like nuts or bolts], depending on the item in question, impact of said item on the engine structure may lead to immediate rupture of a vital engine component. The fan disk (fan blade attached to this disk, which in turn is attached to the LP shaft; commonly made from a single titanium alloy) may fail under low cycle fatigue (accumulated loading at lower frequency in the higher plastic-deformation stress regime), or high cycle fatigue (accumulated fatigue at higher frequency from resonant vibrational loading at lower stresses in the elastic regime). When a fan disk fails suddenly in a disk burst (rupture) incident, the damage to the surrounding engine and airframe can be significant, depending on the level of containment of the resulting fragments. Moving to the engine's compressor section, a compressor stator can fail under high cycle fatigue or through the mechanism of runaway aerodynamically-induced resonant fluid-structural vibration (flutter). A compressor rotor blade can fail under high cycle fatigue, flutter, erosion or corrosion. A compressor disk can fail in the same manner as a fan disk, including bursting at the time of complete failure. Moving to the engine's combustor, a combustor liner structure can fail due to thermal fatigue (creep) or overheating. In the engine's turbine section, stators, rotor blades and disks can fail in a manner analogous to the compressor section, but with additional concerns of thermal fatigue and over-heating to address with the hot gas flow passing through.

7.6 Environmental Issues

7.6.1 Environmental Issues: Noise

Perceivable noise generated by combustion instability in the combustor of a jet engine was noted earlier, as a symptom of a flaw in the combustion process. In general, jet engine noise itself, produced from various locations within the engine for the most part as a function of the engine's normal operation, depending on its frequency and strength, can range from annoying to painful to those people without hearing protection who might be located nearby during takeoffs and landings (see Figs. 7.18, 7.19). Transport authorities continue to push for quieter engines (Fig. 7.20, [7]), and certain airports have especially stringent regulations as to the allowed noise that can be produced by airplanes using that particular location (note: the airframe moving through the air can in some instances produce as much, if not more, noise than the engines; the sonic boom produced by a fighter aircraft flying overhead at supersonic speed will attest to that fact).

Common sources of jet engine noise are:

1. the rotating fan blades of a turbofan engine (comparable to noise produced by a rotating propeller, although ducting of the fan helps to reduce the level of noise)
2. exit stator vanes of a turbofan engine's fan duct emit noise at a characteristic blade passage frequency (referred to as tone noise; one can use Helmholtz and quarter-wave resonator approaches to help suppress [damp] targeted noise frequencies)
3. compressor rotor/stator stages generate characteristic noise as air moves from one row of blades or vanes to the next
4. similarly to the compressor, turbine stator/rotor stages generate noise
5. core engine nozzle exhaust jet will generate significant noise (noise largely generated by turbulence, referred to as broadband noise since across a range of frequencies, which makes it more difficult to suppress); directing the core jet's direction upward a few degrees, while penalizing thrust, may be effective for reducing noise detected on the ground; a chevron nozzle (with a serrated or sawtooth edge at the core nozzle's exit plane (see Fig. 7.21), comparable to fluted or lobed forced exhaust mixers referred to earlier [4, 5]) shows promise in acting to cancel out some noise frequencies
6. fan exhaust from turbofan engine fan duct interacting with the core engine exhaust will generate a characteristic noise, and the dual inner-outer exhaust streams may in fact be quieter than the inner core exhaust jet by itself (shielding

Fig. 7.18 Typical noise field produced by a stationary jet engine close to the ground, spreading in still air [2]. The curved lines represent lines of constant sound level as noted in decibels, with 140 dB experienced very close to the engine exhaust. Note that this noise field is for an older jet engine. Reprinted with permission of The McGraw-Hill Companies, Inc

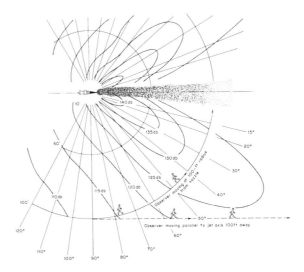

Fig. 7.19 Noise levels produced by various components of a turbofan engine during takeoff. Courtesy of NASA (graph from 1999 NASA Glenn Research Center fact sheet [1999-07-003-GRC])

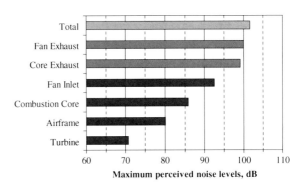

effect of the outer fan stream); possible use of an ejector duct or shroud aft of the nozzle exit, for more active noise suppression; chevron-shaped fan duct exit nozzle (Fig. 7.21, [8]) can act to lower generated noise (to be employed for the Boeing 787 airliner's engines)

7.6.2 Environmental Issues: Air Pollution

Present-day jet fuel combustion products that are exhausted by the engine into the atmosphere do contribute to various elements of air pollution. While the amount of pollution contributed is small compared to other sources (like automobiles), the location of the exhaust delivery, specifically at high altitude, may have a magnifying effect. One jet exhaust product, carbon dioxide (CO_2, a major component under the pollution category of *carbon emissions*) is a greenhouse gas (GHG), so-called in that

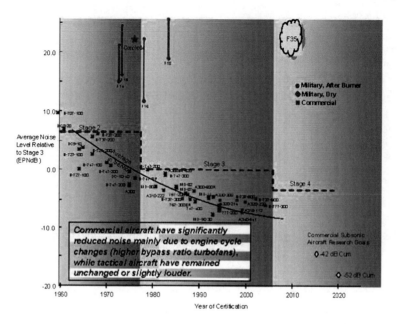

Fig. 7.20 Trend chart showing decreasing noise levels produced from various jet engines over the years [7]. Chart courtesy of U.S. Navy

it contributes to global warming by acting to inordinately trap heat in the lower atmosphere. Other exhaust products like water vapor (H_2O), nitrogen oxides (NOx), sulfur oxides (SOx), unreacted hydrocarbons, and particulate matter (soot) are GHG contributors. The above exhaust constituents also enable the increased production, by sunlight, of ground-level ozone, an undesirable element associated with smog and corresponding human respiratory system health issues. Sulfur oxides like sulfur dioxide (SO_2) have the notoriety of also contributing to acid rain.

It has been observed that NOx production in jet engine combustors goes up as one increases the combustor exit gas temperature (T_{04}). Earlier, it was noted that a heightened T_{04} helps to increase specific thrust ($F/\dot{m}_a$), but at lower flight speeds, increasing T_{04} leads to a higher TSFC (worsening fuel economy). As a result, there may be some benefit in obtaining a higher fuel economy (lower TSFC) in combination with a lower NOx level, if one can design or allow for a lower specific thrust over a substantial portion of a given flight mission.

7.6.3 Environmental Issues: Dwindling Jet Fuel Supply

While there are substantially differing estimates of the amount of the remaining liquid petroleum reserves that can be used for anticipated conventional jet fuel production (ranging from 50 years to as long as 200 or more years), there is a

Fig. 7.21 Views of a chevron fan duct nozzle exit, and a chevron core nozzle exit as well, for a high bypass turbofan engine, to help reduce noise [8]. Note the core engine tail cone extending substantially beyond the nominal core nozzle exit plane. Upper photo courtesy of NASA

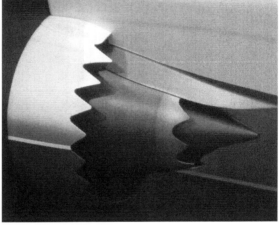

general consensus that research and development should be conducted presently towards evaluating alternative fuels and their use down the road for air transportation. Synthetic fuels produced predominantly from bituminous coal via the Fischer–Tropsch process have the longest track record (since the 1920s), with a comparable energy content to conventional jet fuel. For those countries with large coal reserves, this is potentially an attractive option. Currently, blends of synthetic and conventional jet fuel are being evaluated, meeting the industry requirement of a simple "drop-in" capability in the short term, i.e., no significant engine or fuel storage modifications need be made for existing aircraft designs. While the exhaust products from synthetic fuels is generally cleaner than that seen for conventional jet fuels, there is the existing stigma of carbon emissions surrounding coal production.

More recently, ethanol has been a widely studied and tested biofuel for transportation applications, used by itself, or blended with conventional fuels. Biofuels use forest biomass or agricultural biomass as the "feedstock" (raw material required for the production of the alternative fuel). Conventional ethanol can be produced from corn in cooler climates, and sugarcane in hotter climates (both being examples of agricultural biomass). One substantial drawback of ethanol is that its energy content (around 30 MJ/kg of ethanol, in reacting with air, as compared to 43 MJ/kg for jet fuel) is 30% less than jet fuel. In a drop-in usage scenario, that means the effective range of existing aircraft would potentially be cut back by as much as 30% (the liquid density of ethanol, 790 kg/m^3, and Jet A-1, 804 kg/m^3, is similar), given the existing tank storage of each aircraft. A substantial amount of energy is expended in the process of producing ethanol from corn, while the cost is significantly less for transforming sugarcane. Critics also note that corn and sugarcane are food for the human population (and for farm livestock), and at some point, shortages will almost surely arise in a conflicting supply-and-demand context. In addressing the latter two criticisms, cellulosic ethanol, cost-effectively made from wood (an example of forest biomass), grasses (like switchgrass, which is not edible by livestock), and inedible parts (stems, etc.) of plants, is being evaluated as a potentially more popular alternative to conventional ethanol.

Algal fuels, which include biodiesel fuels derived from algae (note: biodiesels are also derived from other substances, like discarded vegetable oil), are more energetic than ethanol, with an energy content of 90% of jet fuel. Since algae-based fuels are at a higher liquid density (around 880 kg/m^3) than jet fuel, they are similar to synthetic fuels in being a one-to-one potential replacement for conventional jet fuel, i.e., provide a similar aircraft range for the same volume of fuel. Like synthetic fuels, the exhaust gas is relatively clean compared to conventional jet exhausts.

With respect to more exotic alternative fuels, hydrogen is often proposed. It has a high energy content, from 120 to 140 MJ/kg of H_2 when reacted with incoming air, if utilizing hydrogen in a combustion process (as opposed to a fuel cell/electrical energy production approach). Hydrogen can be produced from water, which is, of course, plentiful in most parts of the world. The major drawback with hydrogen is its low density in storage, as a pressurized gas certainly, but even if stored as a cryogenic liquid (70 kg/m^3). For widespread civil usage, there are a number of safety and handling issues with hydrogen as a gas or liquid that would need to be addressed. Certainly, there is no "drop-in" capability in going from conventional jet fuel to hydrogen; one would need a new aircraft engine and corresponding fuel storage arrangement in the aircraft.

7.7 Example Problems

7.1. Perform a design-point cycle analysis for a non-mixing turbofan engine at its cruise Mach number of 0.85 and altitude of 12,200 m. Take note of the following design parameters:

diffuser, $\eta_d = 0.97$; $\gamma_d = 1.4$; fan, $\eta_f = 0.85$, $\gamma_f = 1.4$, $\pi_f = 1.5$, $\gamma_{fn} = 1.4$, $\eta_n = 0.97$, BPR = 5.0
compressor, $\eta_c = 0.95$, $\gamma_c = 1.37$, $\pi_c = 20$;
burner, $\eta_b = 1.0$, $\gamma_b = 1.35$, $\pi_b = 0.95$, $T_{max} = 1,500$ K

q_R = 45 MJ/kg of fuel; turbine, γ_t = 1.33; η_t = 0.90; η_m = 0.99
core nozzle, η_n = 0.98; γ_n = 1.36

Given the above flight and engine information, determine the specific thrust, TSFC, propulsive efficiency and overall efficiency.

7.2. Take note of the following performance parameters for an aircraft's non-mixing (separate-stream) turbofan engine:

diffuser, η_d = 0.96; γ_d = 1.4; fan, η_f = 0.89, γ_f = 1.4, π_f = 1.7, γ_{fn} = 1.4,
η_{fn} = 0.97, BPR = 7.0
compressor, η_c = 0.93, γ_c = 1.38, π_c = 27;
burner, η_b = 0.97, $C_{pb,exit}$ =1131 J/kg-K, π_b = 0.98, T_{max} = 1,400 K
q_R = 42 MJ/kg of fuel; turbine, γ_t = 1.35; η_t = 0.90; η_m = 0.99
core nozzle, η_n = 0.98; γ_n = 1.37
engine monitoring instruments: core air intake, 90 kg/s; T_{02} = 246.5 K,
p_{02} = 32 kPa, T_{03} = 638.2 K, p_{03} = 864 kPa, T_{04} = T_{max}, p_{04} = 846.7 kPa
current flight conditions at 11,585 m altitude: outside air pressure = 20.67 kPa,
outside air temperature = 216.7 K, outside air density = 0.332 kg/m³,
flight Mach number = 0.83

Given the above flight and engine information, determine the net thrust being delivered by the engine at this time.

7.3. Evaluate the performance of the engine of Problem 7.1, but for a lower fan pressure ratio of 1.35 (vs. 1.5).

7.4. Perform a design-point cycle analysis for a low-bypass mixed-stream after-burning turbofan engine at its cruise Mach number of 0.85 and altitude of 12,200 m. Take note of the following design parameters:

diffuser, η_d = 0.97; γ_d = 1.4; fan, η_f = 0.85, γ_f = 1.4, π_f = 3.5, γ_{fn} = 1.4,
η_{fn} = 0.97, BPR = 0.70
compressor, η_c = 0.95, γ_c = 1.37, π_c = 20;
burner, η_b = 0.99, γ_b = 1.35, π_b = 0.95, $T_{max,B}$ = 1,500 K
q_R = 45 MJ/kg of fuel; turbine, γ_t = 1.33; η_{nt} = 0.90; η_m = 0.99
turbine/afterburner mixer, η_{mixer} = 0.99
afterburner, η_{AB} = 0.9, γ_{AB} = 1.33, π_{AB} = 0.98, q_R = 45 MJ/kg of fuel,
$T_{max,AB}$ = 2,000 K
core nozzle, η_n = 0.98; γ_n = 1.36

Given the above flight and engine information, determine the specific thrust and TSFC.

7.5. An aircraft is presently flying at a flight Mach number of 0.7 at 10,000 m ISA altitude. This airplane employs three Pratt & Whitney JT9D-59A turbofan engines (see engine performance chart, Fig. 7.9). The engine throttle setting is presently set at 95% of maximum continuous N_1. Establish the rate at which the airplane is removing fuel from its tank storage (kg/h).

7.6. *Computer-Based Project*: Perform an off-design cycle analysis for a large commercial non-mixing turbofan engine at its cruise throttle setting, at altitudes from 0 to 15,000 m ISA, and flight Mach numbers from zero to 0.9. Take note of the following design parameters:

diffuser, $\eta_d = 0.98$; $\gamma_d = 1.4$; fan, $\eta_f = 0.965$, $\gamma_f = 1.4$, $\pi_f = 1.5$, $\gamma_{fn} = 1.4$, $\eta_{fn} = 0.985$, BPR = 7.50, fan nozzle fixed exit area $A_{ef} = 4.0$ m^2
compressor, $\eta_c = 0.985$, $\gamma_c = 1.375$, $\pi_c = 30.1$;
burner, $\eta_b = 0.99$, $\gamma_b = 1.35$, $\pi_b = 0.985$, $T_{max} = 1350$ K
$q_R = 44$ MJ/kg of fuel; turbine, $\gamma_t = 1.335$; $\eta_t = 0.955$; $\eta_m = 0.99$
core nozzle, $\eta_n = 0.98$; $\gamma_n = 1.36$

Plot thrust F vs. flight Mach number Ma_∞ at the 4 principal altitudes of 0, 5,000, 10,000 and 15,000 m ISA (generate enough data points so that each curve appears smooth in appearance). Ascertain the TSFC data points when values of 0.04, 0.05, 0.06 and 0.07 kg/h-N are attained, and graph these cross-hatch TSFC curves onto the main $F(h_{ASL})$ versus Ma_∞ graph (the TSFC curves will have 4 data points each, and thus may not be particularly smooth in appearance as a result). Comment on the characteristics of this engine's thrust profile at different flight conditions.

7.8 Solutions to Problems

7.1. Looking at a turbofan engine, at a flight Mach no. Ma_∞ of 0.85, at 12,200 m altitude ($p_\infty = 18.75$ kPa, $T_\infty = 216.7$ K, $\rho_\infty = 0.30$ kg/m^3). Bypass ratio B of 5.
Starting from intake (diffuser), as in Problem 6.2 (same numbers as input):
$T_{02} = T_\infty[1 + \frac{\gamma_d - 1}{2} Ma_\infty^2] = 216.7[1 + 0.2(0.85^2)] = 248$ K, at diffuser outlet
$p_{02} = p_\infty[1 + \eta_d(\frac{T_{02}}{T_\infty} - 1)]^{\frac{\gamma_d}{\gamma_d - 1}} = 18.75[1 + 0.97(1.144 - 1)]^{3.5} = 29.63$ k Pa, at diffuser outlet

Moving through fan:
$p_{08} = \pi_f p_{02} = 1.5(29.63) = 44.5$ kPa, at fan outlet;

$T_{08} = T_{02}[1 + \frac{1}{\eta_f}[\pi_f^{\frac{\gamma_f - 1}{\gamma_f}} - 1]] = 248[1 + 1/0.85[1.5^{0.286} - 1]] = 284$ K, at fan outlet.

$\frac{p_{08}}{p_\infty} = \frac{44.5}{18.75} = 2.27$ and choking criterion $\left(\frac{\gamma_f + 1}{2}\right)^{\frac{\gamma_f}{\gamma_f - 1}} = 1.89 < \frac{p_{08}}{p_\infty}$, so fan nozzle flow is *choked*.

$$V_{ef} = \sqrt{2\eta_{fn} C_{p,fn} T_{08} \left[\frac{\gamma_{fn} - 1}{\gamma_{fn} + 1}\right]} = \sqrt{2(0.97)1004(284)[0.4/2.4]} = 304 \text{ m/s}$$

$$p_{\text{ef}} = p_{08} \left[\frac{T_{\text{ef}}}{T_{08}} \right]^{\frac{\gamma_{\text{fn}}}{\gamma_{\text{fn}}-1}} = p_{08} \left[\frac{2}{\gamma_{\text{fn}}+1} \right]^{\frac{\gamma_{\text{fn}}}{\gamma_{\text{fn}}-1}} = p_9 = 23.5 \text{ kPa}$$

$$\frac{A_{\text{ef}}}{\dot{m}_a} = B \left[\frac{\gamma_{\text{fn}}+1}{2} \right]^{\frac{1}{\gamma_{\text{fn}}-1}} \left[\frac{\gamma_{\text{fn}}+1}{2\gamma_{\text{fn}}RT_{08}} \right]^{0.5} \frac{RT_{08}}{p_{08}} = 5 \left[\frac{2.4}{2} \right]^{2.5} \left[\frac{2.4}{2(1.4)287(284)} \right]^{0.5} \frac{287(284)}{44500}$$
$$= 0.047$$

$$T_9 = \frac{2}{\gamma_{fn}+1} T_{08} = 237 \text{ K}$$

Moving through central core compressor (see Problem 6.2):

$$T_{03} = T_{02} \left[1 + \frac{1}{\eta_c} [\pi_c^{\frac{\gamma_c-1}{\gamma_c}} - 1] \right] = 248[1 + 1/0.95 [20^{0.27} - 1]] = 573 \text{ K}, \text{ at compressor outlet}$$

$$p_{03} = \pi_c p_{02} = 20(29.63) = 592.6 \text{ kPa}, \text{ at compressor outlet};$$

Moving through combustor (burner):

$$T_{04} = T_{\text{max}} = 1500 \text{ K}, \quad p_{04} = \pi_b p_{03} = 0.95(592.6) = 563 \text{ kPa}, \text{ at burner exit.}$$

$$f = \frac{\dfrac{T_{04}}{T_{03}} - \dfrac{C_{p,c}}{C_{p,b,\text{exit}}}}{\dfrac{\eta_b q_R}{C_{p,b,\text{exit}} T_{03}} - \dfrac{T_{04}}{T_{03}}} = \frac{\dfrac{1500}{573} - \dfrac{1063}{1151}}{\dfrac{1.0(45 \times 10^6)}{1151(573)} - \dfrac{1500}{573}} = 0.0258, \text{ fuel–air ratio (same}$$

as Problem 6.2)

Moving through the turbine section of engine:

$$T_{05} \approx T_{04} - \frac{1}{\eta_m}(T_{03} - T_{02}) - \frac{B}{\eta_m}(T_{08} - T_{02})$$
$$= 1500 - 1/0.99 \cdot (573 - 248) - 5/0.99 \cdot (284 - 248)$$

$$= 990 \text{ K (vs. 1172, Problem 6.2)}$$

$$p_{05} = p_{04}[1 - \frac{1}{\eta_t}(1 - \frac{T_{05}}{T_{04}})]^{\frac{\gamma_t}{\gamma_t-1}} = 563[1 - 1/0.9 \cdot (1 - 990/1500)]^{4.03} =$$

83.2 kPa, at turbine outlet

No afterburner, so $T_{06} = T_{05} = 990$ K, and $p_{06} = p_{05} = 83.2$ kPa. Now, approaching subsonic converging nozzle, need to check for choking:

$\frac{p_{06}}{p_\infty} = \frac{83.2}{18.75} = 4.44$ and choking criterion $\left(\frac{\gamma_n+1}{2} \right)^{\frac{\gamma_n}{\gamma_n-1}} = 1.87 < \frac{p_{06}}{p_\infty}$, so nozzle flow is *choked.*

$$V_e = \sqrt{2\eta_n C_{p,n} T_{06} \left[\frac{\gamma_n-1}{\gamma_n+1} \right]} = \sqrt{2(0.98)1084(990)[0.36/2.36]} = 566.5 \text{ m/s}$$

$T_7 = \frac{2}{\gamma_n+1} T_{06} = 2/2.36 \cdot 990 = 839$ K, exit nozzle static temperature at sonic flow condition.

$$p_7 = p_{06}\left[\tfrac{T_7}{T_{06}}\right]^{\frac{\gamma_n}{\gamma_n-1}} = p_{06}\left[\tfrac{2}{\gamma_n+1}\right]^{\frac{\gamma_n}{\gamma_n-1}} = 83.2[2/2.36]^{3.778} = 44.5 \text{ kPa} = p_e > p_\infty. \quad \text{For}$$
choked flow,

$$\frac{A_e}{\dot{m}_a} = \frac{A_e(1+f)}{\dot{m}} = \left[\frac{\gamma_n+1}{2}\right]^{\frac{1}{\gamma_n-1}}\left[\frac{\gamma_n+1}{2\gamma_n RT_{06}}\right]^{0.5}\frac{RT_{06}}{p_{06}}(1+f)$$

$$= [2.36/2]^{2.778}[2.36/(2(1.36)287(990))]^{0.5}287(990)/83200.\ (1 +$$
$$0.0258) = 0.00969 \text{ m}^2\text{s/kg}.$$

For specific thrust,

$$\frac{F}{\dot{m}_a} = (1+f)V_e + \mathcal{B} \cdot V_{ef} - (1+\mathcal{B})V_\infty + \frac{A_e}{\dot{m}_a}(p_e - p_\infty) + \frac{A_{ef}}{\dot{m}_a}(p_{ef} - p_\infty)$$

$$= (1 + 0.0258)566.5 + 5(304) - 6(251) + 0.00969(44500-18750)$$
$$+ 0.047(23500-18750)$$
$$= 1068 \text{ N s/kg or } 1.07 \text{ kN s/kg (vs. 0.76 for the TJ)}$$

$$\text{TSFC} = \frac{\dot{m}_f}{F} = \frac{f}{\frac{F}{\dot{m}_a}} = 0.0258/1068$$
$$= 2.42 \times 10^{-5} \text{ kg/(s N) or } 0.0871 \text{ kg/(h N) or } 0.85 \text{ lb/(h lb)}$$

vs. 1.2 for the TJ of Problem 6.2.

$$\eta_o = \frac{1}{\text{TSFC}}\frac{V_\infty}{q_R} = (1/2.42 \times 10^{-5})(251/45 \times 10^6) = 0.23, \quad \text{overall   efficiency}$$
(vs. 0.16 for the TJ).

Given $V_{e,\infty} = \sqrt{2\eta_n C_{p,n}T_{06}[1 - (\frac{p_\infty}{p_{06}})^{\frac{\gamma_n-1}{\gamma_n}}]}$

$$= \sqrt{2(0.98)(1084)990[1 - (\tfrac{18.75}{83.2})^{0.265}]} = 828 \text{ m/s}, \quad \text{and} \quad V_{ef,\infty} =$$

$$\sqrt{2\eta_{fn}C_{p,fn}T_{08}[1 - (\tfrac{p_\infty}{p_{08}})^{\frac{\gamma_{fn}-1}{\gamma_{fn}}}]} = \sqrt{2(0.97)(1004)284[1 - (\tfrac{18.75}{44.5})^{0.286}]} = 348 \text{ m/s, so}$$

$$\eta_p = \frac{\frac{F}{\dot{m}_a}V_\infty}{(1+f)\frac{V_{e,\infty}^2}{2} + \mathcal{B}\frac{V_{ef,\infty}^2}{2} - (1+\mathcal{B})\frac{V_\infty^2}{2}}$$
$$= \frac{1068(251)}{1.0258(828^2)/2 + 5(348^2)/2 - (6)251^2/2} = 0.57,$$

the TF's propulsive efficiency (vs. 0.35 for the TJ). Here, the fan is delivering approximately 46% of the overall net thrust delivered, while the core exhaust jet is delivering 54%. Increasing the fan pressure ratio π_f will increase the fan's percentage of the overall thrust delivered, up to a threshold limit value for π_f, above which the percentage will begin to drop.

7.2. Looking at a turbofan engine, at a flight Mach no. Ma_∞ of 0.83, at 11,585 m altitude ($p_\infty = 20.67$ kPa, $T_\infty = 216.7$ K, $\rho_\infty = 0.332$ kg/m^3). Bypass ratio B of 7. Core flow of 90 kg/s. Information provided already for front part of core engine.

Core:

$C_{p,c} = 1042$ J/(kg K)

$$f = \frac{\dfrac{T_{04}}{T_{03}} - \dfrac{C_{p,c}}{C_{p,b,\text{exit}}}}{\dfrac{\eta_b q_R}{C_{p,b,\text{exit}} T_{03}} - \dfrac{T_{04}}{T_{03}}} = \frac{\dfrac{1400}{638.2} - \dfrac{1042}{1131}}{\dfrac{0.97(42 \times 10^6)}{1131(638.2)} - \dfrac{1400}{638.2}} = 0.0235, \text{ fuel–air ratio}$$

Moving through fan:

$p_{08} = \pi_f p_{02} = 1.7(32) = 54.4$ kPa, at fan outlet;

$T_{08} = T_{02}[1 + \frac{1}{\eta_f}[\pi_f^{\frac{\gamma_f - 1}{\gamma_f}} - 1]] = 246.5[1 + 1/0.89 \cong [1.7^{0.286} - 1]] = 292$ K,   at fan outlet.

$\frac{p_{08}}{p_\infty} = \frac{54.4}{20.67} = 2.63$ and choking criterion $\left(\frac{\gamma_f + 1}{2}\right)^{\frac{\gamma_f}{\gamma_f - 1}} = 1.89 < \frac{p_{08}}{p_\infty}$ so fan nozzle flow is *choked*. $C_{p,fn} = 1{,}005$ J/(kg K).

$$V_{ef} = \sqrt{2\eta_{fn} C_{p,fn} T_{08}\left[\frac{\gamma_{fn} - 1}{\gamma_{fn} + 1}\right]} = \sqrt{2(0.97)1005(292)[0.4/2.4]} = 308 \text{ m/s}$$

$$p_{ef} = p_{08}\left[\frac{T_{ef}}{T_{08}}\right]^{\frac{\gamma_{fn}}{\gamma_{fn} - 1}} = p_{08}\left[\frac{2}{\gamma_{fn} + 1}\right]^{\frac{\gamma_{fn}}{\gamma_{fn} - 1}} = p_9 = 28.74 \text{ kPa}$$

$$\frac{A_{ef}}{\dot{m}_a} = B\left[\frac{\gamma_{fn} + 1}{2}\right]^{\frac{1}{\gamma_{fn} - 1}}\left[\frac{\gamma_{fn} + 1}{2\gamma_{fn} RT_{08}}\right]^{0.5}\frac{RT_{08}}{p_{08}}$$

$$= 7\left[\frac{2.4}{2}\right]^{2.5}\left[\frac{2.4}{2(1.4)287(292)}\right]^{0.5}\frac{287(292)}{54400} = 0.0544$$

Moving through central core turbine:

$$T_{05} \approx T_{04} - \frac{1}{\eta_m}(T_{03} - T_{02}) - \frac{B}{\eta_m}(T_{08} - T_{02}) = 1400 - 1/0.99$$
$$\cong (638.2 - 246.5) - 7/0.99 \cong (292 - 246.5) = 682.6 \text{ K}$$

$p_{05} = p_{04}[1 - \frac{1}{\eta_t}(1 - \frac{T_{05}}{T_{04}})]^{\frac{\gamma_t}{\gamma_t - 1}} = 846.7[1 - 1/0.9 (1 - 682.6/1400)]^{3.857}$
$= 32.85$ kPa, at turbine outlet

No afterburner, so $T_{06} = T_{05} = 990$ K, and $p_{06} = p_{05} = 83.2$ kPa. Now, approaching subsonic converging nozzle, need to check for choking:

$\frac{p_{06}}{p_\infty} = \frac{32.85}{20.67} = 1.59$ and choking criterion $(\frac{\gamma_n+1}{2})^{\frac{\gamma_n}{\gamma_n-1}} = 1.875 > \frac{p_{06}}{p_\infty}$, so nozzle flow is *unchoked*. $p_7 = p_e = p_\infty$. ($\gamma_n = 1.37$, $C_{p,n} = 1.37(287)/0.37 = 1{,}063$ J/(kg K).

$$V_e = \sqrt{2\eta_n C_{p,n} T_{06}\left[1 - \left(\frac{p_\infty}{p_{06}}\right)^{\frac{\gamma_n-1}{\gamma_n}}\right]} = \sqrt{2(0.98)1063(682.6)\left[1 - \left(\frac{20.67}{32.85}\right)^{0.27}\right]}$$

$$= 408.9\,\text{m/s}$$

$$V_\infty = a_\infty Ma_\infty = \sqrt{\gamma_{\text{air}} R_{\text{air}} T_\infty} \cdot Ma_\infty 295(0.83) = 245\,\text{m/s}$$

For thrust,

$$F = \dot{m}_a\{(1+f)V_e + \mathcal{B} \cdot V_{ef} - (1+\mathcal{B})V_\infty + \frac{A_e}{\dot{m}_a}(p_e - p_\infty) + \frac{A_{ef}}{\dot{m}_a}(p_{ef} - p_\infty)\}$$

$$= 90\{(1 + 0.0235)408.9 + 7(308) - 8(245) + 0 + 0.0544(28740 - 20670)\}$$
$$= 94817\,\text{N or } 94.8\,\text{kN}$$

Here, the fan is delivering approximately 83% of the overall net thrust delivered, while the core exhaust jet is delivering 17%.

7.3. Looking at a turbofan engine, at a flight Mach no. Ma_∞ of 0.85, at 12,200 m altitude ($p_\infty = 18.75$ kPa, $T_\infty = 216.7$ K, $\rho_\infty = 0.30$ kg/m^3). Bypass ratio $\mathcal{B}$ of 5.

Starting from intake (diffuser), as in Problem 6.2 (same numbers as input):
$T_{02} = T_\infty[1 + \frac{\gamma_d-1}{2}Ma_\infty^2] = 216.7[1 + 0.2(0.85^2)] = 248$ K, at diffuser outlet
$p_{02} = p_\infty[1 + \eta_d(\frac{T_{02}}{T_\infty} - 1)]^{\frac{\gamma_d}{\gamma_d-1}} = 18.75[1 + 0.97(1.144 - 1)]^{3.5} = 29.63$ kPa, at diffuser outlet

Moving through fan:

$p_{08} = \pi_f p_{02} = 1.35(29.63) = 40.0$ kPa, at fan outlet;
$$T_{08} = T_{02}\left[1 + \frac{1}{\eta_f}\left[\pi_f^{\frac{\gamma_f-1}{\gamma_f}} - 1\right]\right] = 248[1 + 1/0.85 \cong [1.35^{0.286} - 1]]$$

$$= 274.2\,\text{K, at fan outlet.}$$

$\frac{p_{08}}{p_\infty} = \frac{40.0}{18.75} = 2.13$ and choking criterion $(\frac{\gamma_f+1}{2})^{\frac{\gamma_f}{\gamma_f-1}} = 1.89 < \frac{p_{08}}{p_\infty}$, so fan nozzle flow is *choked*.

$$V_{ef} = \sqrt{2\eta_{fn} C_{p,fn} T_{08}\left[\frac{\gamma_{fn}-1}{\gamma_{fn}+1}\right]} = \sqrt{2(0.97)1004(274.2)[0.4/2.4]} = 298.4\,\text{m/s}$$

$$p_{ef} = p_{08} \left[\frac{T_{ef}}{T_{08}} \right]^{\frac{\gamma_{fn}}{\gamma_{fn} - 1}} = p_{08} \left[\frac{2}{\gamma_{fn} + 1} \right]^{\frac{\gamma_{fn}}{\gamma_{fn} - 1}} = p_9 = 21.1 \, \text{kPa}$$

$$\frac{A_{ef}}{\dot{m}_a} = B \left[\frac{\gamma_{fn} + 1}{2} \right]^{\frac{1}{\gamma_{fn} - 1}} \left[\frac{\gamma_{fn} + 1}{2\gamma_{fn} RT_{08}} \right]^{0.5} \frac{RT_{08}}{p_{08}}$$

$$= 5 \left[\frac{2.4}{2} \right]^{2.5} \left[\frac{2.4}{2(1.4)287(274.2)} \right]^{0.5} \frac{287(274.2)}{40000} = 0.0512$$

$$T_9 = \frac{2}{\gamma_{fn} + 1} T_{08} = 228.5 \, \text{K}$$

Moving through central core compressor (see Problem 6.2):

$T_{03} = T_{02}[1 + \frac{1}{\eta_c} [\pi_c^{\frac{\gamma_c - 1}{\gamma_c}} - 1]] = 248[1 + 1/0.95 \, [20^{0.27} - 1]] = 573 \, \text{K}$,   at   compressor outlet

$p_{03} = \pi_c p_{02} = 20(29.63) = 592.6 \, \text{kPa}$, at compressor outlet;

Moving through combustor (burner):

$T_{04} = T_{max} = 1,500 \, \text{K}, \; p_{04} = \pi_b p_{03} = 0.95(592.6) = 563 \, \text{kPa}$, at burner exit.

$$f = \frac{\dfrac{T_{04}}{T_{03}} - \dfrac{C_{p,c}}{C_{p,b,exit}}}{\dfrac{\eta_b q_R}{C_{p,b,exit} T_{03}} - \dfrac{T_{04}}{T_{03}}} = \frac{\dfrac{1500}{573} - \dfrac{1063}{1151}}{\dfrac{1.0(45 \times 10^6)}{1151(573)} - \dfrac{1500}{573}} = 0.0258, \; \text{fuel–air ratio (same}$$

as Problem 6.2)

Moving through the turbine section of engine:

$T_{05} \approx T_{04} - \frac{1}{\eta_m}(T_{03} - T_{02}) - \frac{B}{\eta_m}(T_{08} - T_{02}) = 1500 - 1/0.99 \, (573 - 248) -$
$5/0.99 \, (274.2 - 248)$

    $= 1,040 \, \text{K} \; (\text{vs. } 1,172, \text{Problem 6.2; } 990, \text{Problem 7.1})$

$p_{05} = p_{04}[1 - \frac{1}{\eta_t}(1 - \frac{T_{05}}{T_{04}})]^{\frac{\gamma_t}{\gamma_t - 1}} = 563[1 - 1/0.9 \, (1 - 1040/1500)]^{4.03} = 105 \, \text{kPa}$,

at turbine outlet

No afterburner, so $T_{06} = T_{05} = 1,040 \, \text{K}$, and $p_{06} = p_{05} = 105 \, \text{kPa}$. Now, approaching subsonic converging nozzle, need to check for choking:

$\frac{p_{06}}{p_\infty} = \frac{105}{18.75} = 5.6$ and choking criterion $(\frac{\gamma_n + 1}{2})^{\frac{\gamma_n}{\gamma_n - 1}} = 1.87 < \frac{p_{06}}{p_\infty}$, so nozzle flow is choked.

$$V_e = \sqrt{2\eta_n C_{p,n} T_{06} \left[\frac{\gamma_n - 1}{\gamma_n + 1} \right]} = \sqrt{2(0.98)1084(1040)[0.36/2.36]} = 580.6 \, \text{m/s}$$

$T_7 = \frac{2}{\gamma_n + 1} T_{06} = 2/2.36.1040 = 881 \, \text{K}$, exit nozzle static temperature at sonic flow condition.

$$p_7 = p_{06} \left[\frac{T_7}{T_{06}}\right]^{\frac{\gamma_n}{\gamma_n-1}} = p_{06} \left[\frac{2}{\gamma_n+1}\right]^{\frac{\gamma_n}{\gamma_n-1}} = 105[2/2.36]^{3.778} = 56.2 \, \text{kPa} = p_e > p_\infty. \quad \text{For}$$

choked flow,

$$\frac{A_e}{\dot{m}_a} = \frac{A_e(1+f)}{\dot{m}} = \left[\frac{\gamma_n+1}{2}\right]^{\frac{1}{\gamma_n-1}} \left[\frac{\gamma_n+1}{2\gamma_n RT_{06}}\right]^{0.5} \frac{RT_{06}}{p_{06}} (1+f)$$

$$= [2.36/2]^{2.778}[2.36/(2(1.36)287(1040))]^{0.5} 287(1040)/105000 \, (1+0.0258)$$

$$= 0.00765 \, \text{m}^2\text{s/kg}$$

For specific thrust

$$\frac{F}{\dot{m}_a} = (1+f)V_e + B \cdot V_{\text{ef}} - (1+B)V_\infty + \frac{A_e}{\dot{m}_a}(p_e - p_\infty) + \frac{A_{\text{ef}}}{\dot{m}_a}(p_{\text{ef}} - p_\infty)$$

$$= (1 + 0.0258)580.6 + 5(298.4) - 6(251) + 0.00765(56200 - 18750) +$$
$$0.0512(21100 - 18750)$$

$$= 988.4 \, \text{N s/kg or } 0.988 \, \text{kN s/kg (vs. 0.76 for the TJ of 6.2, 1.07 for the TF of}$$
Problem 7.1)

Conforms with trend observed in Fig. 7.10.

$TSFC = \frac{\dot{m}_f}{F} = \frac{f}{\frac{F}{\dot{m}_a}} = 0.0258/988.4 = 2.61 \times 10^{-5} \, \text{kg/(s N) or } 0.094 \, \text{kg/(h N) or}$

0.92 lb/(h lb)

vs. 1.2 for the TJ of Problem 6.2, 0.85 for TF of Problem 7.1.

Conforms with trend observed in Fig. 7.10.

$\eta_o = \frac{1}{TSFC} \frac{V_\infty}{q_R} = (1/2.61 \times 10^{-5})(251/45 \times 10^6) = 0.214$, overall efficiency

(vs. 0.16 for the TJ of 6.2, 0.23 for the TF of Problem 7.1)

Given

$$V_{e,\infty} = \sqrt{2\eta_n C_{p,n} T_{06}\left[1 - \left(\frac{p_\infty}{p_{06}}\right)^{\frac{\gamma_n-1}{\gamma_n}}\right]}$$

$$= \sqrt{2(0.98)(1084)1040\left[1 - \left(\frac{18.75}{105}\right)^{0.265}\right]} = 900 \, \text{m/s},$$

$$V_{\text{ef},\infty} = \sqrt{2\eta_{\text{fn}} C_{p,\text{fn}} T_{08}\left[1 - \left(\frac{p_\infty}{p_{08}}\right)^{\frac{\gamma_{\text{fn}}-1}{\gamma_{\text{fn}}}}\right]} = \sqrt{2(0.97)(1004)274.2[1 - \left(\frac{18.75}{40}\right)^{0.286}]}$$

$$= 322.6 \, \text{m/s, so}$$

$$\eta_p = \frac{\frac{F}{\dot{m}_a} V_\infty}{(1+f)\frac{V_{e,\infty}^2}{2} + B\frac{V_{\text{ef},\infty}^2}{2} - (1+B)\frac{V_\infty^2}{2}}$$

$$= \frac{988.4(251)}{1.0258(900^2)/2 + 5(322.6^2)/2 - (6)251^2/2} = 0.51$$

the TF's propulsive efficiency (vs. 0.35 for the TJ of Problem 6.2, 0.57 for TF of Problem 7.1). Here, the fan is delivering approximately 36% of the overall net thrust delivered, while the core exhaust jet is delivering 64%.

7.4. Looking at a mixed-stream, afterburning turbofan engine, at a flight Mach no. Ma_∞ of 0.85, at 12,200 m altitude $(p_\infty = 18.75$ kPa, $T_\infty = 216.7$ K, $\rho_\infty = 0.30$ kg/m^3). Bypass ratio B of 0.7.

Starting from intake (diffuser), as in Problem 6.2 (same numbers as input):

$$T_{02} = T_\infty\left[1 + \frac{\gamma_d - 1}{2}Ma_\infty^2\right] = 216.7\left[1 + 0.2\left(0.85^2\right)\right] = 248 \text{ K at diffuser outlet}$$

$$p_{02} = p_\infty\left[1 + \eta_d\left(\frac{T_{02}}{T_\infty} - 1\right)\right]^{\frac{\gamma_d}{\gamma_d - 1}} = 18.75[1 + 0.97(1.144 - 1)]^{3.5}$$
$$= 29.63 \text{ kPa, at diffuser outlet}$$

Moving through fan:
$p_{08} = \pi_f p_{02} = 3.5(29.63) = 103.7$ kPa, at fan outlet; assume no stagnation pressure losses in fan duct, $p_{08.5} \approx p_{08}$

$$T_{08} = T_{02}[1 + \frac{1}{\eta_f}[\pi_f^{\frac{\gamma_f - 1}{\gamma_f}} - 1]] = 248[1 + 1/0.85\cdot[3.5^{0.286} - 1]] = 374 \text{ K}, \quad \text{at fan}$$
outlet; assume no stagnation temperature losses in fan duct, $T_{08.5} \approx T_{08}$

Moving through central core compressor (see Problem 6.2):
$$T_{03} = T_{02}\left[1 + \frac{1}{\eta_c}\left[\pi_c^{\frac{\gamma_c - 1}{\gamma_c}} - 1\right]\right] = 248[1 + 1/0.95\cdot[20^{0.27} - 1]] = 573 \text{ K}, \quad \text{at}$$
compressor outlet
$p_{03} = \pi_c p_{02} = 20(29.63) = 592.6$ kPa, at compressor outlet;

Moving through combustor (burner):
$T_{04} = T_{\max,B} = 1,500$ K, $p_{04} = \pi_b p_{03} = 0.95(592.6) = 563$ kPa, at burner exit.

$$f = \frac{\dfrac{T_{04}}{T_{03}} - \dfrac{C_{p,c}}{C_{p,b,\text{exit}}}}{\dfrac{\eta_b q_R}{C_{p,b,\text{exit}} T_{03}} - \dfrac{T_{04}}{T_{03}}} = \frac{\dfrac{1500}{573} - \dfrac{1063}{1151}}{\dfrac{0.99(45\times10^6)}{1151(573)} - \dfrac{1500}{573}} = 0.0261, \text{ fuel–air ratio}$$

Moving through the turbine section of engine:
$$T_{05} \approx T_{04} - \frac{1}{\eta_m}(T_{03} - T_{02}) - \frac{B}{\eta_m}(T_{08} - T_{02})$$
$$= 1500 - 1/0.99\cdot(573 - 248) - 0.7/0.99\cdot(374 - 248) = 1083 \text{ K}$$

$$p_{05} = p_{04}[1 - \frac{1}{\eta_t}(1 - \frac{T_{05}}{T_{04}})]^{\frac{\gamma_t}{\gamma_t - 1}} = 563[1 - 1/0.9\cdot(1 - 1083/1500)]^{4.03} =$$
127 kPa, at turbine outlet
In mixing zone between turbine exit and afterburner entry:

$$T_{05.5} \approx \frac{(1+f)}{(1+f+B)}T_{05} + \frac{B}{(1+f+B)}T_{08.5}$$

$$= \frac{1+0.0261}{1+0.0261+0.7} \cdot 1083 + \frac{0.7}{1+0.0261+0.7} \cdot 374 = 795.5 \text{ K}$$

$$p_{05.5} = p_{05}\left[1 - \frac{1}{\eta_{mixer}}\left(1 - \frac{T_{05.5}}{T_{05}}\right)\right]^{\frac{\gamma_t}{\gamma_t-1}} = 127[1 - 1/0.99(1 - 795.5/1083)]^{4.03}$$

$$= 36.1 \text{ kPa}$$

At exit of the afterburner, $T_{06} = T_{\max, \text{A/B}} = 2{,}000$ K, and $p_{06} = \pi_{AB}\,p_{05.5} = 0.98(36.1) = 35.4$ kPa.

Given $\gamma_t = 1.33$ and$_{AB} = 1.33$, $C_{p, \text{A/B,exit}} = C_{p,t} = 1.33(287)/0.33 = 1157$ J/(kg K), so that:

$$f_{AB} = \frac{\dfrac{T_{06}}{T_{05.5}} - \dfrac{C_{p,t}}{C_{p,\text{AB,exit}}}}{\dfrac{\eta_{AB}q_R}{C_{p,\text{AB,exit}}T_{05.5}} - \dfrac{T_{06}}{T_{05.5}}} = \frac{\dfrac{2000}{795.5} - \dfrac{1157}{1157}}{\dfrac{0.9(45 \times 10^6)}{1157(795.5)} - \dfrac{2000}{795.5}} = 0.0365 = \frac{\dot{m}_{f.AB}}{\dot{m}_a + \dot{m}_f + \dot{m}_{by}}$$

One checks that $f_d + f_{AB} = 0.0261 + 0.0365 = 0.063 < f_{\text{stoich}}$ (0.067 for this air/fuel mixture), so can sustain the afterburner combustion process.

$\frac{p_{06}}{p_\infty} = \frac{35.4}{18.75} = 1.89$ and choking criterion $\left(\frac{\gamma_n+1}{2}\right)^{\frac{\gamma_n}{\gamma_n-1}} = 1.87 < \frac{p_{06}}{p_\infty}$, so nozzle flow is *choked*.

$$V_e = \sqrt{2\eta_n C_{p,n}T_{06}\left[\frac{\gamma_n-1}{\gamma_n+1}\right]} = \sqrt{2(0.98)1084(2000)[0.36/2.36]} = 805.1 \text{m/s}$$

$T_7 = \frac{2}{\gamma_n+1}T_{06} = 2/2.36\,2000 = 1695$ K, exit nozzle static temperature at sonic flow condition.

$p_7 = p_{06}[\frac{T_7}{T_{06}}]^{\frac{\gamma_n}{\gamma_n-1}} = p_{06}[\frac{2}{\gamma_n+1}]^{\frac{\gamma_n}{\gamma_n-1}} = 35.4[2/2.36]^{3.778} = 18.94 \text{ kPa} = p_e > p_\infty$. For choked flow,

$$\frac{A_e}{\dot{m}_a} = \frac{A_e(1+f+B+[1+f+B]f_{AB})}{\dot{m}}$$

$$= \left[\frac{\gamma_n+1}{2}\right]^{\frac{1}{\gamma_n-1}}\left[\frac{\gamma_n+1}{2\gamma_n RT_{06}}\right]^{0.5}\frac{RT_{06}}{p_{06}}(1+f+B+[1+f+B]f_{AB})$$

$$= [2.36/2]^{2.778}[2.36/(2(1.36)287(2000))]^{0.5}287(2000)/35400\cdot$$

$$(1+0.0261+0.7+[1+0.0261+0.7]0.0365) = 0.0565 \text{ m}^2\text{s/kg}$$

For specific thrust

$$\frac{F}{\dot{m}_a} = (1+f+\mathcal{B}+[1+f+\mathcal{B}]f_{AB})V_e - (1+\mathcal{B})V_\infty + \frac{A_e}{\dot{m}_a}(p_e - p_\infty)$$

$$= (1+0.0261+0.7+[1+0.0261+0.7]0.0365)805.1 - (1+0.7)(251)$$

$$+ 0.0565(18940 - 18750)$$

$$= 1024.4\,\text{N s/kg} \quad \text{or} \quad 1.024\,\text{kN s/kg(vs.1.095 for the a/b TJ of Prob.6.3)}$$

$$\text{TSFC} = \frac{\dot{m}_f + \dot{m}_{f,AB}}{F} \approx \frac{f + [1+f+\mathcal{B})]f_{AB}}{\frac{F}{\dot{m}_a}}$$

$$= (0.0261 + [1+0.0261+0.7]0.0365)/1024.4 = 8.7 \times 10^{-5}\text{kg/sN}$$

or 0.313 kg/(h N) or 3.07 lb/(h lb), vs. 1.64 for the a/b TJ of Problem 6.3

7.5. An aircraft is presently flying at a flight Mach number of 0.7 at 10000 m ISA altitude. This airplane employs three Pratt & Whitney JT9D-59A turbofan engines (see engine performance chart for 10,000 m, Fig. 7.9). The engine throttle setting is presently set at 95% of maximum continuous N_1. Establish the rate at which the airplane is removing fuel from its tank storage (kg/h).

From Fig. 7.9, the thrust being delivered by one engine at that flight condition is around 50 kN. The TSFC at that flight condition is around 0.59 kN/(h kN), or 0.06 kg/(h N). Total thrust being delivered by the airplane is:

$$F = 3\,F_{\text{eng}} = 3\,(50,000) = 150,000\,\text{N or 150 kN, so for fuel consumption:}$$

$$\dot{m}_f = F \cdot \text{TSFC} = 150000\,(0.06) = 9000\text{kg/h}$$

7.6. Perform an off-design cycle analysis for a large commercial non-mixing turbofan engine at its cruise throttle setting, at altitudes from 0 to 15,000 m ISA, and flight Mach numbers from zero to 0.9. Take note of the following design parameters:

diffuser, $\eta_d = 0.98$; $\gamma_d = 1.4$; fan, $\eta_f = 0.965$, $\gamma_f = 1.4$, $\pi_f = 1.5$, $\gamma_{fn} = 1.4$, $\eta_{fn} = 0.985$, BPR $= 7.50$, fan nozzle fixed exit area $A_{ef} = 4.0\,\text{m}^2$
compressor, $\eta_c = 0.985$, $\gamma_c = 1.375$, $\pi_c = 30.1$;
burner, $\eta_{nb} = 0.99$, $\gamma_b = 1.35$, $\pi_b = 0.985$, $T_{max} = 1350$ K
$q_R = 44$ MJ/kg of fuel; turbine, $\gamma_t = 1.335$; $\eta_t = 0.955$; $\eta_m = 0.99$
core nozzle, $\eta_n = 0.98$; $\gamma_{\gamma n} = 1.36$

Plot thrust F vs. flight Mach number Ma_∞ at the 4 principal altitudes of 0, 5000, 10000 and 15000 m ISA (generate enough data points so that each curve appears smooth in appearance). Ascertain the TSFC data points when values of 0.04, 0.05, 0.06 and 0.07 kg/(h N) are attained.

As a check, sample set of calculations, $Ma_\infty = 0.7$ at $h_{ASL} = 10,000$ m:
$p_\infty = 26.5$ kPa, $T_\infty = 223.3$ K, $\gamma_\infty = 0.414$ kg/m^3, $a_\infty = 299.5$ m/s
$A_{ef} = 4$ m^2

Starting from intake (diffuser):

$T_{02} = T_\infty[1 + \frac{\gamma_d-1}{2}Ma_\infty^2] = 223.3[1 + 0.2(0.7^2)] = 245.2$ K, at diffuser outlet

$p_{02} = p_\infty[1 + \eta_d(\frac{T_{02}}{T_\infty} - 1)]^{\frac{\gamma_d}{\gamma_d-1}} = 26.5[1 + 0.98(1.098 - 1)]^{3.5} = 36.5$ kPa,   at diffuser outlet

Moving through fan:

$p_{08} = \pi_f p_{02} = 1.5(36.5) = 54.8$ kPa, at fan outlet;

$T_{08} = T_{02}\left[1 + \frac{1}{\eta_f}\left[\pi_f^{\frac{\gamma_f-1}{\gamma_f}} - 1\right]\right] = 245.2[1 + 1/0.965\,[1.5^{0.286} - 1]] = 276.5$ K,

at fan outlet.

$\frac{p_{08}}{p_\infty} = \frac{54.8}{26.5} = 2.07$ and choking criterion $(\frac{\gamma_f+1}{2})^{\frac{\gamma_f}{\gamma_f-1}} = 1.89 < \frac{p_{08}}{p_\infty}$ so fan nozzle flow is *choked*.

$$V_{ef} = \sqrt{2\eta_{fn}C_{p,fn}T_{08}\left[\frac{\gamma_{fn}-1}{\gamma_{fn}+1}\right]} = \sqrt{2(0.985)1004(276.5)[0.4/2.4]} = 302 \text{ m/s}$$

$$p_{ef} = p_{08}\left[\frac{T_{ef}}{T_{08}}\right]^{\frac{\gamma_{fn}}{\gamma_{fn}-1}} = p_{08}\left[\frac{2}{\gamma_{fn}+1}\right]^{\frac{\gamma_{fn}}{\gamma_{fn}-1}} = p_9 = 29\text{kPa}$$

$$\frac{A_{ef}}{\dot{m}_a} = B\left[\frac{\gamma_{fn}+1}{2}\right]^{\frac{1}{\gamma_{fn}-1}}\left[\frac{\gamma_{fn}+1}{2\gamma_{fn}RT_{08}}\right]^{0.5}\frac{RT_{08}}{p_{08}}$$
$$= 7.5\left[\frac{2.4}{2}\right]^{2.5}\left[\frac{2.4}{2(1.4)287(276.5)}\right]^{0.5}\frac{287(276.5)}{54800} = 0.056$$

$\dot{m}_a = A_{ef}/0.056 = 4/0.056 = 71.43$ kg/s$(158$ lb/s$)$

$$T_9 = \frac{2}{\gamma_{fn}+1}T_{08} = 230.4 \text{ K}$$

Moving through central core compressor:

$T_{03} = T_{02}\left[1 + \frac{1}{\eta_c}\left[\pi_c^{\frac{\gamma_c-1}{\gamma_c}} - 1\right]\right] = 245.2[1 + 1/0.985\cdot[30.1^{0.273} - 1]] = 627$ K, at compressor outlet

$p_{03} = \pi_c p_{02} = 30.1(36.5) = 1,099$ kPa, at compressor outlet;

$C_{p,c} = \frac{\gamma_c R_{air}}{\gamma_c-1} = 1.375(287)/0.375 = 1052$ J/kg K,$C_{p,b} = 1107$ J/kg K

Moving through combustor (burner):

$T_{04} = T_{max} = 1,350$ K, $p_{04} = \pi_b p_{03} = 0.985(1099) = 1,083$ kPa,   at burner exit.

$C_{p,b,exit} \approx 2\bar{C}_{p,b} - C_{p,c} = 2(1107) - 1052 = 1162$ J/kg $\cdot$K .

$$f = \frac{\frac{T_{04}}{T_{03}} - \frac{C_{p,c}}{C_{p,b,\text{exit}}}}{\frac{\eta_b q_R}{C_{p,b,\text{exit}} T_{03}} - \frac{T_{04}}{T_{03}}} = \frac{\frac{1350}{627} - \frac{1063}{1162}}{\frac{0.99(44 \times 10^6)}{1162(627)} - \frac{1350}{627}} = 0.0215, \text{ fuel–air ratio}$$

Moving through the turbine section of engine:

$T_{05} \approx T_{04} - \frac{1}{\eta_m}(T_{03} - T_{02}) - \frac{B}{\eta_m}(T_{08} - T_{02}) =$ 1350–1/0.99·(627–245.2)–7.5/0.99

(276.5– 245.2) = 727.2 K

$$p_{05} = p_{04}\left[1 - \frac{1}{\eta_t}\left(1 - \frac{T_{05}}{T_{04}}\right)\right]^{\frac{\gamma_t}{\gamma_t-1}} = 1083[1 - 1/0.955\,(1 - 727.2/1350)]^{3.99}$$

= 77.9 kPa, at turbine outlet

No afterburner, so $T_{06} = T_{05} = 727.2$ K, and $p_{06} = p_{05} = 77.9$ kPa. Now, approaching subsonic converging nozzle, need to check for choking:

$\frac{p_{06}}{p_\infty} = \frac{77.9}{26.5} = 2.94$ and choking criterion $\left(\frac{\gamma_n+1}{2}\right)^{\frac{\gamma_n}{\gamma_n-1}} = 1.87 < \frac{p_{06}}{p_\infty}$ so nozzle flow is *choked*.

$$V_e = \sqrt{2\eta_n C_{p,n} T_{06}\left[\frac{\gamma_n - 1}{\gamma_n + 1}\right]} = \sqrt{2(0.98)1084(727.2)[0.36/2.36]} = 485.5 \text{ m/s}$$

$T_7 = \frac{2}{\gamma_n+1} T_{06} = 2/2.36 \cdot 727.2 = 616.3$ K, exit nozzle static temperature at sonic flow condition.

$p_7 = p_{06}\left[\frac{T_7}{T_{06}}\right]^{\frac{\gamma_n}{\gamma_n-1}} = p_{06}\left[\frac{2}{\gamma_n+1}\right]^{\frac{\gamma_n}{\gamma_n-1}} = 77.9[2/2.36]^{3.778} = 41.7 \text{ kPa} = p_e > p_\infty.$ For choked flow,

$$\frac{A_e}{\dot{m}_a} = \frac{A_e(1+f)}{\dot{m}} = \left[\frac{\gamma_n + 1}{2}\right]^{\frac{1}{\gamma_n-1}}\left[\frac{\gamma_n + 1}{2\gamma_n RT_{06}}\right]^{0.5}\frac{RT_{06}}{p_{06}}(1+f)$$

$= [2.36/2]^{2.778}[2.36/(2(1.36)287(727.2))]^{0.5}287(727.2)/77,900 \cdot (1 + 0.0215) = 0.00884$ m²s/kg.

$V_\infty = 0.7 (299.5) = 210$ m/s

For specific thrust,

$$\frac{F}{\dot{m}_a} = (1+f)V_e + B \cdot V_{\text{ef}} - (1+B)V_\infty + \frac{A_e}{\dot{m}_a}(p_e - p_\infty) + \frac{A_{\text{ef}}}{\dot{m}_a}(p_{\text{ef}} - p_\infty)$$

= (1 + 0.0215)485.5 + 7.5(302)–8.5(210) + 0.00884(41700–26500) + 0.056 (29000–26500)

= 1250 N·s/kg or 1.25 kN·s/kg

$F = \frac{F}{\dot{m}_a} \cdot \dot{m}_a = 1250\,(71.43) = 89290$ N (20,060 lbf)

$TSFC = \frac{\dot{m}_f}{F} = \frac{f}{\frac{F}{\dot{m}_a}} = 0.0215/1250 = 1.72 \times 10^{-5}$ kg/s ·N or 0.062 kg/h ·N   or

0.607 lb/h·lb

References

1. Archer RD, Saarlas M (1996) An introduction to aerospace propulsion. Prentice-Hall, Upper Saddle River
2. Treager IE (1970) Aircraft gas turbine engine technology. McGraw-Hill, New York
3. Anonymous (2004) Airplane flying handbook. FAA-H-8083-3A, Airman Testing Standards Branch, Federal Aviation Administration, U.S. Department of Transportation, Oklahoma City
4. Hu H, Kobayashi T, Saga T et al (1999) Research on the rectangular lobed exhaust/ejector mixer systems. T Jpn Soc Aeronaut S 41:87–194
5. Mengle VG, Dalton WN, Bridges JC, Boyd KC (1997) Noise reduction with lobed mixers: nozzle-length and free-jet speed effects. NASA TM 97-206221
6. Mattingly JD (1996) Elements of gas turbine propulsion. McGraw-Hill, New York
7. Anonymous (2009) Report on jet engine noise reduction. Naval Research Advisory Committee Panel on Jet Engine Noise Reduction, U.S. Navy
8. Schlinker RH, Simonich JC, Shannon DW et al (2009) Supersonic jet noise from round and chevron nozzles: experimental studies. 15th AIAA/CEAS Aeroacoustics Conference, AIAA Paper No. 2009-3257, Miami (Florida)

Chapter 8
Turboprop and Turboshaft Engines

8.1 Introduction

The first turboprop engine appeared in 1938 (Cs-1 engine developed and tested by György Jendrassik's team in Hungary in the later 1930s and early 1940s), almost in conjunction with the introduction elsewhere of the functional turbojet engine. Rolls-Royce and General Electric separately produced functional turboprop engines for flight usage in 1945 [1]. The first turboshaft engine appeared on the scene in 1948 (the Turbomeca TT 782, forerunner to Turbomeca's 400-hp Artouste II, which would drive the first production turboshaft-powered helicopter, the Sud Aviation S.E. 3130 Alouette II, beginning in 1955; prior to this advance, all previous production helicopters of significance, from 1936's Focke-Aghelis FA-61, had been piston engined; [2]). Both turboprop [TP] and turboshaft [TS] engines use a gas turbine core engine to drive an output power shaft for a propeller or helicopter rotor. The main difference between the two variants is that a TP engine might also produce a fraction of its overall thrust via a hot core exhaust jet, while a conventional TS engine will have a lower exhaust velocity but correspondingly somewhat higher shaft power as the tradeoff. TPs commonly employ a free power turbine to drive the speed reduction gear that in turn drives the propeller at a lower rotation speed. This is typically also true for TS engines, in driving the main rotor at a significantly lower rotation speed. It is less common that the free power turbine would also drive, on the same shaft, an LP compressor section; if so, it would no longer technically be "free". With respect to shaft rotation speeds, the high-pressure (HP) core compressor/turbine shaft can rotate at say 35,000 rpm, the free LP (low-pressure) turbine at around 20,000 rpm, and the propeller shaft at around 1,200 rpm. If a single compressor/turbine rotor combination is used for the core engine (note: for more flexibility and power, one could go to a LP/HP two-spool core setup), this TP variant would be called a "one-spool/two-shaft" setup, as shown in Fig. 8.1 (the second shaft, from the free turbine, drives the reduction gear). The mechanical independence of the free turbine from the core rotor elements allows for more flexibility in managing the propeller power

D. R. Greatrix, *Powered Flight*, DOI: 10.1007/978-1-4471-2485-6_8,
© Springer-Verlag London Limited 2012

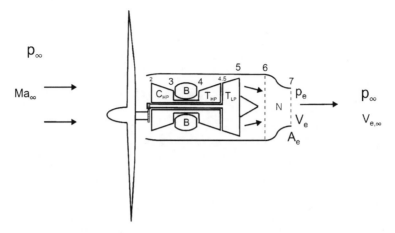

Fig. 8.1 Schematic diagram illustrating the various turboprop engine components and associated station numbering. The *above* shows a "one-spool/two-shaft" configuration

demands, without the potential for adversely affecting core engine operational efficiency. See Fig. 8.2 for an example engine using this split-shaft approach, and Figs. 8.3 and 8.4 for engines only using a single fixed shaft [3, 4].

The overall thrust given by a TP engine driving a propeller can be described by the following:

$$F_{\text{overall}} = F_{\text{prop}} + F_{\text{jet}}|_{\text{net}} = \frac{\eta_{\text{pr}} P_s}{V_{\infty}} + \dot{m}_a([1 + f]V_{e,\infty} - V_{\infty}) \tag{8.1}$$

We know that propeller propulsive efficiency η_{pr} will have a relatively low value at lower flight speeds (e.g., takeoff, early climbout from runway), and only reach its nominal peak value around cruise conditions, say attaining a value approaching 0.85. The thrust contribution from the core exhaust jet will usually be delivered from an unchoked nozzle exit, as reflected by Eq. 8.1. With respect to the thrust from the propeller, we note that flight airspeed V_{∞} goes to zero faster than η_{pr}, so that in the limit case where V_{∞} is around zero, one would usually have the peak thrust value possible from the engine/prop combination (a good thing for takeoffs).

One can define an equivalent power P_{eq} that nominally comprises both the shaft and jet power contributions:

$$P_{\text{eq}} = P_s + \frac{F_{\text{jet}} V_{\infty}}{\eta_{\text{pr}}} = P_s + \frac{\dot{m}_a V_{\infty}}{\eta_{\text{pr}}} \left([1 + f]V_{e,\infty} - V_{\infty}\right) \tag{8.2}$$

In the static case ($V_{\infty} = 0$), Eq. 8.2 becomes:

$$P_{\text{eq}} = P_{s,o} + F_{\text{jet},o} = P_{s,o} + \dot{m}_a(1 + f)V_{e,\infty} \tag{8.3}$$

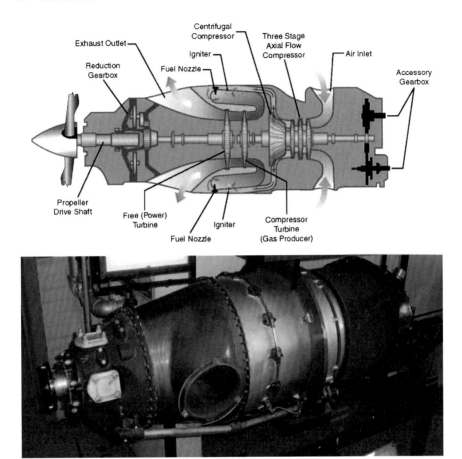

Fig. 8.2 Schematic cutaway diagram of a free-turbine two-spool (LP/HP core)/three-shaft turboprop engine [3]. Note the reverse flow arrangement relative to the propeller at *left*, as a means of avoiding foreign object damage to the compressor, combustor or turbine, and a means for more convenient on-wing hot-section (combustor, turbine) maintenance. *Lower* photo of Pratt & Whitney Canada PT6 turboprop engine on display at the Western Canada Aviation Museum (this engine has a comparable configuration to that indicated by the schematic engine diagram *above* it)

With the removal of V_∞ and η_{pr} in Eq. 8.3, one does not have a singularity issue. Along these lines, for a convenient expression estimating P_{eq} rather than the more elaborate Eq. 8.2, utilizing a somewhat conservative (low, by modern standards for conventional propellers) conversion of 2.5 lb of thrust per horse-power (14.9 N/kW) as the mean proportionality factor, the following has been adopted by the turboprop community:

$$P_{eq} = P_s(\text{hp}) + \frac{F_{\text{jet}}(\text{lb})}{2.5}, \text{ hp} \qquad (8.4a)$$

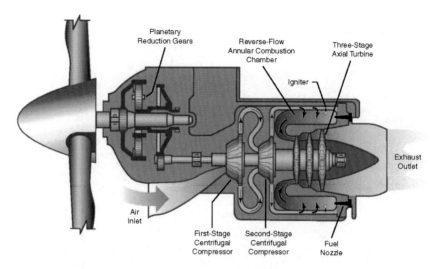

Fig. 8.3 Schematic cutaway diagram of a fixed single-shaft turboprop engine [3]. The example engine drive shaft may run at around 40,000 rpm, while the propeller shaft rotates at around 2,000 rpm

or

$$P_{eq} = P_s(\text{kW}) + \frac{F_{jet}(N)}{14.9}, \text{kW} \tag{8.4b}$$

Note that the core jet power can be as much as 20% of the equivalent power in the static sea-level case, but will drop to less than 5% at cruise conditions, for those TPs that utilize a significant portion of the exhaust jet energy for thrust.

8.2 Cycle Analysis of Conventional Free-Turbine Turboprop Engine

As noted for previous gas turbine variants, some of the baseline equations for turbojet cycle analysis can be utilized for the turboprop (and turboshaft) case. We will look at the one-spool/two-shaft TP engine of Fig. 8.1 for introducing the cycle analysis that can be applied for this category of engine. The approach for a turboshaft engine will be similar, with differences noted below. To begin, we note that the power output of the LP free turbine (to the speed reduction gearbox, as per Fig. 8.1) can be estimated as follows:

$$P_s = \eta_m \dot{m}_t C_{p,t}(T_{04.5} - T_{05}) \tag{8.5}$$

Station 4.5 is the inlet to the free turbine, and η_m is the turbomachinery mechanical efficiency (around 0.99 in value). The upstream HP turbine must drive the main compressor, such that:

$$\dot{m}_a C_{p,c}(T_{03} - T_{02}) = \eta_m \dot{m}_t C_{p,t}(T_{04} - T_{04.5}) \qquad (8.6)$$

Given the approximation

$$\dot{m}_a C_{p,c} \approx \dot{m}_t C_{p,t} \qquad (8.7)$$

then Eq. 8.6 can be reduced to:

$$T_{04.5} \approx T_{04} - \frac{1}{\eta_m}(T_{03} - T_{02}) \qquad (8.8)$$

and correspondingly,

$$p_{04.5} \approx p_{04}\left[1 - \frac{1}{\eta_t}\left(1 - \frac{T_{04.5}}{T_{04}}\right)\right]^{\frac{\gamma_t}{\gamma_t - 1}} \qquad (8.9)$$

Further downstream,

$$T_{05} = T_{04.5} - \frac{P_s}{\eta_m \dot{m}_t C_{p,t}} \approx T_{04.5} - \frac{P_s/\dot{m}_a}{\eta_m C_{p,t}} \qquad (8.10)$$

where $P_s/\dot{m}_a$ is defined as the specific power, and

$$p_{05} \approx p_{04.5}\left[1 - \frac{1}{\eta_t}\left(1 - \frac{T_{05}}{T_{04.5}}\right)\right]^{\frac{\gamma_t}{\gamma_t - 1}} \qquad (8.11)$$

As noted in Chap. 3, the shaft power demanded by the propeller can be ascertained via

$$P_s = C_p \rho n^3 d^5 \qquad (8.12)$$

Here, C_p is the propeller power coefficient (a function of the number of blades, B, pitch angle, β, and advance ratio J), d is the propeller diameter and n is the propeller shaft rotation speed (revolutions per second, unless indicated otherwise). The shaft power being demanded, in the case of a turboprop, would be associated with generating the desired forward thrust F_{prop} from the propeller, which is tied to C_p, etc., via the propeller's corresponding thrust coefficient C_T :

$$F_{\text{prop}} = C_T \rho n^2 d^4 = \frac{\eta_{\text{pr}} C_p}{J} \rho n^2 d^4 = \frac{\eta_{\text{pr}} P_S}{V_\infty} \qquad (8.13)$$

In the case of a turboshaft engine desiring the ideal maximum shaft power output,

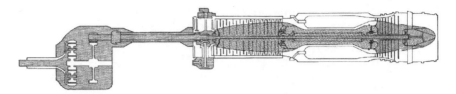

Fig. 8.4 Schematic cutaway diagram of an older single-shaft Allison T56 turboprop engine. The engine drive shaft runs at around 14,000 rpm, while the propeller shaft at the far left rotates at 1,020 rpm. Courtesy of Rolls-Royce plc

$$p_{05} \rightarrow p_{\infty}$$

$$Ma_e \rightarrow 0$$

$$T_{05} \rightarrow T_e$$

such that

$$T_{e,\text{ideal TS}} = T_{04.5}\left\{1 + \eta_t\left[\left(\frac{p_{\infty}}{p_{04.5}}\right)^{\frac{\gamma_t-1}{\gamma_t}} - 1\right]\right\} \tag{8.14}$$

and for specific power,

$$\frac{P_s}{\dot{m}_a}\Big|_{\text{ideal TS}} = \eta_m C_{p,t}(T_{04.5} - T_e) \tag{8.15}$$

In practice, one will likely need Ma_e to be at least 0.3 in order for adequate positive throughput of flow (avoiding transient backflow). In the general (non-ideal) case for either a TP or TS, re-arranging Eq. 8.10 gives

$$\frac{P_s}{\dot{m}_a}\Big|_{\text{TP, TS}} = \eta_m C_{p,t}(T_{04.5} - T_{05}) \tag{8.16}$$

As noted on previous occasions, for power-based fuel economics, one uses brake specific fuel consumption:

$$\text{BSFC} = \frac{\dot{m}_f}{P_s} = \frac{f}{\frac{P_s}{\dot{m}_a}} \tag{8.17}$$

In the typical case that a TP (or TS) would not be using an afterburner, one can transfer flow property values immediately from stations 5 to 6 for the nozzle entry. The TP's core exhaust jet is usually unchoked, such that for a simple convergent nozzle with station 7 as the exit plane,

$$p_7 \rightarrow p_{\infty}$$

$$\frac{T_7}{T_{06}} = \left(\frac{p_\infty}{p_{06}}\right)^{\frac{\gamma_n - 1}{\gamma_n}}$$

$$\rho_7 = \frac{p_\infty}{RT_7}$$

$$V_{e,\infty} = \sqrt{2\eta_n C_{p,n} T_{06}\left[1 - \left(\frac{p_\infty}{p_{06}}\right)^{\frac{\gamma_n - 1}{\gamma_n}}\right]} \tag{8.18}$$

$$A_{e,\infty} \approx A_e$$

The specific thrust delivered by a TP with a substantial core jet thrust contribution is given by

$$\frac{F}{\dot{m}_a} = \frac{\eta_{pr}}{V_\infty} \cdot \frac{P_s}{\dot{m}_a} + (1+f)V_{e,\infty} - V_\infty \tag{8.19}$$

If one were to find it convenient for some part of an analysis, or for comparison to another propulsion system, one could estimate a TP's fuel economics in terms of thrust-based terms, i.e.,

$$\text{TSFC} = \frac{f}{\frac{F}{\dot{m}_a}} \tag{8.20}$$

and for overall efficiency,

$$\eta_o = \frac{FV_\infty}{\dot{m}_f q_R} = \frac{1}{\text{TSFC}}\frac{V_\infty}{q_R} \tag{8.21}$$

Recall that in terms of power (not thrust), a turboprop or turboshaft engine's overall efficiency may be determined via

$$\eta_o = \eta_{pr}\eta_{th} \tag{8.22}$$

where thermal efficiency is available from

$$\eta_{th} = \frac{P_s}{\dot{m}_f q_R} = \frac{1}{\text{BSFC} \cdot q_R} \tag{8.23}$$

Moving from specific power to actual shaft power, an example power chart for a medium-sized two-spool/three-shaft turboprop engine is provided in Fig. 8.5. The throttle setting for these curves is maximum continuous power (a bit below the takeoff power throttle setting, so as to save on engine wear). One can observe that at low altitude, the output shaft power is more or less flat (constant), regardless of forward airspeed. In this case, the engine throttle is likely brought down progressively at higher airspeeds, to maintain this maximum power output at its "flat-rated" value. At higher altitudes, shaft power tends to

Fig. 8.5 Shaft power chart
for example turboprop engine
(maximum cruise power
throttle setting). Engine
performance comparable to
Pratt & Whitney PW120,
compression ratio π_c of 12

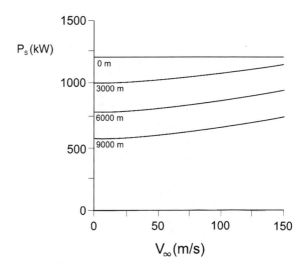

Fig. 8.6 Shaft static power
chart for example turboprop
engine (full takeoff power
throttle setting) for takeoff at
different outside air
temperatures and airfield
altitudes. Engine performance
comparable to Pratt &
Whitney PW120 at takeoff
throttle setting

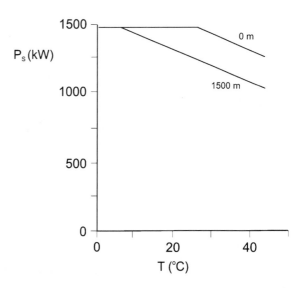

drop in general (lower ambient air pressure) so that at some point flat-rating of
the engine is no longer required. As a result, presumably due to aerodynamic
ram compression, one sees the shaft power going up a bit with increasing airspeed for a given altitude. The engine's corresponding power capability at the
higher takeoff throttle setting can be seen in Fig. 8.6, showing static shaft power
(P_{S0}) as a function of outside air temperature and altitude. As one would expect,
beyond the flat-rated power threshold, power drops off at higher outside air
temperatures.

Fig. 8.7 Photos of wing-mounted Kuznctsov NK-12 turboprop engines installed on a Tupolev Tu-95MS Bear. Note the use of contra-rotating (4 + 4 blades) propellers, and rearward-directed exhaust, to maximize thrust from each engine

Please refer to the solution for Example Problem 8.1 at the end of this chapter, for a completed sample turboprop engine cycle analysis. For students learning this material as part of a course, I would encourage them to attempt the given example problem first, and then check the solution to see if any mistakes were made.

8.3 Installation Issues

Integration requirements will differ for mating a turboprop engine to the wing or fuselage of an airplane, and mating a turboshaft engine to the fuselage of a helicopter. For the turboprop case, a wing mounting usually allows for a clear exhaust path for hot gas exiting the TP engine's core, in conjunction with a relatively unobstructed propeller slipstream (see example of Fig. 8.7). In the case where the core exhaust is not expected to deliver a thrust component, one will sometimes see the exhaust gas directed at a significant angle to the main airflow, to ensure sufficient clearance to avoid damaging the surrounding airframe. Ease of maintenance is an issue when on-wing (for quicker tasks that do not usually require the removal of the engine from the wing), and off-wing (see Fig. 8.8).

In the case of positioning a turboshaft engine on the fuselage of a helicopter, a number of considerations may come into play. The engine intake must be capable of bringing in the required airflow for effective engine operation, at any number of attitudes and speeds expected for a highly maneuverable flight vehicle. In a similar fashion, the hot engine exhaust gas must adequately clear the surrounding airframe and rotor(s), under a wide spectrum of flight conditions. The combined flow of the

Fig. 8.8 Rolls-Royce T56 turboprop engine being operated on an outdoor test rig. The reduction gear unit that ultimately rotates the propeller is positioned well ahead of the main body of the turboprop engine, as in Fig. 8.4. *Left* photo courtesy of New Zealand Aviation News, and *right* diagrams courtesy of Rolls-Royce plc

main rotor downwash and the air passing over the helicopter's body when in translation would need to be examined. Under expected (albeit infrequent) engine failure conditions, like a turbine blade separation, the flight crew and passengers should not be inordinately endangered. A benefit of having the engine positioned on the roof of the helicopter's main body, as shown in Figs. 8.9 and 8.10, is that it puts the engine intake well away from picking up and sucking in ground debris (an FOD consideration).

8.4 Example Problems

8.1. Consider an airplane's free-turbine turboprop engine operating at standard sea-level conditions at a flight Mach number of 0.3. The propeller is demanding 1,600 shp from the free turbine.

diffuser, $\eta_d = 0.97$; $\gamma_d = 1.4$;

compressor, $\eta_c = 0.95$, $\gamma_c = 1.37$, $\pi_c = 11$;

burner, $\eta_b = 1.0$, $\gamma_b = 1.35$, $\pi_b = 0.95$, $T_{\max} = 1,400$ K

$q_R = 45$ MJ/kg of fuel; turbine, $\gamma_t = 1.33$; $\eta_t = 0.9$; $\eta_m = 0.99$

core nozzle, $\eta_n = 0.98$; $\gamma_n = 1.36$

engine monitoring instruments: core air intake, 5 kg/s;

Fig. 8.9 Pratt & Whitney Canada PT6T twinned turboshaft engine in position on the upper fuselage of a Bell CH-146 Griffon helicopter

Ascertain the shaft and equivalent BSFC. If there were no minimum exhaust Mach number required for proper throughput, what would be the maximum possible shaft horsepower available?

8.2. Consider an airplane's free-turbine turboprop engine operating at an ISA altitude of 5,488 m (18,000 ft). The 4-bladed propeller has a diameter of 2.39 m.

diffuser, $\eta_d = 0.96$; $\gamma_d = 1.4$;

compressor, $\eta_c = 0.94$, $\gamma_c = 1.38$, $C_{pc} = 1,042$ J/(kg K), $\pi_c = 11.6$;

burner, $\eta_b = 0.97$, $C_{pb,\ exit} = 1,144$ J/(kg K), $\pi_b = 0.98$, $T_{max} = 1,450$ K

$q_R = 42.4$ MJ/kg of fuel; turbine, $\gamma_t = 1.335$; $\eta_t = 0.89$; $\eta_m = 0.985$

core nozzle, $\eta_n = 0.96$; $\gamma_n = 1.365$

engine monitoring instruments: core air intake, 0.65 kg/s; compressor inlet, $T_{02} = 266$ K, $p_{02} = 60.4$ kPa; propeller setting indicates $C_T = 0.04$, $C_P = 0.15$ at 1,700 rpm prop shaft rotation

current flight conditions at 5,488 m altitude: outside air pressure = 50.56 kPa, outside air temperature = 252.3 K, outside air density = 0.698 kg/m^3, flight Mach number = 0.52

(a) What is the shaft BSFC (kg/(h kW)) of the engine at this flight condition?
(b) Does the engine have sufficient throughput (core nozzle exit $Ma_e > 0.3$) for satisfactory operation?

8.3. An aircraft employing four PW120 turboprop engines is flying steady and level at 6,000 m altitude (ISA), at a true flight speed of 135 m/s. The aerodynamic drag acting on the airplane is 19,570 N. Determine the percentage of maximum

Fig. 8.10 Photo of a small conventional helicopter (Bell UH-1C Iroquois). The single turboshaft engine positioned on the roof of the helicopter fuselage is an 1,100-hp Lycoming T53-L-11 (*lower right* photo of engine on static display)

cruise power from the engines at 6,000 m that is required to maintain this flight condition. Refer to Fig. 8.5, for a relevant engine performance chart for a Pratt & Whitney PW120 turboprop engine. The 4-bladed 2.8-m-diameter propellers on each engine are rotating at 1,900 rpm (PSRU used to reduce the engine shaft speed down to the propeller speed). Refer to Fig. 3.6 for the relevant 4-bladed propeller charts.

8.4. *Computer-Based Project*: Perform an off-design cycle analysis for a commercial turboprop engine at its cruise throttle setting, at altitudes from 0 to 9,000 m ISA, at flight true airspeeds from zero to 150 m/s. Take note of the following design parameters:

diffuser, $\eta_d = 0.97$; $\gamma_d = 1.4$;

compressor, $\eta_c = 0.95$, $\gamma_c = 1.37$, $C_{pc} = 1062$ J/(kg K), $\pi_c = 11.0$;

burner, $\eta_b = 1.0$, $\gamma_b = 1.35$, $C_{pb,exit} = 1157$ J/(kg K), $\pi_b = 0.95$, $T_{max} = 1400$ K

$q_R = 45.0$ MJ/kg of fuel; turbine, $\gamma_t = 1.33$; $\eta_t = 0.9$; $\eta_m = 0.99$

core nozzle, $\eta_n = 0.98$; $\gamma_n = 1.36$; core nozzle fixed exit diameter $d_e = 0.3$ m;

fixed core exhaust downstream velocity (for positive throughput) $V_{e,\infty} = 170$ m/s

Plot shaft power P_S (kW) vs. true airspeed V_∞ (m/s) at the 4 principal altitudes of 0, 3,000, 6,000 and 9,000 m ISA (generate enough data points so that the curves appear smooth). In a second graph, plot fuel flow rate $\dot{m}_f$ (kg/h) vs. V_∞ (m/s) for the corresponding curves. Comment on the engine's power and fuel consumption profiles at different flight conditions.

8.5 Solutions to Problems

8.1. Looking at a free-turbine turboprop engine, at a flight Mach no. Ma_∞ of 0.3, at 0 m altitude ($p_\infty = 101.325$ kPa, $T_\infty = 288.2$ K, $\rho_\infty = 1.225$ kg/m^3). Inflow of air, $\dot{m}_a = 5$ kg/s, and required shaft power is 1,600 shp (1,190 kW).

Starting from intake (diffuser):

$$T_{02} = T_\infty \left[1 + \frac{\gamma_d - 1}{2} Ma_\infty^2 \right] = 288.2[1 + 0.2(0.3^2)] = 293 \text{ K, at diffuser outlet}$$

$$p_{02} = p_\infty \left[1 + \eta_d \left(\frac{T_{02}}{T_\infty} - 1 \right) \right]^{\frac{\gamma_d}{\gamma_d - 1}} = 101.3[1 + 0.97(1.017 - 1)]^{3.5} = 107 \text{ kPa,}$$

at diffuser outlet

Moving through compressor:

$$T_{03} = T_{02} \left[1 + \frac{1}{\eta_c} \left[\pi_c^{\frac{\gamma_c - 1}{\gamma_c}} - 1 \right] \right] = 293[1 + 1/0.95 \cdot [11^{0.27} - 1]] = 574 \text{ K,}$$

at compressor outlet

$$p_{03} = \pi_c \, p_{02} = 11(107) = 1,177 \text{ kPa, at compressor outlet;}$$

$$C_{p,c} = \frac{\gamma_c R_{air}}{\gamma_c - 1} = 1.37(287)/0.37 = 1,063 \text{ J/(kg K)}.$$

Moving through combustor (burner):

$$\bar{C}_{p,b} \approx \frac{C_{p,c} + C_{p,b,\text{exit}}}{2} = \frac{\gamma_b R_{\text{air}}}{\gamma_b - 1} = 1.35(287)/(1.35 - 1) = 1,107 \text{ J/(kg K), so that}$$

at burner exit :

$$C_{p,b,\text{exit}} \approx 2\bar{C}_{p,b} - C_{p,c} = 2(1107) - 1063 = 1151 \text{ J/(kg K)}.$$

$$T_{04} = T_{\text{max}} = 1400 \text{ K}, p_{04} = \pi_b p_{03} = 0.95(1177) = 1,118 \text{ kPa, at burner exit.}$$

$$f = \frac{\dfrac{T_{04}}{T_{03}} - \dfrac{C_{p,c}}{C_{p,b,\text{exit}}}}{\dfrac{\eta_b q_R}{C_{p,b,\text{exit}} T_{03}} - \dfrac{T_{04}}{T_{03}}} = \frac{\dfrac{1400}{574} - \dfrac{1063}{1151}}{\dfrac{1.0(45 \times 10^6)}{1151(574)} - \dfrac{1400}{574}} = 0.023, \text{fuel--air ratio}$$

(a reasonable value)

Moving through the two turbine sections of the engine:

$$T_{04.5} \approx T_{04} - \frac{1}{\eta_m}(T_{03} - T_{02}) = 1400 - 1/0.99 \cdot; (574 - 293) = 1,116 \text{ K},$$

at outlet of comp. turbine

$$p_{04.5} = p_{04}\left[1 - \frac{1}{\eta_t}\left(1 - \frac{T_{04.5}}{T_{04}}\right)\right]^{\frac{\gamma_t}{\gamma_t - 1}} = 1118[1 - 1/0.9 \cdot (1 - 1116/1400)]^{4.03}$$

$$= 399.4 \text{ kPa, at comp. turbine outlet.}$$

$$C_{p,t} = \frac{\gamma_t R_{\text{air}}}{\gamma_t - 1} = 1157 \text{ J/(kg K)},$$

$$T_{05} \approx T_{04.5} - \frac{P_s/\dot{m}_a}{\eta_m C_{p,t}} = 1116 - (1190 \times 10^3/5.0)/[0.99(1157)] = 908 \text{ K},$$

at outlet of free turbine

$$p_{05} = p_{04.5}\left[1 - \frac{1}{\eta_t}\left(1 - \frac{T_{05}}{T_{04.5}}\right)\right]^{\frac{\gamma_t}{\gamma_t - 1}} = 399.4[1 - 1/0.9 \cdot (1 - 908/1116)]^{4.03}$$

$$= 156.8 \text{ kPa, at outlet of free turbine}$$

No afterburner, so $T_{06} = T_{05} = 908$ K, and $p_{06} = p_{05} = 156.8$ kPa. Now, approaching subsonic converging nozzle, need to check for choking:

$$\frac{p_{06}}{p_\infty} = \frac{156.8}{101.3} = 1.55 \text{ and choking criterion} \left(\frac{\gamma_n + 1}{2}\right)^{\frac{\gamma_n}{\gamma_n - 1}} = 1.87 > \frac{p_{06}}{p_\infty},$$

so nozzle flow is *unchoked*

$$C_{p,n} = \frac{\gamma_n R_{\text{air}}}{\gamma_n - 1} = 1084 \text{ J/(kg K)},$$

$$V_{e,\infty} = \sqrt{2\eta_n C_{p,n} T_{06}\left[1 - \left(\frac{p_\infty}{p_{06}}\right)^{\frac{\gamma_n - 1}{\gamma_n}}\right]}$$

$$= \sqrt{2(0.98)1084(908)[1 - (101.3/156.8)^{0.265}]} = 460.7 \text{ m/s}$$

$T_7 = T_{06}\left(\frac{p_\infty}{p_{06}}\right)^{\frac{\gamma_n - 1}{\gamma_n}} = 908(101.3/156.8)^{0.265} = 808$ K, exit nozzle static temperature at subsonic flow condition: $Ma_e = \frac{V_e}{\sqrt{\gamma_n R T_7}} = 460.7/562 = 0.82 > 0.3$, so some margin available for more shaft power output.

$V_\infty = a_\infty Ma_\infty = \sqrt{\gamma_{\text{air}} R_{\text{air}} T_\infty} \cdot Ma_\infty = 340.2(0.3) = 102$ m/s, aircraft flight speed. For net thrust produced by the core jet exhaust,

$F_{\text{jet,net}} = (1 + f)\dot{m}_a V_{e,\infty} - \dot{m}_a V_\infty = (1 + 0.023)\ 5.0\ (460.7) - 5.0\ (102) = 1846$ N or 415 lbf

$\text{BSFC}_{\text{shaft}} = \frac{f}{\frac{P_s}{\dot{m}_a}} = 0.023/(1190 \times 10^3/5.0) = 9.7 \times 10^{-8}$ kg/(s W) or 0.35 kg/ (h kW) or 0.575 lb/h·shp

$$P_{\text{eq}} = P_s + \frac{F_{\text{jet,net}}}{2.5(\text{lbf/hp})} = 1600 + 415/2.5 = 1766 \text{ ehp or } 1318 \text{ kW}$$

$$\text{BSFC}_{\text{eq}} = \frac{f}{\frac{P_{\text{eq}}}{\dot{m}_a}} = 0.023/(1318 \times 10^3/5.0)$$
$$= 8.7 \times 10^{-8} \text{ kg/s} \cdot \text{W or } 0.313 \text{ kg/hr} \cdot \text{kW or } 0.52 \text{ lb/hr} \cdot \text{ehp}$$

Also, $\dot{m}_f = f\dot{m}_a = 0.023(5.0) = 0.12$ kg/s or 414 kg/h.

For maximum possible shaft power output,

$P_{s,\text{max}} = \dot{m}_a \eta_m C_{p,t}(T_{04.5} - T_e) = 5.0(0.99)1157(1116 - 808) = 1.67 \times 10^6$ W or 1.67 MW or 2250 hp, about 40% more shaft horsepower than the previous value. However, this result is based on a zero throughput assumption, i.e., likely not very practical for actual operations, where one needs substantial positive throughput ($Ma_e > 0.3$) for stable engine operation.

8.2. Looking at a free-turbine turboprop engine, at a flight Mach no. Ma_∞ of 0.52, at 5,488 m altitude ($p_\infty = 50.56$ kPa, $T_\infty = 252.3$ K, $\rho_\infty = 0.698$ kg/m^3).

Inflow of core engine air, $\dot{m}_a = 0.65$ kg/s, and required shaft power needs to be determined.

(a) $V_\infty = a_\infty Ma_\infty = \sqrt{\gamma_{air} R_{air} T_\infty} \cdot Ma_\infty = 318.4(0.52) = 165.6$ m/s, $n = 1700/60 = 28.33$ rps

$$J = V_\infty/(n \cdot d) = 165.6/(28.33 \cdot 2.39) = 2.45$$

$P_S = C_p \rho_\infty n^3 d^5 = 0.15(0.698)28.33^3(2.39^5) = 185642$ W or 185.6 kW (approx. 249 hp)

Given $T_{02} = 266$ K, $p_{02} = 60.4$ kPa.

Moving through compressor:

$$T_{03} = T_{02}\left[1 + \frac{1}{\eta_c}\left[\pi_c^{\frac{\gamma_c-1}{\gamma_c}} - 1\right]\right] = 266[1 + 1/0.94; [11.6^{0.275} - 1]] = 538.3 \text{ K},$$

at compressor outlet

$$p_{03} = \pi_c p_{02} = 11.6(60.4) = 700.6 \text{ kPa, at compressor outlet}$$

Moving through combustor (burner):

$$T_{04} = T_{max} = 1450 \text{ K}$$

$$f = \frac{\dfrac{T_{04}}{T_{03}} - \dfrac{C_{p,c}}{C_{p,b,exit}}}{\dfrac{\eta_b q_R}{C_{p,b,exit} T_{03}} - \dfrac{T_{04}}{T_{03}}} = \frac{\dfrac{1450}{538.3} - \dfrac{1042}{1144}}{\dfrac{0.97(42.4 \times 10^6)}{1144(538.4)} - \dfrac{1450}{538.4}} = 0.0278, \text{ fuel–air ratio}$$

$$\text{BSFC}_{shaft} = \frac{f}{\frac{P_s}{\dot{m}_a}} = 0.0278/(185642/0.65) = 9.736 \ 10^{-8} \text{kg}/(\text{s W}) = 0.35 \text{kg}/(\text{h kW})$$

(b) Moving through the two turbine sections of the engine:

$p_{04} = \pi_b p_{03} = 0.98(700.6) = 686.6$ kPa, at burner exit/turbine inlet.

$T_{04.5} \approx T_{04} - \frac{1}{\eta_m}(T_{03} - T_{02}) = 1450 - 1/0.985 \cdot (538.3 - 266) = 1{,}173.6$ K, at outlet of comp. turbine

$$p_{04.5} = p_{04}\left[1 - \frac{1}{\eta_t}\left(1 - \frac{T_{04.5}}{T_{04}}\right)\right]^{\frac{\gamma_t}{\gamma_t-1}} = 686.6[1 - 1/0.89. \ (1 - 1173.6/1450)]^{3.985} =$$
262.8 kPa, at comp. turbine outlet.

$$C_{p,t} = \frac{\gamma_t R_{air}}{\gamma_t - 1} = 1144 \text{ J}/(\text{kg K}),$$

$T_{05} \approx T_{04.5} - \frac{P_s/\dot{m}_a}{\eta_m C_{p,t}} = 1173.6 - (185642/0.65)/[0.985(1144)] = 920.2$ K, at outlet of free turbine

$$p_{05} = p_{04.5}\left[1 - \frac{1}{\eta_t}\left(1 - \frac{T_{05}}{T_{04.5}}\right)\right]^{\frac{\gamma_t}{\gamma_t - 1}} = 262.8[1 - 1/0.89 \cdot (1 - 920.2/1173.6)]^{3.985} =$$

86.8 kPa, at outlet of free turbine

No afterburner, so $T_{06} = T_{05} = 920.2$ K, and $p_{06} = p_{05} = 86.8$ kPa. Now, approaching subsonic converging nozzle, need to check for choking:

$$\frac{p_{06}}{p_\infty} = \frac{86.8}{50.56} = 1.72 \text{ and choking criterion} \left(\frac{\gamma_n + 1}{2}\right)^{\frac{\gamma_n}{\gamma_n - 1}} = 1.87 > \frac{p_{06}}{p_\infty},$$

so nozzle flow is *unchoked*.

$$C_{p,n} = \frac{\gamma_n R_{air}}{\gamma_n - 1} = 1073.3 \text{ J/(kg K)},$$

$$V_{e,\infty} = \sqrt{2\eta_n C_{p,n} T_{06}\left[1 - \left(\frac{p_\infty}{p_{06}}\right)^{\frac{\gamma_n - 1}{\gamma_n}}\right]}$$

$$= \sqrt{2(0.96)1073.3(920.2)[1 - (50.56/86.8)^{0.2674}]} = 505 \text{ m/s}$$

$T_7 = T_{06}\left(\frac{p_\infty}{p_{06}}\right)^{\frac{\gamma_n - 1}{\gamma_n}} = 920.2(50.56/86.8)^{0.265} = 796$ K, exit nozzle static tempera-

ture at subsonic flow condition: $Ma_e = \frac{V_e}{\sqrt{\gamma_n R T_7}} = 505/558.4 = 0.9 > 0.3$, so throughput criterion is met.

8.3. An aircraft employing four PW120 turboprop engines is flying steady and level at 6000 m altitude (ISA), at a true flight speed of 135 m/s. The aerodynamic drag acting on the airplane is 19570 N. Determine the percentage of maximum cruise power from the engines at 6000 m that is required to maintain this flight condition. Refer to Fig. 8.5, for a relevant engine performance chart for a Pratt & Whitney PW120 turboprop engine. The 4-bladed 2.8-m-diameter propellers on each engine are rotating at 1900 rpm (PSRU used to reduce the engine shaft speed down to the propeller speed). Refer to Fig. 3.6 for the relevant 4-bladed propeller charts.

$h_{ASL} = 6,000$ m, $V_\infty = 135$ m/s, $a_\infty = 316.5$ m/s, $Ma_\infty = 0.43$

$n = 1900/60 = 31.7$ revs/s; $\rho_\infty = 0.66$ kg/m^3; $J = V_\infty/(nd) = 135/(31.7 \cdot 2.8) = 1.52$

$F_{eng} = F/4 = D/4 = 19570/4 = 4,893$ N (1,100 lbf of thrust from each engine)

$F_{eng} = C_T \rho n^2 d^4 = C_T (0.66)(31.7^2)2.8^4 = 4893$ N, $C_T = 0.12$; from Fig. 3.6, $\beta = 34°$

$C_P = 0.235$, from corresponding chart; $\eta_{pr} = J C_T/C_P = 0.78$

$P_S = C_P \rho n^3 d^5 = 0.235(0.66)31.7^3 2.8^5 = 850300$ W or 850 kW, or 1,140 hp

At 6,000 m ISA and 135 m/s, $P_{S,\text{max}} = 900$ kW, so percentage of max. cruise power is around $850/900 \times 100\% = 94.5\%$

8.4. Perform an off-design cycle analysis for a commercial turboprop engine at its cruise throttle setting, at altitudes from 0 to 9000 m ISA, at flight true airspeeds from zero to 150 m/s. Take note of the following design parameters:

diffuser, $\eta_d = 0.97$; $\gamma_d = 1.4$;

compressor, $\eta_c = 0.95$, $\gamma_c = 1.37$, $C_{pc} = 1062$ J/(kg K), $\pi_c = 11.0$;

burner, $\eta_b = 1.0$, $\gamma_b = 1.35$, $C_{pb,\text{exit}} = 1157$ J/(kg K),

$\pi_b = 0.95$, $T_{\text{max}} = 1400$ K

$q_R = 45.0$ MJ/kg of fuel; turbine, $\gamma_t = 1.33$; $\eta_t = 0.9$; $\eta_m = 0.99$

core nozzle, $\eta_n = 0.98$; $\gamma_n = 1.36$; core nozzle fixed exit diameter $d_e = 0.3$ m;

fixed core exhaust downstream velocity (for positive throughput) $V_{e,\infty} = 170$ m/s

Plot shaft power P_S (kW) vs. true airspeed V_∞ (m/s) at the 4 principal altitudes of 0, 3,000, 6,000 and 9,000 m ISA (generate enough data points so that the curves appear smooth). In a second graph, plot fuel flow rate $\dot{m}_f$ (kg/h) vs. V_∞ (m/s) for the corresponding curves.

As a check, sample set of calculations, $V_\infty = 120$ m/s at $h_{\text{ASL}} = 6000$ m:

$p_\infty = 47.2$ kPa, $T_\infty = 249.2$ K, $\rho_\infty = 0.66$ kg/m^3, $a_\infty = 316.5$ m/s

$A_e = \pi d_e^2/4 = 0.0707$ m^2, $Ma_\infty = 120/316.5 = 0.38$

Starting from intake (diffuser):

$$T_{02} = T_\infty \left[1 + \frac{\gamma_d - 1}{2} Ma_\infty^2\right] = 249.2[1 + 0.2(0.38^2)] = 256.4\,\text{K},$$

at diffuser outlet

$$p_{02} = p_\infty \left[1 + \eta_d \left(\frac{T_{02}}{T_\infty} - 1\right)\right]^{\frac{\gamma_d}{\gamma_d - 1}} = 47.2[1 + 0.97(1.029 - 1)]^{3.5} = 52\,\text{kPa},$$

at diffuser outlet

Moving through compressor:

$$T_{03} = T_{02} \left[1 + \frac{1}{\eta_c}\left[\pi_c^{\frac{\gamma_c - 1}{\gamma_c}} - 1\right]\right] = 256.4[1 + 1/0.95[11^{0.27} - 1]] = 502.2\,\text{K},$$

at compressor outlet

$$p_{03} = \pi_c p_{02} = 11(52) = 572\,\text{kPa, at compressor outlet;}$$

$$C_{p,c} = \frac{\gamma_c R_{\text{air}}}{\gamma_c - 1} = 1.37(287)/0.37 = 1,063\,\text{J/(kg K)}.$$

Moving through combustor (burner):

$$C_{p,b,\text{exit}} = 1,157\,\text{J/(kg K), given.}$$

$T_{04} = T_{\max} = 1,400\,\mathrm{K}, p_{04} = \pi_b p_{03} = 0.95(572) = 543.4\,\mathrm{kPa}$, at burner exit.

$$f = \dfrac{\dfrac{T_{04}}{T_{03}} - \dfrac{C_{p,c}}{C_{p,b,\mathrm{exit}}}}{\dfrac{\eta_b q_R}{C_{p,b,\mathrm{exit}} T_{03}} - \dfrac{T_{04}}{T_{03}}} = \dfrac{\dfrac{1400}{502.2} - \dfrac{1063}{1157}}{\dfrac{1.0(45 \times 10^6)}{1157(502.2)} - \dfrac{1400}{502.2}} = 0.025,\ \text{fuel} - \text{air ratio}$$

Moving through the two turbine sections of the engine:

$$T_{04.5} \approx T_{04} - \frac{1}{\eta_m}(T_{03} - T_{02}) = 1400 - 1/0.99 \cdot (502.2 - 256.4) = 1152\,\mathrm{K},$$

at outlet of comp. turbine

$$p_{04.5} = p_{04}\left[1 - \frac{1}{\eta_t}\left(1 - \frac{T_{04.5}}{T_{04}}\right)\right]^{\frac{\gamma_t}{\gamma_t - 1}} = 543.4[1 - 1/0.9 \cdot (1 - 1152/1400)]^{4.03}$$

$$= 436.5\,\mathrm{kPa},\ \text{at comp. turbine outlet.}$$

$$C_{p,t} = \frac{\gamma_t R_{\mathrm{air}}}{\gamma_t - 1} = 1,157\,\mathrm{J/(kg\,K)},$$

$$T_{05} \approx T_{04.5} - \frac{P_s/\dot{m}_a}{\eta_m C_{p,t}} = 1152 - (P_s \dot{m}_a)/[0.99(1157)]$$

$$= 1152 - 8.73 \times 10^{-4} P_s \dot{m}_a,\ \text{at outlet of free turbine.}$$

Note, fixed core exhaust downstream velocity (for positive throughput) $V_{e,\infty} = 170$ m/s .

$$p_{05} = p_{04.5}\left[1 - \frac{1}{\eta_t}\left(1 - \frac{T_{05}}{T_{04.5}}\right)\right]^{\frac{\gamma_t}{\gamma_t - 1}} = 436.5[1 - 1/0.9(1 - T_{05}/1152)]^{4.03},$$

at outlet of free turbine.

No afterburner, so $T_{06} = T_{05}$, and $p_{06} = p_{05}$. Now, approaching subsonic converging nozzle, assume flow is unchoked.

$$C_{p,n} = \frac{\gamma_n R_{\mathrm{air}}}{\gamma_n - 1} = 1084\,\mathrm{J/kg \cdot K},$$

$$V_{e,\infty} = \sqrt{2\eta_n C_{p,n} T_{06}\left[1 - \left(\frac{p_\infty}{p_{06}}\right)^{\frac{\gamma_n - 1}{\gamma_n}}\right]}$$

$$= \sqrt{2(0.98)1084(T_{06})[1 - (47.2/p_{06})^{0.265}]} = 170\,\mathrm{m/s}$$

$$\frac{V_{e,\infty}^2}{2\eta_n C_{p,n}} = T_{05}\left[1 - \left(\frac{p_\infty}{p_{04.5}\left[1 - \frac{1}{\eta_t}\left(1 - \frac{T_{05}}{T_{04.5}}\right)\right]^{\frac{\gamma_t}{\gamma_t - 1}}}\right)^{\frac{\gamma_n - 1}{\gamma_n}}\right]$$

$$\frac{(170)^2}{2(0.98)1084} = T_{05}\left[1 - \left(\frac{47.2}{436.5\left[1 - \frac{1}{0.9}\left(1 - \frac{T_{05}}{1152}\right)\right]^{4.03}}\right)^{0.265}\right] = 13.6;$$

$$\text{iterating}, T_{05} = 723\,\text{K} = T_{06}$$

$$p_{05} = 436.5[1 - 1/0.9\,(1 - 723/1152)]^{4.03} = 50.7\,\text{kPa} = p_{06}$$

$$T_7 = T_{06}\left(\frac{p_\infty}{p_{06}}\right)^{\frac{\gamma_n - 1}{\gamma_n}} = 723(47.2/50.7)^{0.265} = 709.4\,\text{K, exit nozzle static temperature}$$

at subsonic flow condition: $Ma_e = \dfrac{V_e}{\sqrt{\gamma_n R T_7}} = 170/526.2 = 0.32 > 0.3$, so margin met for throughput.

$$\rho_7 = \rho_e = \frac{p_{e,\infty}}{RT_e} = \frac{47200}{287(709.4)} = 0.232\,\text{kg/m}^3,$$

$$\dot{m}_e = \rho_e V_{e,\infty} A_e = 0.232(170)0.0707 = 2.79\,\text{kg/s}$$

$$P_s/\dot{m}_a = C_{p.t}\eta_m(T_{04.5} - T_{05}) = 1157(0.98)(1152 - 723) = 486426\,\text{W/(kg/s)}$$

$$P_S = \dot{m}_a \cdot \frac{P_S}{\dot{m}_a} = \frac{\dot{m}_e}{1+f} \cdot \frac{P_S}{\dot{m}_a} = 2.79/(1 + 0.025) \cdot 486426$$
$$= 1324\,\text{kW}(1,774\,\text{hp}), \dot{m}_a = 2.72\,\text{kg/s}$$

For net thrust produced by the core jet exhaust,

$$F_{\text{jet,net}} = (1 + f)\dot{m}_a V_{e,\infty} - \dot{m}_a V_\infty = (1 + 0.025)2.72(170) - 2.72(120)$$
$$= 148\,\text{N or 33 lbf}$$

$$\text{BSFC}_{\text{shaft}} = \frac{f}{\frac{P_s}{\dot{m}_a}} = 0.025/(1324 \times 10^3 2.72)$$
$$= 5.14 \times 10^{-8}\text{kg/s} \cdot \text{W or } 0.185\,\text{kg/h} \cdot \text{kW or } 0.304\,\text{lb/(h shp)}$$

$$P_{\text{eq}} = P_s + \frac{F_{\text{jet,net}}}{2.5(lbf/hp)} = 1774 + 33/2.5 = 1,787\,\text{ehp or } 1,333\,\text{kW}$$

$$\text{BSFC}_{\text{eq}} = \frac{f}{\frac{P_{\text{eq}}}{\dot{m}_a}} = 0.025/(1333 \times 10^3 2.72) = 5.1 \times 10^{-8}\text{kg/(s W) or}$$

$0.184\,\text{kg/(h kW) or } 0.302\,\text{lb/(h ehp)}$

Also, $\dot{m}_f = f\dot{m}_a = 0.025(2.72) = 0.068$ kg/s or 245 kg/h.

References

1. Anonymous (1990) Eight decades of progress–a heritage of aircraft turbine technology. General Electric Company, Cincinnati
2. Young WR (1982) The epic of flight–the helicopters. Time-Life, Chicago
3. Anonymous (2004) Airplane flying handbook. FAA-H-8083-3A, Airman Testing Standards Branch, Federal Aviation Administration, U.S. Department of Transportation, Oklahoma City
4. Treager IE (1970) Aircraft gas turbine engine technology. McGraw-Hill, New York

Part II
Rocket Propulsion

Chapter 9
Introduction to Space Flight

9.1 Introduction to Propulsion for Rocket and Space Vehicles

From an atmospheric flight performance background, thrust is one of four principal force components (the others being lift, drag and weight) that we need to establish for a given flight application to succeed. From a general aerospace propulsion background, we realize that the selection of the most appropriate thrust delivery system may be determined by a number of important factors, including: range of altitudes (h) and attitudes (α, β incidence angles of the flight vehicle relative to airflow), range of vehicle flight Mach numbers (Ma_∞) and ability to intake sufficient air. Other factors, such as flight economy, are important, but may be kept in the background early on in the design process, unless cost is in fact a predominant issue for the given application.

If one does not have a sufficient amount of air coming into the engine at a given Ma_∞ or h, and one has a substantial thrust (F) requirement, one would typically consider a chemical rocket engine for the given application. A rocket is generally (traditionally) defined as a propulsion system that carries both fuel and oxidizer as storage within the vehicle, burning the propellant as required to produce a high-speed exhaust jet that delivers the needed thrust. Chemical systems, such as solid-propellant, liquid-propellant and hybrid rocket engines, certainly fall under this definition. Air-breathing rocket engines may bend the above traditional definition to some degree, by using outside air to deliver at least some of the needed oxidizer content to the engine. Other propulsion systems, such as electric, solar or nuclear, may not even employ a combustion process to produce the high-speed exhaust jet. We will have a look at all of the above in the following chapters, and in doing so, develop a good general understanding of rocket propulsion for atmospheric and space flight.

D. R. Greatrix, *Powered Flight*, DOI: 10.1007/978-1-4471-2485-6_9,
© Springer-Verlag London Limited 2012

9.2 Mission Requirements

A number of rocket propulsion applications exist, including the largest thrust delivery application to date, for propelling heavy space launch vehicles. There are various factors that go into choosing one system over another. The mission profile for a given vehicle's flight application will strongly influence propulsion system design choices. More elaborate mission profiles for flight through the atmosphere and/or space may necessitate the use of a more comprehensive six-degree-of-freedom (3 degrees in translation, 3 degrees in rotation) numerical flight simulation program early on in the design process, in order to better assess the propulsion requirements.

By way of introduction, let us consider one relatively straightforward flight mission example, and see how one would use the associated analysis to help establish the rocket propulsion system requirements. Consider the case of a rocket vehicle launched in a vertical ascent trajectory from a horizontal surface (flat, non-rotating Earth being assumed, for simplicity), as per Fig. 9.1. For now, we will assume the vehicle has only one engine, but later, we can consider the benefits of multi-staging (sequenced use of multiple engines). Note that gravity (gravitational acceleration g) decreases with increasing altitude via Newton's law of gravity:

$$g = g_o \left(\frac{R_E}{R_E + h_{ASL}} \right)^2 \tag{9.1}$$

noting that g_o is 9.81 m/s^2 at sea level altitude, R_E is a mean Earth radius value of around 6,378 km and h_{ASL} is altitude above sea level. The general ordinary differential equation describing this vertical flight is:

$$\frac{dV}{dt} = \frac{F - D}{m} - g \tag{9.2}$$

where all the variables on the right-hand side may vary with time, i.e., given $g = f(h)$ as noted earlier, vehicle mass $m = f(t)$ as propellant is burned away via

$$\dot{m} = \frac{F}{I_{sp} g_o} \tag{9.3}$$

where specific impulse I_{sp} can vary from 200 to 400 s for chemical systems, and correspondingly thrust F can be constant ("flat" or "neutral") or vary with time ("progressive" if ramping up, "regressive" if ramping down). Aerodynamic drag acting on the flight vehicle will vary in the lower atmosphere as the vehicle accelerates upward, via:

$$D = C_D \frac{1}{2} \rho V^2 S \tag{9.4}$$

where drag coefficient $C_D = f(Ma)$, air density $\lambda = f(h)$, airspeed $V = f(t)$ and S is a reference area, typically taken as the maximum body (fuselage) cross-section,

Fig. 9.1 Schematic diagram
of rocket vehicle at sea level
launch, quadrant elevation
angle $\theta_o = 90°$

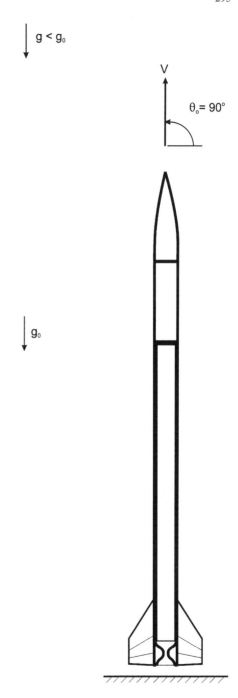

rather than wing or fin area, for a rocket or missile. For a general solution, we would have to solve Eq. 9.2 numerically.

However, for illustrative purposes, let us simplify the above by neglecting vehicle drag, and prescribing the thrust as being constant at a value F_{01}, where the subscript 0 refers to the thrust value at the segment start time, i.e., at $t = 0$ s, and the subscript 1 refers to the first motor stage (if a multi-staged vehicle):

$$\frac{dV}{dt} = \frac{F_{01}}{m_o - \left(\frac{F_{01}}{I_{sp,1}g_o}\right)t} - g \tag{9.5}$$

where m_o is the initial vehicle mass at launch. One can integrate the above with respect to time to arrive at the vehicle's forward speed in its vertical ascent:

$$V(t) = -g_o I_{sp,1} \ln\left(1 - \frac{F_{01}}{I_{sp,1}m_o g_o}t\right) - \bar{g}_1 t,\ 0 < t < t_{B1} \tag{9.6}$$

Here, $\bar{g}_1$ is the mean gravitational acceleration over the integration period, and t_{B1} is the first motor's burn time, i.e., the time for first-stage propellant m_{P1} to be completely consumed. Given $V = dy/dt$ in this example, height attained becomes:

$$y = \int_0^t V dt = g_o I_{sp,1}\left[\left(\frac{m_o g_o I_{sp,1}}{F_{01}} - t\right)\ln\left(1 - \frac{F_{01}}{m_o g_o I_{sp,1}}t\right) + t\right] - \bar{g}_1 \frac{t^2}{2},\ 0 < t < t_{B1} \tag{9.7}$$

Multistaging allows for the jettisoning of some dead structure (weight) over the course of the flight, dropping off empty motor stages that would otherwise slow the acceleration of the flight vehicle in subsequent phases of the flight. In this fashion, one can produce a lighter overall flight vehicle at launch, to meet a given final target speed and altitude. The more motor stages, the lighter the overall vehicle to meet the objective.

Let us consider that we had a second motor stage for our example vehicle above, as per Fig. 9.2, and that we start the second motor at the instant that the first motor burns out and is separated from the flight vehicle (for simplicity of calculations). Given the first motor's burn time t_{B1} may be ascertained via

$$t_{B1} = \frac{I_{sp,1}m_{P1}g_o}{F_{01}} \tag{9.8}$$

and with the jettisoning of the empty first motor (mass of m_{E1}), the flight vehicle's new mass would be

$$m_1 = m_o - m_{P1} - m_{E1}. \tag{9.9}$$

We can then estimate the vehicle's velocity at the end of the second-stage burn (note: at $t = t_{B1} + t_{B2}$) as follows:

Fig. 9.2 Schematic diagram
of two-stage rocket vehicle

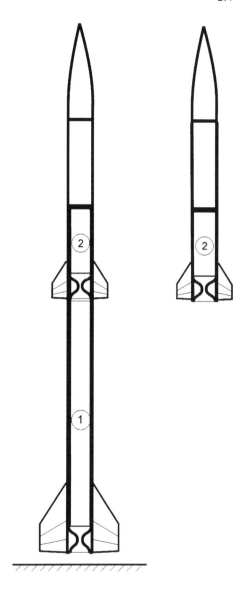

$$V_2 = -g_o I_{\mathrm{sp},1} \left\{ \ln\left(1 - \frac{m_{P1}}{m_o}\right) + \bar{g}_1 \frac{m_{P1}}{F_{o,1}} + \left(\ln\left(1 - \frac{m_{P2}}{m_1}\right) + \bar{g}_2 \frac{m_{P2}}{F_{o,2}} \right) \frac{I_{\mathrm{sp},2}}{I_{\mathrm{sp},1}} \right\}$$

(9.10)

Note that $\bar{g}_2$ is the mean gravitational acceleration over the integration period of
the second flight segment. The vertical height attained at second-stage burnout, h_2,
can then be determined via

$$h_2 = \frac{g_o^2 I_{\mathrm{sp},1}^2}{F_{o,1}} \left[(m_o - m_{P1}) \ln\left(1 - \frac{m_{P1}}{m_o}\right) + m_{p1} \right] - \frac{\bar{g}_1}{2} \left[\frac{I_{\mathrm{sp},1} m_{P1} g_o}{F_{o,1}} \right]^2$$
$$+ V_1 \left[\frac{I_{\mathrm{sp},2} m_{P2} g_o}{F_{o,2}} \right] + \frac{g_o^2 I_{\mathrm{sp},2}^2}{F_{o,2}} \left[(m_1 - m_{P2}) \ln\left(1 - \frac{m_{P2}}{m_1}\right) + m_{P2} \right] - \frac{\bar{g}_2}{2} \left[\frac{I_{\mathrm{sp},2} m_{P2} g_o}{F_{o,2}} \right]^2$$

$$(9.11)$$

where vehicle velocity at first-stage burnout, V_1, is established from

$$V_1 = -g_o I_{\mathrm{sp},1} \left\{ \ln\left(1 - \frac{m_{P1}}{m_o}\right) + \bar{g}_1 \frac{m_{P1}}{F_{o,1}} \right\} \qquad (9.12)$$

The peak height attained (apogee) in this example is reached when all the vehicle's kinetic energy is transferred to potential energy, i.e.,

$$h_{\max} = h_2 + \frac{V_2^2}{2\bar{g}_3} = \frac{1}{\frac{1}{R_E + h_2} - \frac{V_2^2}{2g_o R_E^2}} - R_E \qquad (9.13)$$

Here, $\bar{g}_3$ is the mean gravity over the unpowered ascent portion of the vertical flight.

Following the approach above, one can undertake a similar derivation for further motor stages, to illustrate the higher $V_{\max}$ and $h_{\max}$ attained, for the same initial launch mass m_o. In practice, one would need to consider other factors in establishing the number of motor stages to be used, given the increased complexity and cost associated with adding each stage, versus going with a heavier rocket vehicle at launch. Presently, it is common to see only two stages being used to reach low Earth orbit, when liquid-propellant rocket engines are used in whole or in part for the two stages (see the Space Shuttle example), while 3–4 stages are commonly used when lower-I_{sp} solid-propellant rocket motors are predominantly employed. See Fig. 9.3 for a multi-stage vehicle example. Four Alliant Techsystems (ATK) GEM-40 strap-on solid rocket booster motors (with graphite *epoxy motor* casings, hence the acronym) are used as part of the overall first-stage thrust delivery, in conjunction with the Rocketdyne RS-27A liquid rocket engine.

Please refer to the solution for Example Problem 9.1 at the end of this chapter, for a completed sample multi-staged rocket performance analysis. For students learning this material as part of a course, I would encourage them to attempt the given example problem first, and then check the solution to see if any mistakes were made.

9.3 Launch Vehicle Trajectory to Orbit

The simplified vertical ascent trajectory analysis covered earlier provides for some quick analytical results, as does the even simpler ideal (no-gravity) rocket equation for ΔV estimation (change in vehicle velocity V as a function of propellant mass), shown below for a given flight segment:

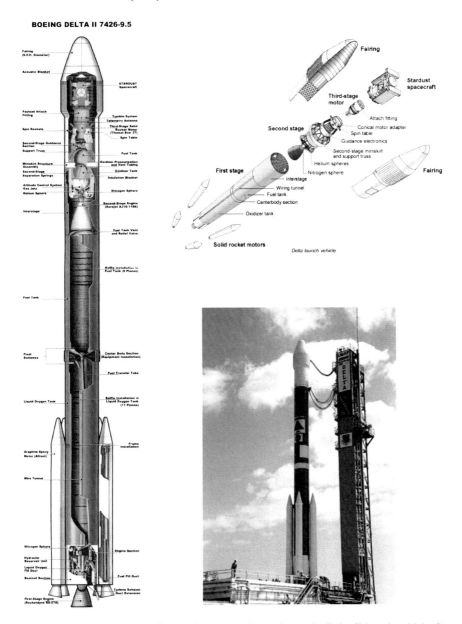

Fig. 9.3 Schematic diagram of the various stages that make up the Delta II launch vehicle, for the Stardust spacecraft/Comet Wild-2 fly-by mission. Launched in February 1999, the Stardust spacecraft passed by the comet in January 2004, successfully collecting and subsequently returning to Earth samples of comet and interstellar dust (January 2006). The nitrogen sphere stores N_2 used for a cold-gas attitude-control thruster array for mid-course flight corrections; the helium sphere stores used He as pressurant for propellant feed of the second-stage Aerojet AJ-10 engine. Courtesy of NASA

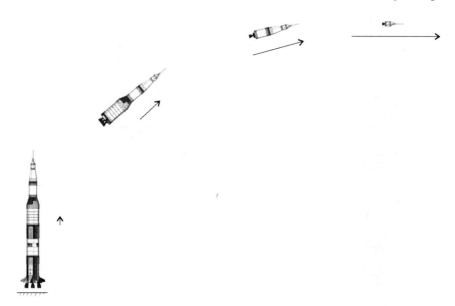

Fig. 9.4 Flight trajectory of multi-stage launch vehicle up to orbital altitude and speed

$$\frac{m_{\text{final}}}{m_{\text{initial}}} = \exp\left(\frac{-\Delta V}{I_{\text{sp}} \cdot g_o}\right) = \exp\left(\frac{-\Delta V}{u_e}\right) \qquad (9.14)$$

Equations like these allow for simple parametric trade off studies on staging (number of stages, thrust for a given stage engine, propellant for a given stage, etc.) versus maximum attained velocity (V_{max}) or maximum altitude (apogee, h_{max}).

Let us look briefly at the more complicated flight profiles that one associates with inserting a payload (e.g., a satellite) into Earth orbit. One can refer to Fig. 9.4, which illustrates a simple multi-stage launch vehicle trajectory to low Earth orbit (LEO). With the launch vehicle at gross liftoff weight (GLOW) on the launch pad, the first-stage booster motor(s) are primarily there for altitude attainment Δh in accelerating the vehicle through the higher-drag lower atmosphere. As one reaches higher altitudes where drag becomes less of a factor, vehicle velocity becomes more paramount. In addition to bringing the vehicle closer to the desired orbital speed, the centrifugal acceleration effect of higher vehicle speeds at lower pitch angles θ relative to the Earth's surface below helps to offset gravitational losses to the vehicle's kinetic energy over the latter portion of the ascent trajectory. Throughout the ascent flight segment, the vehicle will be moving from a near-vertical pitch attitude ($\theta \sim 90°$) toward an attitude ultimately in parallel with the Earth's surface ($\theta \sim 0°$). Rather than using a simple, gradual "gravity-turn" ballistic trajectory approach in nosing from 90 to 0° while picking up speed from upper-stage thrust, one commonly sees a more aggressive pitchdown at one or more portions of the ascent to hasten the dropping of the flight vehicle's nose [1].

Pitching down when the vehicle is still at a relatively low forward speed bu
clear of the launch site is a common choice, since the amount of propellant ε
expended for pitching down is relatively less versus later at higher forward vehicle
speeds. This approach somewhat aligns with the well-known Hohmann minimum-
energy transfer ellipse approach, bearing in mind that one needs to consider such
factors as aerodynamic drag at lower altitudes and angle-of-attack g-loading
limitations of the flight vehicle's structure. At or approaching the desired orbital
altitude (in essence, the apogee of a transitional ellipse), an upper-stage "apogee
kick" motor (AKM) can provide the final velocity increment to bring the payload
to the required orbital speed, as well as "circularizing" the orbit in conjunction
(bringing the vehicle from an elliptic to circular orbit, in essence).

Consider the Space Transporation System (STS)/Space Shuttle's overall
velocity budget (ΔV) to attain a 200-km altitude Earth orbit:

Desired (nominal) circular orbital velocity	7,790 m/s
Gravity losses	1,220 m/s
Pitch angle trajectory adjustment	360 m/s
Atmospheric drag losses	118 m/s
Final orbital insertion	145 m/s
Minor correction maneuvres	62 m/s
Inertial assist from Earth rotation, lat. $\lambda = 28.5°$	−408 m/s
Total required mission velocity (ΔV)	9,347 m/s

One would use the total ΔV shown to estimate via the ideal rocket equation the
amount of propellant and its distribution through the two principal motor stages (as
per Fig. 9.5, first stage comprised of the thrust from two outboard solid-propellant
SRBs (solid rocket boosters) and three central liquid-propellant SSMEs (Space
Shuttle main engines); after jettisoning the two SRBs, second stage is thrust
delivered from the remaining three SSMEs; after shutting down the SSMEs and
jettisoning the external liquid propellant tank, a *de facto* third stage is thrust
delivered from the Shuttle's orbital maneuvring system (OMS) liquid-propellant
rocket engines to bring the Shuttle into the final-phase correct orbital position and
speed). The overall mass fraction (m_p/m_o, i.e., total propellant mass over vehicle's
gross liftoff mass) of the STS is 0.82, which significantly influences the value of
the above ΔV.

9.4 Gasdynamics and Thermodynamics of Internal Flow

The principal means of delivering thrust F by a chemical rocket is the production
of a high-velocity exhaust jet. Heat energy input via combustion produces a high-
temperature gas having a flame temperature T_F from 2,000 to 4,000 K in higher

Fig. 9.5 Photo of the Space Transportation System (STS; Space Shuttle *Discovery*) being slowly rolled out from the Vehicle Assembly Building toward the launch pad at Cape Canaveral, Florida. The large copper-colored body beneath the *Discovery* orbiter flight vehicle is the main external propellant tank structure (ET) containing an upper liquid oxygen tank and a much larger lower liquid hydrogen tank, for feeding the 3 SSME liquid-propellant rocket engines positioned at the rear of the orbiter. Photos courtesy of NASA

performance chemical engine combustion chambers. This high-temperature internal flow will pick up speed in the combustion chamber as the gas moves toward and into the exhaust nozzle convergence. The gas will continue to accelerate within the nozzle, reaching sonic speed in the proximity of the nozzle throat, and expanding supersonically (in the ideal case) through the nozzle expansion (divergence) to the nozzle exit plane. In order to deliver thrust at substantial levels, the combustion chamber pressure p_c must also be substantial, typically from 500 to 2,000 psi (3.45–13.8 MPa). This correlation can be seen with the use of the thrust coefficient C_F:

$$F = C_F p_c A_t \qquad (9.15)$$

where A_t is the nozzle throat area, and $C_F = f(\gamma, p_c, p_e)$, with p_e being the static pressure at the nozzle exit. It is not unusual to have a second phase accompanying the gas flow for at least some length of the flow in the chamber, e.g., liquid droplets entering via head-end injectors, or solid particulates entering from the burning solid-propellant surface.

In the more general case, one may in practice need a more sophisticated internal flow model than the one-dimensional models commonly used in this book. In modeling the movement of gas (and possibly a second phase within it, i.e., solid or liquid), one will need to assess the appropriate differential equations for conservation of mass, linear momentum and energy as discussed above, and perhaps a fourth conservation equation for angular momentum of the gas, and possibly a fifth conservation equation, for mixture fraction (i.e., chemical species) within a reacting gas flow, or even a sixth conservation equation, for mass diffusion within a medium. These differential equations would need to be applied in the three principal spatial directions [x, y, z in a cartesian (rectangular) 3D system; conservation of mass in 3D can be framed by a single differential equation, rather than three equations], and by analogy, an additional set of conservation equations would need to be applied for the second phase, if present.

The effects of laminar and turbulent viscosity on the gas flow are firstly felt in the momentum equation(s), in their general form referred to as the Navier–Stokes equations. A typical turbulence modeling approach is to incorporate two additional differential equations into the flow solution model, one for transport of turbulence kinetic energy (k) and transport of turbulence dissipation (ε, i.e., conversion of kinetic energy to heat energy) as done in the two-equation k–ε approach. In this case, the two turbulence parameters k and ε are incorporated into the Reynolds-averaged (incompressible flow) or Favre-averaged (compressible flow) conservation equations for linear momentum and energy.

In the most general case, the partial differential equation(s) for conservation of energy will equate the increase in the gas's internal energy with time to the sum of the heat influx from conduction and radiation, mass diffusion, pressure-related flow work (e.g., pressure acting to move mass through the system), laminar and turbulent dissipation of kinetic energy to heat via viscous stresses, heat input Q due to combustion and any body-force work.

Combustion can in practice be modeled in different ways and through various equations to meet the needs for a given application or propulsion system analysis, without necessarily requiring intensive computations as outlined above. In a similar fashion, on occasion we can simplify the gasdynamic flow equations to some degree, and in conjunction with one or a few combustion equations, still produce a reasonably accurate estimate of a given rocket's performance.

9.5 Rocket Nozzle Flow

The standard one-dimensional thrust equation, for thrust generated by a rocket's propulsive exhaust nozzle, is given by:

$$F = \dot{m}_e u_e + (p_e - p_\infty)A_e \tag{9.16}$$

The first term on the righthand side of the equation is the momentum flux at the nozzle exit plane, while the second term is the net pressure force acting at the exit plane of cross-sectional area A_e. A conventional chemical rocket would employ a convergent–divergent nozzle, with sonic flow in the vicinity of the throat (of area

Fig. 9.6 a Flow characteristics in convergent–divergent nozzle as chamber pressure is progressively increased relative to constant outside air pressure. Case (*1*): subsonic flow throughout. Case (*2*): flow has become choked, with flow ahead of upstream-facing standing normal shock S_2 being supersonic, and subsonic downstream (overexpanded nozzle). Case (*3*): standing normal shock S_3, with bigger pressure increase across it than S_2, is positioned very near to the nozzle exit plane (overexpanded nozzle; [2]). Inviscid flow assumed; **b** flow characteristics in convergent/divergent nozzle as chamber pressure is progressively increased relative to constant outside air pressure. Case (*4*): supersonic flow throughout internal nozzle region; upstream-facing oblique shock S_4 with supersonic flow upstream and downstream to bring pressure up toward ambient level (overexpanded nozzle). Case (*5*): flow has reached design point, exit-plane exhaust at ambient air pressure. Case (*6*): exit-plane exhaust pressure now exceeds outside air pressure, thus producing an upstream-facing Prandtl–Meyer rarefaction (expansion) wave to bring pressure down (underexpanded nozzle; [2]). Inviscid flow assumed

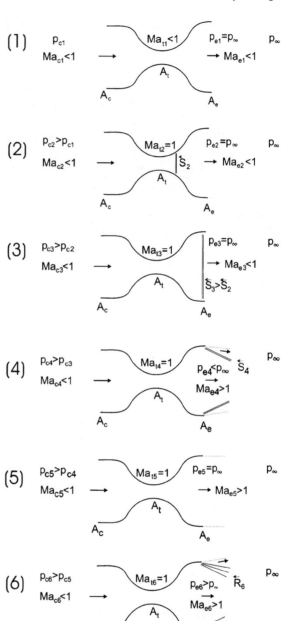

A_t) and ideally, supersonic flow downstream through the expansion (divergence). If the flow becomes over-expanded in the nozzle divergence, a stationary normal shock may form there in order to decelerate (diffuse) the flow. Figure 9.6 illustrates the flow characteristics that one would expect through the nozzle under ideal

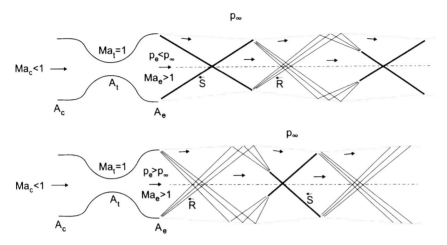

Fig. 9.7 Nominal exhaust flow patterns for an overexpanded supersonic nozzle (*upper* diagram: Case (*4*) of Fig. 9.6b) and an underexpanded supersonic nozzle (*lower* diagram: Case (*6*) of Fig. 9.6b) [2]

inviscid compressible flow conditions, from a low chamber pressure setting toward a high chamber pressure value, relative to a constant outside air pressure [2]. The result would be analogous to the case of a constant high chamber pressure, and having the outside air pressure drop progressively from Case (1) through to Case (6), i.e., emblematic of the rocket vehicle gaining altitude. Further information for Case (4) and (6) with respect to the downstream flow and standing diamond-shaped exhaust plume wave pattern may be observed in Fig. 9.7. In practice, as illustrated by Fig. 9.8, viscous boundary layer behavior might play a significant role in adjusting the above flow patterns. Depending on the nozzle geometry, at lower nozzle pressure ratios, where the flow is unchoked or just marginally choked and a good portion of the flow in the nozzle divergence is subsonic, the effective flow exit area for thrust delivery calculations may be closer to the nozzle throat area if the flow is separating from the bounding nozzle wall, in the vicinity of the nozzle throat, and expanding down to the outside air pressure level as an exhaust jet of relatively constant cross-sectional area.

Referring again to Fig. 9.7, one can briefly review some supersonic 2D planar-flow equations that have some relevance, with respect to doing analyses of flows containing standing oblique shocks and rarefaction waves. For example, referring now to Fig. 9.9 for supersonic flow over a wedge, if the effective wedge angle is shallow enough [2], one will have an attached oblique shock whose angle may be established via:

$$\tan\delta = \frac{\frac{2}{\tan\theta}\left(Ma_1^2\sin^2\theta - 1\right)}{(\gamma + 1)Ma_1^2 - 2\left(Ma_1^2\sin^2\theta - 1\right)} \tag{9.17}$$

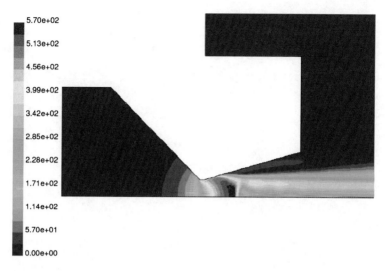

Fig. 9.8 Example flow contour diagram (contours of velocity magnitude in m/s) of steady channel gas flow passing through a choked 2D-axisymmetric convergent-divergent nozzle moving from left to right into the open atmosphere; viscous-flow CFD simulation via FLUENT V5.4. Diagram shows upper half of flow field, with flow centerline along the bottom boundary. A standing normal shock is evident in the nozzle divergence section, indicative of an overexpanded nozzle. The flow is separated from the nozzle expansion wall downstream of the nozzle throat, resulting in an exhaust jet that is of relatively constant cross-sectional area as it extends and expands downstream

When θ goes to 90°, one arrives at the normal shock solution, with δ going to 0°. Conversely, when $Ma \cdot \sin\theta$ goes to unity, one has the weak standing acoustic Mach wave solution [2], which also produces going to 0°. In solving Eq. 9.17, where δ is less than the detachment limit δ_{max}, one will have two possible solutions for a given Ma_1 and δ, namely the strong-shock solution (large θ) and the weak-shock solution (smaller θ). At the shock detachment limit, when δ equals δ_{max}, there is one unique solution, with $dMa_1/d\theta$ being zero for a given δ, or $d\delta/d\theta$ being zero for a given Ma_1:

$$\tan\delta_{max} = \frac{4Ma_1^2\sin\theta_{max} \cdot \cos^3\theta_{max}}{(\gamma + 1)Ma_1^2 - 2(Ma_1^2\sin^2\theta_{max} - 1) - 4Ma_1^2\sin^2\theta_{max} \cdot \cos^2\theta_{max}} \quad (9.18)$$

Barring an unusual application, in the case where there are two possible solutions, the weaker solution is generally the better estimate. Once θ is ascertained, one can estimate downstream flow Ma_2 via [2]:

$$Ma_2\sin(\theta - \delta) = \left\{ \frac{Ma_1^2\sin^2\theta + \frac{2}{\gamma-1}}{\frac{2\gamma}{\gamma-1}Ma_1^2\sin^2\theta - 1} \right\}^{1/2} \quad (9.19)$$

Fig. 9.9 Oblique shock wave patterns for steep and shallow wedge angle δ. In the steep wedge case (*upper diagram*), a bow shock will be produced, with some standoff distance at the centerline between the shock front and the starting point (*apex*) of the wedge. The flow streamlines will be curved in the subsonic flow region, with the flow accelerating toward the sonic line boundary and beyond. For a wedge angle shallower than δ_{max}, as is the case for the *lower* diagram, one observes an attached oblique shock [2]

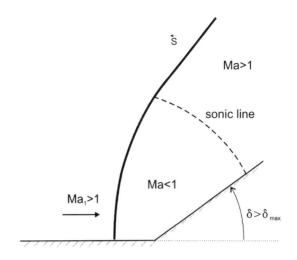

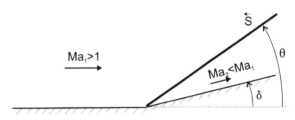

Additionally, the following can be determined:

$$\frac{T_2}{T_1} = \frac{\left[2 + (\gamma - 1)Ma_1^2\sin^2\theta\right]\left(2\gamma Ma_1^2\sin^2\theta - \gamma + 1\right)}{(\gamma + 1)^2 Ma_1^2\sin^2\theta} \tag{9.20}$$

$$\frac{p_2}{p_1} = \frac{2\gamma Ma_1^2\sin^2\theta - \gamma + 1}{\gamma + 1} \tag{9.21}$$

$$\frac{\rho_2}{\rho_1} = \frac{(\gamma + 1)Ma_1^2\sin^2\theta}{(\gamma - 1)Ma_1^2\sin^2\theta + 2} \tag{9.22}$$

Referring to Fig. 9.10 for supersonic flow over a convex corner, one can correlate the following for estimating the downstream supersonic flow Mach number, Ma_2, given the convex angle $v_2 - v_1$ is known:

$$v_2 - v_1 = \sqrt{\frac{\gamma + 1}{\gamma - 1}} \cdot \left[\tan^{-1}\sqrt{\left(\frac{\gamma - 1}{\gamma + 1}\right)(Ma_2^2 - 1)} - \tan^{-1}\sqrt{\left(\frac{\gamma - 1}{\gamma + 1}\right)(Ma_1^2 - 1)}\right]$$

$$+ \tan^{-1}\sqrt{Ma_1^2 - 1} - \tan^{-1}\sqrt{Ma_2^2 - 1}$$

$$\tag{9.23}$$

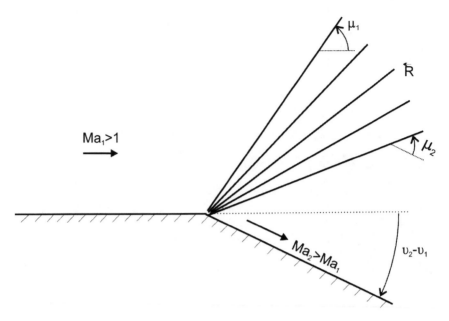

Fig. 9.10 Centered Prandtl–Meyer expansion (rarefaction) wave fan positioned at a convex corner [2]

For each wavelet (Mach wave) of the Prandtl–Meyer expansion wave fan [2], the following applies:

$$\sin\mu = \frac{1}{Ma} \tag{9.24}$$

Since the expansion process is isentropic, one can employ the standard isentropic flow relations, e.g.,

$$\frac{p_2}{p_1} = \left[\frac{1 + \frac{\gamma-1}{2} \cdot Ma_1^2}{1 + \frac{\gamma-1}{2} \cdot Ma_2^2} \right]^{\frac{\gamma}{\gamma-1}} \tag{9.25}$$

Note that the above coverage applies to 2D planar flows. The solutions will be somewhat different when examining 2D axisymmetric flows, and in general a bit more involved. The Taylor–MacColl solution [2] for an oblique shock over a cone-shaped nose (rather than a wedge as discussed earlier) is a well-known example. Some flow conditions by necessity can only be solved numerically.

Assuming no mass injection in the nozzle divergence, the mass flow $\dot{m}_e$ exiting the nozzle is equal to the mass flow through the nozzle throat, $\dot{m}_t$. From gasdynamics, choked mass flow of an ideal gas at the throat can be estimated as a function of chamber (or more specifically, head-end) pressure p_c:

$$\dot{m}_t = \dot{m}_e = \frac{1}{c^*} A_t p_c = \left[\frac{\gamma}{RT_F} \left(\frac{2}{\gamma+1} \right)^{\frac{\gamma+1}{\gamma-1}} \right]^{1/2} A_t p_c \tag{9.26}$$

Stagnation reservoir conditions are assumed at the point of measuring p_c, an approximation if considering some actual rocket combustion chambers. The parameter c^* is called the characteristic exhaust velocity (a.k.a, 'c-star'), which is connected to the effective (fully expanded) exhaust velocity c by the following:

$$c = \frac{F}{\dot{m}_e} = I_{sp} g_o = c^* C_F \tag{9.27a}$$

and in space,

$$c = \frac{F_{vac}}{\dot{m}_e} = I_{sp,vac} g_o = c^* C_{F,v} \tag{9.27b}$$

The vacuum thrust coefficient $C_{F,v}$ noted above will be discussed in more detail later. If the nozzle is choked and not over-expanded (no shock in the nozzle expansion), the following equation applies for the exit gas velocity, where $u_e \leq c$:

$$u_e = a_e \cdot Ma_e = \left[RT_F \left(\frac{2\gamma}{\gamma - 1} \right) \left\{ 1 - \left(\frac{p_e}{p_c} \right)^{\frac{\gamma-1}{\gamma}} \right\} \right]^{1/2} \tag{9.28}$$

The nozzle exit-plane area A_e governs the ideal (no-shock) static exit pressure p_e via the isentropic area-Mach number relation:

$$u_e = a_e \cdot Ma_e = \left[RT_F \left(\frac{2\gamma}{\gamma - 1} \right) \left\{ 1 - \left(\frac{p_e}{p_c} \right)^{\frac{\gamma-1}{\gamma}} \right\} \right]^{1/2} \tag{9.29}$$

From here, given Ma_t is 1 in value, thus establishing Ma_e, one can ascertain the ideal exit-plane static pressure:

$$p_e = p_c \left[1 + \frac{\gamma - 1}{2} Ma_e^2 \right]^{\frac{-\gamma}{\gamma-1}} \tag{9.30}$$

Substituting from Eqs. 9.26 and 9.28 into Eq. 9.25, one can produce the following estimate of thrust for a conventional rocket with an ideal or underexpanded choked nozzle flow condition:

$$F = C_F A_t p_c = C_{F,v} \left[1 - \left(\frac{p_e}{p_c} \right)^{\frac{\gamma-1}{\gamma}} \right]^{1/2} A_t p_c + (p_e - p_\infty) A_e \tag{9.31}$$

The parameter $C_{F,v}$ is the vacuum thrust coefficient, determined via:

$$C_{F,v} = \left[\frac{2\gamma^2}{\gamma - 1} \left(\frac{2}{\gamma + 1} \right)^{\frac{\gamma+1}{\gamma-1}} \right]^{1/2} \tag{9.32}$$

which is a function of the gas ratio of specific heats.

In an actual engine case, one may likely need to adjust predicted thrust F downward a little bit via a nozzle thrust delivery efficiency factor η_F (where actual $F_{act} = \eta_F F$), whose value might range from as low as 0.8 for a rudimentary nozzle well away from its design flight condition (e.g., significantly over- or under-expanded), to as high as 0.99 for a well-designed nozzle at near-optimal design flight conditions. In the case that the nozzle is choked but over-expanded (p_e via Eq. 9.30 less than p_∞), a better estimate of thrust using Eq. 9.31 would be to substitute p_e with p_∞.

Specific impulse I_{sp} is a measure of thrust performance and efficiency, defined by:

$$I_{sp} = \frac{F}{\dot{m}g_o} \text{ s} \tag{9.33}$$

This parameter tends to go up in value (a good thing) with a higher R (thus a lower molecular mass $\mathscr{M}$), a lower γ and a higher T_F. A rocket motor's average specific impulse $\bar{I}_{sp}$ may be estimated experimentally via:

$$\bar{I}_{sp} = \frac{1}{g_o m_p} I_{tot} = \frac{1}{g_o m_p} \int_0^{t_b} F dt \tag{9.34}$$

where I_{tot} is the total impulse (N·s) from the motor firing, and m_p is the mass of propellant consumed over burn time t_b. Specific impulse is in essence the inverse of thrust specific fuel consumption (TSFC). Not including the mass of air being consumed (since it is free as it is collected from the atmosphere), a turbojet engine has a typical I_{sp} of around 3,000 s, while a turbofan engine has a value that can approach 5,000 s. A conventional air-breathing ramjet engine can have an I_{sp} of around 1,500–2,000 s at its design cruise condition, and a well-designed pulsejet engine somewhat lower than that. A typical solid rocket will have a value of around 240 s at sea level, and up to 280 s at very high altitude (thrust performance improves with lower outside air pressures).

The shape of the nozzle divergence for maximum thrust attainment would in general be bell-shaped (classical de Laval contour nozzle profile, as shown in Fig. 9.11; [3]). For example, a bell (contour) nozzle would possibly deliver an η_F of approximately 0.98, while a comparably sized conical nozzle (simple cone expansion section) might have an η_F closer to 0.85. In practice, the length of a nozzle may be restricted by overall vehicle design concerns, concerns that might include keeping the nozzle's structural mass below a certain value; this generally results in a "truncated optimum" nozzle. Nozzle expansion profiles typically need to be adjusted under these circumstances of having a shorter nozzle than otherwise desired, in order to deliver the highest thrust. The method of characteristics (characteristic or weak wave theory) and CFD are two common mathematical tools for nozzle analysis and design. If one ignores viscous boundary layer effects, the method of characteristics would suggest that an ideal minimum-length maximum-thrust nozzle would have the shape as shown by the right nozzle in Fig. 9.11 [3]. The sharp-edged nozzle throat allows for a Prandtl–

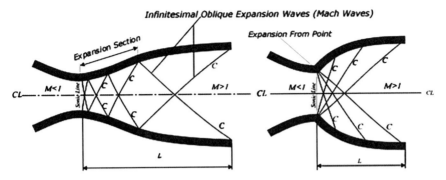

Fig. 9.11 Classical de Laval contour nozzle (*left*) and ideal minimum-length nozzle (*right*) [3]

Meyer expansion wave fan's origin to be ideally positioned; however, with actual viscous flows (that can be modelled by CFD), such an abrupt geometry change for the nozzle throat (versus a smooth geometric transition) would likely result in a separation of the boundary layer at that location, with associated flow momentum losses. Addtionally, sharp discontinuities in structure can lead to stress loading, heating and erosion issues. Thus in practice, one would tend to provide some radius of curvature to the throat area.

As a further alternative for nozzle selection, an aerospike nozzle (Fig. 9.12; falls in the category of a radial or annular-flow design) does have the potential advantage of lower structural mass (employing a center-body, but no peripheral wall structure), and a passive (rather than mechanical) means for allowing the nozzle exit area to increase as one goes up in altitude. However, at any one design flight condition, the aerospike is inferior to a comparable walled conventional contour nozzle designed for that condition, especially at lower flight speeds and altitudes. Aerospikes are potentially competitive, overall, for flight missions that take the vehicle from low to high altitude at predominantly high vehicle speeds (supersonic to hypersonic). Plug, expansion-deflection or other annular-flow nozzles can be designed to mechanically compensate for altitude (e.g., axially moving a central plug or center-body back-and-forth as required in the nozzle throat and expansion regions), as an active mechanical alternative to the aerospike approach. The drawback with this active mechanical approach is higher design complexity, higher viscous drag losses (in some cases) and extra weight.

For multi-staged space launch applications, one typically would see higher and higher exit-to-throat area ratios in the upper-stage motor nozzles. Given that outside air pressure drops with increased altitudes above the Earth's surface and has a significant effect on increasing thrust and specific impulse, it is commonly found that it is beneficial to accept the additional structural weight of the larger nozzle expansion sections in order to attain the desired performance. For storage while enclosed within an upper-stage structure, one may see the use of an extensible nozzle expansion that is compacted initially, and once free to do so, the

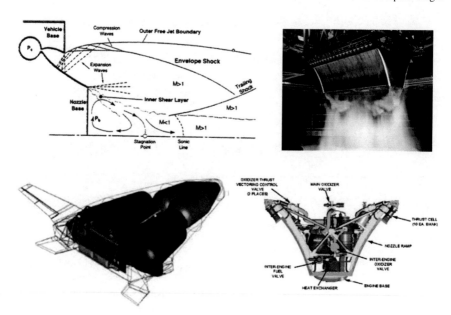

Fig. 9.12 Schematic diagram at *upper left* of aerospike nozzle and exhaust flow pattern. Linear Aerospike XRS-2200 LH_2/LO_2 liquid rocket engine by Rocketdyne being test-fired at *upper right*; two side-by-side engines were proposed for powering the NASA/Lockheed Martin X-33 hypersonic/SSTO flight vehicle at *lower left*. Government funding for the program ceased in 2001. Courtesy of NASA

nozzle expansion is mechanically extended to provide the desired extra length and exit area A_e. A stepped nozzle is a simpler fixed-geometry design compromise for flight over a wide altitude range, whereby an intermediate step in the nozzle expansion contour allows for a smaller exiting exhaust jet at low altitude (flow separating cleanly from the step location as it moves downstream through the remainder of the nozzle expansion), and at high altitude (lower outside air pressure), the exiting flow passes by the step more smoothly (re-attaches downstream with the positive pressure gradient) and exits the nozzle expansion exit plane relatively cleanly.

In the case of two-phase flow through the nozzle, the mass of the particles or droplets may be substantial enough to include in the determination of thrust, i.e., allow for particle momentum flux $\dot{m}_p u_p$ at the nozzle exit plane (and also upstream of the exit plane, allow for gas-particle drag and heat exchange which may act to counter net thrust delivery a little bit). Due to slip (difference in velocity) and the temperature difference between the particles (under high acceleration through the nozzle throat) and the surrounding gas, one would not use the equilibrium estimate for various two-phase flow properties. To further illustrate this point, let us first define some relevant parameters. If one defines average two-phase density in a given volume as

$$\rho_{2ph} = \frac{m}{\Psi} \approx \frac{N \cdot m_p + \rho\Psi}{\Psi} = \frac{\rho_p \Psi + \rho\Psi}{\Psi} = \alpha_p \rho_{2ph} + \rho \quad (9.35)$$

where α_p is the particle loading fraction in the flow, and N is the number of particles of average mass m_p in the elemental volume above, then one can show the correlation between particle density ρ_p and gas density ρ:

$$\rho_p = \frac{N \cdot m_p}{\Psi} = \alpha_p \rho_{2ph} = \frac{\alpha_p}{1 - \alpha_p} \cdot \rho \quad (9.36)$$

Note that in the above relations, we are assuming the particle loading is low enough in value such that the gas phase void fraction, defined by

$$\alpha_g = \frac{\Psi - \Psi_p}{\Psi} = 1 - \frac{\Psi_p}{\Psi} \quad (9.37)$$

is relatively close to unity (Ψ_p is the volume occupied by the particles in an elemental total volume of Ψ). Under equilibrium conditions (no slip [$u = u_p$], no ΔT [$T = T_p$]), the two-phase value for ratio of specific heats is:

$$\gamma_e = \frac{\gamma + \frac{\rho_p}{\rho}\gamma\frac{C_m}{C_p}}{1 + \frac{\rho_p}{\rho}\gamma\frac{C_m}{C_p}} < \gamma \quad (9.38)$$

where C_m is the specific heat of the solid particle or liquid droplet. Similarly, for the two-phase equilibrium sound speed,

$$a_e = \sqrt{\frac{\gamma_e RT}{1 + \frac{\rho_p}{\rho}}} < a \quad (9.39)$$

As noted earlier, one would not be advised to assume the two-phase flow properties would be in equilibrium under appreciable flow acceleration, and certainly not in the vicinity of the nozzle throat. It is known that the two-phase flow in fact chokes at the nozzle throat, which is to say $u/b = 1$ at that location, where b is the non-equilibrium sound speed of the two-phase mixture [4]:

$$b = a \left[\frac{1 + \beta_g \frac{C_m}{C_p}\frac{T_p}{T}}{\alpha_g^2 \left(1 + \beta_g \frac{u_p}{u}\right)\left(1 + \gamma\beta_g \frac{C_m}{C_p}\frac{T_p}{T}\right) + \alpha_g \beta_g (\gamma - 1)\frac{u_p}{u}\left(1 - \frac{u_p}{u}\right)} \right]^{1/2} \quad (9.40)$$

where

$$\beta_g = \frac{\rho_p u_p}{\rho u \alpha_g} \quad (9.41)$$

Since one would expect $u_p < u$ at or near the throat, it turns out that $b < a$ at that location. If that is the case, downstream of the nozzle throat, one will have $u/a = Ma = 1$.

On occasion, in the case of solid propellants, one may see the use of two or three sets of particles of differing nominal diameters, composed of the same material (e.g., aluminum). If such is the case, one needs to keep track of each set. With Eq. 9.40 as the point of interest, let us look at the example of two particle sets. In that case, the following would apply:

$$\rho_p = \frac{N_1 m_{p1} + N_2 m_{p2}}{\mathcal{V}} = \rho_{p1} + \rho_{p2} \tag{9.42}$$

$$\beta_g = \frac{\dot{m}_p}{\dot{m}_g} = \frac{\rho_{p1} u_{p1} + \rho_{p2} u_{p2}}{\rho u \alpha_g} \tag{9.43}$$

$$\beta_g \frac{C_m}{C_p} \frac{T_p}{T} = \frac{C_m}{C_p T} \cdot \frac{\left[\rho_{p1} u_{p1} T_{p1} + \rho_{p2} u_{p2} T_{p2}\right]}{\rho u \alpha_g} = \Sigma_1 \tag{9.44}$$

$$\beta_g \frac{u_p}{u} = \frac{\rho_{p1} u_{p1}^2 + \rho_{p2} u_{p2}^2}{\rho u^2 \alpha_g} = \Sigma_2 \tag{9.45}$$

$$\beta_g \frac{u_p^2}{u^2} = \frac{\rho_{p1} u_{p1}^3 + \rho_{p2} u_{p2}^3}{\rho u^3 \alpha_g} = \Sigma_3 \tag{9.46}$$

By substitution then, Eq. 9.40 would become:

$$b = a \left[\frac{1 + \Sigma_1}{\alpha_g^2 (1 + \Sigma_2)(1 + \gamma \Sigma_1) + \alpha_g (\gamma - 1)(\Sigma_2 - \Sigma_3)} \right]^{1/2} \tag{9.47}$$

Note the effect of particle mass loading ΔM_p into a solid propellant that originally was of solid density $\rho_{s,o}$, such that the loading mass fraction into a solid volume $\mathcal{V}$ is:

$$\alpha_{p,s} = \frac{\Delta M_p}{\Delta M_{s,o} + \Delta M_p} = \frac{\rho_m \mathcal{V}_m}{\rho_{s,o}(\mathcal{V} - \mathcal{V}_m) + \rho_m \mathcal{V}_m} = \frac{1}{\frac{\rho_{s,o}}{\rho_m}\left(\frac{\mathcal{V}}{\mathcal{V}_m} - 1\right) + 1} \tag{9.48}$$

The new effective solid propellant overall density becomes:

$$\rho_{s,new} = \frac{\Delta M_p + \Delta M_{s,o}}{\mathcal{V}} = \frac{\rho_{s,o}(\mathcal{V} - \mathcal{V}_m) + \rho_m \mathcal{V}_m}{\mathcal{V}} = \rho_{s,o}\left(1 - \frac{\mathcal{V}_m}{\mathcal{V}}\right) + \rho_m \frac{\mathcal{V}_m}{\mathcal{V}} \tag{9.49}$$

By substitution, one can show that

$$\rho_{s,new} = \frac{\rho_{s,o}}{1 - \alpha_{p,s} + \alpha_{p,s} \frac{\rho_{s,o}}{\rho_m}} \tag{9.50}$$

Boundary layer development at and beyond the nozzle throat may also be substantial. This will bring a loss in flow momentum. In some analyses where one needs to be accurate, boundary layer losses and the effective reduction in flow cross-sectional area would need to be accounted for in establishing a good prediction for thrust.

In an analogy similar to two-phase flow not being in equilibrium in the vicinity of the nozzle throat and beyond, the molecular characteristics of the gas flow may be changing substantially in this region (i.e., being appreciably in thermochemical nonequilibrium, such that values for γ, C_p, etc., are varying noticeably), in transitioning from a frozen flow condition to an equilibrium flow condition further downstream. One might retain an ideal-gas framework with frozen or mean flow property assumptions, for convenient preliminary calculations.

9.6 Combustion Considerations for Chemical Rockets

In establishing the performance of a chemical rocket engine, we are aware that the combustion process and how it is managed will determine the effective mean molecular mass $\mathcal{M}$ of the resulting gas, the ratio of specific heats of the gas and the flame temperature. All of these parameters have a strong influence on thrust delivery. There are some basic considerations that are applicable to all conventional combustors which will be discussed here in this section, while there are other considerations that will be specific to the system in question.

Consider a simple, ideal combustion application bringing the fuel, hydrogen (H_2), and the oxidizer, oxygen (O_2), together as reactants, producing water (H_2O) and heat energy as the products of the resulting stoichiometric reaction:

$$H_2 + \frac{1}{2}O_2 \rightarrow H_2O + \Delta Q_r \tag{9.51}$$

In the above relation, note the molar basis for mass conservation. If one defines r as the oxidizer/fuel mixture ratio of the initial reactants, then

$$r = \frac{m_O}{m_F} = \frac{\dot{m}_O}{\dot{m}_F} = \frac{\frac{1}{2}(32\,\text{kg per mole})}{1(2\ \text{kg per mole})} = 8{:}1 \tag{9.52}$$

and the resulting products would have a molecular mass of

$$\mathcal{M} = 2 + 16 = 18\,\text{amu} \tag{9.53}$$

The net heat of reaction ΔQ_r can be determined from equilibrium thermochemistry calculations that will also establish the flame (combustion) temperature T_F (referred to as the adiabatic flame temperature when there is no heat into or out of the system, and the combustion is complete; maximum adiabatic flame temperature occurs when the mixture is stoichiometric), for a given combustor pressure p_c. Thermochemical equilibrium for establishing molar concentrations n_j for a jth molecule is based on the minimization of the chemical potential (also known as the Gibbs free energy G in J/(kg mole), which is a function of enthalpy h and entropy s; [5, 6]) of the various reactants, i.e., minimize:

$$G = h - Ts = u + pv - Ts = \frac{\sum n_j G_j}{\sum n_j} = \frac{\sum n_j (h - Ts)_j}{\sum n_j} \quad (9.54)$$

In the above calculations, in using the existing thermochemical tables for a given pressure, one will balance contributions from the heats of formation of the various reactants and products, required heat of vaporization, etc.

Given the above introduction, one can now look at the hydrogen–oxygen combustion process in a more realistic (and less ideal) context:

$$a\mathrm{H}_2 + b\mathrm{O}_2 \rightarrow n_{\mathrm{H}_2\mathrm{O}}\mathrm{H}_2\mathrm{O} + n_{\mathrm{H}_2}\mathrm{H}_2 + n_{\mathrm{O}_2}\mathrm{O}_2 + n_{\mathrm{O}}\mathrm{O} + n_{\mathrm{H}}\mathrm{H} + n_{\mathrm{OH}}\mathrm{OH} + \Delta Q_r \quad (9.55)$$

In practice, liquid rocket engines (LREs) run a fuel-rich mixture (equivalence ratio $\phi > 1$, where $\phi = (\dot{m}_f)_{\mathrm{act}}/(\dot{m}_f)_{\mathrm{stoich}}$) with oxygen–hydrogen, usually around or a bit less than 6:1 for mixture ratio r (thus, given $a = 1$, then b 3/8 rather than 1/2, for Eq. 9.55 above). The unreacted H_2 in the exhaust products acts to lower the mean molecular mass of the gas, which helps thrust performance, even though the flame temperature will drop a bit with the need to heat the unreacted hydrogen (in fact, the peak thrust delivery occurs at r around 3.5, but then storage of liquid hydrogen would be too lopsided at such a low r, versus the oxygen stored, in terms of available tank volume). Molecular mass $\mathcal{M}$ can be found via:

$$\mathcal{M} = \frac{\sum n_j \mathcal{M}_j}{\sum n_j} = \frac{18 n_{\mathrm{H}_2\mathrm{O}} + 2 n_{\mathrm{H}_2} + 32 n_{\mathrm{O}_2} + 16 n_{\mathrm{O}} + n_{\mathrm{H}} + 17 n_{\mathrm{OH}}}{n_{\mathrm{H}_2\mathrm{O}} + n_{\mathrm{H}_2} + n_{\mathrm{O}_2} + n_{\mathrm{O}} + n_{\mathrm{H}} + n_{\mathrm{OH}}} \quad (9.56)$$

The molar concentrations n_j can be established via Eq. 9.56. Corresponding to the above calculations, one can find the constant-pressure specific heat of the exhaust gas:

$$C_p = \frac{\sum n_j C_{p,j}}{\sum n_j} \quad (9.57)$$

The specific heat ratio of the exhaust gas may be estimated via:

$$\gamma = \frac{C_p}{C_v} = \frac{C_p}{C_p - R} = \frac{C_p}{C_p - \frac{\mathscr{R}}{\mathscr{M}}} \quad (9.58)$$

While typically the calculations are less voluminous for LREs, a similar approach is applicable for SRM propellant reactants and products, with a typically large number of constituents to be accounted for in the thermochemical calculations.

With respect to the physical structure and mechanism(s) of burning within a given system's flame, combustion science tends to identify flames under 2 general categories: (1) premixed laminar, and (2) turbulent diffusion. Chemical kinetics (rates of reaction of the various molecular constituents) are enhanced by increasing chamber pressure, and thus will dominate the premixed flame's behavior (fuel and oxidizer mixed before burning, or some portion thereof). Mixing turbulence, due to swirling, recirculation, etc., would tend to dominate the diffusion flame's behavior. In practice, and certainly in the case of most SRM and LRE combustors, some aspects of both flame categories will influence the combustion process and its corresponding effectiveness.

9.7 Example Problems

9.1. A sounding rocket has two motor stages. The effective payload (including upper vehicle structure) to be carried in a vertical flight path to its peak altitude is 20 kg in mass. Characteristics of the first motor stage include:

mean thrust $F_{o1} = 34.6$ kN, mean specific impulse $I_{sp1} = 252$ s, propellant mass $m_{P1} = 35$ kg, empty motor structure mass $m_{E1} = 0.15 \, m_{P1}$

Characteristics of the second (upper) motor stage include:

mean thrust $F_{o2} = 9$ kN, mean specific impulse $I_{sp2} = 265$ s, propellant mass $m_{P2} = 17$ kg, empty motor structure mass $m_{E2} = 0.12 \, m_{P2}$

Estimate the maximum ideal vertical apogee for the vehicle's flight, neglecting aerodynamic drag. You may assume a mean gravitational acceleration over the first motor stage's burn time of 9.80 m/s^2, 9.79 m/s^2 over the second motor stage's burn time and 9.15 m/s^2 over the final flight phase.

9.2. A launch vehicle, having a gross vehicle liftoff mass of 4,000 kg, has three motor stages. In order to approximate a curved trajectory to orbit from a low-atmosphere launch, assume a vertical ascent for the flight performance calculations over the three stages, with the mean gravity per flight phase adjusted for non-vertical, lower/higher-drag flight portions (values as shown below). The effective payload (including upper vehicle structure) to be carried in an initial vertical flight path is 45 kg in mass. Desired orbital speed is 8 km/s. Characteristics of the first motor stage (SRM booster) include:

mean thrust $F_{o1} = 120$ kN, mean specific impulse $I_{sp1} = 225$ s, propellant mass $m_{P1} = ?$ kg, empty motor structure mass $m_{E1} = 0.11 \, m_{P1}$, mean $g_1 = 9.80$ m/s^2 for 1st flight phase

Characteristics of the second (intermediate) motor stage (SRM) include:

mean thrust $F_{o2} = 50$ kN, mean specific impulse $I_{sp2} = 250$ s, propellant mass $m_{P2} = ?$ kg, empty motor structure mass $m_{E2} = 0.11 \, m_{P2}$, mean $g_2 = 4.0$ m/s^2 for 2nd flight phase

Characteristics of the third (upper) motor stage (SRM) include:

mean thrust $F_{o3} = 20$ kN, mean specific impulse $I_{sp3} = 275$ s, propellant mass $m_{P3} = ?$ kg, empty motor structure mass $m_{E3} = 0.11 \, m_{P3}$, mean $g_3 = 0.0$ m/s^2 for 3rd flight phase

(a) One engineer in your team notes one can acquire off-the-shelf a first-stage motor having a propellant mass of 2,000 kg, and a second stage motor having a propellant mass of 1,000 kg, at a relative cost savings. The third motor will have to be custom-built, at a greater relative cost, to meet the stated requirements outlined earlier. Once all the information is established, cost aside, can this setup technically do the job (reach an orbital speed of at least 8 km/s)?

(b) A second engineer with more flight optimization experience suggests one might have to bite the bullet, and custom-build all three motor stages to give a better driven payload ratio among the three phases, to reach the minimum 8 km/s target. Based on her calculations so far, a higher first-stage propellant mass of 2,500 kg and a lower second stage propellant mass of 800 kg makes more sense, given the liftoff mass of 4,000 kg. Complete her calculations, to confirm or refute the meeting of the speed target.

9.3. Provide a general equation for solving thrust coefficient C_F, if all other pertinent parameters are known.

9.4. Show that u_e goes to c when the exhaust flow is fully expanded in an ideal fashion.

9.5. Show why the circular orbital speed V of the Shuttle orbiter at 200 km altitude is around 7,790 m/s. To aid in your calculations, you may use Eq. 9.1 for the law of gravity, and from dynamics, the knowledge that the centrifugal acceleration to maintain a circular path is V^2/r, where r is the radial distance between the object and the center of rotation.

9.8 Solutions to Example Problems

9.1. Looking at a two-stage flight vehicle, vertical ascent with gravity, but no drag.
$$m_o = m_{E1} + m_{P1} + m_{E2} + m_{P2} + m_{PL} = 5.25 + 35 + 2.04 + 17 + 20 = 79.3 \text{ kg}$$

Acceleration off launch rail (liftoff), $a_{LO} = \frac{F_{o,1}}{m_o} = \frac{34600}{79.3} = 436$ m/s^2 or 44.5 g

$$m_1 = m_o - m_{P1} - m_{E1} = 39.05 \text{ kg}$$

$$V_2 = -g_o I_{sp,1} \left\{ \ln\left(1 - \frac{m_{P1}}{m_o}\right) + \bar{g}_1 \frac{m_{P1}}{F_{o,1}} + \left(\ln\left(1 - \frac{m_{P2}}{m_1}\right) + \bar{g}_2 \frac{m_{P2}}{F_{o,2}}\right) \frac{I_{sp,2}}{I_{sp,1}} \right\}$$
$$= -9.81(252)\{\ln(1 - 35/79.3) + 9.80(35)/34600 + (\ln(1 - 17/39.05) + 9.79(17)/9000)265/252\}$$
$$= 1415 + 1438 = 2853 \text{ m/s}$$

$$h_2 = \frac{g_o^2 I_{sp,1}^2}{F_{o,1}} \left[(m_o - m_{P1}) \ln\left(1 - \frac{m_{P1}}{m_o}\right) + m_{p1} \right] - \frac{\bar{g}_1}{2} \left[\frac{I_{sp,1} m_{P1} g_o}{F_{o,1}} \right]^2$$

$$+ V_1 \left[\frac{I_{sp,2} m_{P2} g_o}{F_{o,2}} \right] + \frac{g_o^2 I_{sp,2}^2}{F_{o,2}} \left[(m_1 - m_{P2}) \ln\left(1 - \frac{m_{P2}}{m_1}\right) + m_{P2} \right] - \frac{\bar{g}_2}{2} \left[\frac{I_{sp,2} m_{P2} g_o}{F_{o,2}} \right]^2$$

$$= (9.81^2)(252^2)/34600[(79.3 - 35)\ln(1 - 35/79.3) + 35] - 9.80/2[2.5]^2$$

$$+ 1415[4.91] + (9.81^2)(265^2)/9000[(39.05 - 17)\ln(1 - 17/39.05) + 17]$$

$$- 9.79/2[4.91]^2$$

$$= 1626 - 31 + 6948 + 3302 - 118 = 11727\,\text{m}$$

$h_{\max} = h_2 + \frac{V_2^2}{2\bar{g}_3} = 11727 + 2853^2/(2 \cdot 9.15).456514$ m   or   457 km   or   284 stat. mi.

Similarly,

$$h_{\max} = \frac{1}{\frac{1}{R_E + h_2} - \frac{V_2^2}{2g_o R_E^2}} - R_E = \frac{1}{\frac{1}{(6378000 + 11727)} - \frac{2853^2}{2(9.81)6378000^2}} - 6378000$$

$$\approx 457000\,\text{m}.$$

9.2. Looking at a three-stage flight vehicle, vertical ascent with gravity, but no drag.

(a) $m_o = m_{E1} + m_{P1} + m_{E2} + m_{P2} + m_{E3} + m_{P3} + m_{PL} = 4000\,\text{kg}$

$4000 = 2000 + 0.11(2000) + 1000 + 0.11(1000) + 1.11 m_{P3} + 45, m_{P3} = 563\,\text{kg}$

$$m_1 = m_o - m_{P1} - m_{E1} = 4000 - 2000 - 0.11(2000) = 1780 \text{ kg}$$
$$m_2 = m_1 - m_{P2} - m_{E2} = 1780 - 1.11(1000) = 670\,\text{kg}$$

$$V_3 = -g_o I_{sp,1} \left\{ \ln\left(1 - \frac{m_{P1}}{m_o}\right) + \bar{g}_1 \frac{m_{P1}}{F_{o,1}} + \left(\ln\left(1 - \frac{m_{P2}}{m_1}\right) + \bar{g}_2 \frac{m_{P2}}{F_{o,2}} \right) \frac{I_{sp,2}}{I_{sp,1}} \right.$$

$$\left. + \left(\ln\left(1 - \frac{m_{P3}}{m_2}\right) + \bar{g}_3 \frac{m_{P3}}{F_{o,3}} \right) \frac{I_{sp,3}}{I_{so,1}} \right\}$$

$$= -9.81(225)\{\ln(1 - 2000/4000) + 9.80(2000)/120000 + (\ln(1 - 1000/1780)$$

$$+ 4(1000/50000))250/225 + \ln(1 - 563/670) + 0(563/2000))275/225\}$$

$$= -2207.25\{-0.5298 - 0.8279 - 2.2421\}$$

$$= 7946\,\text{m/s} < 8000 \text{ m/s}$$

required. No.

(b) Change mass distribution, $4000 = 2500(1.11) + 800(1.11) + m_{P3}(1.11) + 45, m_{P3} = 263\,\text{kg}$

$$m_1 = m_o - m_{P1} - m_{E1} = 4000 - 2500 - 0.11(2500) = 1225 \text{ kg}$$
$$m_2 = m_1 - m_{P2} - m_{E2} = 1225 - 1.11(800) = 337 \text{ kg}$$

$$V_3 = -9.81(225)\{\ln(1 - 2500/4000) + 9.80(2500)/120000 + (\ln(1 - 800/1225)$$
$$+ 4(800/50000))250/225 + \ln(1 - 263/337) + 0(263/2000))275/225\}$$
$$= -2207.25\{-0.7767 - 1.1051 - 1.8529\}$$
$$= 8244 \text{ m/s} > 8000 \text{ m/s}$$

required. Yes.

Alternative solution approach: set V_3 to 8,000 m/s from the start, and establish needed m_{P3}.

9.3.

$$C_F = F/(A_t p_c) = \left\{ C_{F,v} \left[1 - \left(\frac{p_e}{p_c} \right)^{\frac{\gamma-1}{\gamma}} \right]^{1/2} A_t p_c + (p_e - p_\infty) A_e \right\} / (A_t p_c)$$

$$= C_{F,v} \left[1 - \left(\frac{p_e}{p_c} \right)^{\frac{\gamma-1}{\gamma}} \right]^{1/2} + \frac{A_e}{A_t} \left[\frac{p_e - p_\infty}{p_c} \right]$$

9.4.

$$u_e = a_e \cdot Ma_e = \left[RT_F \left(\frac{2\gamma}{\gamma - 1} \right) \left\{ 1 - \left(\frac{p_e}{p_c} \right)^{\frac{\gamma-1}{\gamma}} \right\} \right]^{1/2}$$

$$= \left[RT_F \left(\frac{2\gamma}{\gamma - 1} \right) \left\{ 1 - \left(\frac{p_\infty}{p_c} \right)^{\frac{\gamma-1}{\gamma}} \right\} \right]^{1/2}, \text{ when fully expanded}$$

$$c = c^* C_F = \left[\frac{\gamma}{RT_F} \left(\frac{2}{\gamma + 1} \right)^{\frac{\gamma+1}{\gamma-1}} \right]^{-1/2} \cdot \left\{ C_{F,v} \left[1 - \left(\frac{p_e}{p_c} \right)^{\frac{\gamma-1}{\gamma}} \right]^{1/2} + \frac{A_e}{A_t} \left[\frac{p_e - p_\infty}{p_c} \right] \right\}$$

$$= \left[\frac{\gamma}{RT_F} \left(\frac{2}{\gamma + 1} \right)^{\frac{\gamma+1}{\gamma-1}} \right]^{-1/2} \cdot \left\{ \left[\frac{2\gamma^2}{\gamma - 1} \left(\frac{2}{\gamma + 1} \right)^{\frac{\gamma+1}{\gamma-1}} \right]^{1/2} \left[1 - \left(\frac{p_\infty}{p_c} \right)^{\frac{\gamma-1}{\gamma}} \right]^{1/2} + \frac{A_e}{A_t} \left[\frac{p_\infty - p_\infty}{p_c} \right] \right\}$$

$$= \left[RT_F \left(\frac{2\gamma}{\gamma - 1} \right) \left\{ 1 - \left(\frac{p_\infty}{p_c} \right)^{\frac{\gamma-1}{\gamma}} \right\} \right]^{1/2} = u_e, \text{ when fully expanded}$$

9.5. From Eq. 9.1, we have:

$$g = g_o \left(\frac{R_E}{R_E + h_{ASL}} \right)^2$$

where g_o is 9.81 m/s^2, R_E is around 6,378 km, and h_{ASL} is 200 km. From dynamics, one can equate the above to the centrifugal acceleration needed to hold that circular path:

$$g = g_o \left(\frac{R_E}{R_E + h_{ASL}} \right)^2 = \frac{V^2}{R_E + h_{ASL}}$$

so that

$$V = \left\{ g_o R_E^2 / (R_E + h_{ASL}) \right\}^{1/2} = \left\{ 9.81 \cdot 6378000^2 / (6378000 + 200000) \right\}^{1/2}$$
$$= 7789 \ \text{m/s}$$

Close enough.

References

1. Greatrix DR, Karpynczyk J (2005) Rocket vehicle design for small payload delivery to orbit. Can Aeronaut Space J 51:123–131
2. John JEA (1984) Gas dynamics, 2nd edn. Prentice-Hall, Saddle River
3. Hill PG, Peterson CR (1992) Mechanics and thermodynamics of propulsion, 2nd edn. Addison-Wesley, New York
4. Forde M (1986) Quasi-one-dimensional gas/particle nozzle flows with shock. AIAA J 24:1196–1199
5. Van Wylen GJ, Sonntag RE (1973) Fundamentals of classical thermodynamics, 2nd edn. Wiley, New York
6. Kuo KK (1986) Principles of combustion. Wiley, New York

Chapter 10
Solid-Propellant Rocket Motors

10.1 Introduction

With regard to specific rocket propulsion systems, let us look first at solid-propellant rocket motors (SRMs), one of the "simpler" chemical rocket systems. They are commonly identified as being simpler because of the typical lack of any moving parts, and generally straightforward operation (see Fig. 10.1 for one well-known large example). With respect to moving parts, while it is true that solid rockets generally do not employ turbomachinery or upstream valving, at the aft downstream end of the motor, one can note that an SRM's nozzle may employ a thrust vector control technique involving actuated movement of the nozzle or a valved injection of fluid therein. With respect to straightforward operation, in a typical SRM, there is no throttling (modulated feed of fuel and/or oxidizer to adjust thrust delivery, as might be done in a hybrid or liquid pro-pellant rocket engine), although on occasion, one sees the use of a pintled nozzle to effectively adjust the nozzle throat area and resulting thrust of an SRM. More commonly, the solid propellant is ignited and allowed to "passively" burn to completion, in a manner pre-arranged by the propellant charge's various design characteristics, including geometry. SRMs in general are a competitive system for simpler one-burn-only applications, sometimes being the least costly of the various alternatives. Thrust delivery can be as low as micro- to milli-newton level for small spacecraft microthrusters, to as high as 10 mega-newtons or more when acting as a booster motor for a space launch vehicle. One solid rocket booster (SRB) that assists in lifting off the Space Shuttle (STS) from the launch pad produces an initial thrust of around 12 MN (see Figs. 10.2 and 10.3), noting that the overall STS flight vehicle weighs 20 MN at liftoff. Specific impulse for higher performance solid rocket motors will be on the order of 240 s at sea level, and approaching 280 s at very high altitudes.

D. R. Greatrix, *Powered Flight*, DOI: 10.1007/978-1-4471-2485-6_10,
© Springer-Verlag London Limited 2012

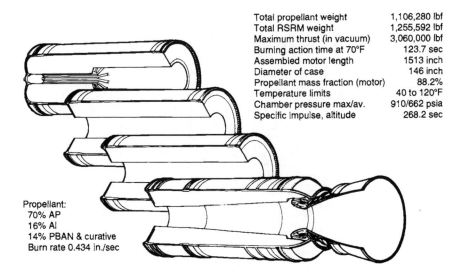

Total propellant weight	1,106,280 lbf
Total RSRM weight	1,255,592 lbf
Maximum thrust (in vacuum)	3,060,000 lbf
Burning action time at 70°F	123.7 sec
Assembled motor length	1513 inch
Diameter of case	146 inch
Propellant mass fraction (motor)	88.2%
Temperature limits	40 to 120°F
Chamber pressure max/av.	910/662 psia
Specific impulse, altitude	268.2 sec

Propellant:
70% AP
16% Al
14% PBAN & curative
Burn rate 0.434 in./sec

Fig. 10.1 Cutaway diagram of a Space Shuttle solid rocket booster (SRB) built by ATK Thiokol. The diagram shows the motor separated into four main segments (which it would be in practice, prior to being assembled). The uppermost propellant segment (closest to the motor's head end) reveals a finocyl design. The lowest propellant segment (closest to the nozzle) reveals a conical flare design, to keep the grain port exit area substantially larger than the nozzle throat area from initial motor ignition. The propellant contains a PBAN polymeric binder (polybutadiene acrylic-acid/acrylonitrile). Observe the submerged nozzle design (nozzle convergence is substantially within the chamber boundaries). Courtesy of NASA

10.2 Performance Considerations

The solid propellant grain (charge) is cast, bonded (to the case) or loaded (as a free-standing cartridge) into place within the combustion chamber. Above a certain size, it becomes more difficult to produce (cast) a single propellant grain as one unit, and may require that the grain comprise several smaller segments brought together when the motor is assembled. This is the case for the aforementioned SRBs for the Space Shuttle. Once the igniter brings the propellant surface above its auto-ignition temperature T_{as} and ignition proceeds, the resulting burning surface will regress toward the motor casing boundary, which would commonly comprise an inner heat-resistant insulation layer and the outer structural wall. The chosen shape of the grain will determine the resulting chamber pressure–time (p_c–t) profile, which will produce a corresponding thrust-time (F–t) profile. It is common to designate the chamber pressure as measured at the motor's head end as the representative p_c.

To produce a relatively constant (flat, neutral) thrust profile, as per Fig. 10.4, one could employ an end-burning (a.k.a., cigarette-burner) design. While end-burners allow for a high propellant loading percentage in occupying most of the available space in the combustion chamber upstream of the nozzle, a disadvantage

Fig. 10.2 Back view of the STS/Space Shuttle on the launch pad, showing the two strap-on solid rocket boosters (SRBs) attached to the central external tank (ET) for carrying liquid oxygen and hydrogen, with the orbiter vehicle's bottom profile visible behind these components. Photo courtesy of NASA

of this approach is the long-period exposure of the casing insulation/wall to the high-temperature flow. This disadvantage is largely avoided by employing a star or wagon-wheel grain design, designs that provide a relatively neutral profile by maintaining a relatively constant burning surface area S throughout most of the firing period. *B*allistic *t*est & *e*valuation *s*ystem (BATES) motors employ segmented, end-face-burning cylinders of solid propellant, with gaps between each cylinder, to produce a relatively neutral profile (see Fig. 10.5 for a five-segment motor example). In the case of shorter length-to-diameter (L/D) motors, e.g., more spherical as opposed to cylindrical, it is common to see the use of fins or slots in the grain, but only partly along the grain's length. These motors are called finocyl designs (see example in Fig. 10.6). When conical slots along part of the grain's length are used, this is referred to as a conocyl design. Upper stage motors are typically shorter (stubbier) due to their location in a space launch vehicle (Fig. 10.7), and are common candidates for the above treatment. A neutral thrust

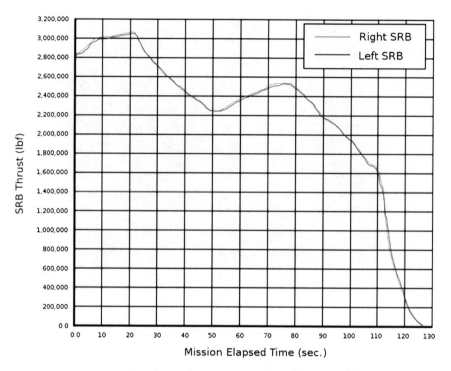

Fig. 10.3 Thrust-time curves for both SRBs used for the Space Shuttle mission STS-107. Courtesy of NASA

Fig. 10.4 Various thrust profiles that can result from differing propellant grain designs

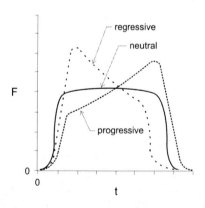

profile is a common flight mission requirement for larger, more sophisticated flight vehicles.

A simple cylindrical grain (a.k.a., circular perforation [CP] grain) produces a progressive (ramping-up) thrust-time profile (see Fig. 10.3), with the burning surface increasing with firing time up until the casing wall is reached, at which

Fig. 10.5 Single-stage
sounding rocket employing
BATES segmented-cylinder
motor

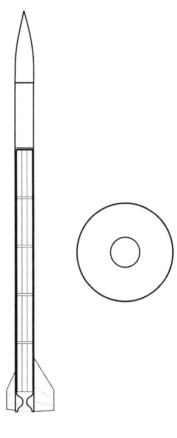

Fig. 10.6 Finocyl propellant
grain design

point the profile experiences a "tail-off" or drop in thrust as the motor approaches
final extinguishment (extinction). One may consider a cylindrical-grain design, if
the flight vehicle and/or its payload can withstand the strongly increasing forward
acceleration during the main motor burn. In addition to the increasing thrust, note
that the flight vehicle will get lighter as propellant is consumed, to further augment
the axial acceleration experience. Like the end-burners noted above, cylindrical
grains can provide a relatively high loading percentage (i.e., volumetric efficiency,
Ψ_p/Ψ_c) of propellant placed in the available volume of the combustion chamber, or
at least higher than conventional star grain designs. An additional advantage with
cylinders is the absence of slivers, which occasionally appear toward the end of

star grain firings. Slivers that remained unburnt by the end of the firing are in essence wasted ("dead") mass carried by the flight vehicle. Also, slivers can collapse, and in the process, significantly alter the expected thrust-time profile, or in the worst case, with pieces of propellant passing through the nozzle throat, potentially initiate a combustion instability episode (transient pressure wave development).

A simple central lengthwise slot in the propellant grain produces a regressive (decreasing) thrust-time profile over the latter part of the firing (see Fig. 10.4), as the burning surface effectively decreases during that time. A combination of a cylindrical grain and a slot can also produce a regressive profile. While not as common as the previously mentioned propellant grain designs, some flight missions do require that thrust decrease during the motor burn, to limit the acceleration experienced by the flight vehicle.

Returning to the issue of igniting the solid propellant, smaller SRMs might employ a pellet-dispersion ignition approach. For example, in the laboratory, a simple bag igniter, containing pyrotechnic igniter pellets (e.g., composed of a monopropellant like boron potassium nitrate [$BKNO_3$; BPN]), has combustion of its pellets initiated by a spark (electric arc) or electrically heated metal wire with pyrotechnic coating ["squib"] that bring some of the pellets' respective surface temperatures above the auto-ignition threshold. Like a lot of igniters, the bag

Fig. 10.8 Basket igniter, typically positioned at the head end, used for igniting solid rocket motor [1]. Diagram courtesy of NASA

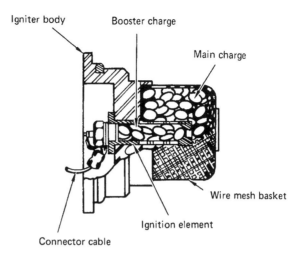

would typically be placed in the central port near the head end of the motor, to allow for a more efficient distribution of pressure and heat downstream toward the nozzle. With pellet combustion initiated, the hot gas evolving from the burning pellets would tend to rapidly pressurize the local space within the chamber as well as locally igniting the solid propellant grain. A series of pressure waves will be generated by this transient overpressure zone, moving axially back-and-forth in the motor chamber to further distribute gas pressure, and send the pellets further downstream in the central port toward the nozzle, convectively and radiatively heating the propellant surface to auto-ignition along the way. At this juncture, with the solid propellant well on its way to complete ignition of its exposed surface (flame spreading becoming complete), the motor chamber will enter the principal filling phase, with pressure continuing to rise in the motor until leveling off to a quasi-steady operational chamber pressure p_c.

Another pellet-dispersion ignition approach is a "shotgun cartridge" system, where for convenience of placement, the igniter may be placed at the aft port exit of the propellant grain (at or just upstream of the nozzle throat), rather than further up the port toward the head end. The cartridge, containing a number of igniter pellets, would upon ignition send the burning pellets upstream toward the head end, heating the solid propellant along the way. The pellets, if not consumed, will eventually reverse direction and move downstream toward the nozzle and exit the motor. A basket igniter is an alternative to the shotgun (pellet launching) approach, where the burning igniter pellets are retained inside the cartridge for much of the hot gas distribution and ignition process, until their size is diminished sufficiently from burning to allow for their exit from the basket into the central port and/or out the nozzle (Fig. 10.8; [1]).

Moving to larger SRMs, a common ignition approach is through the use of a pyrogen [heat-producing] jet igniter typically placed at or near the motor head end (see Fig. 10.9; [1]). This type of igniter is in essence a miniature solid rocket

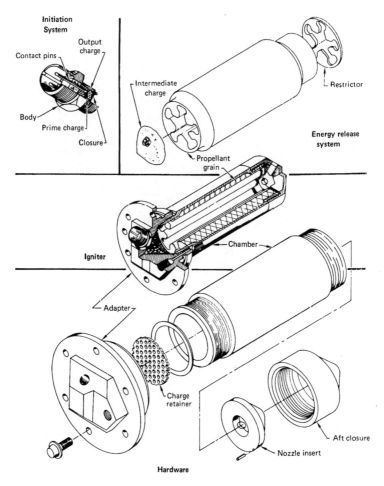

Fig. 10.9 Pyrogen igniter, typically positioned at the head end, used for igniting solid rocket motor [1]. Diagram courtesy of NASA

motor. Once ignited, it sends a hot high-speed exhaust gas flow down the central motor port, convectively heating the solid propellant in a progressive left-to-right fashion to auto-ignition.

10.3 Solid Propellant Types

The most common solid propellant type today is the composite, where the elastomeric polymer binder acts as the principal fuel component binding together crystals of oxidizer and additional powdered fuel constituents like aluminum and catalysts (e.g., ferric oxide Fe_2O_3 or manganese dioxide MnO_2, occasionally used

Table 10.1 Characteristics of various solid propellants at nominal operating conditions

Propellant	ρ_s (kg/m^3)	T_F (K)	T_S(K)	M (amu)	γ	r_b(cm/s)
NC/NG	1630	2300	760	22	1.26	0.7
AP/PS/additives	1635	2500	780	25	1.23	0.8
AP/PU/additives	1620	2400	670	21	1.25	0.5
AP/PBAA/EPON	1600	2300	700	22	1.24	0.8
AP/PBAN/Al	1750	2600	800	24	1.24	1.0
AP/CTPB/additives	1600	2300	800	22	1.25	1.1
AP/HTPB/Al	1750	3050	950	26	1.21	1.1

6.89 MPa chamber pressure; values are typical, although may vary depending on the given propellant formulation

as burning rate modifiers). When fuel and oxidizer components are separate compounds as in this case, this is also referred to as a heterogeneous mixture (or propellant). A common multi-purpose composite propellant might for example use 15% by mass a rubber, hydroxyl-terminated polybutadiene [HTPB], to hold together 75% ammonium perchlorate [NH_4ClO_4; a.k.a., AP] crystals, and 10% aluminum powder (see Table 10.1 for various propellants and their corresponding characteristics). One can note that smaller AP crystals typically result in an increase in pressure-dependent burning rate, while larger AP crystals can contribute to an increase in flow-dependent burning (a.k.a., erosive burning). The addition of aluminum powder is occasionally for energy enhancement (depending on the base polymer/oxidizer combination being used), but usually for restraining the development of combustion instability symptoms in the transverse direction (tangential and radial pressure waves) and the axial direction (axial pressure waves primarily, but possibly axial and transverse waves in combination). While effective in the transverse direction at relatively low loading percentages (1–3% by mass), aluminum powder's effectiveness in the suppression of combustion instability in the axial direction is a little more difficult to predict (even at loading up to 20%). Of course, one factor affecting this suppression ability is the burning down of a reactive fuel powder like aluminum, so that the mean diameter of an aluminum particle is changing as it moves down the core flow toward the nozzle. Inert particles like aluminum oxide Al_2O_3 may be used as a substitute, if one desires a potentially more effective constant particulate size through the firing. In some cases, it is also desirable to avoid excessive aluminum slag buildup in the aft chamber region, a phenomenon associated with reactive aluminum particles and submerged nozzles (where the nozzle convergence entry is substantially within the combustion chamber to save on overall motor length, with a resulting peripheral annular pocket that allows for molten aluminum to accumulate at the aft chamber position rather than being ejected from the motor; see Fig. 10.1 for an example of a submerged nozzle).

Considerable research continues to be done in identifying and assessing more energetic binders that may one day replace HTPB as the popular binder choice. For example, glycidyl azide polymer [GAP] binders do demonstrate higher energy content than HTPB. However, other factors related to reliability and robustness, like predictable, stable burning rates even after months or years in storage before firing, are potentially just as important, and remain to be proven for these new-comer compounds.

For some military applications, a reduced smoke content in the SRM's exhaust plume is desirable for not giving away the position of launch in a combat situation. Reduced-smoke, minimum-signature or smokeless propellants have been developed over the years, but to date, have not seen wide usage. Reasons for this can be similar to that seen with higher energy materials noted in the previous paragraph: lack of proven performance stability after a long period of storage, too costly to produce, etc. A relatively new oxidizer, ammonium dinitramide (ADN), has shown promise as a potential low-smoke replacement for ammonium perchlorate.

Homogeneous solid propellants (so-called because they have both fuel and oxidizer in the same chemical constituent) are historically still referred to as double-base [DB] propellants. The origins of the latter name arise from an early popular combination of solid nitrocellulose (NC, the original single-base propellant, which contains its own fuel and oxidizer agents within its chemical structure) and liquid nitroglycerine absorbed within it (NG, which also contains its own fuel and oxidizing agents in its molecular compound). Newer composite propellants proved to be more effective overall for most flight applications than the older NC/NG solid propellant variants (see Table 10.1), and thus have supplanted these and derivative DB propellants as the most common present choice. As for composite binders, research continues on DB or homogeneous propellants, in search of potential replacements providing more energy, and eventually, more robustness (if possible). Triple-base [TB] propellants historically include nitroguanidine [NQ or NGD] to NC and NG, for better characteristics. RDX (cyclotrimethylene-trinitramine, $C_3H_6N_6O_6$) and HMX (cyclotetramethylene-tetranitramine, $C_4H_8N_8O_8$) are two chemicals that have some heritage as high-energy explosives and standalone homogeneous solid propellants, and presently are being tried as additives to more robust solid propellants. In a similar fashion, well-proven polymers like HTPB and oxidizers like AP are being added to DB propellants, in order to provide an overall improvement with respect to energy, robustness, etc. A new class of solid propellants have arisen from this work, called composite-modified double-base [CMDB] propellants.

10.4 Solid-Propellant Burning Rate Models

In order to evaluate and predict the internal ballistic combustion and flow behavior of SRMs, we need practical models for establishing the solid propellant's burning rate at a given time and location. In being introduced to a topic, or undertaking preliminary design calculations, it is better to deal initially with a few equations with some physical meaning behind them, versus conducting a voluminous number-crunching black-box exercise where trends and mechanisms are not so

Fig. 10.10 Pressure-dependent burning rate behavior of three propellant categories

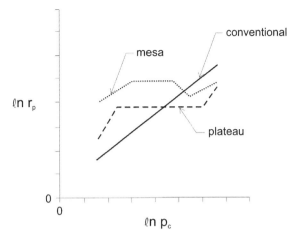

evident. Fortunately, the dependence of most propellants' burning rate on local static pressure over a useful range of pressures is governed by de St. Robert's or Vieille's empirical law:

$$r_p = Cp^n \tag{10.1}$$

where coefficient C and exponent n can be determined by a number of test burns at different chamber static pressures p_c. If a conventional propellant, the resulting curve for those data points will follow a straight line on a log–log graph, as shown in Fig. 10.10, and as expressed by the following equation:

$$\ell n(r_p) = \ell n(C) + n \cdot \ell n(p_c) \tag{10.2}$$

The exponent n needs to be less than one for stable burning (slow deflagration, versus rapid explosive rate). For common composite propellants, n can range from 0.2 to 0.5. Coefficient C happens to be a function of the propellant's initial temperature T_i (around room or outside air temperature), such that:

$$C = C_o exp\left[\sigma_p(T_i - T_{io})\right] \tag{10.3}$$

where C_o and T_{io} are at reference conditions (e.g., standard sea-level values), and σ_p is the pressure-dependent burning-rate temperature sensitivity, which can range from 0.001 to 0.009 K^{-1}. A lower outside air temperature in winter will lower C, thus lowering $p_c(t)$ during the motor firing while extending the burn period. A lower chamber pressure and base burning rate may render the motor susceptible to developing combustion instability symptoms. The resulting lower thrust in turn may adversely affect the vehicle's mission, e.g., endangering the ability to clear the launch rail at a sufficient speed for flight stability.

Solid propellants can exhibit pressure-dependent burning rate behavior different from the conventional profile described above. Referring again to Fig. 10.10, one on occasion may encounter a propellant displaying a plateau characteristic, where the burning rate remains relatively constant over a range of pressures. From a modeling viewpoint, this burning rate phenomenon is quite easy to incorporate, as long as the pressure range for neutral behavior is known. A further example of less typical behavior is also illustrated in Fig. 10.10, for a propellant displaying a mesa characteristic. As the term suggests, the pressure-dependent burning rises with increasing pressure up to a threshold value, at which point r_p remains constant for a range of pressures, and then subsequently drops in value as pressure is further increased. At an even higher pressure, the burning rate recovers and begins to rise again.

The auto-ignition temperature T_{as} at which propellant burning can commence may be estimated via [2]:

$$T_{as} \approx T_i + \frac{Q_s}{C_s} \qquad (10.4)$$

The symbol Q_s represents the net near-surface heat release of the propellant (not to be confused with a later parameter, ΔH_s, the net surface heat of reaction, which typically has a much lower value than Q_s). Values for Q_s that are positive are exothermic (heat is being released to its surroundings), while values that are negative are endothermic (heat is being absorbed from its surroundings). Consider the following example for a common AP/HTPB/Al composite propellant:

$$\begin{aligned} Q_s &\approx \frac{m_{AP}}{m_p} \cdot Q_{s,AP} + \frac{m_{HTPB}}{m_p} \cdot Q_{s,HTPB} + \frac{m_{Al}}{m_p} \cdot Q_{s,Al} \\ &\approx 0.75(+1045,000) + 0.15(-1813,000) + 0.10(-280,000) \\ &\approx +483,800\,\text{J/kg} \end{aligned} \qquad (10.5)$$

where the mass fractions are in terms of the total propellant mass m_p. Similarly, one can find the propellant's solid specific heat:

$$\begin{aligned} C_s &\approx \frac{m_{AP}}{m_p} \cdot C_{s,AP} + \frac{m_{HTPB}}{m_p} \cdot C_{s,HTPB} + \frac{m_{Al}}{m_p} \cdot C_{s,Al} \\ &\approx 0.75(1420) + 0.15(2100) + 0.10(900) \\ &\approx 1470\,\text{J/(kg K)} \end{aligned} \qquad (10.6)$$

Referring to Eq. 10.4, one can continue the above numerical example to show the following for auto-ignition temperature:

$$T_{as} \approx 294 + \frac{483800}{1470} \approx 623\ \text{K}$$

Once ignited, the solid propellant burning surface temperature T_S will reach an approximate equilibrium value that can be estimated via Summerfield's relation [3]:

$$T_S \approx T_{as} + \frac{1}{2\sigma_p} \qquad (10.7)$$

Continuing the above numerical example, using a typical value for σ_p for that type of propellant:

$$T_S \approx 623 + \frac{1}{2(0.0016)} = 935 \ K$$

The above value is a reasonable estimate for this type of propellant, given available experimental information.

10.5 Erosive Burning in SRMs

The propellant burning rate may also be affected by the hot gas flow passing above the propellant surface; this component of burning is referred to as erosive burning. The positive erosive burning augmentation on the base burning rate (r_o) has been observed experimentally over the years to be more evident at lower port-to-throat area ratios (A_p/A_t, with port referring in general to any location along the central core flow of the propellant grain, but sometimes more specifically to mean the central core at the exit from the grain). A common dividing line on the area ratio is 5:1 or less, which corresponds to a flow Mach number of around 0.2 or more, when erosive burning is clearly evident. The original assumption was that this augmentation in surface regression rate was somehow due to some non-combustive mechanical shearing (ablation) effect, hence the origin of the "erosive" adjective.

Through various experiments and a variety of different models, it has become (relatively) clear that erosive burning is in fact due to heightened convective heat transfer (which is a function of predominantly turbulent flow) to the burning surface engendering a reciprocating feedback from the gasification/combustion process. The convective heat transfer model by Lenoir and Robillard [4] was a significant first step in producing a useful predictive tool for estimating the increase in burning rate under core flow. The framework of the L-R model included a transpired flat-plate boundary layer, representative of x-dependent boundary layer buildup under blowing in moving from the head end of the motor to the nozzle. Later, it was clear that for a lot of longer motors with a larger effective L/D, the x-dependence was at some point being replaced or modified by an internal port diameter dependence, suggesting that the boundary layer was reaching a fully developed state earlier than anticipated. As a result, an ad hoc correction for d_p was incorporated into the L-R model, as shown in the equation for overall burning rate r_b which includes de St. Robert's pressure-dependent component:

$$r_b = r_o + r_e = Cp^n + \alpha G^{0.8} d_p^{-0.2} exp(-\beta r_b \rho_s / G) \tag{10.8}$$

where the coefficient α may be estimated from

$$\alpha = \frac{0.0288 C_p \mu^{0.2} Pr^{-2/3}}{\rho_s C_s} \frac{(T_F - T_S)}{(T_S - T_i)} \tag{10.9}$$

and the coefficient β was experimentally determined (from a limited set of firing data available to Lenoir and Robillard at that juncture, the mid-1950s) to be around 53 in value. Here, the symbol G represents the mean or bulk axial mass flux (ρu_∞) of the core flow (influencing two components of the model: fluid shear stress, and reduction thereof due to transpiration), ρ_s is the propellant's solid density, μ is the absolute viscosity of the gas and Pr is the effective Prandtl number ($\mu C_p / k$) of the gas, where k is the thermal conductivity of the gas. In the case where the port is non-circular, one can use the hydraulic diameter for d_p, namely $4A_p/s_p$, where s_p is the peripheral distance around the core. In practice, both α and β can vary considerably, in trying to match up to newer experimental firing data.

A lack of universality is a weakness in a predictive model. Greatrix and Gottlieb [5] introduced a more comprehensive convective heat transfer feedback model that had a lesser reliance on "roving" correction factors. In the G-G energy-conservation model, the positive erosive burning component r_e can be estimated via:

$$r_e = \frac{h(T_F - T_S)}{\rho_s\left[C_s(T_S - T)_i - \Delta H_s\right]} \tag{10.10}$$

Here, we see the appearance of ΔH_s, the net surface heat of reaction (positive if exothermic; typically, low in value). By way of Reynolds' analogy between friction and heat transfer, one establishes the convective heat transfer coefficient h under transpiration through:

$$h = \frac{C_p \tau_s}{Pr^{2/3} u_\infty} \tag{10.11}$$

assuming turbulent flow. One can correlate fluid-wall shear stress τ_s to the Darcy-Weisbach friction factor f by the following:

$$\tau_s = \rho u_\infty^2 (f/8) \tag{10.12}$$

One typically finds the zero-transpiration quantity first, in this case h^* (from previous evaluations of propellant erosive burning, this approach provides a more consistent predictive result in comparing to experimental data; using Eq. 10.11 with an estimation of the friction factor under transpiration to arrive at h, one tends to produce an underprediction of erosive burning). The value for h^* may be found as a function of the zero-transpiration Darcy-Weisbach friction factor f^*, following Eq. 10.11:

$$h* = \frac{k^{2/3} C_p^{1/3}}{\mu^{2/3}} \frac{G f*}{8} \tag{10.13}$$

Here, the value for f^* may be found for fully developed turbulent flow via Colebrook's expression:

$$(f^*)^{-1/2} = -2\log_{10}\left[\frac{2.51}{Re_d (f^*)^{1/2}} + \frac{\varepsilon/d_p}{3.7}\right] \tag{10.14}$$

Fig. 10.11 Schematic
diagram of approximation
approach for prediction of the
stretching of the combustion
zone

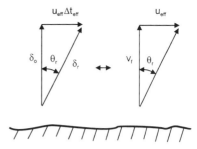

where ε is the effective propellant surface roughness height. In the case of developing flow, e.g., flat-plate flow as a function of axial distance x from the motor's head end, the following empirical expression (see [6]), for fully turbulent flow, might prove useful (if in the applicable Reynolds number regime):

$$f^* = 4\left[1.89 - 1.62\log_{10}\left(\frac{\varepsilon}{x}\right)\right]^{-2.5} - 7.04\left[1.89 - 1.62\log_{10}\left(\frac{\varepsilon}{x}\right)\right]^{-3.5} \quad (10.15)$$

Depending on the local flow Mach number Ma_∞, one may want to account for compressibility effects on the friction factor via the well-known recovery-factor equation for turbulent flow:

$$f_{comp} = f_{incomp}/\left[1 + Pr^{1/3}\left(\frac{\gamma - 1}{2}\right)Ma_\infty^2\right] \quad (10.16)$$

In turn, from film theory (Greatrix and Gottlieb [4]; Mickley et al. [7]; Moffat and Kays [8]), one can correct for transpiration:

$$h = \frac{\rho_s r_b C_p}{exp\left(\frac{\rho_s r_b C_p}{h^*}\right) - 1} \quad (10.17)$$

One should note the use of overall burning rate r_b in solving for the convective heat transfer coefficient. Thus one can find the positive augmentation effect of the core flow, without overly relying on empirical corrections.

Occasionally, it has been observed experimentally that the burning rate appears to drop below the expected base burning rate, at lower flow speeds, and tends to recover at higher core flow speeds. This phenomenon has been referred to as negative erosive burning. Greatrix [9] has proposed that such a reduction in burning rate may be due to a laminar-type stretching or sliding of the effective combustion zone with the local core flow, as illustrated in Fig. 10.11. With increased turbulence, this largely laminar effect would disappear (at higher flow speeds). With this approach in mind, the above positive erosive burning rate component r_e from Eq. 10.10 needs to be added to the burning-rate-reducing effect of combustion zone stretching, to arrive at the overall burning rate:

Fig. 10.12 Theoretical and experimental [10] data for burning rate augmentation as a function of mass flux, double-base propellant A [9]

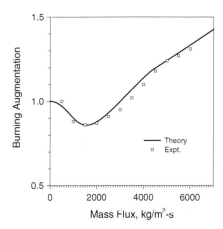

$$r_b = \frac{r_b}{r_o}\Big|_{\delta_r} \cdot r_o + r_e \tag{10.18}$$

where via Greatrix's analysis:

$$\frac{r_b}{r_o}\Big|_{\delta_r} = \cos\left[\tan^{-1}\left(\frac{u_{\text{eff}}}{v_f}\right)\right] = \cos\left[\tan^{-1}\left(K_\delta \delta_o\left[1 - (f/f_{\text{lim}})^{1/2}\right]\frac{\rho u_\infty}{\rho_s r_o}\right)\right], \quad f < f_{\text{lim}} \tag{10.19}$$

Here, δ_r is the effective stretched combustion zone thickness under core flow, that under base conditions (like pressure-dependent burning only) would be at a thickness δ_o (see Eq. 10.25). To date, the demonstrated value for the limit friction factor f_{lim} under transpiration is around 2.5×10^{-4} (no negative effect on burning rate at higher flow speeds, where $f > f_{\text{lim}}$). From film theory (Mickley et al. [7], Moffat and Kays [8]), the effects of transpiration on f, based on f^* calculated earlier, can be incorporated as follows:

$$f = 8\frac{\rho_s r_b}{G}\Big/\left(exp\left[\frac{8\rho_s r_b}{G \cdot f^*}\right] - 1\right) \tag{10.20}$$

One can note the dependence of f and h on G via Eqs. 10.13 and 10.20. The value for the shear layer coefficient K_δ in Eq. 10.19 is around 2,600 m^{-1}, based on available experimental data.

One can refer to Fig. 10.12 [9, 10] for an example profile of burning rate augmentation (r_b/r_o) as a function of the core flow's mass flux (G), where the comparison between the theoretical model above and experimental firing data is quite good. While not always as readily evident, here one can readily see at low G the effect of negative erosive burning on reducing the overall r_b below the base (in this case, pressure-dependent) burning rate ($r_b/r_o < 1.0$). At higher flow speeds, as G increases further, turbulence increases such that eventually the negative erosive burning trough is ultimately replaced by the almost-linear augmentation

Fig. 10.13 Theoretical and
experimental [10, 11] data for
burning rate augmentation as
a function of mass flux,
composite propellant E [9]

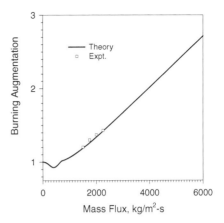

effect of positive erosive burning. Negative erosive burning is not always so strong
in the low flow-speed regime, as may be seen by the example profile provided in
Fig. 10.13 [9–11] for a different propellant. For this example, one might be for-
given for assuming a plateau region on burning rate ($r_b/r_o \approx 1.0$) until a threshold
value for G of around 650 kg/s m^2 is exceeded.

   Please refer to the solution for Example Problem 10.3 at the end of this chapter,
for a completed sample solid rocket motor erosive burning analysis. For students
learning this material as part of a course, I would encourage them to attempt the
given example problem first, and then check the solution to see if any mistakes
were made.

10.6 Acceleration Effects on Burning

In some situations, like the motor spinning about its longitudinal axis (i.e., rolling,
as seen in Fig. 10.14) or experiencing radial vibration of the motor propellant/
casing assembly (see Fig. 10.15), normal acceleration a_n may act to augment the
solid propellant burning rate. The resultant (overall) acceleration vector, if containing some lateral and/or longitudinal component(s) as represented by a_ℓ, has to
be within about $\pm 20°$ from the vertical (angle ϕ relative to the propellant surface
beneath the compressive acceleration field, as shown in Fig. 10.16), in order to
observe an appreciable augmentation effect on overall r_b. As a result, a spinning
motor on a test stand in the laboratory may see a large increase in burning rate and
chamber pressure, but the same motor on-board a flight vehicle in accelerating free
flight may see only a small increase in r_b and p_c above the nominal operating
values, for the same spin rate.

   A number of different models have been put forward over the years, for predicting the effect of normal acceleration, and for predicting the corresponding
reduction thereof when orientation angle ϕ, found from

Fig. 10.14 Schematic
diagram of SRM undergoing
spinning about its
longitudinal axis

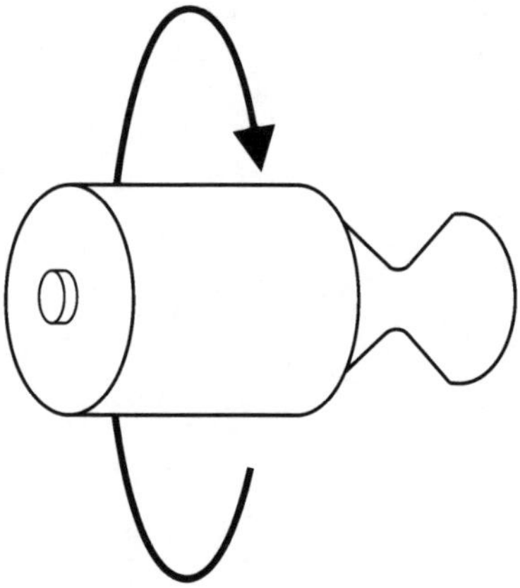

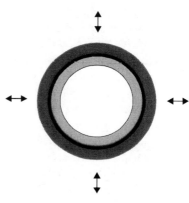

Fig. 10.15 Schematic
diagram of cylindrical-grain
SRM undergoing radial
vibration

Fig. 10.16 Schematic
diagram of acceleration
vectors acting near the
propellant surface

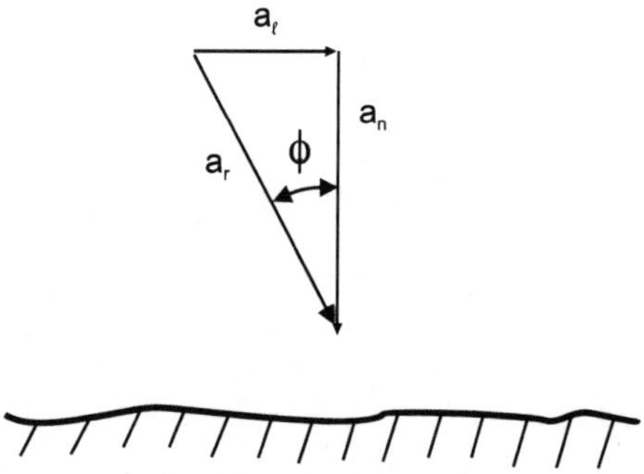

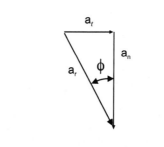

$$\phi = \tan^{-1}\left(\frac{a_\ell}{a_n}\right) \tag{10.21}$$

is of an appreciable value (note: a_ℓ can be a combined vector of lateral and
longitudinal acceleration components). The empirical model by Crowe [12] has

been a popular choice for predictive purposes, or at least for correlating existing experimental data:

$$\frac{r_b}{r_o} = 1 + C_1 \left\{ \frac{(a_n p_c)^{1/4}}{r_o} - C_2 \right\} \tag{10.22}$$

Coefficients C_1 and C_2 need to be established for a given propellant via experimental data, e.g., resulting from various values of a_n, p_c and r_b/r_o obtained from spin-motor test firings under different conditions. The common correction for a_ℓ on a_n effects is a simple sinusoidal assumption along the lines of:

$$\frac{r_b}{r_o} = \left(\frac{r_b}{r_o} \right)_{\phi=0^\circ} \cos \phi \tag{10.23}$$

While Eq. 10.22 may correlate reasonably well with some firing data, one can readily establish that Eq. 10.23 will not provide a good estimate of a_ℓ reduction on a_n effects beyond a very small value for ϕ.

Based on the representation of the combustion zone as being compressed under an acceleration field, Greatrix [13, 14] has put forward an alternative model for acceleration augmentation of burning rate. The principal equation of this model is:

$$r_b = \left[\frac{C_p(T_F - T_S)}{C_s(T_S - T_i) - \Delta H_s} \right] \frac{(r_b + G_a/\rho_s)}{exp\left[C_p \delta_o (\rho_s r_b + G_a)/k \right] - 1} \tag{10.24}$$

where the reference energy film (combustion zone) thickness δ_o can be estimated via

$$\delta_o = \frac{k}{\rho_s r_o C_p} \cdot \ell n \left[1 + \frac{C_p(T_F - T_S)}{C_s(T_S - T_i) - \Delta H_s} \right] \tag{10.25}$$

In the above, r_o is the base burning rate due to factors other than acceleration, e.g., due to pressure and core flow. The accelerative mass flux G_a, negative when a_n is directed into the combustion zone (i.e., compressing said zone), is determined from

$$G_a = \left\{ \frac{a_n p}{r_b} \frac{\delta_o}{RT_F} \frac{r_o}{r_b} \right\}_{\phi=0^\circ} \cos^2 \phi_d \tag{10.26}$$

When the acceleration is directed away from the combustion zone, experimental observation suggests that G_a goes to zero in value. The accelerative mass flux is reduced by increasing values of the total lateral/longitudinal displacement angle (a.k.a., augmented orientation angle) ϕ_d:

$$\phi_d = \tan^{-1} \left[K \left(\frac{r_o}{r_b} \right)^3 \tan \phi \right] \tag{10.27}$$

From Eqs. 10.26 and 10.27, we observe that the reduction (mitigation) effect of a_ℓ increases with increasing values of ϕ_d, which corresponds to an increase in ϕ or a_ℓ,

Fig. 10.17 Predicted
burning rate augmentation of
example composite
propellant due to normal
acceleration

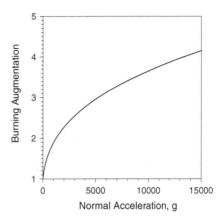

and a decrease in augmentation ratio. The value for correction factor K appears to be around 8, given experimental data available at this date.

In practice, some propellants experience a constant augmentation response over a certain range of normal acceleration. For example, one AP/PBAN composite propellant at a given chamber pressure has an augmentation ratio r_b/r_o sit at 1.3, between 100 and 450 g of a_n. Other propellants display a zero response to acceleration, below a certain threshold value for a_n. One hypothesis, that attempts to explain both of the phenomena above, is that there is an associated flooding + cooling effect occurring with the liquid melt layer at the base of the combustion zone being held in place longer within a certain range of values for a_n .

Augmentation results predicted by the Greatrix model for an increasing a_n on a given solid propellant's burning rate are shown in Fig. 10.17. The relationship can be seen to be nonlinear, becoming more asymptotic at higher normal acceleration levels. For the same propellant at 500 g of normal acceleration, Fig. 10.18 illustrates the marked reduction in burning augmentation as ϕ is increased. It turns out that the normal acceleration effect on burning is comparable to erosive burning; the lower the base burning rate r_o, the higher the augmentation effect.

Please refer to the solution for Example Problem 10.5 at the end of this chapter, for a completed sample solid rocket motor acceleration-dependent burning rate analysis.

10.7 Internal Ballistic Analyses for Steady and Nonsteady Operation of SRMs

Internal ballistic analysis involves the evaluation of the combined combustion and flow behavior within the combustor, and the thrust delivered by the exiting mass flow. In those cases where the bulk flow over the propellant surface is at a relatively low Mach number ($Ma < 0.2$) and acceleration is not a factor, one can

Fig. 10.18 Predicted
burning rate augmentation
reduction as a function of
acceleration orientation angle
ϕ, for a_n of 500 g acting on
example composite
propellant

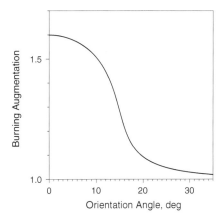

produce some straightforward relations for predicting chamber pressure p_c as a
function of the propellant burning surface area S, via the previously derived
Eq. 9.16:

$$\dot{m}_{in} = \rho_s S r_b \approx \rho_s S C p_c^n \approx \dot{m}_t \approx \left[\frac{\gamma}{RT_F}\left(\frac{2}{\gamma+1}\right)^{\frac{\gamma+1}{\gamma-1}}\right]^{1/2} A_t p_c \tag{10.28}$$

such that

$$p_c = \left\{\left[\frac{\gamma}{RT_F}\left(\frac{2}{\gamma+1}\right)^{\frac{\gamma+1}{\gamma-1}}\right]^{1/2}\frac{A_t}{\rho_s SC}\right\}^{\frac{1}{n-1}}, \text{ Pa} \tag{10.29}$$

One should ensure that the de St. Robert coefficient C is in the proper units for
solving this equation (fundamental case, m/(s Pa^n)), where C originally may be
given in alternative units like cm/(s · kPa^n)). For example, for $n = 0.30$:

$$0.063 \text{ cm}/\left(s \cdot kPa^{0.30}\right) = 7.93 \times 10^{-5} \text{ m}/\left(s \, Pa^{0.30}\right)$$

Computationally, one would need to update $S(t)$ as the propellant grain burns
back, and thus calculate $p_c(t)$ accordingly until burnout. Given $p_c(t)$ is known, one
can use the pertinent relations in Chap. 9 to then establish the thrust delivered by
the nozzle exhaust flow. Please refer to the solution for Example Problem 10.1 at
the end of this chapter, for a completed sample solid rocket motor performance
analysis.

In those cases where local bulk flow Mach number may exceed approximately
0.2 over the propellant surface, for accuracy one would be inclined to employ the
conservation equations of motion for the gas and/or particles within the flow, and
establish burning rate r_b at given locations on the propellant surface as a function
of local flow conditions (and acceleration, if applicable). In the case of motors

Fig. 10.19 **a** Schematic diagram of SRM, showing reference x-direction **b** Schematic diagram of SRM, positioned on a static test stand

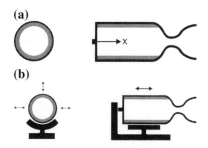

having a healthy L/D (length-to-diameter) ratio (>10:1) at least early on in the firing when flow Mach numbers tend to be higher (and thus erosive burning likely more significant), reasonable results can be obtained with the one-dimensional (1D) equations of motion along the length of the core flow (moving in the x-direction from the motor's head end toward the nozzle, as per Fig. 10.19). Let us consider first the mass continuity (conservation) equation for the gas:

$$\frac{\partial \rho}{\partial t} + \frac{\partial (\rho u)}{\partial x} = -\frac{1}{A}\frac{\partial A}{\partial x}\rho u + (1 - \alpha_p)\rho_s \frac{4r_b}{d} - \left(\frac{4r_b}{d} + \kappa\right)\rho \tag{10.30}$$

noting that $\kappa = 1/A \cdot (\partial A/\partial t)$, which is a radial dilatation term beyond that due to propellant grain regression, i.e., cross-sectional port area change due to vibration. The symbol α_p is the fraction of particle mass in the solid propellant of overall density ρ_s (propellant and particles; here, α_p corresponds to $\alpha_{p,s}$ of the previous chapter, and ρ_s corresponds to $\rho_{s,\text{new}}$ of the previous chapter). The inert (non-reactive, therefore constant diameter) monodisperse particle phase mass-conservation equation is given by:

$$\frac{\partial \rho_p}{\partial t} + \frac{\partial (\rho_p u_p)}{\partial x} = -\frac{1}{A}\frac{\partial A}{\partial x}\rho_p u_p + \alpha_p \rho_s \frac{4r_b}{d} - \left(\frac{4r_b}{d} + \kappa\right)\rho_p \tag{10.31}$$

The subscript p denotes that the flow property is applicable to the particle phase, e.g., u_p is the local particle velocity, while ρ_p is the density of particles in a given core flow segment volume $\Delta\mathcal{V}\left(\rho_p = \Sigma m_p/\Delta\mathcal{V}\right.$, where m_p is the mean mass of a particle$\left.\right)$. Returning to the gas phase, conservation of linear momentum in the x-direction gives:

$$\frac{\partial (\rho u)}{\partial t} + \frac{\partial}{\partial x}\left(\rho u^2 + p\right) = -\frac{1}{A}\frac{\partial A}{\partial x}\rho u^2 - \left(\frac{4r_b}{d} + \kappa\right)\rho u - \rho a_\ell - \frac{\rho_p}{m_p}D$$
$$+ \frac{4r_b}{d}(1 - \alpha_p)\rho_s u_i - \frac{\rho u^2}{2}\frac{f}{d} \tag{10.32}$$

where a_ℓ is the longitudinal acceleration of the gas resulting from the motor's motion in the x-direction (negative when motor acceleration is leftward), and u_i is the horizontal component of the mass injection velocity $v_w = \rho_s r_b/\rho \cdot (1 - \alpha_p)$.

The symbol f here is the Darcy-Weisbach friction factor ($f = 4C_f$), which would be ascertained under transpiration (blowing) if the burning surface is beneath the flow. The symbol D here is the drag force exerted on a particle of mean diameter d_m, e.g., for a spherical particle:

$$D = \frac{\pi d_m^2}{8} C_d \rho (u - u_p) |u - u_p| \tag{10.33}$$

given C_d is the drag coefficient at the pertinent Reynolds number $Re_p = \rho d_m / \mu \cdot |u - u_p|$ and relative flow Mach number $Ma_{rel} = |u - u_p|/a$. The corresponding momentum equation for the particles is given by:

$$\frac{\partial (\rho_p u_p)}{\partial t} + \frac{\partial \left(\rho_p u_p^2 \right)}{\partial x} = -\frac{1}{A} \frac{\partial A}{\partial x} \rho_p u_p^2 - \left(\frac{4 r_b}{d} + \kappa \right) \rho_p u_p - \rho_p a_\ell + \frac{\rho_p}{m_p} D$$
$$+ \frac{4 r_b}{d} \left(\alpha_p \rho_s u_i \right) \tag{10.34}$$

Defining $E = p/[(\gamma - 1)\rho] + u^2/2$ as the total specific energy of the gas at a given location, we have the following for the conservation of energy:

$$\frac{\partial (\rho E)}{\partial t} + \frac{\partial}{\partial x} (\rho u E + u p) = -\frac{1}{A} \frac{\partial A}{\partial x} (\rho u E + u p) - \left(\frac{4 r_b}{d} + \kappa \right) \rho E$$
$$+ (1 - \alpha_p) \rho_s \frac{4 r_b}{d} \left(C_p T_f + \frac{v_f^2}{2} \right) - \rho u a_\ell - \frac{\rho_p}{m_p} (u_p D + Q) \tag{10.35}$$

where the heat transfer Q from the gas to a given particle may be determined for a spherical particle via:

$$Q = \pi d_m k \cdot Nu \cdot (T - T_p) \tag{10.36}$$

where k is the thermal conductivity of the gas. The symbol Nu is the Nusselt number ($h_c d_m / k$), which can be found as a function of Re_p and Ma_{rel}. The corresponding energy conservation equation for the particle phase is given by:

$$\frac{\partial \left(\rho_p E_p \right)}{\partial t} + \frac{\partial \left(\rho_p u_p E_p \right)}{\partial x} = -\frac{1}{A} \frac{\partial A}{\partial x} \left(\rho_p u_p E_p \right) - \left(\frac{4 r_b}{d} + \kappa \right) \rho_p E_p$$
$$+ \alpha_p \rho_s \frac{4 r_b}{d} \left(C_m T_f + \frac{v_f^2}{2} \right) - \rho_p u_p a_\ell + \frac{\rho_p}{m_p} (u_p D + Q) \tag{10.37}$$

where $E_p = C_m T_p + u_p^2/2$, noting that C_m is the particle material specific heat. The symbol T_p is the mean particle temperature, for a particle at a given location.

10.8 Finite-Difference Model for Quasi-Steady Internal Ballistic Case

For steady (or quasi-steady, given the grain is burning back with time, albeit relatively slowly) operation of the aforementioned motor having an appreciable L/D ratio, one can drop the partial derivatives with respect to time, i.e., $\partial/\partial t$ and $\kappa(t)$ terms, so that one would have $3 + 3$ ordinary differential equations in terms of x for the two-phase flow. In integrating with respect to x from station ① to station ② along an element of length Δx (or if infinitesimally short, dx) in moving from the head end toward the nozzle (see Fig. 10.19a for a simple motor schematic diagram, and 10.19b for a static test setup), the resulting conservation of mass relation for the gas in a finite-difference format becomes:

$$\rho_2 u_2 A_2 = \rho_1 u_1 A_1 + \overline{r_b \Delta S}\big[(1 - \alpha_p)\rho_s - \rho\big] = B_1 \tag{10.38}$$

where ΔS is the segment peripheral surface area ($\pi d \cdot \Delta x$ for a cylindrical segment). The overbar symbol represents a mean evaluation between the two stations, but given first order accuracy is typically sufficient for this application, one can in that case simply utilize the given property on the left-hand side of the segment (①). For the particles, continuity provides

$$\rho_{p2} u_{p2} A_2 = \rho_{p1} u_{p1} A_1 + \overline{r_b \Delta S}\big[\alpha_p \rho_s - \overline{\rho}_p\big] = B_{1p} \tag{10.39}$$

Conservation of linear momentum for the gas gives

$$\rho_2 u_2^2 A_2 + p_2\left(\frac{A_1 + A_2}{2}\right) = \rho_1 u_1^2 A_1 + p_1\left(\frac{A_1 + A_2}{2}\right) + \overline{\Delta S}\left[(1 - \alpha_p)\rho_s \overline{r_b u_i} - \overline{r_b \rho u} - \frac{\overline{\rho u^2 f}}{8}\right]$$
$$- \left(\frac{A_1 + A_2}{2}\right)\left[\frac{\overline{\rho_p}}{m_p}\overline{D} + \overline{\rho} a_\ell\right]\Delta x = B_2 \tag{10.40}$$

Momentum conservation for the particle phase gives

$$\rho_{p2} u_{p2}^2 A_2 = \rho_{p1} u_{p1}^2 A_1 + \alpha_p \rho_s \overline{r_b u_i} \Delta S - \overline{r_b \rho u} \Delta S$$
$$+ \left(\frac{A_1 + A_2}{2}\right)\left[\frac{\overline{\rho_p}}{m_p}\overline{D} - \overline{\rho_p} a_\ell\right]\Delta x = B_{2p} \tag{10.41}$$

Conservation of energy for the gas provides

$$\rho_2 u_2 A_2 \left(C_p T_2 + \frac{u_2^2}{2} \right) = \rho_1 u_1 A_1 \left(C_p T_1 + \frac{u_1^2}{2} \right) + \overline{\Delta S} \left[\overline{r_b} \rho_s (1 - \alpha_p) \left(C_p T_F + \frac{\overline{v_w^2}}{2} \right) \right.$$

$$\left. - h_c (\overline{T} - T_w) - \overline{r_b \rho} \left(C_v \overline{T} + \frac{\overline{u^2}}{2} \right) \right] - \left(\frac{A_1 + A_2}{2} \right)$$

$$\frac{\overline{\rho_p}}{m_p} \left(\overline{u_p D} + \overline{Q} \right) \Delta x - \left(\frac{A_1 + A_2}{2} \right) \overline{\rho u \, a_\ell} \Delta x$$

$$= B_3$$

$$(10.42)$$

where the convective heat transfer term, as the friction loss term in Eq. 10.40, is presumed to be significant only over a non-burning surface. Conservation of energy for the particles gives

$$\rho_{p2} u_{p2} A_2 \left(C_m T_{p2} + \frac{u_{p2}^2}{2} \right) = \rho_{p1} u_{p1} A_1 \left(C_m T_{p1} + \frac{u_{p1}^2}{2} \right) + \left[\overline{r_b} \alpha_p \rho_s \left(C_m T_F + \frac{\overline{v_w^2}}{2} \right) \right.$$

$$\left. - \overline{r_b \rho_p} \left(C_m \overline{T_p} + \frac{\overline{u_p^2}}{2} \right) \right] + \left(\frac{A_1 + A_2}{2} \right) \frac{\overline{\rho_p}}{m_p} \left(\overline{u_p D} + \overline{Q} \right) \Delta x$$

$$- \left(\frac{A_1 + A_2}{2} \right) \overline{\rho_p u_p a_\ell} \Delta x$$

$$= B_{3p}$$

$$(10.43)$$

Completing the set of equations needed for a closed solution is the equation of state for the gas:

$$p_2 = \rho_2 R T_2 \qquad (10.44)$$

There are seven pertinent equations, and seven unknown variables: $\rho_2, u_2, p_2, T_2, \rho_{p2}, u_{p2}$ and T_{p2}. Defining $\lambda = 2A_2/(A_1 + A_2)$, and applying variable substitutions and some algebraic manipulation, one can arrive at a quadratic equation for gas velocity at the right-hand side of a given segment (station ②), the solution being:

$$u_2 = \frac{- \frac{C_p B_2 \lambda}{R} \pm \left\{ \left(\frac{C_p B_2 \lambda}{R} \right)^2 + 4 \left(\frac{B_1}{2} - \frac{C_p B_1 \lambda}{R} \right) B_3 \right\}^{1/2}}{2 \left(\frac{B_1}{2} - \frac{C_p B_1 \lambda}{R} \right)} \qquad (10.45)$$

Note that there are two possible solutions for a quadratic equation, as may be observed by the ± symbol in the numerator. The more typical solution for flow

Fig. 10.20 Propellant grain at differing times during firing, as a result of burnback (star grain initial internal port configuration)

over the propellant grain is subsonic, which is arrived at using a plus. In the less typical solution for supersonic flow over the propellant grain, one would employ a minus. Once u_2 is solved, one can move to the following equations for the other gas flow variables:

$$\rho_2 = B_1/(A_2 u_2) \tag{10.46}$$

$$p_2 - 2(B_2 - B_1 u_2)/(A_1 + A_2) \tag{10.47}$$

and

$$T_2 = \lambda u_2(B_2 - B_1 u_2)/(R B_1) = p_2/(R \rho_2) \tag{10.48}$$

Moving to the particle phase,

$$\rho_{p2} = B_{1p}/(A_2 u_{p2}) \tag{10.49}$$

$$u_{p2} = B_{2p}/B_{1p} \tag{10.50}$$

and

$$T_{p2} = \frac{B_{3p}}{B_{1p}C_m} - \frac{u_{p2}^2}{2C_m} \tag{10.51}$$

If one has a single gas phase application (no particles), closure for the overall solution of head-end pressure p_c could be attained by the isentropic flow relation of Eq. 9.30, i.e., substitute the grain end-port values (at the Nth node, center port area A_N, flow Mach number Ma_N) for the nozzle exit-plane values. As the solution converges for p_c, one should observe the throat flow Mach number Ma_t approaching unity. If one has a two-phase application (gas + particles), or when friction losses are substantial in the nozzle contraction section for the single-phase non-isentropic case, one would continue the finite-difference flow calculations further, from the end port ($i = N$) to the nozzle throat. If a two-phase flow case, solution convergence would require that u/b becomes unity at the nozzle throat, referring to Eq. 9.40, recalling that b is the non-equilibrium two-phase sound speed. Upon establishing p_c, one can estimate thrust F by applying isentropic flow

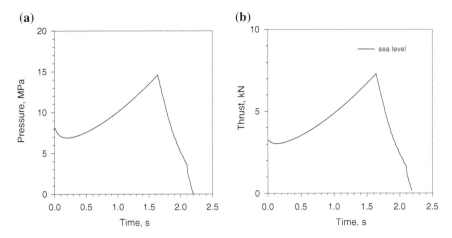

Fig. 10.21 **a** Predicted head-end pressure profile for small cylindrical-grain SRM **b** Predicted thrust profile for small cylindrical-grain SRM. Note the progressive thrust profile for the main part of the firing (before tailoff commences with propellant burnout)

calculations (or non-isentropic finite-difference numerical calculations) from the nozzle throat to the nozzle exit plane, as per Eq. 9.31. As noted previously, one would need to incrementally allow for the propellant grain to burn back, in order to produce an overall profile for $p_c(t)$ and $F(t)$. In Fig. 10.20, one can observe the configuration of a star-grain propellant motor at differing times into its firing, showing the level of burnback as the firing progresses.

Using the numerical approach above, an example predicted pressure–time profile for a small 70-mm outer diameter, 0.9-m long cylindrical-grain SRM is shown in Fig. 10.21a, while the corresponding sea-level thrust profile may be found in Fig. 10.21b. The initial peak that is evident in the pressure profile is due to the high erosive burning level early on in the firing. As the grain burns back (regresses), the burning rate along the grain becomes predominantly pressure-dependent for the remainder of the motor firing. In practice, on occasion one observes that the experimental pressure–time profile is a bit more robust than the predicted profile over the central part of the firing. This is sometimes referred to as a "humping" or "rainbowing" effect. For better prediction comparisons, it has been common practice to incorporate a corrective ballistic anomaly rate factor (BARF) modeling function to bring the predicted profile closer to the experimental result when such humping is observed. Over the years, a number of researchers have proposed different mechanisms that might be lying behind behind some of this anomalous behavior (e.g., erosive burning), in addition to the popular belief that it can be attributed to a variation in propellant characteristics as one moves through the propellant web (thickness) from the initial internal port surface to the outer propellant grain surface.

In the example results of Fig. 10.21, for simplicity of calculations, it has been assumed that the chamber has filled to quasi-equilibrium at $t = 0$ s. In practice, for a motor of this size, it would take from 50 to 100 ms for the chamber to fill during the ignition phase. The ignition of the solid propellant surface would in general not be uniform nor instantaneous, but progressive, say igniting from left-to-right (head end to nozzle end). However, having said that, one has the option of incorporating a simple uniform chamber-filling algorithm into a predictive internal ballistic model such as described in the following section, in order to produce predicted pressure and thrust profiles that have a finite filling phase present.

One can potentially get a better comparison to experimental firing data by allowing the propellant and surrounding motor casing to deform under loading in one's calculations. Propellants are typically treated as incompressible (Poisson's ratio v of 0.5), but even if true, as the surrounding casing deflects outward under pressure, the propellant will move outward as well to take up that deformation volume increment. Propellant deflection will be even greater when its value for v is less than 0.5 (propellant being somewhat compressible). The deflection of a cylindrical-grain motor and casing under pressure loading is relatively straight-forward to establish, using thick-wall deformation theory, but more complicated grain configurations, like star grains, etc., will likely require a finite-element modeling approach to determine the local grain deformation. These points are discussed in further detail a bit later in this chapter.

10.9 Simpler Nonsteady Internal Ballistic Cases

With respect to nonsteady motor conditions, there are some applications where one can estimate $p_c(t)$ without necessarily solving the complete set of partial differential equations for continuity, momentum and energy. One example would be the estimation of the time required in going from a uniform propellant grain surface ignition to reach a quasi-steady chamber pressure p_c (as per Eq. 10.29). Equation 10.30 for mass conservation can be reduced to a simpler chamber filling model:

$$\Psi_C \frac{d\rho}{dt} = \dot{m}_{in} - \dot{m}_e \tag{10.52}$$

where Ψ_C is the effective chamber volume. For the isothermal case ($T = T_F = $ constant), and thus having chamber gas density $\rho = p_c/(RT_F)$, then one arrives at:

$$\frac{dp_c}{dt} \approx \frac{RT_F}{\Psi_C} \left[\rho_s S C p_c^n - \left\{ \frac{\gamma}{RT_F} \left(\frac{2}{\gamma+1} \right)^{\frac{\gamma+1}{\gamma-1}} \right\}^{1/2} A_t p_c \right] \tag{10.53}$$

and

$$t = \int dt = \frac{\Psi_C}{RT_F} \int_{p_0}^{p_1} \left[\rho_s S C p_c^n - \left\{ \frac{\gamma}{RT_F} \left(\frac{2}{\gamma+1} \right)^{\frac{\gamma+1}{\gamma-1}} \right\}^{1/2} A_t p_c \right]^{-1} dp_c \tag{10.54}$$

Similarly, for simple isothermal emptying of the gas from the chamber upon burnout (extinguishment), one can show the following:

$$\frac{dp_c}{dt} \approx -\frac{RT_F}{\Psi_C}\dot{m}_e \approx -\frac{RT_F}{\Psi_C}\left\{\frac{\gamma}{RT_F}\left(\frac{2}{\gamma+1}\right)^{\frac{\gamma+1}{\gamma-1}}\right\}^{1/2} A_t p_c \qquad (10.55)$$

Integrating the above, one finds the following for chamber pressure:

$$p_c(t) = p_o exp\left[-\frac{RT_F}{\Psi_C}\left\{\frac{\gamma}{RT_F}\left(\frac{2}{\gamma+1}\right)^{\frac{\gamma+1}{\gamma-1}}\right\} A_t \cdot t\right] \qquad (10.56)$$

where in this case, p_0 is the chamber pressure at the instant of uniform propellant burnout.

Some unsteady conditions will require the solution of the partial differential equations as described by Eqs. 10.30–10.37 for a better assessment of the situation. Referring to the previous ignition (chamber filling) example, as noted earlier, it is common to expect that propellant ignition will develop progressively along the grain, rather than uniformly as assumed above. Initial flow speeds and corresponding Mach numbers can also be quite high during the ignition transient phase, with additional igniter-generated gas being added to the flow. Sometimes one observes a noticeable "ignition spike" on the pressure–time profile, as a result of this increased early erosive burning level, which can be substantially more severe than the initial peak one observes in Fig. 10.21a.

10.10 Transient Burning of Solid Propellants

There are a number of applications where one commonly needs to include transient effects on the burning rate of a given solid propellant (rather than just assuming a quasi-steady, fast kinetic-rate response of the propellant to the local flow conditions). The most common applications to date have been for evaluating the ignition transient (chamber filling) phase of an SRM firing, the propellant burnout and chamber emptying (tailoff) phase of an SRM firing, and combustion instability symptom development. Experimental studies of solid propellants, e.g., via T-burner [4] studies utilizing different effective chamber lengths (see Fig. 10.22), have confirmed the existence of a frequency dependence on the transient combustion response of most propellants.

In the case of SRM ignition and chamber filling studies, the KTSS (Krier-T'ien-Sirignano-Summerfield; [15]) pressure-coupled flame model, or some variant thereof, has been employed on numerous occasions by researchers. In regard to SRM combustion instability, the A-B (Denison-Baum; [16]) pressure-coupled response model, or some variant thereof, has seen relatively frequent use over the years by researchers. While both of the above models are well known, they are not

Fig. 10.22 Schematic
diagram of conventional T-
burner for frequency-
dependent burning
evaluation, showing different
standing axial wave modes
(first to third harmonic) [4].
Reprinted with permission of
John Wiley & Sons, Inc.

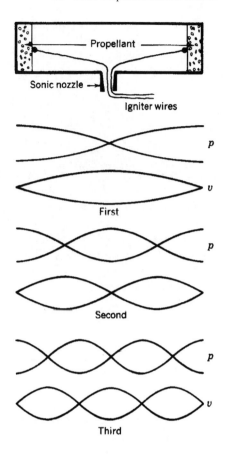

in general easy to apply to new motors or propellants, given the models, for accurate predictive capability, require some essential pieces of information from experimental firings for the specific motor in question for setting certain parameter values.

A general numerical model developed recently by Greatrix [17], based on the Zeldovich-Novozhilov [18] solid-phase energy conservation result for unsteady solid-propellant burning, is one means for getting around the inconvenience of the above models. Unlike past Z-N models, the integrated temperature distribution in the solid phase is utilized directly for estimating instantaneous burning rate (rather than the thermal gradient at the burning surface). The burning model is general in the sense that the model may be incorporated for various propellant burning-rate mechanisms, and not just for pressure-dependent burning (a limitation of the above models).

The Z-N solid-phase energy conservation result is commonly presented in the following time-dependent temperature-based relationship:

$$T_{i,\text{eff}} = T_i - \frac{1}{r_b^*}\frac{\partial}{\partial t}\int_{-\infty}^{0}\Delta T\,dy \qquad (10.57)$$

where $T_{i,\text{eff}}$ is the effective initial propellant temperature for instantaneous burning rate estimation, T_i is the actual initial propellant temperature and in this context, $\Delta T = T(y,t) - T_i$ is the temperature distribution in moving from the propellant surface at $y = 0$ (and $T = T_s$) to that spatial location in the propellant where the temperature reaches T_i. An alternative representation of the above result is in fact more convenient, for a more general approach:

$$r_b^* = r_{b,\text{qs}} - \frac{1}{(T_s - T_i - \Delta H_s/C_s)}\frac{\partial}{\partial t}\int_{-\infty}^{0}\Delta T\,dy \qquad (10.58)$$

In Eq. 11.58, r_b^* is the nominal (unconstrained) instantaneous burning rate, and its value at a given propellant grain location is solved at each time increment via numerical integration of the temperature distribution through the heat penetration zone of the solid phase. One can note the inclusion of the net surface heat of reaction, ΔH_s (sign convention: positive when exothermic output of heat). The quasi-steady burning rate may be ascertained as a function of such parameters as static pressure of the local flow, for example through de St. Robert's law:

$$r_{b,\text{qs}}(p) = Cp^n \qquad (10.59)$$

Note that $r_{b,\text{qs}}$ is not restricted to the mechanism of pressure in this model, and can also be a combined function of flow velocity, acceleration, radiation, etc. In the solid phase, the transient heat conduction is governed by:

$$k_s\frac{\partial^2 T}{\partial y^2} = \rho_s C_s\frac{\partial T}{\partial t} \qquad (10.60)$$

An additional equation limiting the transition of the instantaneous burning rate r_b with time is required to physically constrain the model. In general numerical modeling, where lagging a parameter's value is a desired objective, a simple empirical means for applying this constraint is as follows:

$$\frac{dr_b}{dt} = K_b\left(r_b^* - r_b\right) \qquad (10.61)$$

where the rate limiting coefficient K_b effectively damps or slows the change in value of the unconstrained burning rate r_b^* with time when:

$$K_b < \frac{1}{\Delta t} \qquad (10.62)$$

The boundary condition at the propellant surface ($y = 0$, $T = T_s$) to first-order accuracy may be applied through

Fig. 10.23 Frequency response of example solid propellant, $r_{b,o} = 0.01$ m/s, $\Delta H_s = 0.0$ J/kg

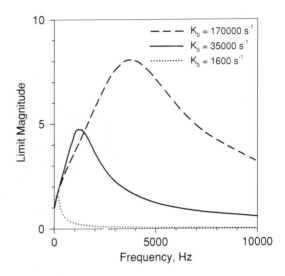

$$T_{t+\Delta t,-\Delta y} = T_{t,-\Delta y} + \frac{\Delta t}{\rho_s C_s \Delta y} \left[\frac{k_s \left(T_{t,-2\Delta y} - T_{t,-\Delta y} \right)}{\Delta y} + \frac{k_s \left(T_s - T_{t,-\Delta y} \right)}{\Delta y} + \Delta q_{\text{eff}} \right]$$

$$(10.63)$$

where Δq_{eff} represents the net heat input from the gas phase into the regressing solid, as determined by:

$$\Delta q_{\text{eff}} = (K_b \Delta t) \rho_s C_s \left(r_{b,\text{qs}} - r_b \right) \left(T_s - T_i - \Delta H_s / C_s \right) \qquad (10.64)$$

The empirical coefficient K_b will typically need to be set below a maximum permissible value for rendering a nondivergent solution for r_b, and adjusted even further downward in order to match up approximately with combustion response behavior for a typical solid propellant. Examining propellants with various characteristics, one means for comparing and potentially aligning the numerical model to actual firing data is to examine the frequency response to cyclic $r_{b,\text{qs}}$ input, at different values for the burn rate limiting coefficient. For example, application of a ± 0.001 m/s sinusoidal cycle on a reference base burning rate $r_{b,o}$ of 0.01 m/s at different driving frequencies on the reference propellant, with a given value for K_b, produces a set of limit-amplitude cycle results. The nondimensional limit magnitude M_ℓ is defined by

$$M_\ell = \frac{r_{b,\text{peak}} - r_{b,o}}{r_{b,\text{qs,peak}} - r_{b,o}} \qquad (10.65)$$

For an example solid propellant, one can observe the effect of K_b's value on the resulting frequency response behavior in Fig. 10.23. As K_b is increased, the peak resonant frequency increases, as does the peak response magnitude. On the other hand, as the net surface heat of reaction ΔH_s is increased in positive value

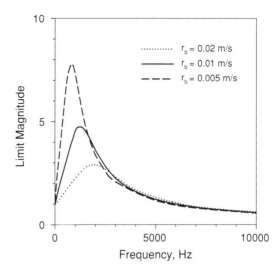

Fig. 10.24 Frequency response of example propellant, $K_b = 35000$ s^{-1}, $\Delta H_s = 0.0$ J/kg

(everything else being held constant, like K_b), the peak response magnitude increases, but the resonant frequency remains relatively unchanged. In Fig. 10.24, one can observe the effect of changing the base burning rate, for the same K_b value. The lower the base burning rate, the higher the peak response magnitude; this is consistent with experimental observation, where instability issues worsen as one lowers a given motor's base burning rate.

10.11 Axial and Transverse Combustion Instability of SRMs

The numerical modeling of nonsteady combustion instability problems is another example of the usage of Eqs. 10.30–10.37. At low pressure wave frequencies, L^* or bulk-mode instability can occasionally be observed occurring in SRMs during the lower pressure periods of the ignition start-up phase or the firing tailoff phase (this instability is sometimes descriptively referred to as chuffing or sputtering; L^* is the motor's characteristic length, namely chamber volume divided by nozzle throat area). Combustion during a chuffing episode is cyclically falling above and below the needed threshold for smooth operation, at pressures temporarily below the nominal operational level.

At higher frequencies, traditional nonlinear axial combustion instability pertains to the appearance of a sustained axial pressure wave moving-back-and forth within the motor core flow, and in more severe cases, the wave will have a steep shock compression front (see Fig. 10.25; [19, 20]). Shock wave pressure magnitudes have been observed to exceed 50% of the base operating pressure (see Fig. 10.26, for example instability behavior). This is not a desirable symptom, and if severe enough, can prematurely end a flight mission. The fundamental axial

Fig. 10.25 Schematic
diagram of SRM axial wave
cavity

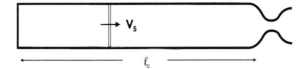

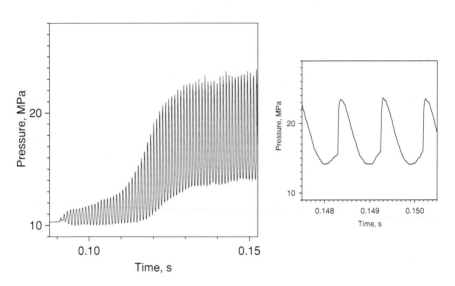

Fig. 10.26 Predicted head-end pressure profile for SRM undergoing nonlinear axial combustion
instability symptoms (sustained 1L pressure wave, dc shift)

acoustic wave frequency is given by $f_{1L} \approx a/(2\ell_c)$, noting the compression shock
front wave speed V_s can in practice be significantly greater than sound speed a, the
weak wave limit. This pressure wave may be initiated by various sources,
including by bits of igniter, propellant or insulation passing through the nozzle
throat. Deliberate pyrotechnic pulses may be applied in static motor tests evaluating the susceptibility of a given motor. The mechanism for sustaining the axial
pressure wave has traditionally been ascribed to an augmented, unsteady, fre-
quency-dependent, pressure- and/or velocity-coupled combustion response, i.e., an
enhanced local propellant burning rate feeding and sustaining the compression
portion of the traveling wave. Hence, the development of such devices as T-
burners (Fig. 10.22, [4]) for testing a propellant's combustion instability
susceptibility.

Based on experimental evidence gathered over the years, researchers have
identified other mechanisms at play that might explain the appearance of c.i.
symptoms that do not necessarily fit the combustion-only thesis. For example,
altering a nozzle's convergence length has been demonstrated to alter the c.i.
behavior. This would suggest a gas dynamic effect related to shock-wave rein-
forcement upon nozzle reflection, a factor clearly independent of combustion

Fig. 10.27 Schematic
diagram of SRM tangential
wave motion in cavity

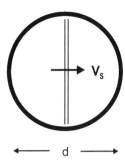

response. As a second example, altering a motor casing's material from aluminum to composite has been demonstrated to alter the c.i. behavior of that motor. There is the correlation between normal acceleration-dependent burning and transient radial vibration of the surrounding structure with the repeating passage of the axial pressure wave, that might suggest the influence of a material change. The c.i.symptom known as the "dc shift" (undesired base pressure rise, as per Fig. 10.26) can potentially be attributed to radial vibration, rather than the traditional assumption that velocity-coupled (erosive burning related) combustion response is playing a role. Given the experimental evidence, it is important to remember that a solid propellant cannot in general be stated as being unstable, if in one experimental setup the motor using that propellant goes unstable; it is just as likely that in a different motor system setup, say a comparable motor but at a different size [21], the propellant's contribution will be benign, and the motor will function adequately.

Transverse combustion instability is commonly associated with the appearance of either or both tangential and radial pressure waves within the motor chamber. As per Fig. 10.27, the fundamental tangential acoustic wave frequency is $f_{1T} \approx 0.59a/d$, for a wave passing back-and-forth laterally in a cylindrical passage (as opposed to a rectangular passage, where the coefficient would be 0.50). As per Fig. 10.28, the fundamental radial acoustic wave frequency is $f_{1R} \approx 1.22a/d$, for a wave passing back and forth between a central focal point and a peripheral wall (as opposed to a rectangular passage, where the coefficient would be 1.00). In past experimental rocket firings, researchers may not have had pressure transducers capable of picking up the higher transverse frequencies, and thus may have failed to consider the potential implications of transverse wave symptoms on motor performance. As noted earlier for axial combustion instability, pressure- and velocity-coupled combustion response is commonly identified as the driving mechanism for transverse c.i. symptoms. Other mechanisms may play a role, however. Radial and tangential structural vibration can be a factor in pressure wave growth, for example.

Experimental and numerical research has shown that about 2% by mass of particles of a suitable mean diameter can act to suppress any appreciable pressure wave development in the transverse direction. Particle loading, even as high as

Fig. 10.28 Schematic
diagram of SRM radial wave
motion in cavity

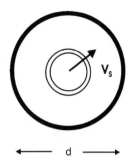

15–20%, will not necessarily be as effective, in the axial direction. Particles may be inert (like aluminum oxide) or reactive (like aluminum, a fuel), and this may have some bearing on the effectiveness of symptom suppression. Further research remains to be done in this respect, in establishing comprehensively and consistently what can be expected from a given motor with respect to nonlinear instability behavior. Other techniques exist for suppressing instability symptoms, including the use of internal rods and various alterations to the propellant grain's geometry (e.g., upstream-to-downstream port expansion or contraction).

On a final note, some larger SRMs, such as used for the boosters of the Ariane 5 launch vehicle, have a low-level non-traditional axial combustion instability problem, commonly called vortex-shedding instability [4]. These motors, because of their large size, utilize a set of smaller propellant cylinder segments separated by dividers, rather than one large propellant grain. As a firing progresses, the gas flow passing over the divider gaps between propellant segments tends to produce vortices which are shed from the gaps, corners or other associated obstacles in the flow, and these vortices pass downstream toward the nozzle. The common result is a low-level pressure oscillation (on the order of 1% of base pressure) superimposed on the nominal base chamber pressure, typically oscillating at f_{1L}. An additional category of weak vortices that may also appear in longer segmented or unsegmented ports as one moves downstream is referred to as being parietal or surface-based in origin, rather than gap- or obstacle-based.

10.12 Structural Issues for SRMs

At the most fundamental level, the metal or composite motor casing of a cylindrical flightweight solid-propellant rocket motor (see Fig. 10.29; [22]) may be treated as a thin-walled pressure vessel. The solid propellant, in general (with the binder being an elastomeric polymer, and a Young's elastic modulus E at least one or two orders of magnitude less than aluminum or steel), bears little of the pressure load. With a Poisson's ratio v close to 0.5, solid propellants are typically close to being incompressible (e.g., while certainly more rigid as a rubber-based material,

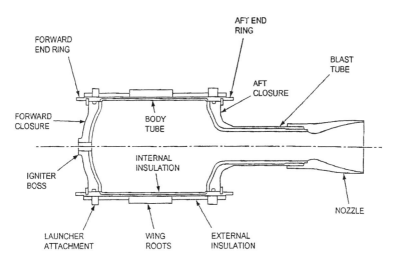

Fig. 10.29 Sample motor structure for a missile motor emplying a blast tube forward of the nozzle to allow for hardware placement in the rear vehicle region [22]. Diagram reprinted with permission of the American Institute of Aeronautics and Astronautics

can be close to the behavior of water with regard to deflecting under load). The propellant, like water, deflects outward to fill the freed-up volume created by the outward deflection of the surrounding casing (if v less than 0.5, will deflect even further). With this in mind, one has the well-known thin-walled cylinder relation for circumferential (hoop) stress:

$$\sigma_h = \frac{pr}{h} \tag{10.66}$$

where r is the mean radius of the casing wall from the motor's centerline, p is the local internal (gauge) gas pressure, and h is the casing wall thickness. Knowing the casing material's yield stress limit σ_y, one can size the casing wall thickness so that the anticipated maximum σ_h falls below the value for σ_y by the required safety margin. That safety margin may incorporate other factors, like repeated impact resistance (shock), or cyclic stress-induced and temperature-induced (creep) fatigue, depending on the expected lifetime and environment of the casing. A second well-known relation applies for longitudinal stress acting on a thin-walled cylindrical pressure vessel:

$$\sigma_\ell = \frac{pr}{2h} \tag{10.67}$$

This indicates that longitudinal stress in the ideal case is half the hoop stress, everything else being equal. For the same material then, hoop stress will be the likely principal criterion for setting the nominal casing wall thickness. Moving to a spherical pressure vessel, as one might see for some upper stage motors, one has the following thin-walled result:

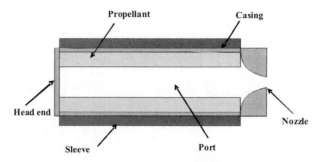

Fig. 10.30 Simple schematic diagram of a cylindrical-grain SRM with a steel static-test sleeve surrounding a flightweight aluminum motor casing

$$\sigma_h = \sigma_\ell = \frac{pr}{2h} \tag{10.68}$$

This confirms the structural weight savings associated with a spherical casing (or later, for a liquid propellant storage tank, for a liquid-propellant rocket engine).

Returning to the more common cylindrical (versus spherical) case for SRMs, one may wish to apply the thick-wall theory to establish the quasi-steady radial deflection of the propellant surface, for a cylindrical-grain motor, so as to establish the influence of the increased central port area on the resulting internal ballistic performance. Assuming one is relatively far away from constraining end effects, e.g., due to the head-end motor cap or the nozzle, the local radial displacement under static loading may be evaluated via the following ordinary differential equation for a thick-walled cylinder (Popov [23]):

$$\frac{d^2\psi}{dr^2} + \frac{1}{r}\frac{d\psi}{dr} - \frac{\psi}{r^2} = 0 \tag{10.69}$$

Here, ψ is the local radial static-load displacement from the no-load position, i.e., at some reference position j,

$$\psi_j = \Delta r_j = r_j - r_{j,o} \tag{10.70}$$

Let us assume a more comprehensive thick-walled problem where there are as many as three layers of differing materials, e.g., propellant (A)/casing (B)/static-test-sleeve (C) as shown in Fig. 10.30, or propellant (A)/insulation (B)/casing (C). The solution of the above equation at the inner surface of the propellant (no-load $r = r_{i,o}$) is

$$\psi_i = A_1^A r_{i,o} + A_2^A / r_{i,o} \tag{10.71}$$

where coefficients A_1 and A_2 are applicable to material A (solid propellant), from r_i to the boundary with the middle layer inner surface, r_m. At the inner propellant surface, under the local gauge static pressure $p(x)$, where x is the local axial position along the propellant grain, the following radial stress (σ_r) boundary condition applies:

$$-p(x) = \frac{E_A}{(1 + v_A)(1 - 2v_A)} \left[A_1^A - (1 - 2v_A) \frac{A_2^A}{r_{i,o}^2} \right] \tag{10.72}$$

Assuming the first three-layer scenario, at the external surface of the static test sleeve wall (no-load $r = r_{e,o}$), for material C:

$$\psi_e = A_1^C r_{e,o} + A_2^C / r_{e,o} \tag{10.73}$$

At the external sleeve wall surface, one has the following σ_r boundary condition, where outside gauge pressure is zero:

$$0 = \frac{E_C}{(1 + v_C)(1 - 2v_C)} \left[A_1^C - (1 - 2v_C) \frac{A_2^C}{r_{e,o}^2} \right] \tag{10.74}$$

Additional boundary conditions for equivalent pressure loading on each side of an interface between two layers (propellant/casing [$r = r_m$], casing/sleeve [$r = r_n$]), that is:

$$\frac{E_A}{(1 + v_A)(1 - 2v_A)} \left[A_1^A - (1 - 2v_A) \frac{A_2^A}{r_{m,o}^2} \right]$$
$$= \frac{E_B}{(1 + v_B)(1 - 2v_B)} \left[A_1^B - (1 - 2v_B) \frac{A_2^B}{r_{m,o}^2} \right] \tag{10.75}$$

and

$$\frac{E_B}{(1 + v_B)(1 - 2v_B)} \left[A_1^B - (1 - 2v_B) \frac{A_2^B}{r_{n,o}^2} \right]$$
$$= \frac{E_C}{(1 + v_C)(1 - 2v_C)} \left[A_1^C - (1 - 2v_C) \frac{A_2^C}{r_{n,o}^2} \right] \tag{10.76}$$

and for equivalent displacement at each side:

$$\psi_m = A_1^A r_{m,o} + A_2^A / r_{m,o} = A_1^B r_{m,o} + A_2^B / r_{m,o} \tag{10.77}$$

$$\psi_n = A_1^B r_{n,o} + A_2^B / r_{n,o} = A_1^C r_{n,o} + A_2^C / r_{n,o} \tag{10.78}$$

allows for a solution of the aforementioned six coefficients, and resulting local displacements as needed. For an example sleeveless cylindrical-grain SRM, Fig. 10.31 provides a series of grain profiles from left-to-right, through the course of a 2-s simulated firing, comparing the no-load curves to the pressure-loaded curves. The differences between the corresponding profile positions are evident, although not relatively substantial to have much effect on the resulting internal ballistic performance of this particular example motor.

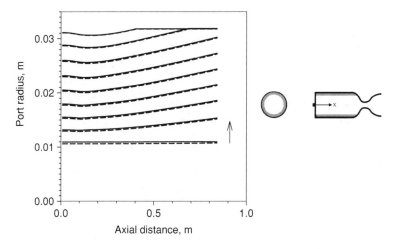

Fig. 10.31 Predicted port grain radius profile for reference cylindrical-grain SRM, with (*solid curves*) and without pressure loading (*dashed curves*), at 0.25-s increments (curves move upward as the grain burns back, from the initial port radius of 1.05 cm, toward the upper insulation/wall boundary at a radius of 3.175 cm)

One can encounter the need to analyze the deflection of the SRM structure (propellant/casing/sleeve) for grain shapes that are more elaborate than the simple cylinder discussed above. Referring to Fig. 10.32a [24] for a two-dimensional section view of a star-grain motor (starting configuration) contained in a static-test sleeve, this example illustrates the need for a more sophisticated numerical structural analysis, in this case via the finite element method (FEM). An example mesh containing approximately 2,500 elements is shown, as the basis for proceeding toward a numerical solution of the deflection of the structure under a given internal pressure loading. In this example, the amount and direction of deflection of the inner propellant surface will depend on the particular position one is located, with more outward radial deflection where the so-called web thickness (distance from propellant surface to casing wall) is greater. Moving to Fig. 10.32b, due to symmetry, one can see that some computational savings can be obtained by undertaking calculations for a pie section, rather than analyzing the entire star grain section. One may be interested in evaluating the structure at various points into the firing, as the grain burns back. An example grain pie-section profile further into the motor firing is also provided in Fig. 10.32b.

With respect to modeling the dynamic motion of the propellant and surrounding structure (e.g., under cyclic vibration in a combustion instability scenario), one should note that most solid propellants, like plastic or rubber, behave as nonlinear viscoelastic materials. This indicates that the relationship between stress and strain (relative displacement under load) is not linear as stress loading is increased on the material, and there is substantial intrinsic damping acting to slow the movement of the material in deflection. In contrast, metals (such as aluminum for the motor casing) are more readily modeled as linear elastic materials, at least at lower

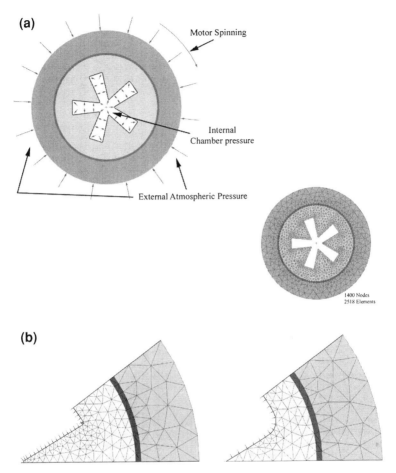

Fig. 10.32 a Schematic diagram (illustrating sources of loading on structure, including centrifugal loading under spin) and corresponding finite element mesh for star-grain SRM full two-dimensional section, with aluminum casing, and steel static-test sleeve (for further information, refer to study reported by Loncaric et al. [24]). **b** Finite element mesh for star-grain SRM, 36° pie section, with aluminum casing, and steel static-test sleeve. At left, start of firing (no burnback), and at right, 28% grain burnback (later into firing)

deflection magnitudes, where the relationship between stress and strain is approximately linear. Viscous damping on dynamically deforming metals in a structure is largely from external sources, rather than intrinsic to the material itself. In practice, working under the implied assumption that the deflection of the solid propellant is not beyond a reasonable threshold, it is common to treat the propellant as a damped linear elastic material in dynamic deformation studies, to save on computational effort (e.g., Montesano et al. [25]).

The exhaust nozzle would typically experience the most severe heat loading of the SRM structure, especially in the vicinity of the nozzle throat. Upstream of the nozzle, the SRM casing structure is largely protected by heat-shielding propellant or insulation material. Convective heat transfer analysis tends to confirm that peak heating of the nozzle occurs at sonic flow conditions. One commonly sees the use of nozzle throat inserts that have superior heat resistance characteristics relative to the surrounding nozzle structure (e.g., high-temperature graphite inserts are popular). Upstream or downstream of the throat insert, the nozzle surface may employ an ablative thermal protection layer or coating that slowly erodes away with time into the firing, but effectively protects the underlying cooler structure [26].

If one wanted to avoid the issues surrounding the use of a nozzle, and can accept lower performance efficiency, the option of a nozzleless SRM configuration is available. In lieu of a convergent–divergent nozzle, one can introduce the shape of an expansion cone at the aft end of the propellant grain. If the generated chamber pressure is sufficiently high, the internal core flow will go supersonic upon entering the expansion and ideally exit the motor as a high-speed exhaust jet. Of course, the expansion entry (effective throat for sonic flow) and downstream cone will both be expanding out with time into the motor's firing, as the propellant in that region burns back. The net result of the burnback of the expansion is generally a regressive chamber pressure profile (the thrust profile may still be progressive, or relatively neutral) and a somewhat lower specific impulse, relative to a comparable conventionally nozzled motor [27].

10.13 Example Problems

10.1. A solid rocket motor has the following characteristics later in its operational firing, with ambient flight static pressure at 83 kPa:

> grain length, $\ell_p = 2.1$ m; effective grain core peripheral distance,
> $s_p = 36$ cm
> nozzle throat dia., $d_t = 4.5$ cm; nozzle exit dia., $d_e = 12.25$ cm;
> composite propellant solid density, $\rho_s = 1710$ kg/m^3
> combustion products, $\gamma = 1.21$; molecular weight 25 amu; $T_F = 2,850$ K
> pressure-based burning rate coeffs., $n = 0.32$, $C = 5.2 \times 10^{-4}$ m/(s kPa^n)

Estimate the chamber pressure, thrust, and specific impulse at this flight condition.

10.2. A solid rocket motor has the following characteristics later in its operational firing, with ambient flight static pressure at 83 kPa:

> grain length, $\ell_p = 2.1$ m; effective grain core peripheral distance,
> $s_p = 36$ cm
> nozzle throat dia., $d_t = 4.5$ cm; nozzle exit dia., $d_e = 12.25$ cm;
> composite propellant solid density, $\rho_s = 1,710$ kg/m^3
> combustion products, $\gamma = 1.21$; molecular weight 25 amu; $T_F = 2,850$ K

pressure-based burning rate coeffs., $n = 0.32$, $C_o = 5.2 \times 10^{-4}$ m/m/(s kPa^n) at ref. temp. of 288 K; pressure-dep. burning-rate temp. sensitivity $\sigma_p = 0.0016$ K^{-1}

Initial temperature of the propellant at time of launch was $-30°$C Estimate the chamber pressure, thrust, and specific impulse at this flight condition. Comment on the change in performance from the reference case of Problem 10.1 above.

10.3. A solid-propellant grain has the following characteristics:

composite propellant solid density, $\rho_s = 1{,}740$ kg/m^3; solid specific heat $C_S = 1{,}490$ J/(kg K); net surface heat release $\Delta H_S = 0$ J/kg
combustion products, $\gamma = 1.2$; specific gas constant $R = 320$ J/(kg K);
$T_F = 3{,}000$ K
pressure-based burning rate coeffs., $n = 0.32$, $C_o = 5.2 \times 10^{-4}$ m/(s kPa^n) at ref. temp. of 288 K; pressure-dep. burning-rate temp. sensitivity $\sigma_p = 0.0016$ K^{-1}; burning surface temp. of 950 K; propellant burning surface roughness $\varepsilon = 200$ μm absolute gas viscosity $\mu = 8.1 \times 10^{-5}$ kg/(m s); gas Prandtl no. $= 0.8$

Estimate the burning rate of the propellant under static pressure of 6.9 MPa and core flow Mach number of 0.6, for a local grain core diameter of 4 cm. Initial temperature of the propellant is 15°C Compare the results of using the Greatrix/ Gottlieb model and that produced by the Lenoir/Robillard model. You may neglect any negative erosive burning influence.

10.4. A solid-propellant grain for a motor spinning on a static test stand has the following characteristics:

composite propellant solid density, $\rho_s = 1{,}740$ kg/m^3; solid specific heat $C_S = 1{,}490$ J/(kg K); net surface heat release $\Delta H_S = 0$ J/kg
combustion products, $\gamma = 1.2$; specific gas constant $R = 320$ J/(kg K);
$T_F = 3{,}000$ K
pressure-based burning rate coeffs., $n = 0.32$, $C_o = 5.2 \times 10^{-4}$ m/(s kPa^n) at ref. temp. of 288 K; pressure-dep. burning-rate temp. sensitivity $\sigma_p = 0.0016$ K^{-1}; burning surface temp. of 950 K; propellant burning surface roughness $\varepsilon = 200$ μm absolute gas viscosity $\mu = 8.1 \times 10^{-5}$ kg/(m s); gas Prandtl no. $= 0.8$

Estimate the burning rate of the propellant under static pressure of 6.9 MPa and normal acceleration of 500 g ($-4{,}905$ m/s^2 into burning surface). Initial temperature of the propellant is 15°C Once the local burning rate has been determined, also ascertain the augmentation ratio r_b/r_o .

10.5. A solid-propellant grain for a motor with the vehicle spinning while in free flight has the following characteristics:

composite propellant solid density, $\rho_s = 1,740$ kg/m³; solid specific heat
$C_S = 1,490$ J/(kg K); net surface heat release $\Delta H_S = 0$ J/kg
combustion products, $\gamma = 1.2$; specific gas constant $R = 320$ J/kg K;
$T_F = 3,000$ K
pressure-based burning rate coeffs., $n = 0.32$, $C_o = 5.2 \times 10^{-4}$ m/(s kPaⁿ)
at ref. temp. of 288 K; pressure-dep. burning-rate temp. sensitivity
$\sigma_p = 0.0016$ K^{-1}; burning surface temp. of 950 K; propellant burning surface roughness $\varepsilon = 200$ μm absolute gas viscosity $\mu = 8.1 \times 10^{-5}$ kg/(m s);
gas Prandtl no. $= 0.8$

(a) Estimate the burning rate of the propellant under static pressure of 6.9 MPa,
normal acceleration of 500 g ($-4,905$ m/s², into burning surface) due to spinning,
and longitudinal acceleration of 100 g due to free flight acceleration. Initial
temperature of the propellant is 15°C Once the local burning rate has been
determined, also ascertain the augmentation ratio r_b/r_o .
(b) Repeat part (a), but with the longitudinal acceleration now being 200 g.

10.6. A solid-propellant grain for a motor has the following characteristics at a
given time in the motor firing:

composite propellant solid density, $\rho_s = 1,740$ kg/m³; solid specific heat
$C_S = 1,490$ J/(kg K); net surface heat release $\Delta H_S = 0$; circular port
diameter of 4.0 cm; no particle loading
combustion products, $\gamma = 1.2$; specific gas constant $R = 320$ J/(kg K);
$T_F = 3,000$ K
pressure-based burning rate coeffs., $n = 0.32$, $C_o = 5.2 \times 10^{-4}$ m/(s kPaⁿ)
at ref. temp. of 288 K; pressure-dep. burning-rate temp. sensitivity
$\sigma_p = 0.0016$ K^{-1}; burning surface temp. of 950 K; propellant burning surface roughness $\varepsilon = 200$ μm absolute gas viscosity $\mu = 8.1 \times 10^{-5}$ kg/(m s);
gas Prandtl no. $= 0.8$

(a) Estimate the flow properties (u, ρ, p, Ma) 0.1 m downstream of the head-end
position, where the head-end pressure is 6.9 MPa. Initial temperature of the propellant is 15°C You may neglect erosive burning (positive or negative), friction,
heat loss via convection, longitudinal acceleration; mass injection velocity horizontal component u_i may be assumed to be zero.
(b) Repeat part (a), but a further 0.1 m downstream from the location of (a).

10.7. A solid rocket motor has the following characteristics:

grain burning surface area $= 1.7$ m²; nozzle throat dia., $d_t = 9.5$ cm;
composite propellant solid density, $\rho_s = 1,650$ kg/m³; $C_s = 1,560$ J/(kg K);
$\Delta H_s = 0$ J/kg
combustion products, $\gamma = 1.2$; molecular weight 25 amu; $T_F = 2,850$ K;
$T_s = 890$ K
gas Pr $= 0.73$; gas $k = 0.23$ W/m–K;
pressure-based burning rate coeffs., $n = 0.33$, $C = 5.833 \times 10^{-5}$ m/(s Paⁿ)

The initial temperature of the propellant at the time of launch is 19°C. Estimate the motor's percentage increase in base chamber pressure, if undergoing a spin rate that delivers 1,200 g to the propellant surface.

You may neglect erosive burning, and assume a large port-to-throat area ratio, to simplify your calculations. Before solving for the augmented burning rate (which will be substantially higher than the base burn rate, over 100% in fact, at this acceleration level) as part of finding this chamber pressure increase, you may first want to establish a relation, via mass conservation (see textbook), for chamber pressure as a function of burning rate (to allow for convenient elimination of variables in the solution process; ultimately, you will still require an iterative solution for the new burning rate under spinning, in an implicit equation for r_b). Finally, note that due to the increase in base pressure with spinning, you should be aware that a new value for base burning rate r_o will result, with implications on your calculations thereof.

10.8. A solid rocket motor's propellant is currently under ambient conditions at 35°C. There is some dispute on the engineering team about the expected burning rate, under static pressure of 6.89 MPa in the motor chamber. One of the engineers believes that the temperature sensitivity σ_p is 0.0016 K^{-1}, while another engineer believes that the Arrhenius relation holds, and that the activation energy E_{as} should be around 40×10^6 J/kg For certain, the following is accurate:

pressure-based burning rate coeffs., n = $0.32, C_o = 5.2 \times 10^{-4}$ m/s · kPa^n at ref. temp. of 288 K; net near-surface heat release $Q_s = 483,800$ J/kg; pro-pellant specific heat $C_s = 1470$ J/(kg K)

The Arrhenius expression for solid pyrolysis states:

$$r_b = A_s exp(-E_{as}/[\mathscr{R}T_s])$$

where $\mathscr{R}$ is the universal gas constant (8,314 J/(kg K)).

(a) Compare the nominal estimates by the standard de St. Robert expression, and the Arrhenius expression.
(b) A third engineer thinks the activation energy is closer to being 11×10^6 J/kg. Comment.
(c) A fourth engineer disagrees with the third engineer, and thinks that it is more likely that the temperature sensitivity is in fact closer to being 0.0065 K^{-1}. Comment.

10.14   Solutions to Example Problems

10.1. Looking at a solid rocket motor, at a flight altitude giving $p_\infty = 83$ kPa.

$$R = \frac{\mathscr{R}}{\mathscr{M}} = \frac{8314}{25} = 333 \ \text{J/(kg K)}$$

Nozzle throat area, $A_t = \frac{\pi d_i^2}{4} = \pi(0.045)^2/4 = 1.59 \times 10^{-3}\,\mathrm{m}^2$

Propellant grain surface area, $S = s_p \ell_p = (0.36)2.1 = 0.756\,\mathrm{m}^2$

Nozzle exit area, $A_e = \frac{\pi d_e^2}{4} = \pi(0.1225)^2/4 = 1.18 \times 10^{-2}\,\mathrm{m}^2$; $n = 0.32$

$C = 5.2 \times 10^{-4}\,\mathrm{m/s\,kPa}^n/1000^{0.32} = 5.7 \times 10^{-5}\,\mathrm{m/(s\,Pa}^n)$, so for chamber pressure:

$$p_c = \left\{ \left[\frac{\gamma}{RT_f} \left(\frac{2}{\gamma+1} \right)^{\frac{\gamma+1}{\gamma-1}} \right]^{1/2} \frac{A_t}{\rho_s SC} \right\}^{\frac{1}{n-1}}$$

$$= \left\{ \left[\frac{1.21}{333(2850)} (0.905)^{10.524} \right]^{0.5} \frac{1.59 \times 10^{-3}}{1710(0.756)5.7 \times 10^{-5}} \right\}^{-1.4706}$$

$$= 13.163 \times 10^6\,\mathrm{Pa} \text{ or } 13.163\,\mathrm{MPa}$$

Area-Mach No. relation, where Ma_t goes to unity:

$$\frac{A_t}{A_e} = \frac{Ma_e}{Ma_t} \left[\frac{2 + (\gamma-1)Ma_t^2}{2 + (\gamma-1)Ma_e^2} \right]^{\frac{\gamma+1}{2(\gamma-1)}} = Ma_e \left[\frac{2.21}{2 + 0.21 Ma_e^2} \right]^{5.262} = 0.135$$

Via iteration, find supersonic solution, $Ma_e = 3.09$. For nozzle exit pressure:

$$p_e = p_c \left[1 + \frac{\gamma-1}{2} Ma_e^2 \right]^{\frac{-\gamma}{\gamma-1}} = 0.241\,\mathrm{MPa}.$$

$$C_{F,v} = \left[\frac{2\gamma^2}{\gamma-1} \left(\frac{2}{\gamma+1} \right)^{\frac{\gamma+1}{\gamma-1}} \right]^{1/2} = 2.208;$$

$$c^* = \left[\frac{\gamma}{RT_f} \left(\frac{2}{\gamma+1} \right)^{\frac{\gamma+1}{\gamma-1}} \right]^{-1/2} = 1498\,\mathrm{m/s}$$

$$F = C_{F,v} \left[1 - \left(\frac{p_e}{p_c} \right)^{\frac{\gamma-1}{\gamma}} \right]^{1/2} A_t p_c + (p_e - p_\infty)A_e$$

$$= 2.208 \left[1 - (0.241/13.163)^{0.174} \right]^{0.5} 1.59 \times 10^{-3} (13.163 \times 10^6)$$

$$+ (241000 - 83000)1.18 \times 10^{-2}$$

$$= 32724 + 1864 = 34588\,\mathrm{N} \text{ or } 34.59\,\mathrm{kN}$$

$$\dot{m}_e = \frac{p_c A_t}{c^*} = 13.163 \times 10^6 (1.59 \times 10^{-3})/1498 = 13.97\,\mathrm{kg/s}$$

$$I_{sp} = \frac{F}{\dot{m}_e g_o} = \frac{34588}{13.97(9.81)} = 252\,\mathrm{s}$$

10.2. Looking at the solid rocket motor from Problem 10.1, with a change in propellant initial temperature T_i from 288 to 243 K, at a flight altitude giving $p_\infty = 83$ kPa.

$C_o = 5.2 \times 10^{-4}$ m/s kPa^n/$1000^{0.32} = 5.7 \times 10^{-5}$ m/(s Pa^n), at ref. 288 K temperature;

$$C = C_o \exp\left[\sigma_p\left(T_i - T_{i,o}\right)\right] = 5.7 \times 10^{-5} \exp[0.0016(-30° - 15°)]$$
$$= 5.304 \times 10^{-5} \text{ m/s Pa}^n;$$

$$p_c = \left\{\left[\frac{\gamma}{RT_f}\left(\frac{2}{\gamma+1}\right)^{\frac{\gamma+1}{\gamma-1}}\right]^{1/2} \frac{A_t}{\rho_s SC}\right\}^{\frac{1}{n-1}}$$

$$= \left\{\left[\frac{1.21}{333(2850)}(0.905)^{10.524}\right]^{0.5} \frac{1.59 \times 10^{-3}}{1710(0.756)5.3 \times 10^{-5}}\right\}^{-1.4706}$$

$$= 11.84 \times 10^6 \text{ Pa or } 11.84 \text{ MPa}$$

Area-Mach No. relation, where Ma_t goes to unity:

$$\frac{A_t}{A_e} = \frac{Ma_e}{Ma_t}\left[\frac{2 + (\gamma-1)Ma_t^2}{2 + (\gamma-1)Ma_e^2}\right]^{\frac{\gamma+1}{2(\gamma-1)}} = Ma_e\left[\frac{2.21}{2 + 0.21Ma_e^2}\right]^{5.262} = 0.135$$

Via iteration, find supersonic solution, $Ma_e = 3.09$, as in Problem 10.1. For nozzle exit pressure:

$$p_e = p_c\left[1 + \frac{\gamma-1}{2}Ma_e^2\right]^{\frac{-\gamma}{\gamma-1}} = 0.217 \text{ MPa.}$$

$$C_{F,v} = \left[\frac{2\gamma^2}{\gamma-1}\left(\frac{2}{\gamma+1}\right)^{\frac{\gamma+1}{\gamma-1}}\right]^{1/2} = 2.208;$$

$$c^* = \left[\frac{\gamma}{RT_f}\left(\frac{2}{\gamma+1}\right)^{\frac{\gamma+1}{\gamma-1}}\right]^{-1/2} = 1498 \text{ m/s}$$

$$F = C_{F,v}\left[1 - \left(\frac{p_e}{p_c}\right)^{\frac{\gamma-1}{\gamma}}\right]^{1/2} A_t p_c + (p_e - p_\infty)A_e$$

$$= 2.208\left[1 - (0.217/11.84)^{0.174}\right]^{0.5} 1.59 \times 10^{-3}\left(11.84 \times 10^6\right)$$
$$+ (217000 - 83000)1.18 \times 10^{-2}$$
$$= 29430 + 1580 = 31010 \text{ N or } 31 \text{ kN, about 10\% lower, vs. Problem 10.1}$$

$$\dot{m}_e = \frac{p_c A_t}{c^*} = 11.84 \times 10^6 \left(1.59 \times 10^{-3}\right)/1498 = 12.58 \text{ kg/s}$$

$$I_{sp} = \frac{F}{\dot{m}_e g_o} = \frac{31010}{12.58(9.81)} = 251.3 \text{ s, a small drop relative to Problem 10.1}$$

10.3. Looking at a solid rocket propellant's burning rate under pressure and core flow.

Consider the Lenoir-Robillard model first:

$$r_b = Cp^n + \alpha G^{0.8} D^{-0.2} \exp(-\beta r_b \rho_s/G);$$
$$\beta = 53 \text{ from exptl. observation}; D = d_p = 0.04 \text{ m}$$

$$\rho \approx \frac{p}{RT_F} = \frac{6.9 \times 10^6}{320(3000)} = 7.19 \text{ kg/m}^3;$$
$$u = a \cdot Ma \approx \sqrt{\gamma RT_F}\, Ma = 1073(0.6) = 644$$

$$G = \frac{\dot{m}}{A_p} = \rho u = 7.19(644) = 4630.4 \text{ kg/m}^2 \text{ s};$$
$$C_p = \frac{\gamma R}{\gamma - 1} = \frac{1.2(320)}{1.2 - 1} = 1920 \text{ J/(kg K)}$$

$$\alpha = \frac{0.0288 C_p \mu^{0.2} Pr^{-2/3}}{\rho_s C_s} \frac{(T_F - T_S)}{(T_S - T_i)}$$
$$= \frac{0.0288(1920)(8.1 \times 10^{-5})0.8^{-0.667}}{1740(1490)} \frac{(3000 - 950)}{(950 - 288)}$$
$$= 1.165 \times 10^{-5}.$$

One can now estimate burning rate via L-R:

$$r_b = 5.2 \times 10^{-4}(6900)^{0.32} + 0.019 \exp(-19.92 r_b) = 0.0088 + 0.0124 = 0.0212 \text{ m/s},$$

through iteration.

Now, let us consider the Greatrix–Gottlieb model:

$$r_b = Cp^n + \frac{h(T_F - T_S)}{\rho_s C_s(T_S - T_i) - \rho_s \Delta H_s};$$

$$Re_d = \frac{\rho u d}{\mu} = \frac{7.19(644)0.04}{8.1 \times 10^{-5}} = 2.287 \times 10^6; \quad \varepsilon = 2 \times 10^{-4} \text{ m}$$

$$\frac{1}{f^{1/2}} = -2 \cdot \log_{10}\left[\frac{2.51}{f^{1/2} Re_d} + \frac{\varepsilon/d}{3.7}\right] = -2 \cdot \log_{10}\left[\frac{1.0975 \times 10^{-6}}{f^{0.5}} + 1.3514 \times 10^{-3}\right]$$

$f = 0.0304$ via iteration. This is the incompressible flow estimate. Accounting for compressibility in a turbulent flow, apply the recovery factor equation:

$$f_{comp} = f_{incomp}/\left[1 + Pr^{1/3}\frac{\gamma - 1}{2} Ma^2\right] = 0.0294;$$

$$k = \frac{\mu C_p}{Pr} = \frac{8.1 \times 10^{-5} \cdot 1920}{0.8} = 0.1944 \;\text{W/(m K)};$$

$$h^* = \frac{k}{d} Re_d Pr^{1/3} \frac{f}{8} = \frac{0.1944}{0.04} 2.287 \times 10^6 (0.8^{0.333}) \frac{0.0294}{8} = 37919 \;\text{W/(m}^2\,\text{K)}; \quad \text{so,}$$

finally,

$$h = \frac{\rho_s r_b C_p}{\exp\left(\frac{\rho_s r_b C_p}{h^*}\right) - 1} = \frac{1740(1920) r_b}{\exp\left(\frac{1740(1920) r_b}{37919}\right) - 1} = \frac{3.3408 \times 10^6 r_b}{\exp(88.104 r_b) - 1}, \quad \text{where}$$

$$r_b = 0.0088 + \frac{h(2050)}{1.7163 \times 10^9} \approx 0.023 \;\text{m/s, via iteration of these two equations on } r_b$$

The results for the L-R and G-G models are similar.

10.4. Looking at a solid rocket propellant's burning rate under pressure and acceleration.

$$C_p = \frac{\gamma R}{\gamma - 1} = \frac{1.2(320)}{1.2 - 1} = 1920 \;\text{J/(kg K)};$$

$$k = \frac{\mu C_p}{Pr} = \frac{8.1 \times 10^{-5} \cdot 1920}{0.8} = 0.1944 \;\text{W/(m K)};$$

$$r_o = Cp^n = 5.2 \times 10^{-4}(6900)^{0.32} = 0.0088 \;\text{m/s}$$

$$\delta_0 = \frac{k}{\rho_s r_o C_p} \ell n \left[1 + \frac{C_p(T_F - T_S)}{C_S(T_S - T_i) - \Delta H_S} \right] = \frac{0.1944}{1740(0.0088)1920} \ell n \left[1 + \frac{1920(2050)}{1490(662) - 0} \right]$$

$$= 10.63 \times 10^{-6} \,\text{m}$$

$$G_a = \frac{a_n p}{r_b} \frac{\delta_o}{RT_F} \frac{r_o}{r_b} = \frac{-4905(6.9 \times 10^6)}{r_b} \frac{10.63 \times 10^{-6}}{320(3000)} \frac{0.0088}{r_b} = -\frac{3.3 \times 10^{-3}}{r_b^2}$$

$$r_b = \frac{C_p(T_F - T_S)}{C_S(T_S - T_i) - \Delta H_S} \cdot \frac{r_b + G_a/\rho_s}{\exp\left[\frac{C_p \delta_o}{k}(\rho_s r_b + G_a)\right] - 1}$$

$$= 3.99 \frac{r_b - \frac{1.9 \times 10^{-6}}{r_b^2}}{\exp\left[0.105\left(1740 r_b - \frac{3.3 \times 10^{-3}}{r_b^2}\right)\right] - 1} = 0.0141 \;\text{m/s, via iteration.}$$

Thus, $\dfrac{r_b}{r_o} = \dfrac{0.0141}{0.0088} = 1.60$

10.5. Looking at the solid rocket propellant for Problem 10.4, burning rate under pressure and normal/longitudinal acceleration.

(a) 100 g longitudinal acceleration in combination with 500 g normal acceleration:

$$r_o = Cp^n = 5.2 \times 10^{-4}(6900)^{0.32} = 0.0088 \;\text{m/s}$$

$$\delta_0 = \frac{k}{\rho_s r_o C_p} \ell n \left[1 + \frac{C_p(T_F - T_S)}{C_S(T_S - T_i) - \Delta H_S}\right] = \frac{0.1944}{1740(0.0088)1920} \ell n \left[1 + \frac{1920(2050)}{1490(662) - 0}\right]$$

$$= 10.6310^{-6}\,\text{m}$$

$\phi = \tan^{-1}\left(\frac{a_\ell}{a_n}\right) = \tan^{-1}\left(\frac{100}{500}\right) = 11.31°$; from Problem 10.4, for 500 g normal accel., $r_b = 0.0141$ m/s

$\phi_d = \tan^{-1}\left[K\left(\frac{r_o}{r_b}\right)^3 \tan\phi\right]$, where default value for K is 8, unless otherwise indicated.

$$G_a = \left\{\frac{a_n p}{r_b} \frac{\delta_o}{RT_F} \frac{r_o}{r_b}\right\}_{\phi=0°} \cos^2\phi_d = \left\{\frac{-4905(6.9 \times 10^6)}{0.0141} \frac{10.63 \times 10^{-6}}{320(3000)} \frac{0.0088}{0.0141}\right\}$$

$$\times \cos^2\left[\tan^{-1}\left[8\left(\frac{0.0088}{r_b}\right)^3 \tan(11.31°)\right]\right]$$

$$= -16.826 \cos^2\left[\tan^{-1}\left[\frac{1.0904 \times 10^{-6}}{r_b^3}\right]\right]$$

$$r_b = \frac{C_p(T_F - T_S)}{C_S(T_S - T_i) - \Delta H_S} \cdot \frac{r_b + G_a/\rho_s}{\exp\left[\frac{C_p \delta_o}{k}(\rho_s r_b + G_a)\right] - 1}$$

$$= 3.99 \cdot \frac{r_b - 9.36 \times 10^{-3} \cos^2\left[\tan^{-1}\left[\frac{1.0904 \times 10^{-6}}{r_b^3}\right]\right]}{\exp\left[0.105\left(1740 r_b - 16.826 \cos^2\left[\tan^{-1}\left[\frac{1.0904 \times 10^{-6}}{r_b^3}\right]\right]\right)\right] - 1}$$

$$= 0.01344\,\text{m/s, via iteration.}$$

Thus, $\dfrac{r_b}{r_o} = \dfrac{0.01344}{0.0088} = 1.53$ when $\phi = 11.3°$, as compared to 1.6 when $\phi = 0°$

(b) 200 g longitudinal acceleration in combination with 500 g normal acceleration:

$$\phi = \tan^{-1}\left(\frac{a_\ell}{a_n}\right) = \tan^{-1}\left(\frac{200}{500}\right) = 21.8°$$

$$G_a = \left\{\frac{a_n p}{r_b} \frac{\delta_o}{RT_F} \frac{r_o}{r_b}\right\}_{\phi=0°} \cos^2\phi_d = \left\{\frac{-4905(6.9 \times 10^6)}{0.0141} \frac{10.63 \times 10^{-6}}{320(3000)} \frac{0.0088}{0.0141}\right\}$$

$$\times \cos^2\left[\tan^{-1}\left[8\left(\frac{0.0088}{r_b}\right)^3 \tan(21.8°)\right]\right]$$

$$= -16.826 \cos^2\left[\tan^{-1}\left[\frac{2.1806 \times 10^{-6}}{r_b^3}\right]\right]$$

$$r_b = \frac{C_p(T_F - T_S)}{C_S(T_S - T_i) - \Delta H_S} \cdot \frac{r_b + G_a/\rho_s}{\exp\left[\frac{C_p \delta_o}{k}(\rho_s r_b + G_a)\right] - 1}$$

$$= 3.99 \cdot \frac{r_b - 9.36 \times 10^{-3} \cos^2\left[\tan^{-1}\left[\frac{2.1806 \times 10^{-6}}{r_b^3}\right]\right]}{\exp\left[0.105\left(1740 r_b - 16.826 \cos^2\left[\tan^{-1}\left[\frac{2.1806 \times 10^{-6}}{r_b^3}\right]\right]\right)\right] - 1}$$

$$= 0.00947 \text{ m/s, via iteration.}$$

Thus, $\dfrac{r_b}{r_o} = \dfrac{0.00947}{0.0088} = 1.075$ when $\phi = 21.8°$, as compared to 1.6 when $\phi = 0°$

10.6. Looking at a solid rocket motor, near its head end (so lower internal port flow speed).

(a) Going 0.1 m downstream from motor's head end, constant-area cylindrical port to begin with. As boundary condition, at head end ($x = 0$ m), can assume $u_1 = 0$ m/s and $p_1 = p_c = 6.9 \times 10^6$ Pa, and $T_1 = T_F = 3{,}000$ K. No particle loading $(\alpha_p = 0)$. Subsequently, then

$$\rho_1 = \frac{p_1}{RT_1} = \frac{6.9 \times 10^6}{320(3000)} = 7.1875 \text{ kg/m}^3, \quad a_1 = \sqrt{\gamma RT_1} = 1073.3 \text{ m/s}, Ma_1 = 0;$$

$$A_p = \frac{\pi d_p^2}{4} = 1.257 \times 10^{-3} \text{ m}^2; \quad C_v = C_p/\gamma = 1920/1.2 = 1600 \text{ J/(kg K)}$$

To first order accuracy, can use upstream values for properties nominally designated as means.

$$\overline{\Delta S} = \bar{s}_p \cdot \Delta x = \pi d_p \Delta x = \pi(0.04)0.1 = 0.01257 \text{ m}^2; \quad \lambda = 1 \quad \text{(constant-area duct... see textbook)}$$

$$r_{b,1} = C p_1^n = 5.2 \times 10^{-4}(6900)^{0.32} = 0.0088 \text{ m/s} = \bar{r}_b;$$
$$v_{w,1} = \frac{\rho_s r_{b,1}}{\rho_1}(1 - \alpha_p) = 2.13 \text{ m/s}$$

$$B_1 = \rho_1 u_1 A_1 + \overline{r_b \Delta S}\left[(1 - \alpha_p)\rho_s - \rho\right]$$
$$= 7.1875(0)(1.257 \times 10^{-3}) + 0.0088(0.01257)[(1 - 0)1740 - 7.1875]$$
$$= 0.1917$$

$$B_2 = \rho_1 u_1^2 A_1 + p_1 \left(\frac{A_1 + A_2}{2}\right) + \overline{\Delta S}\left[(1 - \alpha_p)\rho_s \overline{r_b u_i} - \overline{r_b \rho u} - \frac{\overline{\rho u^2 f}}{8}\right]$$

$$- \left(\frac{A_1 + A_2}{2}\right)\left[\frac{\bar{p}_p}{m_p}\bar{D} + \bar{\rho}a_\ell\right]\Delta x$$

$$= 7.1875(0)^2\left(1.257 \times 10^{-3}\right) + 6.9 \times 10^6\left(1.257 \times 10^{-3}\right)$$
$$+ 0.01257[(1 - 0)1740(0.0088)(0) - 0.0088(7.1875)(0)$$
$$- 7.1875(0)(0)/8] - \left(1.257 \times 10^{-3}\right)[(0)(0) + 7.1875(0)]0.1$$
$$= 8673.3$$

$$B_3 = \rho_1 u_1 A_1 \left(C_p T_1 + \frac{u_1^2}{2}\right) + \overline{\Delta S}\left[\overline{r_b \rho_s}(1 - \alpha_p)\left(C_p T_F + \frac{v_w^2}{2}\right)\right.$$

$$- h_c(\bar{T} - T_w) - \overline{r_b \rho}\left(C_v \bar{T} + \frac{\bar{u}^2}{2}\right)\Big] - \left(\frac{A_1 + A_2}{2}\right)\frac{\bar{p}_p}{m_p}\left(\overline{u_p D} + \bar{Q}\right)\Delta x - \left(\frac{A_1 + A_2}{2}\right)\overline{\rho u}a_\ell \Delta x$$

$$= 7.1875(0)\left(1.257 \times 10^{-3}\right)(1920(3000) + 0^2/2) + 0.01257[0.0088(1740)$$
$$(1 - 0)(1920(3000) + 2.13^2/2)$$
$$- (0)(3000 - 3000) - 0.0088(7.1875)(1600(3000) + 0^2/2)] \quad 0 - 0$$
$$= 1,104,822$$

$$u_2 = \frac{-\frac{C_p B_2 \lambda}{R} \pm \left\{\left(\frac{C_p B_2 \lambda}{R}\right)^2 + 4\left(\frac{B_1}{2} - \frac{C_p B_1 \lambda}{R}\right)B_3\right\}^{1/2}}{2\left(\frac{B_1}{2} - \frac{C_p B_1 \lambda}{R}\right)}$$

$$= \frac{-52039.8 + \left\{52039.8^2 + 4(-1.05435)1104822\right\}^{0.5}}{2(-1.05435)}$$

$$= 24678.62 - 24657.38 = 21.24 \text{ m/s}$$

$$\rho_2 = B_1/(A_2 u_2) = 0.1917/\left(1.257 \times 10^{-3} \times 21.24\right) = 7.1801 \text{ kg/m}^3$$

$$p_2 = 2(B_2 - B_1 u_2)/(A_1 + A_2) = (8673.3 - 0.1917(21.24))/1.257 \times 10^{-3}$$
$$= 6896761 \text{ Pa}$$

$$T_2 = \lambda u_2(B_2 - B_1 u_2)/(RB_1) = p_2/(R\rho_2) = 6896761/(320 \times 7.1801)$$
$$= 3001.67 \text{ K}$$

$$a_2 = \sqrt{\gamma R T_2} = 1073.6 \text{ m/s}, \quad Ma_2 = \frac{u_2}{a_2} = 21.24/1073.6 = 0.0198$$

(b) Go a further 0.1 m downstream from above. Old station 2 at right is now new station 1 at left.

$$r_{b,1} = Cp_1^n = 5.2 \times 10^{-4}(6896.761)^{0.32} = 0.008798 \text{ m/s} = \bar{r}_b$$

$v_{w,1} = \frac{\rho_s r_{b,1}}{\rho_1}(1 - \alpha_p) = 2.1313$ m/s, $\lambda = 1$ (constant port area at this time in the firing)

$$B_1 = 7.1801(21.24)(1.257 \times 10^{-3}) + 0.008798(0.01257)[(1 - 0)1740 - 7.1801]$$
$$= 0.38327$$

$$B_2 = 7.1801(21.24)^2(1.257 \times 10^{-3}) + 6.89676 \times 10^6(1.257 \times 10^{-3})$$
$$+ 0.01257[(1 - 0)1740(0.008798)(0)$$
$$- 0.008798(7.1801)(21.24) - 7.1875(0)(0)/8] - (1.257 \times 10^{-3})$$
$$[(0)(0) + 7.1801(0)]0.1$$
$$= 8673.3$$

$$B_3 = 7.1801(21.24)(1.257 \times 10^{-3})(1920(3000) + 21.24^2/2)$$
$$+ 0.01257[0.008798(1740)(1 - 0)(1920(3000) + 2.1313^2/2) - (0)(3000 - 3000)$$
$$- 0.008798(7.1801)(1600(3001.7) + 21.24^2/2)] - 0 - 0$$
$$= 2,209,041$$

$$u_2 = \frac{-52039.8 + \{52039.8^2 + 4(-2.108)2209041\}^{0.5}}{2(-2.108)} = 12343.41 - 12300.88$$
$$= 42.53 \text{ m/s}$$

$$\rho_2 = B_1/(A_2 u_2) = 0.38327/(1.257 \times 10^{-3} \cdot 42.53) = 7.1693 \text{ kg/m}^3$$

$$p_2 = 2(B_2 - B_1 u_2)/(A_1 + A_2) = (8673.3 - 0.38327(42.53))/1.257 \times 10^{-3}$$
$$= 6887032 \text{ Pa}$$

$$T_2 = \lambda u_2(B_2 - B_1 u_2)/(RB_1) = p_2/(R\rho_2) = 6887032/(320 \cdot 7.1693)$$
$$= 3001.98 \text{ K}$$

$$a_2 = \sqrt{\gamma R T_2} = 1073.7 \text{ m/s}, \quad Ma_2 = \frac{u_2}{a_2} = 42.53/1073.7 = 0.0396$$

10.7. Looking at a solid rocket propellant's internal ballistics under pressure and acceleration.

$$R = \frac{\mathscr{R}}{\mathscr{M}} = 8315/25 = 333 \text{ J/(kg K)}; \quad C_p = \frac{\gamma R}{\gamma - 1} = \frac{1.2(333)}{1.2 - 1} = 1996 \text{ J/(kg K)};$$

$$A_t = \frac{\pi d_t^2}{4} = \pi(0.095)^2/4 = 7.09 \times 10^{-3} \text{ m}^2$$
$$= 0.1917$$

$$p_{c,o} = \left\{ \left[\frac{\gamma}{RT_f} \left(\frac{2}{\gamma+1} \right)^{\frac{\gamma+1}{\gamma-1}} \right]^{1/2} \frac{A_t}{\rho_s SC} \right\}^{\frac{1}{n-1}}$$

$$= \left\{ \left[\frac{1.2}{333(2850)} (2/2.2)^{11} \right]^{0.5} \frac{7.09 \times 10^{-3}}{1650(1.7)5.833 \times 10^{-5}} \right\}^{-1.4925}$$

$$= 5.97 \times 10^6 \text{ Pa or } 5.97 \text{ MPa, reference baseline pressure}$$

$r_o = Cp_{c,o}^n = 5.833 \times 10^{-5}(5.97 \times 10^6)^{0.33} = 0.01$ m/s, reference baseline burning rate.

Need combustion zone thickness at new base pressure $p_{c,a}$, due to spinning.

$$\delta_0 = \frac{k}{\rho_s \left(Cp_{c,a}^n \right) C_p} \ell n \left[1 + \frac{C_p(T_F - T_S)}{C_S(T_S - T_i) - \Delta H_S} \right]$$

$$= \frac{0.23}{1650 \left(Cp_{c,a}^n \right) 1996} \ell n \left[1 + \frac{1996(2850 - 890)}{1560(890 - 293) - 0} \right] = \frac{1.1505 \times 10^{-7}}{Cp_{c,a}^n}, \text{ m}$$

$$G_a = \frac{a_n p_{c,a}}{r_b} \frac{\delta_o}{RT_F} \frac{Cp_{c,a}^n}{r_b} = \frac{-1200(9.81)(p_{c,a})}{r_b} \frac{1.1505 \times 10^{-7}}{333(2850)Cp_{c,a}^n} \frac{Cp_{c,a}^n}{r_b}$$

$$= -\frac{1.428 \times 10^{-9} p_{c,a}}{r_b^2}$$

From mass conservation:

$$\rho_s S r_b = \dot{m}_t = \frac{A_t p_{c,a}}{c^*} = \left[\frac{\gamma}{RT_F} \left(\frac{2}{\gamma+1} \right)^{\frac{\gamma+1}{\gamma-1}} \right]^{1/2} A_t p_{c,a}$$

$$1650(1.7)r_b = 6.661 \times 10^{-4}(7.09 \times 10^{-3})p_{c,a}, \text{ or } p_{c,a} = 5.942 \times 10^8 r_b$$

As a result, substituting into an earlier result,

$$G_a = -\frac{1.428 \times 10^{-9} p_{c,a}}{r_b^2} = \frac{-0.849}{r_b}$$

Finally then,

$$r_b = \frac{C_p(T_F - T_S)}{C_S(T_S - T_i) - \Delta H_S} \cdot \frac{r_b + G_a/\rho_s}{\exp\left[\frac{C_p \delta_o}{k}(\rho_s r_b + G_a)\right] - 1}$$

$$= 4.194 \cdot \frac{r_b - \dfrac{0.849}{1650 r_b}}{\exp\left[\dfrac{9.9844 \times 10^{-4}}{5.833 \times 10^{-5}(5.942 \times 10^8 r_b)^{0.33}}\left(1650 r_b - \dfrac{0.849}{r_b}\right)\right] - 1}$$

$= 0.0253$ m/s, via iteration.

$p_{c,a} = 5.942 \times 10^8 r_b = 15.03\,\mathrm{MPa}$, and new $r_o = C_{p_{c,a}}^n = 5.833 \times 10^{-5}(15.03 \times 10^6)^{0.33} = 0.0136$ m/s with respect to augmentation of burning rate due to normal acceleration relative to pressure-dependent burning, and $\Delta p_c = \dfrac{p_{c,a} - p_{c,o}}{p_{c,o}} \times$

$100\% = \dfrac{15.03 - 5.97}{5.97} \times 100\% = 152\%$ increase in base pressure from the original no-spin case.

10.8. Solid propellant initial temp. at 308 K, substantially above the ref. initial temp. of 288 K.

(a) $r_{b,o} = C_o p^n = 5.2 \times 10^{-4}(6890)^{0.32} = 0.8795$ cm/s, ref. burning rate at 288 K

$$C = C_o \exp\left[\sigma_p(T_i - T_{i,o})\right] = 5.2 \times 10^{-4}\exp[0.0016(308 - 288)]$$
$$= 5.369 \times 10^{-4} \text{ m/(s kPa}^n)$$

$r_b = C p^n = 5.369 \times 10^{-4}(6890)^{0.32} = 0.9081$ cm/s, via the standard de St. Robert eqn.

$$T_{S,ref} = T_{i,o} + \frac{Q_S}{C_S} + \frac{1}{2\sigma_p} = 288 + \frac{483800}{1470} + \frac{1}{2(0.0016)} = 930 \text{ K}$$

$$T_{S,new} = T_{i,new} + \frac{Q_S}{C_S} + \frac{1}{2\sigma_p} = 308 + \frac{483800}{1470} + \frac{1}{2(0.0016)} = 950 \text{ K}$$

Need to get first estimate of A_s coefficient to be used in the Arrhenius eqn.
$$A_s = \frac{r_{b,o}}{\exp\left(\frac{-E_{as}}{\mathcal{R}T_{S,ref}}\right)} = \frac{0.008795}{\exp\left(\frac{-40\times10^6}{8314 \times 930}\right)} = 1.5523 \text{ m/s, such that}$$

$$r_b = A_s \exp\left(\frac{-E_{as}}{\mathcal{R}T_{S,new}}\right) = 1.5523\exp\left(\frac{-40\times10^6}{8314\cdot950}\right) = 0.00981 \text{ m/s or 0.981 cm/s,}$$

thus predicting a higher burning rate than the standard estimate (0.9081 cm/s).

(b) Assume $E_{as} = 11 \times 10^6$ J/kg, rather than 40×10^6.

$$A_s = \frac{r_{b,o}}{\exp\left(\frac{-E_{as}}{\mathcal{R}T_{S,ref}}\right)} = \frac{0.008795}{\exp\left(\frac{-11\times10^6}{8314\cdot930}\right)} = 0.0365 \text{ m/s}$$

$r_b = A_s \exp\left(\dfrac{-E_{as}}{\mathcal{R}T_{S,\text{new}}}\right) = 0.0365\exp\left(\dfrac{-11 \times 10^6}{8314 \cdot 950}\right) = 0.00907$ m/s or 0.907 cm/s, thus similar to the standard de St. Robert estimate above (0.908 cm/s).

(c) $C = C_o\exp\left[\sigma_p\left(T_i - T_{i,o}\right)\right] = 5.2 \times 10^{-4}\exp[0.0065(308 - 288)] = 5.922 \times 10^{-4}$ m/(s kPa^n)

$r_b = Cp^n = 5.922 \times 10^{-4}(6890)^{0.32} = 1.0$ cm/s, via the standard de St. Robert eqn.

$T_{S,\text{ref}} = T_{i,o} + \dfrac{Q_S}{C_S} + \dfrac{1}{2\sigma_p} = 288 + \dfrac{483800}{1470} + \dfrac{1}{2(0.0065)} = 694$ K, a bit cool for this propellant type

$$T_{S,\text{new}} = T_{i,\text{new}} + \frac{Q_S}{C_S} + \frac{1}{2\sigma_p} = 308 + \frac{483800}{1470} + \frac{1}{2(0.0065)} = 714 \text{ K}$$

$$A_s = \frac{r_{b,o}}{\exp\left(\dfrac{-E_{as}}{\mathcal{R}T_{S,\text{ref}}}\right)} = \frac{0.008795}{\exp\left(\dfrac{-40 \times 10^6}{8314 \cdot 694}\right)} = 9.0 \text{ m/s, such that}$$

$$r_b = A_s\exp\left(\frac{-E_{as}}{\mathcal{R}T_{S,\text{new}}}\right) = 9.0\exp\left(\frac{-40 \times 10^6}{8314 \cdot 714}\right) = 0.0107 \text{ m/s or 1.07 cm/s,}$$

thus predicting a higher burn rate than the standard estimate of 1.00 cm/s.

References

1. Barrett DH (1971) Solid rocket motor igniters. NASA SP-8051
2. T'ien JS, Sirignano WA, Summerfield M (1970) Theory of L-star combustion instability with temperature oscillations. AIAA J 8:120–126
3. Steinz JA, Summerfield M (1969) Low pressure burning of composite solid propellants. In: Boyars C, Klager K (eds.) Propellants manufacture, hazards and testing. American Chemical Society, Washington, DC
4. Sutton GP, Biblarz O (2001) Rocket propulsion elements, 7th edn. Wiley, New York
5. Greatrix DR, Gottlieb JJ (1987) Erosive burning model for composite-propellant rocket motors with large length-to-diameter ratios. Can Aeronaut Space J 33:133–142
6. Munson BR, Young DF, Okiishi TH (1994) Fundamentals of fluid mechanics, 2nd edn. Wiley, New York
7. Mickley HS, Ross RC, Squyers AL, Stewart WF (1954) Heat, mass, and momentum transfer for flow over a flat plate with blowing or suction. NACA TN 3208
8. Moffat RJ, Kays WM (1968) The turbulent boundary layer on a porous plate: experimental heat transfer with uniform blowing and suction. Int J Heat Mass Tran 11:1547–1566
9. Greatrix DR (2007) Model for prediction of negative and positive erosive burning. Can Aeronaut Space J 53:13–21
10. Godon JC, Duterque J, Lengellé G (1987) Solid propellant erosive burning. In: Proceedings of 23rd AIAA/ASME/SAE/ASEE joint propulsion conference, San Diego, June 29–July 2
11. Godon JC, Duterque J, Lengellé G (1991) Erosive burning in solid propellant motors. In: Proceedings of 27th AIAA/ASME/SAE/ASEE joint propulsion conference, Sacramento, 24–26 June
12. Crowe CT (1972) A unified model for the acceleration-produced burning rate augmentation of metalized solid propellant. Combust Sci Technol 5:55–60

13. Greatrix DR, Gottlieb JJ (1988) Normal acceleration model for composite-propellant combustion. Trans Can Soc Mech Eng 12:205–211
14. Greatrix DR (1994) Parametric analysis of combined acceleration effects on solid-propellant combustion. Can Aeronaut Space J 40:68–73
15. Krier H, T'ien JS, Sirignano WA, Summerfield M (1968) Nonsteady burning phenomena of solid propellants: theory and experiments. AIAA J 6:278–285
16. Culick FEC (1968) A review of calculations for unsteady burning of a solid propellant. AIAA J 6:2241–2255
17. Greatrix DR (2008) Transient burning rate model for solid rocket motor internal ballistic simulations. Int J Aerosp Eng 2008:1–10. doi:10.1155/2008/826070
18. Novozhilov BV (1992) Theory of nonsteady burning and combustion instability of solid propellants by the Zeldovich-Novozhilov method. In: De Luca L, Price EW, Summerfield M (eds.) Nonsteady burning and combustion stability of solid propellants. AIAA, Washington, DC
19. Blomshield FS (2001) Historical perspective of combustion instability in motors: case studies. In: Proceedings of AIAA/ASME/SAE/ASEE 37th joint propulsion conference, Salt Lake City, 8–11 July
20. Baum JD, Levine JN, Lovine RL (1988) Pulsed instability in rocket motors: a comparison between predictions and experiments. J Propul Power 4:308–316
21. Greatrix DR (2011) Scale effects on solid rocket combustion instability behaviour. Energies 4:90–107
22. Chase M, Thorp GP (1996) Solid rocket case design. In: Jensen GE, Netzer DW (eds.) Tactical missile propulsion. AIAA, Reston (Virginia)
23. Popov EP (1976) Mechanics of materials, 2nd edn. Prentice-Hall, Upper Saddle River (New Jersey)
24. Loncaric S, Greatrix DR, Fawaz Z (2004) Star-grain rocket motor–nonsteady internal ballistics. Aerosp Sci Technol 8:47–55
25. Montesano J, Behdinan K, Greatrix DR, Fawaz Z (2008) Internal chamber modeling of a solid rocket motor: effects of coupled structural and acoustic oscillations on combustion. J Sound Vib 311:20–38
26. Heister S (1995) Solid rocket motors. In: Humble RW, Henry GN, Larson WJ (eds.) Space propulsion analysis & design. McGraw-Hill, New York
27. Davenas A (1996) Solid rocket motor design. In: Jensen GE, Netzer DW (eds.) Tactical missile propulsion. AIAA, Reston (Virginia)

Chapter 11
Liquid-Propellant Rocket Engines

11.1 Introduction

We noted earlier that liquid-propellant rocket engine (LRE) operation allows for the selection of propellant constituents that can deliver a lower gas molecular mass $\mathcal{M}$, a bonus in thrust delivery. This advantage comes with more complexity of system construction and operation versus the simpler SRM, given the need for tank storage, feed (pumping) systems, cooling systems and an effective spray injection system for delivery of the propellant to the combustion (thrust) chamber. The higher specific impulse (I_{sp}, with values ranging from around 340 s at low altitude to 420 s or more at high altitude) with higher-energy bipropellant combinations such as LO_2/LH_2, possibly augmented due to a higher flame temperature T_F allowed by effective cooling of the surrounding structure and higher chamber pressure p_c allowed by a thicker-walled combustion chamber, may in some flight applications warrant the selection of an LRE, in spite of the heightened cost and complexity. The ready ability to throttle or modulate thrust at different points in the flight mission, or shut down entirely and restart at a later time in the mission, are additional factors that favor LREs for some applications.

11.2 Propellant Types

There are various means for categorizing propellants. Bipropellants utilize separate mixtures of oxidizer and fuel, and are typically more energetic relative to monopropellants. Bipropellants as a result are more in use for higher thrust applications where a higher I_{sp} is more essential for flight mission success. On occasion, one may see the use of a tripropellant approach, e.g., vehicle attitude control thruster using kerosene–oxygen for the first part of the flight mission (at low altitude, denser kerosene fuel giving a storage volume advantage over this early mission period; the parameter density specific impulse I_d or I_ρ, the product of

D. R. Greatrix, *Powered Flight*, DOI: 10.1007/978-1-4471-2485-6_11,
© Springer-Verlag London Limited 2012

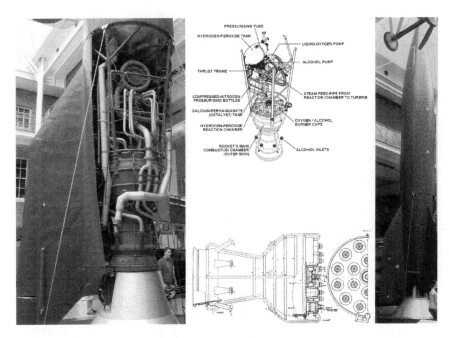

Fig. 11.1 View of the liquid-propellant rocket engine used for powering the flight of the 14-meter-high V-2 (A-4) ballistic missile. Developed by Wernher von Braun's team during World War II for the German war effort (first successful flight, October 3, 1942), some four thousand V-2 missiles were launched against targets during WWII [1]. Beneath the *copper-colored* combustion chamber and nozzle expansion is a *silver-colored* conical display stand used by the Imperial War Museum for mounting the vehicle. The above rocket engine and flight vehicle were effectively the baseline by which manned space flight by Russia (April 12, 1961), and then the United States (May 5, 1961, through the work of Dr. von Braun's team working predominantly out of Huntsville, Alabama, at that time) would eventually be first accomplished, 19 years later. (Engine diagrams courtesy of V2Rocket.com)

average propellant density or specific gravity and I_{sp}, is sometimes quoted in this regard). The kerosene fuel is removed and replaced by hydrogen–oxygen for the latter part of the flight mission (in space) in completing this tripropellant example. Liquid oxygen, LO_2 (a.k.a., LOx or LOX), is a common high-performance choice as an oxidizer. The infamous V-2 rocket [1] of WWII utilized LO_2 in combination with a storable liquid fuel comprised of a water-diluted ethyl alcohol mixture (see Fig. 11.1). The alcohol was diluted to keep the flame temperature below the heating threshold of the surrounding structure of the thrust chamber. However, as a cryogenic propellant with an approximate boiling point of 90 K at 1 atm and 130 K at 10 atm, LO_2 must be stored at a very low temperature to remain liquid in storage (at a relatively low tank pressure, a weight-saving bonus with cryogenic propellants). Refer to the solid–liquid–vapor phase diagram of Fig. 11.2, for an example substance, showing the influence of pressure and temperature. Earth storable oxidizers that remain liquid under ambient room conditions (e.g., 294 K at 1 atm) include hydrogen peroxide (H_2O_2), nitric acid (HNO_3) and nitrogen

Fig. 11.2 Phase diagram for an example substance. Critical pressure and critical temperature occur at the critical point, as shown

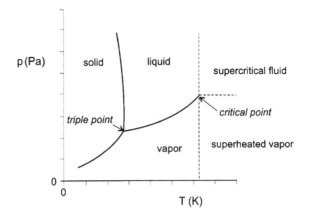

tetroxide (N$_2$O$_4$; one sometimes sees the use of *mixed oxides of nitrogen*, such as MON-3, which is 3% nitric oxide, NO, and MON-25, at 25% NO, in an N$_2$O$_4$ base mixture). Earth storables require less plumbing and care than cryogenic propellants, but in general, they are less energetic. Fluorine atoms can act as an oxidizer in place of oxygen atoms, such as seen with the space storable (liquid in simple storage in the temperatures of space; "mildly cryogenic") oxidizer, chlorine pentafluoride (ClF$_5$; boils at 260 K at 1 atm). Chlorine trifluoride, ClF$_3$, is hypergolic with H$_2$ (will spontaneously ignite when brought together, unlike LO$_2$/LH$_2$, which requires an initial heat source). Liquid fluorine, F$_2$, is a cryogenic oxidizer. While certainly energetic, fluorine-based oxidizers are not commonly used, since they have issues related to toxicity and corrosion that must be dealt with. See Table 11.1 for a comparison of propellant and performance characteristics of various propellant combinations, including fluorine [2]. Note that the mixture ratio for maximum specific impulse does not usually coincide with the stoichiometric value (which gives the peak adiabatic flame temperature), since, for example, it is common to get better thrust delivery with a portion of the reactants uncombusted to give an exhaust gas at a lower molecular mass, even though the flame temperature is somewhat lower.

On the fuel side of bipropellant systems, we have noted the use of LH$_2$ as a high-energy choice with an advantageous low $\mathscr{M}$ as used by the Space Shuttle orbiter's three main engines (SSMEs; see Figs. 11.3, 11.4). Other cryogenic fuels include the hydrocarbon fuel methane (CH$_4$), and ammonia (NH$_3$), which can also be placed under the space storable category (boils at 240 K at 1 atm). LH$_2$ does have the drawback of lower density (70 kg/m^3 compared to liquid H$_2$O at 1,000 kg/m^3 and liquid O$_2$ at 1,140 kg/m^3) in storage, necessitating larger tanks (larger volume and corresponding diameter). With the diameter and exposed surface area issues as relates to overall vehicle aerodynamic drag in the lower atmosphere, it is not uncommon to see H$_2$ being substituted by a less energetic but denser liquid fuel for first-stage booster engines. Earth storable fuels include hydrocarbon mixture such as RP-1 (kerosene-based *rocket propellant no. 1*) which like cryogenic fuels can also be used as a coolant to keep the combustor/nozzle

Table 11.1 Maximum-I_{sp} oxidizer-fuel mixture ratios, gas flame temperatures, gas molecular mass, gas ratio of specific heats (frozen) and sea-level I_{sp} (frozen) for various liquid propellant combinations under nominal operating conditions [2]

Oxidizer	Fuel	r (max-I_{sp})	T_F, K	$\mathcal{M}$, amu	γ	$I_{sp,SL}$, sec
Fluorine	Hydrazine	1.9	4,550	18.5	1.33	360
Fluorine	Hydrogen	4.6	3,100	8.9	1.33	390
Hydrogen peroxide	RP-1	7.0	2,750	21.7	1.19	300
IRFNA	Aerozine[a]	1.8	3,000	20.6	1.22	270
IRFNA	Hydyne[b]	3.1	3,200	24.1	1.22	270
IRFNA	RP-1	4.1	3,200	24.6	1.22	260
Nitrogen tetroxide	Hydrazine	1.1	3,250	19.5	1.26	285
Nitrogen tetroxide	Aerozine	1.7	3,250	21.0	1.24	280
Nitrogen tetroxide	RP-1	3.4	3,300	24.1	1.23	300
Nitrogen tetroxide	MMH	2.1	3,400	22.3	1.23	280
Oxygen	Methane	3.2	3,500	20.6	1.25	300
Oxygen	Hydrazine	0.8	3,300	18.3	1.25	300
Oxygen	Hydrogen	3.4	2,950	8.9	1.26	390
Oxygen	RP-1	2.3	3,600	21.9	1.24	300
Oxygen	UDMH	1.4	3,550	19.8	1.25	295

Reprinted with permission of John Wiley & Sons, Inc.
[a] Aerozine = 50% UDMH/50% hydrazine
[b] Hydyne = 60% UDMH/40% DETA

(thrust chamber) structure below the maximum temperature allowed (if fuel subsequently burned in preburners or the main combustion chamber, referred to as a regenerative cooling method). Hydrocarbon compounds can be combined with hydrazine (N_2H_4) as unsymmetrical dimethylhydrazine (UDMH, $(CH_3)_2NHH_2$) or monomethylhydrazine(MMH, CH_3NHNH_2) as multi-purpose storable fuels for high-end low-thrust applications, where hypergolic (spontaneous upon contact) ignition is advantageous. For example, MMH is hypergolic with N_2O_4 (used in the Space Shuttle's two large orbital maneuvering system thrusters).

Monopropellants such as Earth storable hydrazine are typically less energetic than bipropellants, but having both oxidizer and fuel present in one chemical compound can be advantageous for lower thrust applications (lower complexity/cost, structural weight) such as vernier thrusters used for satellite attitude control. While monopropellants, like bipropellants, can be ignited by electrical or flame-induced heat, in the case of hydrazine, ignition is attained by passing it through a suitable catalyst, e.g., an iridium/aluminum oxide wire mesh.

11.3 Propellant Storage

The means by which one would store the applicable propellant constituents will depend not only on the category of propellant discussed in the previous section, but also on the method of feeding the liquid propellant to its ultimate destination at

Fig. 11.3 One engine (of three Space Shuttle liquid-propellant main rocket engines [SSMEs] by Rocketdyne (Pratt and Whitney) positioned at the rear of the orbiter flight vehicle) being moved into place within the orbiter. *Upper* photo courtesy of NASA. Positioning and size of the SSMEs relative to the two SRBs and central external tank in a bottom view of the overall Space Shuttle vehicle from launch are shown in the *lower* photos

the injector face of the thrust chamber (feed systems will be discussed in the next section). Note that the higher the propellant liquid density, the lower one can make the propellant tanks in size and weight, everything else being equal. If one propellant constituent's density is very different from the other, tank sizing may force a design compromise (ideally, the oxidizer tank and the fuel tank would be comparable in size, and the tanks' contents would be of comparable liquid

Fig. 11.4 View of SSME being fired on a test rig. *Left* photo courtesy of NASA Stennis Space Center. *Right* photos show different views of an SSME on static display at the U.S. Space and Rocket Center in Huntsville, Alabama

densities; if one constituent is substantially denser, one would more likely consider a compromise on bringing more of that substance, and less of the lighter). For example, LO_2/LH_2 has an optimum I_{sp} delivery at an oxidizer-fuel mixture ratio r of around 3.5, due to the beneficial effects of the low molecular mass of unreacted H_2 on the exhaust products (rather than running at stoichiometric r_{st}, which is 8). In practice, one sees LO_2/LH_2 run at an r of 5 or 6, to allow for more storage of the denser O_2 at 1,140 kg/m^3 and a bit less of the lighter H_2 at 70 kg/m^3. This is the case for the Space Shuttle, for its liquid propellant storage (see Fig. 11.5 for a photo of the external tank). For first-stage booster applications, tank sizing versus I_{sp} performance may force the use of a higher density fuel such as Earth storable RP-1 (liquid density, around 800 kg/m^3), rather than LH_2, in combination with cryogenic LO_2.

One also wants to keep tank internal pressures as low as possible, in order to reduce the required tank wall thicknesses and therefore structural weight. For

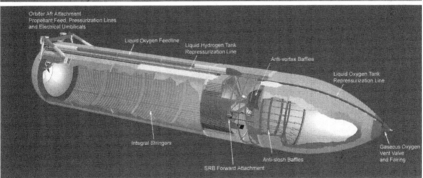

Fig. 11.5 The main external liquid propellant tank (ET) for feeding the Space Shuttle's liquid-propellant main rocket engines (SSMEs), containing a smaller *upper* liquid oxygen tank and a larger *lower* liquid hydrogen tank within the above aluminum-alloy structure. (Constructed by Lockheed Martin. Courtesy of NASA)

example, at the low end, cryogenic tank pressures, with a turbopump-feed system downstream, can range from 10 to 50 psi (70–350 kPa). One-step pressure-feed systems require a feed pressure in the order of 20% greater than the downstream combustion chamber pressure p_c, thus tank pressures in those cases would be substantially higher. Cryogenic propellants will slowly heat up, vaporizing ("boiling off") with a resulting increase in tank internal pressure, unless this gas is vented from the tank. The term *ullage* refers to the extra volume of the tank required for accommodating various factors that include: (1) allowance for some gas release from the liquid propellant, (2) thermal expansion of the liquid as its temperature rises and (3) allowance for some tank structure contraction due to varying external and internal temperatures (this last item applies to both cryogenics and storables). Ullage can impose 3–10% volume increases on the base tank size.

Typical tank materials are aluminum, stainless steel and titanium, all metals with good corrosion resistance (corrosion more often associated with the oxidizers). Composite tanks are gaining popularity due to their relative lightness, but typically require a metal inner liner to prevent leakage through the microscale pores in the composite structure. For the least amount of structural weight, one would choose a spherical shape for the tank. However, in practice, for larger tank requirements, the flight vehicle's body diameter may force a cylindrical main tank body, with ellipsoidal end caps, in order to fit inside the vehicle.

The required total volume of a propellant tank, V_t, is the sum of the following:

$$V_t = V_u + \Delta V_t + \Delta V_b + \Delta V_u + \Delta V_c + \Delta V_{\text{aux}} \tag{11.1}$$

Here, V_u is the usable propellant volume determined from propulsion system requirements. For example:

$$\dot{m} = \dot{m}_O + \dot{m}_F = (1+r)\dot{m}_F = \frac{F}{I_{sp}g_o} = \frac{p_c A_t}{c^*} \tag{11.2}$$

For a nominal burn time t_B, the usable fuel tank volume would be

$$V_{u,\text{fuel}} = \frac{\dot{m}_F N_{\text{eng}} t_B}{\rho_F} \tag{11.3}$$

Here, N_{eng} is the number of operating engines for that burn time period. The usable oxidizer tank volume would be

$$V_{u,\text{oxid}} = \frac{\dot{m}_O N_{\text{eng}} t_B}{\rho_O} \tag{11.4}$$

Note that ρ_F and ρ_O are the nominal liquid densities of the fuel and oxidizer, respectively. The symbol ΔV_t in Eq. 11.1 refers to the trapped or residual propellant volume that is not able to be used for the particular flight mission segment, and ΔV_b refers to the boiled-off propellant volume expected (for a cryogenic). As discussed earlier, ΔV_u applies to the ullage volume (portion of tank not ever occupied by liquid, nominally). ΔV_c refers to the cooling bleed-off requirements (less corrosive fuels, rather than oxidizers, more commonly used for this purpose, with the fuel bled-off from the storage tank). ΔV_{aux} allows for auxiliary usage of the given propellant, e.g., bleed-off fuel and oxidizer for use in gas generation needed for turbopump-feed operation.

Fig. 11.6 Schematic
diagram of pressure-fed
bipropellant LRE, utilizing
high-pressure helium to move
the oxidizer and fuel from
storage

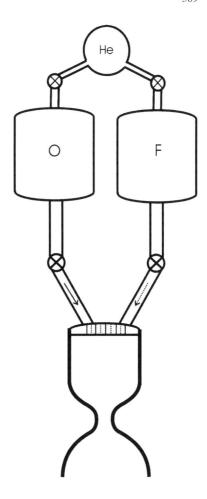

11.4 Propellant Feed System

11.4.1 Pressure-Feed

With respect to propellant feed from storage through piping and valves to the injectors, LREs fall under two general categories: (1) gas-pressure-fed, and (2) turbopump-fed. The simpler of the two approaches, a typical pressure-feed system, as illustrated in Fig. 11.6, will use an inert gas such as helium (He) (or less reactive gas such as nitrogen (N_2) in combination with a separating diaphragm or membrane) at high pressure (say up to 4,000 psi [28 MPa] at the start of a flight) to move the liquid propellant from the respective storage tank where they are at a considerably lower pressure [3]. The pressurant gas may be heated during the latter part of the flight to help prolong its usefulness. In a simpler one-step pressure-feed system, the storage tank pressure would be around 20% higher than the maximum expected combustion chamber pressure p_c, to ensure positive throughput of propellant through to the injectors (no backflow), and avoid feed instability issues related to transient operation. Thrust (combustion) chamber pressures for LREs can range from as low as around 1 MPa (for low-level thrusters) to as high as around 12 MPa for high-performance engines. Not surprisingly then, those systems with a lower p_c requirement (such as thrusters, or low-thrust upper stage motors for the final phase of satellite orbital insertion) will more likely employ a pressure-feed approach, given in part due to the lower storage pressure requirements and lighter resulting tank structural mass, in combination with the lower cost and complexity associated with pressure-feed systems.

For higher performance, some efforts are being directed towards developing multi-step pressure-feed systems, which would potentially allow for storage of propellant at pressures below p_c. With this approach, between the storage tanks and the injectors would be an intermediate mechanism for stepping up the feed-line pressure while avoiding backflow, e.g., electric pumps of varying design, or valved heating segments in the feed line for pressurizing the liquid flow. The highest attained thrust-to-weight ratio (commonly quoted in terms of F_{vac}/W_{dry}, peak thrust at vacuum conditions over dry engine weight, i.e., no propellant) using a pressure-feed approach for an LRE has to date been on the order of 40, for a second- or third-stage engine utilized for a launch vehicle application. A typical pressure-fed thruster for vehicle attitude control has a thrust-to-weight ratio of 0.6 or less, as a measure of comparison. A modern large high-performance turbopump-fed LRE may have a thrust-to-weight ratio exceeding 100.

11.4.2 Turbopump-Feed

The alternative to pressure-feed systems is the use of turbopumps [3]. As turbomachinery, turbopumps, comprised of a rotodynamic pump and a driving turbine, share a number of similarities with respect to turbosuperchargers, and by extension, gas turbine engines. A typical turbopump-feed system, as illustrated in Fig. 11.7, uses centrifugal pumps (Fig. 11.8; [4]), employing an impeller + diffuser, to pressurize the propellant in a one-way (no backflow) fashion. The aforementioned V-2 rocket employed a circular-shaped centrifugal (radial) turbopump-feed system (Fig. 11.1), powered in a gas-generator approach by steam (H_2O in vapor form) produced from the catalysis of hydrogen peroxide and calcium permanganate. Axial pumps (Fig. 11.9; see Table 11.2 for comparing the performance of the two J-2 engine pumps; [3]), employing a number of pressurizing rotor + stator stages, have also been used on occasion, as an alternative. In either case, storage tank pressures can be well below p_c (cryogenic propellants as low as 1.5–3 atm in storage), while downstream of the pumps, the line pressure will be of the order of 20% higher than p_c. The relatively low structural mass of the storage tanks provides a major weight advantage over pressure-fed alternatives,

Table 11.2 Comparison of Performance of J-2 Engine's Fuel and Oxidizer Pumps [4]

Pump characteristic	Liquid oxygen, 70.8 lbm/ft³	Liquid hydrogen, 4.4 lbm/ft³
Pump type	Centrifugal	Axial
Number of stages	1	7 + inducer
Weight flowrate (lbm/s)	460.4	83.6
Volume flowrate (gpm)	2,920	8,530
Rotational speed (rpm)	8,753	27,130
Pressure rise (psi)	1,075	1,208
Inducer headrise (ft)	(no inducer)	5,050
Pump headrise (ft)	2,185	38,000
NPSH (ft)	18	75
Impeller discharge tip speed (ft/s)	390	865
Turbopump weight (lbm)	305	369
Power (hp)	2,358	7,977

Fig. 11.7 Schematic diagram of turbopump-fed bipropellant LRE, utilizing a gas generator (GG) to drive a turbine (T) that in turn drives the pumps (P) that move and pressurize the fuel (F) and oxidizer (O) out of storage

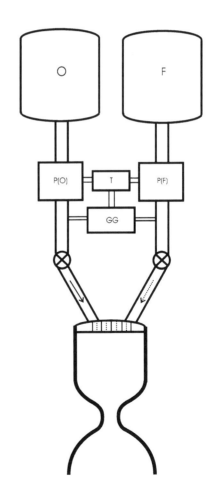

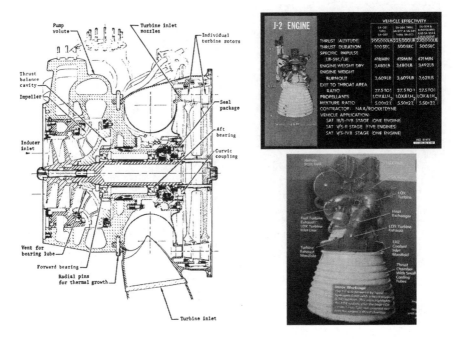

Fig. 11.8 Cutaway diagram at *left* of the centrifugal oxidizer pump used by the LH$_2$/LO$_2$ J-2 engine [4]. The North American Aviation [NAA]/Rocketdyne J-2 was used for several stages of the Apollo/Saturn V vehicle, as noted in the NASA brochure at the *upper right*

and thus is commonly the choice for higher performance flight missions. The pumps are driven typically by axial turbines, which are themselves driven by the expansion of hot gases. Different approaches for providing these driving gases are illustrated in Fig. 11.10 [4]. For example, these gases can be provided by pre-burning some fuel and oxidizer bled off from the main storage tanks (1–5% bleed off from the total propellant load is typical for a flight mission). The preburning is usually done at fuel-rich conditions, to keep the gas cooler than peak stoichiometric flame levels (e.g., oxidizer-fuel r at 1.5 rather than 8 for LO$_2$/LH$_2$) and thus keep the surrounding structure below the heat-damage threshold temperature. With cryogenic-propellant engines, this is called a staged combustion cycle approach. With engines using storable propellants, this is called a gas-generator cycle approach. A fuel preburner and oxidizer preburner in combination with the main combustion chamber (for thrust delivery) are together commonly referred to as the engine's powerhead (see example of Fig. 11.11; [5]). An alternative to preburning is the use of heated fuel, fuel that has been bled off from storage and passed through a coolant flow-through cycle in cooling various engine structures, for the turbine working fluid. This approach is called an expander cycle system. An example chart comparing the performance of the various driving approaches for a given application is provided in Fig. 11.12 [4].

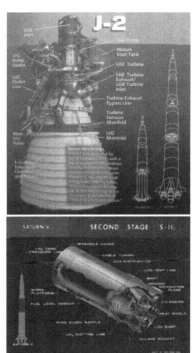

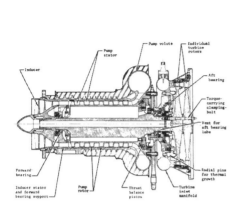

Fig. 11.9 Cutaway diagram at *right* of the axial fuel pump used by the LH₂/LO₂ J-2 engine [4]. Five Rocketdyne J-2 engines were used for the second stage of the Saturn V vehicle, as displayed in the museum diagram at the above *right* and the NASA brochure at *lower right*. A gas-generator approach was used for driving the respective pumps of the J-2 (see Fig. 11.7)

Movement of heated fuel resulting from a coolant run-through can also be fed to the main combustion chamber, in a regenerator/regenerative-cooling approach (Figs. 11.13 and 11.14; [6]). The V-2 rocket of Fig. 11.1 used a regenerative-cooling approach. Open-cycle turbopump systems dump their turbine exhaust flow overboard, while closed-cycle (topping-cycle) turbopump systems inject the turbine exhaust flow into the thrust chamber for a little better I_{sp}. Propellant flow rates in the piping (lines) are at velocities around 10 m/s, a value kept low to keep frictional pressure losses on the order of 50 kPa before reaching the injectors.

Rocket engine turbopumps do have similar design considerations with respect to other hydraulic pumps, including having a cavitation (deleterious bubble formation and collapsing) limit, where net positive suction head [NPSH] stipulates this limit. NPSH is the difference between the total (stagnation = static + dynamic; the front inducer speeds up the flow to produce a dynamic suction effect with fluid static pressure dropping, while downstream, the impeller acts to pressurize the flow) fluid pressure and that due to the local vapor pressure (at the point of cavitation, referred to as the *required* NPSH), as measured at the pump intake [suction side], with pressure Δp in terms of head Δh as related to the liquid

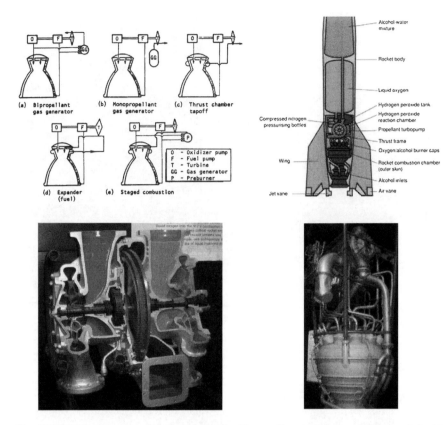

Fig. 11.10 Different approaches for driving liquid propellant turbopumps [4] at top left: (a) bipropellant gas generator, (b) monopropellant gas generator, (c) thrust chamber tapoff, (d) expander (fuel), and (e) staged combustion. As shown at lower left, the V-2 missile used method (**b**) to drive its respective centrifugal fuel and oxidizer turbopumps (*lower* photos). The SSME of Fig. 11.4 uses method (**e**). *Right upper* diagram courtesy of V2Rocket.Com

propellant, i.e., $\Delta p = \rho_{\text{Liq}} \cdot g_o \cdot \Delta h$; [7]). One needs to keep the *available* head difference value *above* the *required* NPSH value to avoid cavitation, e.g., the higher the liquid's pressure and the lower the vapor pressure, the better. An example performance chart for a turbopump's rotational speed relative to the delivered pressurization level is given in Fig. 11.15 [4].

Pump specific speed N_s is an important performance parameter that arises from dimensional analysis:

$$N_s = \frac{N \cdot Q^{1/2}}{H^{3/4}} \tag{11.5}$$

where N is the pump shaft rotational speed, Q is the volumetric flow rate, and H is the headrise (pressurization level reached, in terms of head). In designing pumps and charting their performance, it is useful to work in terms of Q and H, since

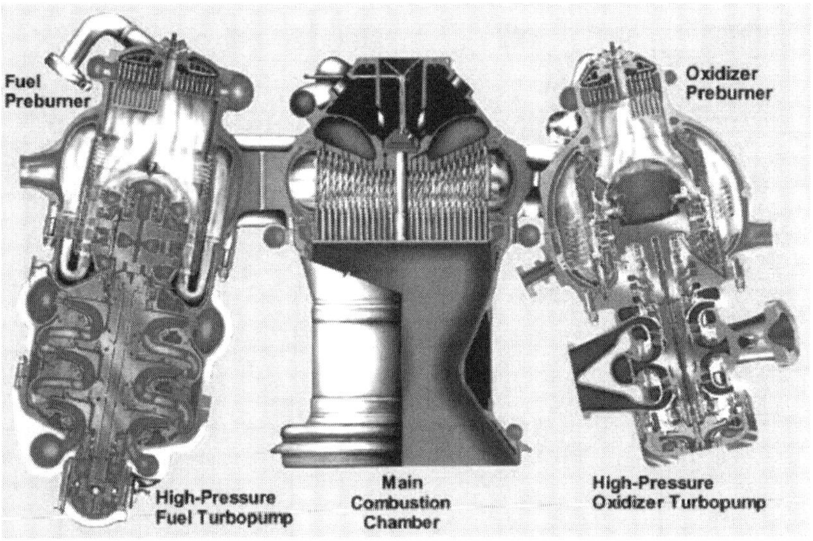

Fuel
Preburner

Oxidizer
Preburner

High-Pressure
Fuel Turbopump

Main
Combustion
Chamber

High-Pressure
Oxidizer Turbopump

© NATO RTO

Fig. 11.11 Powerhead for SSME (preburners, turbopumps and thrust chamber assembly) built by Rocketdyne. Diagram from Haidn [5]

Fig. 11.12 Example performance graph for various turbopump driving cycle approaches, for an LH$_2$/ LO$_2$ engine running at an oxidizer-to-fuel mixture ratio $r = 6:1$, showing pump discharge pressure relative to the operating main combustion chamber pressure [4]

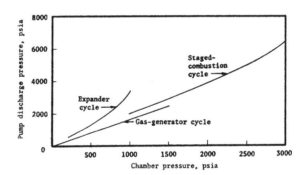

these parameters are typically independent of the approximately incompressible liquid propellant density, and thus the charts, once established, can be used for a wide range of propellants. As per Fig. 11.15, there may be a value (in this case, around 2,000 rpm·gpm$^{1/2}$/ft$^{3/4}$), or relatively narrow value range, for which N_s best fits the system requirements, e.g., in terms of peak efficiency. The pump-developed head (pressurization, H) is the difference in pressure head in going from the suction (incoming) head to the discharge (outgoing) head. The discharge head (see discharge pressure curves of Fig. 11.12) must be high enough to overcome the hydraulic resistance faced by the fluid as it moves from the pump outlet through

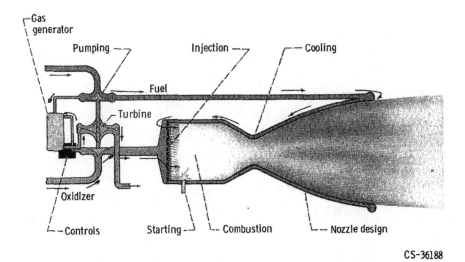

Fig. 11.13 Schematic diagram of a regeneratively cooled liquid rocket engine (method (**a**) of Fig. 11.10; [6])

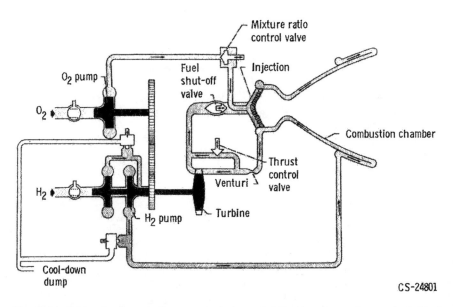

Fig. 11.14 Schematic diagram of a regeneratively cooled liquid rocket engine (method (**d**) of Fig. 11.10; [6])

the required piping before it arrives at the combustion chamber injector plate at a design pressure significantly higher than the operating combustion chamber pressure p_c (thus ensuring a proper forward spray velocity through the injector

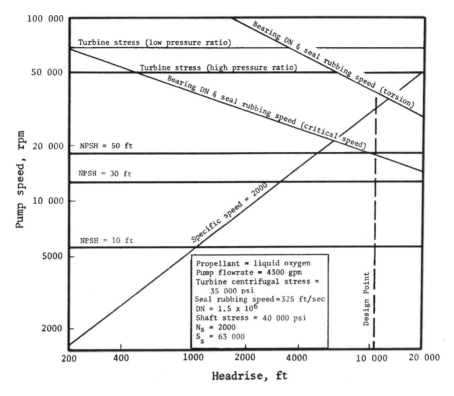

Fig. 11.15 Pump rotational speed limits for a 4,300 gallons/minute (16,280 liters/min) LO$_2$ pump [4]. A headrise H of 10,000 ft corresponds here to 34.09 MPa (4,950 psia). DN is the product of bearing bore diameter (mm) and shaft rotation rate (rpm, revs./min). N_s is the pump-specific speed (rpm gpm$^{1/2}$/ft$^{3/4}$). S_s is the suction-specific speed (rpm·gpm$^{1/2}$/ft$^{3/4}$)

ports). For example, in the case of a regeneratively cooled engine employing a gas-generator cycle, the discharge pressure at the fuel pump outlet must equal the sum of the fuel line losses, the pressure drop through the cooling jacket (primarily in the nozzle structure), the required pressure drop through the injector port array and the nominal operating combustion chamber pressure.

Suction speed S_s is another significant pump design parameter that arises from dimensional analysis, defined by:

$$S_s = \frac{N \cdot Q^{1/2}}{(\text{NPSH})^{3/4}} \tag{11.6}$$

In comparing to N_s, we see the correlation of S_s is now in terms of NPSH, rather than headrise H. The influence of S_s is seen in designing the pump's inducer section, to avoid cavitation as one lowers fluid pressure during the initial suction process, before pressurization.

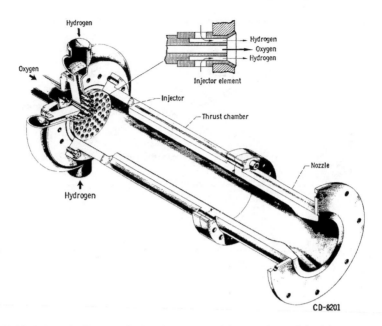

Fig. 11.16 Schematic diagram of subscale heavy-walled LH$_2$/LO$_2$ LRE for laboratory tests [8].
A concentric-tube (*coaxial*) hollow-post injector is shown in the *inset*

11.5 Injectors

As illustrated in Figs. 11.16 [8] and 11.17 [9], the injectors atomize the incoming
propellant (break liquid into small droplets as part of the spraying process) and
distribute/mix the fuel and oxidizer in a suitable pattern downstream of the injector
plate at the head end of the thrust chamber. The substantial number of individual
injector exit ports on an injector plate may give the appearance of a showerhead (a
larger number of ports tends to reduce the appearance of combustion instability;
see example of Fig. 11.18). There are a number of alternative injector designs
(Fig. 11.19; [2, 6, 10]). Impinging spray sheet injectors are commonly associated
with storable propellants. Coaxial hollow-post injectors are often used with
cryogenic propellants. Initial ignition of the mixture is accomplished by various
options, including a spark plug or flame tube approach. For one-start-only applications, a pyrotechnic cartridge might be employed. An occasional ignition-related
problem, a hard start, results from unburned fuel accumulating in an unusually
prolonged starting process, and then subsequently being ignited in conjunction
with significant transient pressure-wave motion. This problem was noted earlier in
Chap. 6 for jet engine afterburners.

The hydraulic injection velocity v_{inj} (note: this quantity includes the upstream
line velocity component) is a useful parameter for various combustor design and
operational considerations. Via Bernoulli's energy conservation equation for an
incompressible fluid:

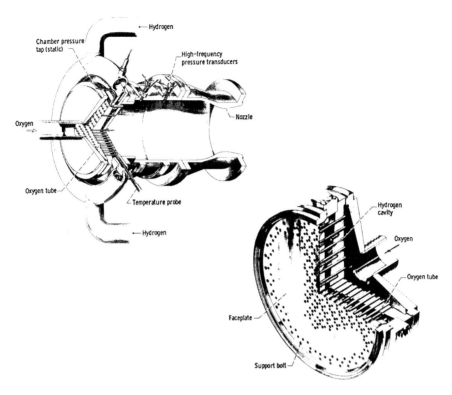

Fig. 11.17 Schematic diagram of subscale heavy-walled LH$_2$/LO$_2$ LRE for laboratory tests [9]. Note the various sensors for firing data acquisition and stability rating

$$v_{\text{inj}} = C_{d,\text{inj}} \sqrt{\frac{2\Delta p_{\text{inj}}}{\rho\left(1 - \left[\frac{A_{\text{inj}}}{A_{\text{line}}}\right]^2\right)}} \qquad (11.7)$$

Here, $C_{d,\text{inj}}$ is the injector's spray orifice discharge coefficient, ρ is the propellant fluid density and Δp_{inj} is the pressure drop through the injector. In Eq. 11.7, note the influence as well of the ratio of the upstream flow cross-sectional area A_{line} and the downstream effective injector orifice area A_{inj}. Values for v_{inj} should be at least 20 m/s for positive throughput (at least double the upstream pipe flow velocity), and preventing feed-related combustion instability symptoms. The associated mass flow rate through the injector(s) may be determined via

$$\dot{m} = Q \cdot \rho \qquad (11.8)$$

where overall volumetric flow rate Q can be found by mass conservation:

$$Q = A_{\text{inj}} v_{\text{inj}} = A_{\text{inj}} C_{d,\text{inj}} \sqrt{\frac{2\Delta p_{\text{inj}}}{\rho\left(1 - \left[\frac{A_{\text{inj}}}{A_{\text{line}}}\right]^2\right)}} \qquad (11.9)$$

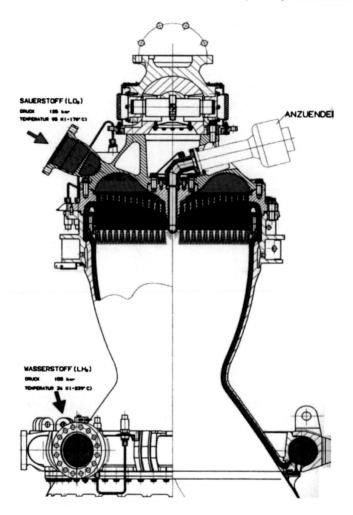

© NATO RTO

Fig. 11.18 Cutaway diagram of the Ariane 5's first-stage Vulcain 2 gas-generator-cycle cryogenic engine's thrust chamber, illustrating the path taken by the liquid oxygen (*blue*) and the liquid hydrogen (*red*), both ultimately passing simultaneously through the injector plate into the combustion chamber. The *center* of the injector plate is occupied by the solid-propellant pyrotechnic igniter (igniter flame delivery tube shown in *yellow*). Diagram from Haidn [5]

An undesirable phenomenon called blanching results from the presence of oxidizer-rich gas mixed with cryogenic liquid droplets (a result of poor mixing/ atomization) coming in contact with the injector face plate or surrounding combustion chamber walls in the vicinity of the injectors, chemically attacking the surface structure in a reduction/oxidation process. Lowering the gas temperature in

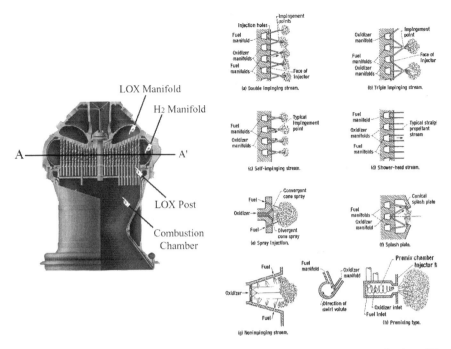

Fig. 11.19 Illustration of various injector designs for liquid rocket engine applications [6]. Schematic at *left* shows typical setup for LO$_2$/LH$_2$ injection [10]

the regions at risk mitigates this problem (via use of a coolant film, or using a richer fuel content [lower oxidizer-fuel mixture ratio r] in the injectors at the periphery of the injector plate; [5]).

11.6 Thrust Chamber

In practice, when one refers to the thrust chamber in LRE terminology, this might commonly be inferred to include both the combustion chamber and the nozzle, given their more contiguous structural pairing, say in relative comparison to SRM chamber-nozzle attachments. In this section, we will focus more on the combustor side of the thrust chamber. In the combustion chamber, one would often strive for a high flame temperature T_F, a value often higher than the allowed surrounding structural temperature limit. As a result, in LREs, cooling of the combustor and nozzle walls with circulating coolant is a common requirement that must be met. Fuel, given its less corrosive nature relative to oxidizers, is typically bled off for use as the coolant, as done for the V-2 vehicle (see Fig. 11.20).

Combustion chambers for LREs can vary in shape, from spherical (the most efficient shape for structural loading) to near-spherical (barrel-shaped with bowed

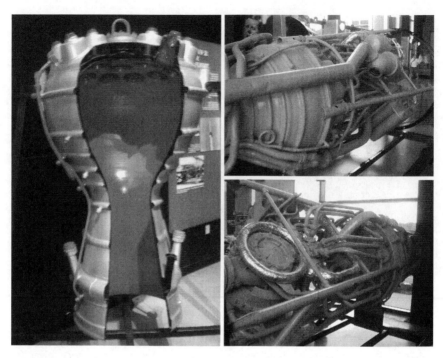

Fig. 11.20 Cutaway view at *left* of the V-2's combustion chamber and nozzle as displayed at the National Museum of the USAF. *Right* photos show alternative views of the outside of the combustion chamber and the "plumbing" above it (U.S. Space & Rocket Center, Alabama)

out wall) to cylindrical (straight wall). The combustion chamber volume $\mathcal{V}_c$ and the corresponding characteristic length L^* defined by [11]:

$$L^* = \frac{\mathcal{V}_c}{A_t} \tag{11.10}$$

must be designed to allow for adequate mixing and vaporization of fuel and oxidizer, and the resulting combustion downstream. The typical range of L^* values for higher performance, larger bipropellant engines is from 0.5 to 3 meters. As reported in [11], the high-performance combination of LO_2/LH_2 with GH_2 injection has a suggested L^* range of 0.55–0.65, while the lower-performance combination of LO_2/RP-1 has a recommended L^* range from 1.0 to 1.25. For lower performance smaller monopropellant engines, the value range is a bit higher than that seen for higher performance bipropellant engines. Referring to Fig. 11.16 for a small laboratory-based bipropellant LRE, the length-to-diameter (L/D) ratio of the chamber is on the higher side (relative to larger flight vehicle LREs), which might allow for more complete combustion (longer reaction time). On the other hand, the LRE of Fig. 11.17 reveals a substantially lower L/D, which might suggest it is a bigger engine relative to that of Fig. 11.16.

The combustor residence or stay time t_c is an additional parameter that is useful for establishing the combustor geometry:

$$t_c = \frac{V_c \bar{\rho}_c}{\dot{m}} \tag{11.11}$$

where $\bar{\rho}_c$ is the mean chamber gas density. The residence time for a molecule passing through the combustor is in the order of 0.001–0.04 s. Referring back to L^*, one can see the following correlation:

$$L^* = \frac{\dot{m} t_c}{\bar{\rho}_c A_t} \tag{11.12}$$

If one sets a mean flow Mach number between 0.2 and 0.6 in the combustor (suitable for flame development and stability), with a relatively constant gas sound speed for the given chamber temperature, then mean gas velocity

$$u = \frac{\dot{m}}{\bar{\rho}_c A_t (A_c/A_t)} \tag{11.13}$$

is relatively constant as well. The chamber-throat contraction area ratio will be relatively set, depending on the design chamber mean flow Mach number, as discussed below. Observation of Eqs. 11.12 and 11.13 then suggests that

$$L^* \approx f(t_c) \tag{11.14}$$

i.e., a direct correlation between the needed combustion time and the resulting thrust chamber geometry.

As one might expect, thrust performance initially climbs with increasing values for L^*, and at some point, levels off and starts to drop with further increases in L^*. In part, frictional flow losses and increased combustor structural mass will contribute to the dropping off in effective performance. The contraction ratio A_c/A_t is typically in the order of 1.3–3 in value for higher performance larger engines, with a higher range of values for lower performance engines. In SRMs, we saw typical values for the comparable A_p/A_t range from 6 to 10, so quite a bit higher (this can be partly explained by the common criterion for low erosive burning over most of the firing of the SRM). Recall the area-Mach number relation for a choked nozzle:

$$\frac{A_c}{A_t} = \frac{1}{Ma_c} \left[\frac{2 + (\gamma - 1)Ma_c^2}{\gamma + 1} \right]^{\frac{\gamma+1}{2(\gamma-1)}} \tag{11.15}$$

where Ma_c is the mean chamber flow Mach number (commonly 0.2–0.6, as noted above), and γ a mean value for the gas ratio of specific heats. A value of 1.2 for γ and a value of 0.4 for Ma_c gives a value for A_c/A_t of 1.6 from Eq. 11.15, while a value of 0.2 for Ma_c gives an A_c/A_t of 3; these are reasonable results.

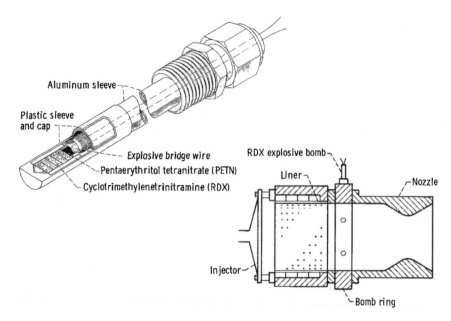

Fig. 11.21 Schematic diagram of bomb used for combustion stability testing [9]

One can note the following correlation between the characteristic length L^* and the actual chamber length L_c (assuming the combustion chamber is close to being cylindrical shaped):

$$L_c \approx \Psi_c/A_c = L^* \cdot \frac{A_t}{A_c} \tag{11.16}$$

If A_c/A_t and L^* are decided upon early on, one can approximately establish L_c via Eq. 11.16. Recall that the nozzle throat area will be determined by the desired thrust delivery level (higher thrust, higher A_t), for a nominal design chamber pressure. Once A_t is established, the approximate A_c will follow. The corresponding chamber length-to-diameter ratio can in turn be found via

$$L_c/D_c \approx L^* \cdot \frac{A_t}{A_c} \cdot \left(\frac{\pi}{4A_c}\right)^{0.5} = L^* \cdot \left(\frac{A_t}{A_c}\right)^{1.5} \cdot A_t^{-0.5} \cdot \left(\frac{\pi}{4}\right)^{0.5} \tag{11.17}$$

This result suggests that chamber L/D will tend to decrease as the engine gets bigger, everything else being equal.

The propellant mass flow $\dot{m}$, with the appropriate mixture ratio for a desired chamber temperature T_F and exhaust gas molecular mass $\mathcal{M}$, will establish the resulting combustion chamber pressure p_c, as covered in Chap. 9. In turn, knowing p_c, one can estimate the resulting thrust delivery produced by the nozzle exhaust flow. Please refer to the solution for Example Problem 11.1 at the end of this chapter, for a completed sample liquid rocket engine performance analysis. For students learning this material as part of a course, I would encourage them to attempt the given example problem first, and then check the solution to see if any mistakes were made.

11.7 Combustion Instability in LREs

As with SRMs, LREs can be susceptible to symptoms of both axial and transverse combustion instability. Rather than bits of material passing through the nozzle, initiation of c.i. in LREs is more commonly as a result of an uncombusted accumulation of reactants suddenly igniting and generating an initial compression wave in the combustion chamber. Deliberate pyrotechnic pulses from charges (so-called "bombs"; Fig. 11.21 [9]) placed at applicable locations in the chamber for c.i. susceptibility tests (called "rating" the LRE's operational roughness or stability) is common practice. Instability symptoms at lower frequencies is commonly attributed to natural resonances in the propellant feed system behind and at the injectors, while symptoms at higher frequencies involve substantial axial and transverse pressure waves. Any fluctuation in the mean chamber pressure p_c above around $\pm 5\%$ is considered "rough" or c.i.-prone. One can compare this threshold to that for SRMs, which is substantially lower at around $\pm 0.5\%$. Experimentally, tangential pressure wave peak-to-trough magnitudes in LREs have been observed to approach 100% the predisturbed p_c. Unlike SRMs, the base pressure does not usually appear to shift much during this strong wave activity (no appreciable dc shift). Due in part to the typically low A_c/A_t for LREs, fundamental axial c.i. is relatively rare in the larger engines (pressure-wave reflection strength is lower at lower contraction ratios; SRMs with their higher A_p/A_t values have a higher incidence of fundamental axial c.i. behavior). At low frequencies, one might see the appearance of a fluid–structure interaction phenomenon referred to as *pogo* instability [2], whereby the rocket vehicle's structure (or lengthwise propellant feed lines) may vibrate axially due to some coupling resonance condition, producing low-level axial pressure-wave motion in the feed lines and combustion chamber. *Chugging* or L^* instability [2] may be somewhat related to pogo instability in some cases (feed-line resonance issues), but is commonly associated with a rough start-up (ignition transient) process, where combustion is cyclically falling above and below the needed threshold for smooth operation, at pressures temporarily below the nominal operational level. Solid rocket motors may also experience a comparable phenomenon during start-up, called *chuffing*.

Tangential pressure waves in the vicinity of the injector plate is a more typical problem area. The placement of baffles in this location is a common although not entirely welcome solution in deterring pressure-wave growth (Fig. 11.22; [9]). One could surmise that combustion response to changes in local pressure and gas velocity would be most intense near the injectors in the primary flame zone, thus the sustaining of tangential pressure waves in this region would clearly be enhanced, rather than further downstream in the combustor. One also sees the use

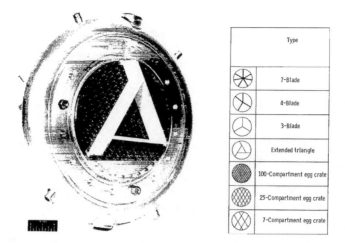

	Type
	7-Blade
	4-Blade
	3-Blade
	Extended triangle
	100-Compartment egg crate
	25-Compartment egg crate
	7-Compartment egg crate

Fig. 11.22 Example of an extended-triangle baffle in position against the injector face plate (*left*), and a table of different baffle patterns (*right*) [9]

of Helmholtz resonators, and quarter-wave tubes/resonators, small side chambers that act to defeat targeted pressure-wave frequencies. The requirement to continually modulate the feed system, the mixture ratio and the resulting combustion process downstream would contribute to a less smooth burning operation, versus say an SRM. Hybrid rocket engines, given their liquid feed system issues, also run rougher than SRMs.

11.8 Note on Thrust Vector Control

Vehicle flight trajectory control in space (i.e., beyond the usable upper atmosphere for a given flight Mach number), or in the lower atmosphere (when vehicle speeds are too low to be aerodynamically effective), dictates the use of thrust vectoring. This approach can be applied for any member of the chemical rocket family (solid, liquid and hybrid). This maneuvering thrust component is generated by either the principal exhaust nozzle, or by smaller, secondary exhaust nozzles. In order to provide directional control through a TVC approach, one common design choice is to allow the principal exhaust nozzle to be swiveled from the central longitudinal axis of the rocket vehicle, possibly as much as 10° or so. The swivel action is commonly accomplished through a combination of gimbal bearings, swivel joints, control actuators and flexible hosing. While being a traditional solution, gimballed nozzles, or some comparable variant (ball & socket nozzle; flexible bearing or joint nozzle, as shown in Fig. 11.23 [12, 13]), can be quite expensive, and for lower cost applications, alternatives are often sought.

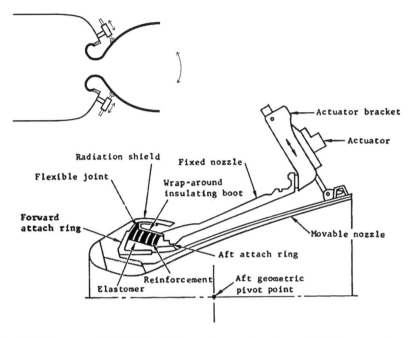

Fig. 11.23 Diagram of flexible joint TVC nozzle apparatus. Lower right diagram courtesy of NASA [12, 13]

An alternative to swiveling the nozzle is the use of exhaust jet deflectors such as jet tabs or jet vanes located at or near the nozzle exit plane. Jet vanes have the advantage of both directional and roll (body torque) control capability (Fig. 11.24). Jet deflectors have the disadvantage of aerodynamic drag losses that penalize the exhaust nozzle's principal thrust delivery, and due to the hot super-sonic gas or gas-particle flow, have a shorter service life. The jetavator approach is a little less intrusive (see Fig. 11.25), with the surrounding exit collar moving in and out of the exhaust flow (at the nozzle exit periphery), as needed [2]. An alternative to mechanical jet deflection is the use of gas or liquid injection [12–14] at appropriate points around the inner surface of the nozzle expansion section (see Fig. 11.26). If one is already carrying liquid fuel and/or oxidizer in storage, one might be able to use one of these liquids as a source for the TVC application. If available, liquid fuels are commonly preferable, since they tend to be more benign in their chemical interaction with the nozzle wall structure (on the issue of injectant compatibility, oxidizers have a reputation for contributing to substantial material erosion; the Titan IV launch vehicle successfully used liquid nitrogen tetroxide, an oxidizer, for TVC purposes). This secondary mass injection approach can also enable both directional and roll control, but its effective range of vectoring is significantly less than that attainable with gimballed nozzles. One can also note that the entry of this additional mass into the exhaust flow generally leads to lower thrust losses, versus the mechanical jet deflection approach.

Fig. 11.24 Diagram of jet
vane TVC apparatus. The V-2
(Fig. 11.10) used this
approach

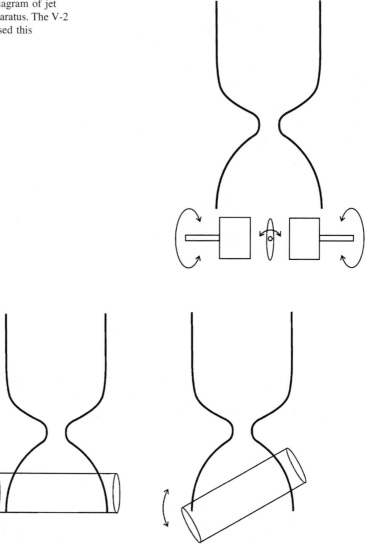

Fig. 11.25 Schematic diagram of jetevator TVC apparatus

If there are two or more principal exhaust nozzles in a clustered engine array, throttling of one or more of the corresponding engines can enable the implementation of vehicle directional control input, versus swiveling said nozzles. Additional secondary (auxiliary) thrusters (also called reaction control system [RCS] thrusters, or vernier thrusters) may nevertheless be required to complement such an approach, in order to adequately meet all attitude control requirements.

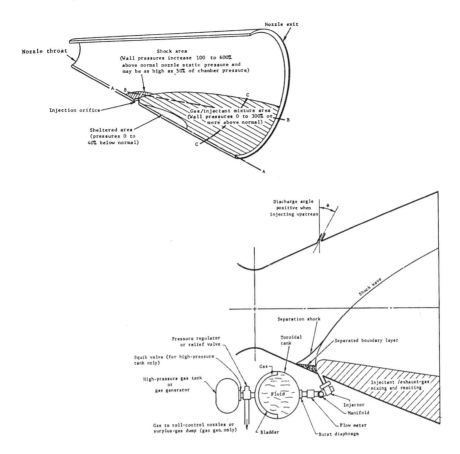

Fig. 11.26 Schematic diagrams illustrating the effects of liquid injection within the nozzle expansion, as a means for TVC. *Upper left* diagram courtesy of NASA [12, 13]. *Lower* diagram [14] reprinted with permission of the American Institute of Aeronautics and Astronautics

11.9 Example Problems

11.1. A liquid-propellant (LH$_2$/LO$_2$) rocket engine has the following characteristics, with a design ambient flight static pressure at 20 kPa:

nozzle throat dia., $d_t = 11.2$ cm; nozzle exit diameter, $d_e = 56$ cm; thrust chamber/throat area ratio, $A_c/A_t = 3.2$; mixture ratio at 4.2:1, running fuel-rich;
design chamber pressure, $p_c = 8.2$ MPa
combustion products, $\gamma = 1.28$; molecular weight, $\mathcal{M} = 12$ amu;
$T_F = 2,900$ K

Estimate the fuel and oxidizer feed rates, thrust and specific impulse at the design point.

11.2 You have been asked to design an appropriate intermediate-stage liquid-propellant rocket engine to meet an interplanetary space flight application. The allowed volume for overall propellant storage (oxidizer and fuel) for this engine is $6\,m^3$. Two oxidizer + fuel combinations are being considered, with characteristics shown below. LO_2/LH_2:

combustion products, $\gamma = 1.2$; gas molecular weight, $\mathcal{M} = 13.5$ amu; $T_F = 3,550$ K
mixture ratio $r = 6:1$; in storage: oxid. liquid density $= 1,140\,kg/m^3$, and fuel liquid density $= 70\,kg/m^3$

LO_2/kerosene:

combustion products, $\gamma = 1.22$; gas molecular weight, $M = 22$ amu; $T_F = 3,600$ K
mixture ratio $r = 2.5:1$; in storage: oxid. liquid density $= 1,140\,kg/m^3$, and fuel liquid density $= 806\,kg/m^3$

Geometry of the vehicle stipulates a nozzle exit diameter of 1.225 m, and thrust performance requirement stipulates that the nozzle throat diameter be 0.19 m, in combination with a chamber pressure of 13.78 MPa. One would add an extra 10% volume to the principal usable (ideal) volume calculated for a given tank size, in allowing for secondary considerations on tank storage. For calculations, assume constant thrust delivery at the above chamber pressure.

1. For the second combination indicated above (LOX/kerosene), what percentage of the overall allowed storage space is for the LOX tank, and what percentage is for the kerosene tank? While not necessary for your calculations, you may wish to note that for the first case, 26.9% is for LOX volume, and 73.1% is for hydrogen volume.
2. What is the specific impulse delivered by the LOX/kerosene engine (assume $p_\infty = 0$ atm)? While not necessary for your calculations, you may wish to note that for the first case, specific impulse for the LOX/hydrogen engine is around 439 s.
3. What is the total impulse (N–s) for the LOX/kerosene engine? Given that the first engine case provides a total impulse of around 8.4×10^6 N–s, which of the two engines would you recommend, if a principal criteria is maximizing the flight vehicle's delta-V (ΔV) for the mission segment?

11.3 A liquid-propellant (LH_2/LO_2) rocket engine has the following characteristics, with a design ambient flight static pressure at 2 kPa:

nozzle throat dia., $d_t = 11.2$ cm; nozzle exit diameter, $d_e = 106$ cm; thrust chamber/throat area ratio, $A_c/A_t = 3.2$; mixture ratio at 4.2:1, running fuel-rich;

design chamber pressure, $p_c = 8.2$ MPa

combustion products, $\gamma = 1.28$; molecular weight, $\mathcal{M} = 12$ amu;

$T_F = 2,900$ K

Estimate the thrust and specific impulse at the design point. Compare results to Problem 11.1 case.

11.4 Establish the oxidizer liquid spray injection velocity and volumetric flow rate for an LRE injector, given the following parameter values: $\rho_{\text{oxid}} = 1,140$ kg/m^3, $d_{\text{inj}} = 5$ mm, upstream $d_{\text{line}} = 8$ mm, $\Delta p_{\text{inj}} = 0.34$ MPa, $C_{d,\text{inj}} = 0.9$

11.5 Establish the approximate chamber length and diameter needed for the engine of Problem 11.1. Assume that the design L^* is 0.8, and the design A_c/A_t of the chamber is 2:1.

11.10 Solutions to Example Problem

11.1. Looking at a liquid rocket engine, at a flight altitude giving $p_\infty = 20$ kPa.

Nozzle throat area, $A_t = \frac{\pi d_t^2}{4} = \pi(0.112)^2/4 = 9.85 \times 10^{-3}$ m^2; $R = \frac{\mathscr{R}}{\mathscr{M}} = 8315/12 = 693$ J/kg·K

Nozzle exit area, $A_e = \frac{\pi d_e^2}{4} = \pi (0.56)^2/4 = 0.2463$ m^2

$$c^* = \left[\frac{\gamma}{RT_f} \left(\frac{2}{\gamma+1} \right)^{\frac{\gamma+1}{\gamma-1}} \right]^{-1/2} = [1.28/[693(2900)] \cdot (2/2.28)^{2.28/0.28}]^{-0.5} = 2136 \text{ m/s}$$

$\dot{m} = \frac{p_c A_t}{c^*} = 8.2 \times 10^6 \ (9.85 \times 10^{-3})/2136 = 37.82$ kg/s

$r = \frac{\dot{m}_O}{\dot{m}_F} = 4.2$ (given), $\dot{m}_F = \frac{\dot{m}}{1+r} = 37.82/(1+4.2) = 7.273$ kg/s, $\dot{m}_O = \dot{m} - \dot{m}_F = r\dot{m}_F = 30.547$ kg/s.

$$C_{F,v} = \left[\frac{2\gamma^2}{\gamma-1} \left(\frac{2}{\gamma+1} \right)^{\frac{\gamma+1}{\gamma-1}} \right]^{1/2} = [2(1.28^2)/(0.28) \cdot (2/2.28)^{2.28/0.28}]^{0.5} = 2.007$$

Area-Mach No. relation, where Ma_t goes to unity:

$$\frac{A_t}{A_e} = \frac{Ma_e}{Ma_t} \left[\frac{2 + (\gamma-1)Ma_t^2}{2 + (\gamma-1)Ma_e^2} \right]^{\frac{\gamma+1}{2(\gamma-1)}} = Ma_e \left[\frac{2.28}{2 + 0.28Ma_e^2} \right]^{4.071} = 0.04$$

Via iteration, find supersonic solution, $Ma_e = 4.3$. For nozzle exit pressure:

$$p_e = p_c \left[1 + \frac{\gamma-1}{2} Ma_e^2 \right]^{\frac{-\gamma}{\gamma-1}} = 0.0238 \text{ MPa.}$$

$$F = C_{F,v} \left[1 - \left(\frac{p_e}{p_c} \right)^{\frac{\gamma-1}{\gamma}} \right]^{1/2} A_t p_c + (p_e - p_\infty)A_e$$

$= 2.007[1-(0.0238/8.2)^{0.219}]^{0.5}9.85 \times 10^{-3}(8.2 \times 10^6) + (23800-20000)0.2463$

$= 137720 + 940 = 138660$ N or 138.7 kN

$$I_{sp} = \frac{F}{\dot{m}_e g_o} = \frac{138660}{37.82(9.81)} = 374 \text{ s}$$

11.2. Compare two oxidizer/fuel combinations for an LRE application.

(a) LOX/kerosene:

$$r = \frac{m_O}{m_F} = 2.5, \; \Psi_{max} = 1.1\Psi_{u,O} + 1.1V_{u,F} = \frac{1.1 r m_F}{\rho_O} + \frac{1.1 m_F}{\rho_F} = 6 \text{ m}^3 \text{ of tank}$$

storage

$$m_F = \frac{\Psi_{max}}{\left(\frac{1.1r}{\rho_O} + \frac{1.1}{\rho_F}\right)} = \frac{6}{\left(\frac{1.1(2.5)}{1140} + \frac{1.1}{806}\right)} = 1588.5 \text{ kg}, \; m_O = r \cdot m_F = 2.5(1588.5)$$

$$= 3971.3 \text{ kg}$$

$$\Psi_O = 1.1\Psi_{u,O} = \frac{1.1 r m_F}{\rho_O} = 3.83 \text{ m}^3 \text{ or } 63.8\% \text{ of storage}, \; \Psi_F = 1.1\Psi_{u,F} = \frac{1.1 m_F}{\rho_F} = 2.17 \text{ m}^3 \text{ or } 36.2\%$$

(b) $R = \dfrac{\mathcal{R}}{\mathcal{M}} = \dfrac{8314}{22} = 377.9 \text{ J/(kg K)}, \; A_t = \dfrac{\pi d_t^2}{4} = \pi(0.19)^2/4 = 0.0284 \text{ m}^2$

$$A_e = \frac{\pi d_e^2}{4} = \pi(1.225)^2/4 = 1.179 \text{ m}^2$$

$$c^* = \left[\frac{\gamma}{RT_f}\left(\frac{2}{\gamma+1}\right)^{\frac{\gamma+1}{\gamma-1}}\right]^{-1/2} = \left[\frac{1.22}{377.9(3600)}\left(\frac{2}{2.22}\right)^{\frac{2.22}{0.22}}\right]^{-0.5} = 1{,}788 \text{ m/s}$$

$$\dot{m} = \frac{p_c A_t}{c^*} = 13.78 \times 10^6 \, (0.0284)/1788 = 218.9 \text{ kg/s}$$

$$\dot{m}_F = \frac{\dot{m}}{1+r} = 62.54 \text{ kg/s}$$

$$t_B = \frac{\rho_F \Psi_{u,F}}{\dot{m}_F N_{eng}} = \frac{806(2.17/1.1)}{62.54(1)} = 25.43 \text{ s}$$

$$C_{F,v} = \left[\frac{2\gamma^2}{\gamma-1}\left(\frac{2}{\gamma+1}\right)^{\frac{\gamma+1}{\gamma-1}}\right]^{1/2} = 2.173$$

$$\frac{A_t}{A_e} = \frac{Ma_e}{Ma_t}\left[\frac{2+(\gamma-1)Ma_t^2}{2+(\gamma-1)Ma_e^2}\right]^{\frac{\gamma+1}{2(\gamma-1)}} = Ma_e\left[\frac{2.22}{2+0.22Ma_e^2}\right]^{5.045} = 0.024$$

Via iteration, find supersonic solution, $Ma_e = 4.39$. For nozzle exit pressure:

$$p_e = p_c\left[1 + \frac{\gamma-1}{2}Ma_e^2\right]^{\frac{-\gamma}{\gamma-1}} = 0.0251 \text{ MPa}.$$

$$F = C_{F,v}\left[1 - \left(\frac{p_e}{p_c}\right)^{\frac{\gamma-1}{\gamma}}\right]^{1/2} A_t p_c + (p_e - p_\infty)A_e$$

$$= 2.173[1 - (0.0251/13.78)^{0.1803}]^{0.5}\, 0.0284(13.78 \times 10^6) + (25100 - 0)1.179$$

$$= 700920 + 29593 = 730513 \text{ N or } 730.5 \text{ kN}$$

$$I_{sp} = \frac{F}{\dot{m}_e g_o} = \frac{730513}{218.9(9.81)} = 340.2 \text{ s}, \text{ substantially lower versus the ref. } O_2/H_2$$

case (439 s)

(c) $I_{tot} = \bar{F}t_B = 730513(25.43) = 18.58 \times 10^6$ N·s, or approximately double the reference oxygen/hydrogen case (8.4×10^6 N·s), due to the higher fuel density here.

With regard to which engine to pick if mission segment Δv is the criterion, the answer is not as clear as one might think (superior I_{tot} of the LOX/kerosene case). The answer depends on the remaining vehicle mass after the propellant (fuel + oxid.) is consumed. From the ideal rocket equation:

$$\Delta v = -g_o I_{sp} \ln\left(1 - \frac{\Delta m_p}{m_o}\right) \approx -g_o I_{sp} \ln\left(1 - \frac{\Delta m_p}{m_{PL} + 1.2\Delta m_p}\right)$$

1. Payload mass of 250 kg: First system (O_2/H_2) provides higher Δv (6,022 vs. 5,423 m/s)
2. Payload mass of 1,250 kg: Second system (O_2/kero.) provides higher Δv (4,036 vs. 3,376 m/s)

11.3. A liquid-propellant (LH_2/LO_2) rocket engine has the following characteristics, with a design ambient flight static pressure at 2 kPa:

nozzle throat dia., $d_t = 11.2$ cm; nozzle exit diameter, $d_e = 106$ cm; thrust chamber/throat area ratio, $A_c/A_t = 3.2$; mixture ratio at 4.2:1, running fuel-rich;
design chamber pressure, $p_c = 8.2$ MPa
combustion products, $\gamma = 1.28$; molecular weight, $\mathcal{M} = 12$ amu; $T_F = 2,900$ K

Estimate the thrust and specific impulse at the design point:

Nozzle throat area, $A_t = \frac{\pi d_t^2}{4} = \pi(0.112)^2/4 = 9.85 \times 10^{-3}$ m^2;
$\mathcal{R} = \frac{\mathcal{R}}{\mathcal{M}} = 8315/12 = 693$ J/(kg K)

Nozzle exit area, $A_e = \frac{\pi d_e^2}{4} = \pi (1.06)^2/4 = 0.8824$ m^2

$c* = \left[\frac{\gamma}{RT_f} \left(\frac{2}{\gamma+1}\right)^{\frac{\gamma+1}{\gamma-1}}\right]^{-1/2} = [1.28/[693(2900)] \cdot (2/2.28)^{2.28/0.28}]^{-0.5} = 2136$ m/s

$\dot{m} = \frac{p_c A_t}{c*} = 8.2 \times 10^6 (9.85 \times 10^{-3})/2136 = 37.82$ kg/s

$C_{F,v} = \left[\frac{2\gamma^2}{\gamma-1} \left(\frac{2}{\gamma+1}\right)^{\frac{\gamma+1}{\gamma-1}}\right]^{1/2} = [2(1.28^2)/(0.28) \cdot (2/2.28)^{2.28/0.28}]^{0.5} = 2.007$

Area-Mach No. relation, where Ma_t goes to unity:

$$\frac{A_t}{A_e} = \frac{Ma_e}{Ma_t}\left[\frac{2 + (\gamma-1)Ma_t^2}{2 + (\gamma-1)Ma_e^2}\right]^{\frac{\gamma+1}{2(\gamma-1)}} = Ma_e \left[\frac{2.28}{2 + 0.28Ma_e^2}\right]^{4.071} = 0.011$$

Via iteration, find supersonic solution, $Ma_e = 5.5$. For nozzle exit pressure:

$p_e = p_c\left[1 + \frac{\gamma-1}{2}Ma_e^2\right]^{\frac{-\gamma}{\gamma-1}} = 8.2[1 + 0.14(5.5^2)]^{-4.57} = 0.00425$ MPa.

$$F = C_{F,v}\left[1 - \left(\frac{p_e}{p_c}\right)^{\frac{\gamma-1}{\gamma}}\right]^{1/2} A_t p_c + (p_e - p_\infty)A_e$$

$= 2.007[1 - (0.00425/8.2)^{0.219}]^{0.5}$ $9.85 \times 10^{-3}(8.2 \times 10^6) + (4250 - 2000)0.8824$

$= 145826 + 1985 = 147{,}811$ N or 147.8 kN, compared to 138,660 N or 138.7 kN for Prob. 11.1.

$I_{sp} = \frac{F}{\dot{m}_e g_o} = \frac{147811}{37.82(9.81)} = 398$ s, compared to 374 s for Prob. 11.1.

11.4. $A_{inj} = \frac{\pi d_{inj}^2}{4} = \pi$ $(0.005)^2/4 = 1.96 \times 10^{-5}$ m^2, $A_{line} = \frac{\pi d_{line}^2}{4} = \pi(0.008)^2/4 = 5.03 \times 10^{-5}$ m^2

$\Delta p_{inj} = 0.34$ MPa, $\rho_{oxid} = 1{,}140$ kg/m^3, $C_{d,inj} = 0.9$

$$v_{inj} = C_{d,inj}\sqrt{\frac{2_{inj}}{\rho\left(1 - \left[\frac{A_{inj}}{A_{line}}\right]^2\right)}} = 0.9\left\{\frac{2(0.34 \times 10^6)}{1140\left(1 - \left[\frac{1.96 \times 10^{-5}}{5.03 \times 10^{-5}}\right]^2\right)}\right\}^{1/2} = 23.9 \text{ m/s}$$

$Q = A_{inj}v_{inj} = 1.96 \times 10^{-5}$ (23.9) $= 4.68 \times 10^{-4}$ m^3/s or 0.028 m^3/min

11.5. $D_c = d_t$ $(A_c/A_t)^{0.5} = 0.112$ $(2)^{0.5} = 0.158$ m, approx. chamber diameter. From Eq. 11.16,

$L_c \approx V_c/A_c = L^* \cdot \frac{A_t}{A_c} = 0.8 \cdot \frac{1}{2} = 0.4$ m, approx. chamber length.

References

1. Gatland KW et al (1981) The illustrated encyclopedia of space technology—a comprehensive history of space exploration. Crown Publishers, New York
2. Sutton GP, Biblarz O (2001) Rocket propulsion rocket propulsion elements, 7th edn. Wiley, New York
3. Huzel DK, Huang DH (1967) Design of liquid propellant rocket engines, NASA SP-125
4. Sobin AJ, Bissell WR (1974) Turbopump systems for liquid rocket engines, NASA SP-8107
5. Haidn OJ (2008) Advanced rocket engines. In: Advances on propulsion technology for high-speed aircraft,aircraft, NATO RTO-EN-AVT-150, Neuilly-sur-Seine (France)
6. Conrad EW (1968) Exploring in aerospace rocketry—7. liquid-propellant rocket systems, NASA TM X-52394
7. Humble RW, Lewis D (1995) Liquid rocket propulsion systems. In: Humble RW, Henry GN, Larson WJ (eds) Space propulsion analysis and design. McGraw-Hill, New York
8. Dankhoff WF, Johnsen IA, Conrad EW, Tomazic, WA (1968) M-1 injector development—philosophy and implementation, NASA TN D-4730
9. Conrad EW, Bloomer HE, Wanhainen JP, Vincent DW (1968) Liquid rocket acoustic-mode-instability studies at a nominal thrust of 20000 pounds, NASA TN D-4968
10. Inoue C, Watanabe T, Himeno T (2007) Numerical study on flow induced vibration of LOX post in liquid rocket engine preburner. Int J Gas Turb Propul Power Syst 1:22–29

11. Archer RD, Saarlas M (1996) An introduction to aerospace propulsion. Prentice-Hall, Upper Saddle River
12. Woodberry RFH, Zeamer RJ (1974) Solid rocket thrust vector control, NASA SP-8114
13. Ellis RA (1975) Solid rocket motor nozzles, NASA SP-8115
14. Prescott BH, Macocha M (1996) Nozzle design. In: Jensen GE, Netzer DW (eds) Tactical missile propulsion. AIAA, Reston

Chapter 12
Hybrid Rocket Engines

12.1 Introduction

Hybrid rocket engines (HREs) are an attempt to exploit some advantages of liquid-propellant rocket engine and solid-propellant rocket motor technology, in a combined package that would (presumably) be more cost-effective for undertaking more complex mission requirements. The traditional arrangement of an HRE is a liquid oxidizer being fed to a solid fuel grain, as per Fig. 12.1. While not common, a "reverse hybrid" utilizes a liquid fuel being fed to a solid oxidizer grain. One clear mission advantage over SRMs is the ability to throttle the oxidizer in-flight upon command (modulate thrust from maximum level ideally right down to zero, as required). Another advantage overall is storage safety, whereby the liquid or gaseous oxidizer is not in direct proximity for inadvertently igniting with the solid fuel grain. In the HRE case, the fuel grain is essentially inert in the absence of an oxidizing agent, unlike an SRM propellant grain. Depending on the chosen liquid oxidizer (a high-energy choice would be oxygen, O_2) and solid fuel (e.g. a higher energy hydrocarbon fuel like paraffin), specific impulse (I_{sp}) for HREs can potentially be higher than SRMs. In a similar fashion, depending on the oxidizer/fuel combination, the resulting exhaust products may be substantially cleaner than that produced by a conventional SRM (which can contain substances like hydrochloric or nitric acid). If the HRE option is capable of fulfilling the given mission requirement, it may be a viable alternative to a higher performance LRE, if lower cost and/or engineering complexity are critical deciding factors. A readier availability of conventional chemical compounds may also move one to choose the HRE approach over either an SRM or LRE approach.

A current disadvantage that is often quoted for HREs is the poorer fuel volumetric loading efficiency (V_f/V_c) relative to SRMs, with the common need for a pre- and post-combustion chamber (where solid fuel is absent) within the engine's overall combustion chamber. Other potential disadvantages that may be more specific to a given HRE design are the need for a lower chamber pressure p_c (noting that p_c must be considerably lower than upstream feed pressure, to prevent

D. R. Greatrix, *Powered Flight*, DOI: 10.1007/978-1-4471-2485-6_12,
© Springer-Verlag London Limited 2012

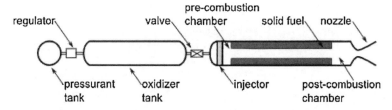

Fig. 12.1 Schematic diagram illustrating a conventional pressure-fed hybrid rocket engine

Table 12.1 Stoichiometric oxidizer-fuel mixture ratios and peak flame temperatures for various propellant combinations under nominal operating conditions

Oxidizer	Fuel	Stoich. ratio	Flame temperature, K
Hydrogen peroxide	HTPB	7.0	2800
Hydrogen peroxide	Paraffin	7.3	2750
Hydrogen peroxide	Polyethylene	8.0	2600
Nitrogen tetroxide	HTPB	3.8	3400
Nitrogen tetroxide	Paraffin	4.2	3350
Nitrous oxide	HTPB	6.5	2800
Nitrous oxide	Paraffin	7.0	2600
Nitrous oxide	Polyethylene	7.0	2600
Oxygen	HTPB	2.7	3650
Oxygen	Paraffin	2.7	3600
Oxygen	Polyethylene	3.4	2800

backflow; structural loading may also be an issue) and a lower effective chamber flame temperature T_F (to avoid heat damage, if cooling capabilities are limited). The common observation of significant residual fuel at the end of HRE firings (in practice, commonly due in part to difficulty in matching up closely to the oxidizer content available; some analogy to slivering seen with SRM star or wagon-wheel propellant grains at the tail-end of their firings) is also cited as a disadvantage. One can observe the stoichiometric mixture ratios (oxidizer-to-fuel) for a variety of HRE propellant combinations in Table 12.1; mixture ratio will have a significant influence on residual fuel values. Ignition of a HRE can sometimes be an issue, depending on the sophistication of the technique, and the design/application requirements for starting. For example, a sequenced approach of introducing a start-up (e.g. lower, to avoid blowout) oxidizer gas flow first into the engine head-end region, followed by a secondary short-term fuel gas flow (alternative: pre-firing placement of metal or plastic fuel mesh or igniter [fuel, or fuel + oxidizer] paste in the forward grain port area, if a one-start-only application), followed by an introduction of a transient heat energy source (e.g. electric arc or electrically-heated wire) to initiate combustion, followed by introducing the main oxidizer delivery flow, can in relative terms consume a considerable amount of time, even if consistent.

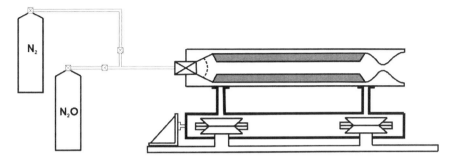

Fig. 12.2 Schematic diagram of laboratory test setup for firing hybrid rocket engine

12.2 Performance Considerations

Unless the liquid oxidizer is self-pressurizing (converting/vaporizing from a liquid into a gas) in the appropriate temperature range for operation (e.g. nitrous oxide, N_2O), then a gas-pressure or turbopump feed system is needed to move the oxidizer from storage to the injectors at the combustor. The injectors are traditionally placed at the head end of the engine combustion chamber (front part of the pre-combustion chamber), but one may also see additional injectors placed aft in the post-combustion chamber in order to add additional oxidizer to the (potentially) fuel-rich reacting core flow. Liquid hydrogen peroxide (H_2O_2) requires a catalyst (commonly accomplished by passing through a silver, platinum or iridium wire mesh) in order to transform it into water vapor (H_2O) and oxygen prior to injection into the combustion chamber. If a gaseous oxidizer is used (e.g. GO_2 or GN_2O; more likely for laboratory tests where storage tank weight is not an issue), then already being at high pressure removes the need for a feed system. Recall that the feed system pressure must be higher than the combustor operational pressure p_c (e.g. to be conservative, a 20–30% drop in pressure through the injectors is not unreasonable, in practice). Refer to Fig. 12.2 for a typical laboratory setup for an HRE firing, in this case using self-pressurizing liquid (or gaseous) N_2O as the oxidizer. Note the high-pressure cylinder of nitrogen (N_2), which is used for two potential purposes: (1) pre- and post-firing flush (purge) of residual oxidizer in the oxidizer delivery system (to avoid inadvertent ignition), and (2) on-command termination (extinguishment) of a firing (in conjunction with oxidizer feed shutdown). Figure 12.3 shows the outdoor static test setup for the firing of Ryerson University's prototype HRE demonstrator, utilizing high-pressure gaseous oxygen as the oxidizer, and low-density polyethylene for the solid fuel [1].

The material for the solid fuel may be harder plastics like polyethylene (PE) or plexiglass (PMMA) for added structural strength, or softer rubbers like hydroxyl-terminated polybutadiene (HTPB), or very soft substances like paraffin wax (an attractive candidate fuel due to a higher regression rate than the above mentioned). If a fuel is considered too soft for a given application, a suitable plasticizer like HTPB may be added to the fuel to give it more rigidity, albeit at the potential

Fig. 12.3 *Upper photo* gives side view of Ryerson University hybrid rocket engine [1] on test stand, prior to test firing at Continuum Aerospace outdoor test facility. *Lower photo* of LDPE/GOx engine firing

expense of a lower regression rate. The fuel grain may have a single central port (as per Fig. 12.4), or multiple ports such as the pie-shaped ports for the other example design of Fig. 12.4. A conventional HRE fuel has a relatively low mean regression rate, say on the order of 0.25 cm/s, as compared to a typical SRM propellant, which may burn at around 1.0 cm/s. As a result, to compensate for a low fuel surface regression rate, one may need to go from a single port to multiple ports to increase the effective burning surface area to get one to the desired chamber pressure p_c.

As noted above, one may have an empty volume called the pre-combustion chamber between the head-end injector plate and the start of the solid fuel grain. This chamber allows for the further vaporization of atomized oxidizer spray before contacting the fuel, to improve combustion efficiency (sometimes defined as a function of characteristic exhaust velocity [*c-star*], $\varepsilon_c = c^*(actual)/c^*(nominal)$) and in turn, I_{sp}. Similarly, one may also have an empty volume aft of the solid fuel grain and before the nozzle entry, called the post-combustion or mixing chamber. This extra volume may allow for an extension of the effective reactive molecule residence time t_c in the overall combustion chamber to potentially allow for more complete combustion of any remaining unreacted gaseous fuel molecules in the aft portion of the HRE combustor. Occasionally, as noted above, one may also see the use of oxidizer injectors at this location, to complement the main head-end injector(s). The use of both a pre- and post-combustion chamber can lower the fuel volumetric loading efficiency to as low as 60%, as compared to a typical SRM

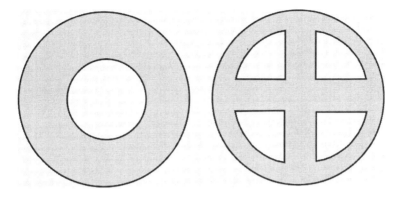

Fig. 12.4 Single (*central*) port and multiple port fuel grain designs

propellant loading efficiency of around 80%. Also on occasion, one may see the use of angled injection ports at the head end or aft to implement a swirl component to the incoming oxidizer flow. This approach has experimentally shown the ability to increase the regression rate of the fuel. Swirl may benefit two possible factors in this regard: increased heat flux in the burning surface's strengthened shear layer, and effectively extending t_c.

12.3 Fuel Regression Rate in HREs

The solid fuel regression rate r_b in conventional (i.e. head-end injection only) hybrid engines is typically found to be primarily a function of local mass flux ($G = \rho u = \dot{m}/A_p$, kg/m^2·s; A_p is local port cross-sectional area):

$$r_b = aG^n, \quad 0.4 < n < 0.85 \tag{12.1}$$

where a is an experimentally determined coefficient (often corrected for port diameter, i.e. lower a at higher diameters), and the exponent n is also largely determined experimentally (originally thought to be around 0.8, which corresponds to turbulent flat-plate boundary layer formulae; a more comprehensive collection of experimental data suggested a wider range of values for n). For preliminary design calculations, one commonly assumes (at least initially) that $G \approx G_O = \dot{m}_O/A_p$, where oxidizer mass flow $\dot{m}_O$ is commonly set at a constant rate throughout a firing. Additionally, it is also common to experimentally establish the values for a and n based on G_O, rather than the actual local value for G (which may be significantly higher, with incoming fuel gas influx upstream of the measurement point). As oxidizer is consumed/reacted in moving down the port from left to right (as per Fig. 12.1) as axial distance x from the head end increases, the local oxidizer/fuel mixture ratio $r(x)$ may very well go from fuel-lean (full possible regression of fuel grain, with stoichiometric reaction enabled; high value for r) to approximately stoichiometric (r near its desired or stoichiometric value), and if the fuel grain length ℓ_f is long enough, to fuel-rich (when oxidizer less available, less than full possible regression and/or heat energy input of the fuel grain, even though aft values for G are significantly larger than forward values; low value for r).

With the above factors at play, it is not uncommon to observe an approximate longitudinal uniformity of regression along the grain (i.e. if one were to take a snapshot of the fuel grain profile at given time into the firing), which supports the earlier assumption that for preliminary design at least:

$$r_b \approx aG_O^n \tag{12.2}$$

With respect to simpler single-port constant-$\dot{m}_O$ HRE firings, it is common to observe that $r(x, t)$ for a given x-location goes up in value as firing time progresses (later, we will see this corresponds to stoichiometric length L_{st} getting bigger with elapsed firing time). For a cylindrical port, one can make the following mathematical correlation. For

$$\dot{m}_f \approx \rho_s S r_b \approx \rho_s S\left(aG_O^n\right) = \rho_s \pi d\ell_f aG_O^n = \rho_s \pi d\ell_f a\left(\frac{\dot{m}_O}{\pi d^2/4}\right)^n \tag{12.3}$$

then for a constant oxidizer mass flow, grouping all the relatively constant parameters into K_1:

$$\dot{m}_f \approx K_1 \cdot d^{1-2n} \tag{12.4}$$

The above result suggests that for $n > 0.5$ (which would be more typical), $\dot{m}_f$ would decrease with firing time as port diameter d grows (which corresponds to r increasing), and for $n < 0.5$, $\dot{m}_f$ would increase.

The apparent independence from pressure in most cases observed suggests that the dominant regression mechanism is via convective heat transfer (which is largely based on mass flux G; [2]), correlating to the erosive burning mechanism observed for SRMs. With this in mind, one can apply the Greatrix–Gottlieb [3] erosive burning model here, namely for overall burning rate r_b:

$$r_b = \frac{h(T_F - T_S)}{\rho_s[C_s(T_S - T_i) - \Delta H_s]} + r_o \tag{12.5}$$

Here, r_o would be the base burning rate due to all other mechanisms, e.g., a_n burn component due to engine spinning, radiation-based burning if sources of radiation exist within the combustion chamber, or in the case of some loading of oxidizer crystals into the fuel (to produce in essence a fuel-rich solid propellant), r_o may incorporate a pressure-dependent term along the lines of de St. Robert's law for SRMs, i.e.

$$r_{o,p} = C_1 p_c^{n_1} \tag{12.6}$$

A principal variable of interest in Eq. 12.5 is the convective heat transfer coefficient under transpiration (i.e. blowing due to propellant gas entering into the core flow), namely h. It can be solved iteratively as a function of incoming mass flow and the zero-transpiration baseline value, $h*$, as discussed for erosive burning modeling in SRMs. In the fundamental case where r_o is not significant, it happens that one can produce a relatively direct solution [4] for r_b:

$$r_b = \frac{h^*}{\rho_s C_p} \ell n\left[1 + \frac{C_p}{C_s}\frac{(T_F - T_S)}{(T_S - T_i - \Delta H_s/C_s)}\right] = \frac{h*}{\rho_s C_p}\ell n[\beta] \qquad (12.7)$$

Via Reynolds' analogy equating shear stress to heat transfer, for a turbulent flow,

$$h^* = \frac{k^{2/3}C_p^{1/3}}{\mu^{2/3}}\frac{Gf^*}{8} \qquad (12.8)$$

In Eq. 12.8, one can observe a direct dependence on local mass flux G. In addition, one can see that there is the influence of the Darcy–Weisbach friction factor (zero-transpiration value, f^*), which for fully developed turbulent flow can be estimated via Colebrook's equation:

$$(f^*)^{-1/2} = -2\log_{10}\left[\frac{2.51}{Re_d(f^*)^{1/2}} + \frac{\varepsilon/d}{3.7}\right] \qquad (12.9)$$

In Eq. 12.9, there is a significant dependence of friction factor on the fuel surface roughness height, ε. Everything else being equal, f^* tends to drop gradually in value in relation to an increasing value for G (incorporated inside Reynolds number Re_d). An example graph showing a comparison of the above model's predictions to experimental data is provided in Fig. 12.5 [4–6].

Work remains to be done on ascertaining the normal acceleration (a_n) sensitivity on HRE burning rates referred to above. The typical low base burning rate would suggest a considerable sensitivity might be expected, whether in spinning (quasi-steady acceleration field) or radial vibration (unsteady oscillatory acceleration field). If at an aft location along the port, a lower oxidizer availability may inhibit the fully possible augmentation of the local burning rate. Later into a firing when oxidizer-fuel $r(x,t)$ is typically higher (constant $\dot{m}_O$), one may see more augmentation with the greater availability of oxidizer when a_n is at appreciable levels.

12.4   Internal Ballistics of HREs

Assuming the flow remains below a moderate subsonic level in the fuel grain port, one can make the following estimation for chamber pressure:

Fig. 12.5 Theoretical and experimental [5, 6] data for burning rate as a function of mass flux, HTPB/GOX propellant A, and paraffin/GOX propellants C and D [4]

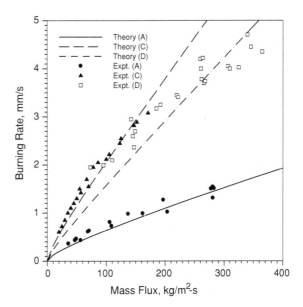

$$p_c = c * \cdot \frac{\dot{m}}{A_t} \approx \left[\frac{\gamma}{RT_f} \left(\frac{2}{\gamma + 1} \right)^{\frac{\gamma + 1}{\gamma - 1}} \right]^{-1/2} \cdot \frac{\rho_s S a \left(\frac{\dot{m}_O}{A_p} \right)^n + \dot{m}_O}{A_t} \qquad (12.10)$$

One needs to be careful to use the proper units for coefficient a of the empirical mass-flux-dependent burning rate equation (see Eq. 12.2), in order to have the units for p_c to come out properly as well (fundamental units for a being m/[s·(kg/[m^2·s])^n]). Once an estimate of p_c is obtained, as per Chap. 9 one can in turn ascertain thrust via:

$$F = C_F A_t p_c = C_{F,v} \left[1 - \left(\frac{p_e}{p_c} \right)^{\frac{\gamma - 1}{\gamma}} \right]^{1/2} A_t p_c + (p_e - p_\infty) A_e \qquad (12.11)$$

where the vacuum thrust coefficient is determined from

$$C_{F,v} = \left[\frac{2\gamma^2}{\gamma - 1} \left(\frac{2}{\gamma + 1} \right)^{\frac{\gamma + 1}{\gamma - 1}} \right]^{1/2} \qquad (12.12)$$

One can see from Eq. 12.10 that increasing the burning surface area S can compensate for a low value for the a coefficient, in getting to a desired chamber pressure. Referring to our earlier parametric evaluation of the effect of n on $\dot{m}_f$ (see Eq. 12.4), for n greater than 0.5, one would expect p_c to drop a bit (depending on

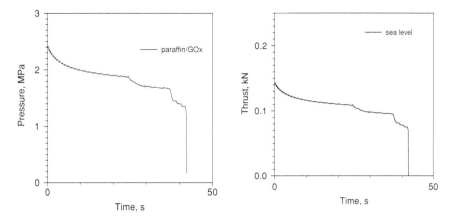

Fig. 12.6 a Predicted head-end pressure–time profile for small cylindrical-grain HRE **b** Predicted head-end sea-level thrust-time profile for small cylindrical-grain HRE

the relative value of $\dot{m}_O$; see Fig. 12.6a for an example profile) as firing time progresses for the single cylindrical port, constant $\dot{m}_O$ scenario. The first distinct trend change in the pressure curve of Fig. 12.6a, at just over 22 s into the example firing, is due to the fuel grain beginning to burn out (reach the inner insulation/ wall) at various locations along the grain. The second distinct drop at around 37 s is due to the progressively growing stoichiometric length having reached the actual fuel grain length at that juncture. A moderately regressive thrust-time profile is the corresponding result of the pressure–time profile's shape, as may be seen in Fig. 12.6b. Please refer to the solution for Example Problem 12.1 at the end of this chapter, for a completed sample hybrid rocket motor performance analysis. For students learning this material as part of a course, I would encourage them to attempt the given example problem first, and then check the solution to see if any mistakes were made.

Realizing that $r(x)$ goes down moving aft, and commonly $r(x,t)$ goes up as the firing progresses in time for a constant $\dot{m}_O$, designers of HREs will typically try to establish the fuel grain length ℓ_f such that at some mean time into the firing, ℓ_f will coincide with the stoichiometric length L_{st}. Based on the empirical law for mass-flux dependent burning, one has the following cylindrical-grain, fixed-oxidizer-rate estimation for L_{st} [7]:

$$L_{st} \approx \frac{G_O d}{4\rho_s r_{st} r_b} \approx \frac{G_O^{1-n} d}{4a\rho_s r_{st}} \approx f\left(d^{2n-1}\right) \tag{12.13}$$

where

$$r_{st} = \left.\frac{\dot{m}_o}{\dot{m}_f}\right|_{\text{stoich}} \approx \frac{G_O A_p}{\rho_s r_b \cdot \pi d L_{st}} \tag{12.14}$$

The stoichiometric mixture ratio r_{st} will vary for different fuel/oxidizer combinations, e.g., about 8:1 for H_2O_2/PE, 2.7:1 for O_2/HTPB. As may be observed by Eq. 12.13, L_{st} grows in value with time into the firing (constant $\dot{m}_O$ case) as d grows with grain burnback for the cylindrical-grain case, for $n > 0.5$. As noted earlier, coefficient a tends to decrease in value as port diameter increases, which would act to further hasten the growth of L_{st}. An alternative expression [4] for stoichiometric length can be generated from the Greatrix regression rate result of Eq. 12.7:

$$L_{st} \approx \frac{Gd}{4\rho_s r_{st} r_b} \approx \frac{2\,Pr^{2/3}\,d}{f * (\ell n[\beta])(r_{st})} \tag{12.15}$$

This expression essentially indicates that L_{st}, for a nominal port diameter d, is an inverse function of the parameter, $f*$, which decreases gradually with increasing mass flux G. Qualitatively, the predictions are comparable between Eqs. 12.13 and 12.15, with L_{st} increasing gradually with increasing G for a given port diameter.

If one wanted to do a more comprehensive steady or nonsteady HRE internal ballistic analysis, especially if aft port flow Mach numbers are getting into the high subsonic levels, one can follow a numerical approach similar to that discussed for SRMs. For the quasi-steady one-dimensional case (parameters not changing too rapidly, in relative terms, e.g., no pressure-wave activity of significance), one has the following ordinary differential equations for conservation of mass, linear momentum and energy of the core gas at a given location along the fuel grain, at a given time in the engine's firing, if moving left to right from the head-end injector toward the exhaust nozzle downstream [8]:

$$\frac{d(\rho u)}{dx} = -\frac{1}{A}\frac{dA}{dx}\rho u + (1 - \alpha_p)\rho_s \frac{4r_b}{d} - \left(\frac{4r_b}{d}\right)\rho \tag{12.16}$$

$$\frac{d}{dx}(\rho u^2 + p) = -\frac{1}{A}\frac{dA}{dx}\rho u^2 - \left(\frac{4r_b}{d}\right)\rho u - \rho a_\ell - \frac{\rho_p}{m_p}D \tag{12.17}$$

$$\frac{d}{dx}(\rho u E + up) = -\frac{1}{A}\frac{dA}{dx}(\rho u E + up) - \left(\frac{4r_b}{d}\right)\rho E$$

$$+ (1 - \alpha_p)\rho_s \frac{4r_b}{d}\left(C_p T_f + \frac{v_f^2}{2}\right) - \rho u a_\ell - \frac{\rho_p}{m_p}(u_p D + Q) \tag{12.18}$$

Here, the total specific energy of the gas is defined for an ideal gas as $E = p/[(\gamma-1)\rho] + u^2/2$. The corresponding quasi-steady equations of motion for a monodisperse inert (non-burning) particle phase (evolving from the fuel grain) within the axial flow may be found as follows [8]:

$$\frac{d(\rho_p u_p)}{dx} = -\frac{1}{A}\frac{dA}{dx}\rho_p u_p + \alpha_p \rho_s \frac{4r_b}{d} - \left(\frac{4r_b}{d}\right)\rho_p \tag{12.19}$$

$$\frac{d\left(\rho_p u_p^2\right)}{dx} = -\frac{1}{A}\frac{dA}{dx}\rho_p u_p^2 - \left(\frac{4r_b}{d}\right)\rho_p u_p - \rho_p a_\ell + \frac{\rho_p}{m_p}D \tag{12.20}$$

$$\frac{d(\rho_p u_p E_p)}{dx} = -\frac{1}{A}\frac{dA}{dx}(\rho_p u_p E_p) - \left(\frac{4r_b}{d}\right)\rho_p E_p$$

$$+ \alpha_p \rho_s \frac{4r_b}{d}\left(C_m T_f + \frac{v_f^2}{2}\right) - \rho_p u_p a_\ell + \frac{\rho_p}{m_p}(u_p D + Q) \tag{12.21}$$

Of course, the boundary condition of fixed (or variable) oxidizer mass injection at the head end would need to be implemented for the HRE. One would also likely want to track the position of L_{st}, so as to possibly implement a non-combustive ablation (if one assumed no oxidizer available) to the solid fuel when local $x > L_{st}$. In the case of ablation, the solid fuel would regress, although possibly at a lower rate, and the decomposition gas entering the central core at a lower effective temperature T_{ds}. Ablation rate e_s can for some materials be assumed to be a function of convective energy feedback in a similar format as Eq. 12.7 above, such that for the simplest case:

$$e_s = \frac{h^*}{\rho_s C_p}\ln\left[1 + \frac{C_p}{C_s}\frac{(T_\infty - T_{ds})}{\{(T_{ds} - T_i) - \Delta H_s/C_s\}}\right] \tag{12.22}$$

As before, zero-transpiration convective heat transfer coefficient h^* may be ascertained via Eq. 12.8. Note the replacement of T_F by local core temperature T_∞ in Eq. 12.16, allowing for the fact that the core temperature will drop significantly as one moves further aft in a non-combusting environment. The value for net surface heat release ΔH_s will likely be small in Eq. 12.16, especially in the absence of combustion.

12.5 Combustion Instability in HREs

With respect to transient unsteady internal ballistics, HREs can on occasion be affected by combustion instability symptoms. These instabilities, like those observed for liquid rocket engines, are typically classified in terms of frequency. It is observed that HREs can sporadically develop intermediate to high-frequency pressure-wave-related ("acoustic") symptoms when fired at colder outside air temperatures. In SRMs, one on occasion observes a dc shift associated with pressure-wave activity, whereby the base chamber pressure rises above the nominal level for a period of time. In HREs, one can observe apparent dc shifts both upward and downward from the nominal p_c. The downward shifts in pressure may be due to a short-term lack of oxidizer to maintain the nominal level of burning in that part of the combustion chamber affected. This may ultimately be an advantage of HREs over SRMs, in having oxidizer starvation prevent transient combustion-driven pressure-wave development from becoming too strong in magnitude.

Lower-frequency instabilities (feed instabilities, or an alternative low-frequency transient phenomenon, chugging, which is also allied to the upstream feed system characteristics) related to feeding the oxidizer to the combustion chamber is a common issue for HREs [9]. A mitigating factor with the low-frequency symptoms is that even though the magnitude of the oscillatory pressure component superimposed on the nominal chamber pressure can be significant, the resulting vibration on the surrounding system is much less compared to that seen for comparable or even lower oscillatory Δp magnitudes at higher frequencies. At higher frequencies, any fluctuation Δp in chamber pressure above about $\pm 3\%$ of the base pressure is commonly characterized as the HRE being under "rough" operation, possibly associated with combustion instability. In liquid rocket engines, this figure is approximately $\pm 5\%$, while for SRMs, the nominal roughness threshold figure is lower by quite an amount, i.e. $\pm 0.5\%$.

12.6 Example Problems

12.1. A small hybrid (LN_2O/PE, liquid nitrous oxide/polyethylene) rocket engine has the following characteristics at a given design point in its flight, for an ambient flight static pressure at 40 kPa:

Nozzle throat diameter, $d_t = 2.5$ cm; nozzle exit diameter, $d_e = 7.5$ cm;
Fuel grain length 0.65 m;
Effective star-grain burning surface distance, $s_p = 22$ cm;
Design port area, $A_p = 1.8\ A_t$;
PE fuel grain solid density, $\rho_s = 1{,}150$ kg/m^{3}; oxidizer feed rate = 2.6 kg/s
Combustion products, $\gamma = 1.22$; molecular mass, M = 24 amu;
$T_F = 2{,}650$ K
Mass-flux based burning rate coefficients, $n = 0.7$, $a = 0.02$ mm/s-(kg/m^2-s)^n;

Estimate the chamber pressure, thrust and specific impulse at the design point.

12.2. For estimating the nominal stoichiometric length L_{st} that designers typically use as a guideline for setting the actual fuel grain length for a conventional hybrid rocket engine [for a nominal stoichiometric oxidizer/fuel mixture ratio (r_{st})], one can show that given

$$r_{st} = \frac{\dot{m}_o}{\dot{m}_f} \approx \frac{G_o A_p}{\rho_s r_b \cdot \pi d L_{st}}$$

then

$$L_{st} \approx \frac{G_o d}{4\rho_s r_{st} r_b} \approx \frac{G_o^{1-n} d}{4a\rho_s r_{st}}$$

via the standard mass-flux-dependent burning rate relationship, $r_b \approx aG_o^n$ for the fuel. This expression essentially indicates that L_{st} increases gradually with increasing G for a given port diameter d. Derive the comparable expression one would get for L_{st} via the newer Greatrix HRE mass-flux-dependent burning rate relationship. Compare the behavior of the new expression for L_{st}, relative to the above one.

12.3. A hybrid rocket engine has the following characteristics:

Fuel grain solid density, $\rho_s = 1100$ kg/m³; solid specific heat $C_s = 2000$ J/kg-K
Net surface heat release $\Delta H_s = 0$ J/kg; $T_s = 800$ K;
Fuel surface roughness $\varepsilon = 7$ μm; ambient temperature $T_i = 288$ K
Combustion products, $\gamma = 1.21$; molecular mass, M = 23 amu;
$T_F = 2725$ K
Absolute gas viscosity $\mu = 7.2 \times 10^{-5}$ kg/m-s; gas Prandtl number 0.73
Mass-flux based burning rate coefficients, $n = 0.75$,
$a = 0.0215$ mm/s-(kg/m²-s)ⁿ;

(a) For a flow static pressure of 8.0 MPa at a Mach number of 0.5, find the fuel's burning rate using the standard mass-flux model (coefficients provided above). You may assume the core gas temperature is roughly the flame temperature.
(b) Alternatively, estimate the burning rate at the above flow conditions using the Greatrix–Gottlieb erosive burning model. Comment: Note that the grain section's hydraulic diameter is 8 cm.

12.4. Assuming that a net normal acceleration level a_n of -500 g resulting from the radial vibration of the local fuel grain surface exists (e.g. due to the passage of a traveling axial shock wave in the combustion chamber) for part (b) of Problem 12.3, estimate the resulting worst-case burning rate, based on the approach covered in Chap.10 for a solid propellant under a normal acceleration field.

12.7 Solutions to Example Problems

12.1. Looking at a hybrid rocket engine, at a flight altitude giving $p_\infty = 40$ kPa.

Nozzle throat area, $A_t = \dfrac{\pi d_t^2}{4} = \pi(0.025)^2/4 = 4.91 \times 10^{-4}$ m²; $R = \dfrac{\mathcal{R}}{\mathcal{M}} = $
$8315/24 = 346.4$ J/kg·K

Nozzle exit area, $A_e = \dfrac{\pi d_e^2}{4} = \pi(0.075)^2/4 = 4.42 \times 10^{-3}$ m²

$$c^* = \left[\frac{\gamma}{RT_f}\left(\frac{2}{\gamma+1}\right)^{\frac{\gamma+1}{\gamma-1}}\right]^{-1/2} = [1.22/[346.4(2650)] \cdot (2/2.22)^{2.22/0.22}]^{-0.5}$$
$$= 1,469 \text{ m/s}$$

$$C_{F,v} = \left[\frac{2\gamma^2}{\gamma-1}\left(\frac{2}{\gamma+1}\right)^{\frac{\gamma+1}{\gamma-1}}\right]^{1/2} = [2(1.22^2)/(0.22) \cdot (2/2.22)^{2.22/0.22}]^{0.5} = 2.173$$

$a = 0.02$ mm/s (kg/m^2·s)^n = 2.0×10^{-5} m/s·(kg/m^2·s)$^{0.7}$; $S = s_p\ell_p = 0.22(0.65)$
$= 0.143$ m^2

$$\dot{m}_e \approx \dot{m}_o + \rho_s Sa\left(\frac{\dot{m}_o}{A_p}\right)^n = 2.6 + 1150(0.143)2.0 \times 10^{-5}(2.6/8.84 \times 10^{-4})^{0.7}$$
$$= 3.481 \text{ kg/s}$$

$$p_c = \frac{c^*\dot{m}_e}{A_t} \approx \frac{\rho_s Sa\left(\dfrac{\dot{m}_o}{A_p}\right)^n + \dot{m}_o}{A_t\left[\dfrac{\gamma}{RT_f}\left(\dfrac{2}{\gamma+1}\right)^{\frac{\gamma+1}{\gamma-1}}\right]^{1/2}}$$

$= (1150(0.143)2.0 \times 10^{-5}(2.6/8.84 \times 10^{-4})^{0.8} + 2.6)/4.91 \times 10^{-4} \cdot 1469 =$
10.42 MPa

Area-Mach No. relation, where Ma_t goes to unity:

$$\frac{A_t}{A_e} = \frac{Ma_e}{Ma_t}\left[\frac{2+(\gamma-1)Ma_t^2}{2+(\gamma-1)Ma_e^2}\right]^{\frac{\gamma+1}{2(\gamma-1)}} = Ma_e\left[\frac{2.22}{2+0.22Ma_e^2}\right]^{5.05} = 0.111$$

Via iteration, find supersonic solution, $Ma_e = 3.26$. For nozzle exit pressure:
$p_e = p_c\left[1 + \frac{\gamma-1}{2}Ma_e^2\right]^{\frac{-\gamma}{\gamma-1}} = 0.142$ MPa.

$$F = C_{F,v}\left[1 - \left(\frac{p_e}{p_c}\right)^{\frac{\gamma-1}{\gamma}}\right]^{1/2}A_t p_c + (p_e - p_\infty)A_e$$

$= 2.173[1-(0.142/10.42)^{0.18}]^{0.5} \, 4.91 \times 10^{-4}(10.42 \times 10^6) + (142000-40000)$
4.42×10^{-3}

$= 8158 + 452 = 8610$ N or 8.61 kN
$$I_{sp} = \frac{F}{\dot{m}_e g_o} = \frac{8610}{3.481(9.81)} = 252 \text{ s}$$

12.2. Derive a comparable expression for L_{st} for HREs, via the Greatrix burning rate result, rather than the standard model.

Assign $\beta = 1 + \frac{C_p}{C_S}\frac{(T_F-T_S)}{(T_S-T_i-\Delta H_S/C_S)}$ so that the Greatrix fuel regression rate model becomes:

$r_b = \frac{h^*}{\rho_s C_p} \ln(\beta)$, noting that for no transpiration, convective heat coefficient becomes

$h^* = \frac{k^{2/3}}{\mu^{2/3}} C_p^{1/3} G \frac{f^*}{8}$, so as a result of substituting this, and letting $G \approx G_O$:

$$r_b \approx \frac{k^{2/3}}{\mu^{2/3}} \frac{C_p^{1/3}}{C_p} \frac{G_O f^*}{\rho_s}{8} \ln(\beta) \approx \frac{1}{Pr^{2/3}} \frac{G_O f^*}{\rho_s}{8} \ln(\beta)$$

From the original derivation of L_{st} :

$$L_{st} \approx \frac{G_O d}{4 \rho_s r_{st} r_b} \approx \frac{G_O d}{4 \rho_s r_{st} \left(\frac{1}{Pr^{2/3}} \frac{G_O f^*}{\rho_s}{8} \ln(\beta) \right)} \approx \frac{2 Pr^{2/3} d}{f^* \ln(\beta) r_{st}}$$

Since f^* decreases gradually as mass flux G increases, then as a result, L_{st} increases gradually with G increasing, for a given port diameter. This trend is consistent with the result using the standard expression of $r_b = a G_O^n$ for the fuel regression rate, where $L_{st} \propto G_O^{1-n}$, with n for an HRE typically being in the 0.7–0.9 range.

12.3. (a) $R = 8314/23 = 361.5$ J/kg·K, $C_p = 1.21(361.5)/0.21 = 2083$ J/kg·K

$$\rho \approx \frac{p}{RT_F} = \frac{8 \times 10^6}{361.5(2725)} = 8.12 \text{ kg/m}^3; \quad u = a \cdot Ma \approx \sqrt{\gamma RT_F} \cdot Ma$$
$$= 1092(0.5) = 546 \text{ m/s}$$

$G = \rho u = 4434$ kg/s·m^2

$r_b = a G^n = 0.0215(4434)^{0.75} = 11.7$ mm/s, or 1.17 cm/s

(b) $k = \frac{\mu C_p}{Pr} = \frac{7.2 \times 10^{-5} \cdot 2083}{0.73} = 0.205$ W/m·K

$Re_d = \frac{\rho u d}{\mu} = \frac{8.12(546)0.08}{7.2 \times 10^{-5}} = 4.926 \times 10^6; \quad \varepsilon = 7 \times 10^{-6}$ m

$$\frac{1}{f^{1/2}} = -2 \cdot \log_{10} \left[\frac{2.51}{f^{1/2} Re_d} + \frac{\varepsilon/d}{3.7} \right] = -2 \cdot \log_{10} \left[\frac{5.0954 \times 10^{-7}}{f^{0.5}} + 2.365 \times 10^{-5} \right]$$

$f = 0.01208$ via iteration. This is the incompressible flow estimate. Accounting for compressibility in a turbulent flow, apply the recovery factor equation:

$$f_{comp} = f_{incomp} / \left[1 + P_r^{1/3} \frac{\gamma - 1}{2} Ma^2 \right] = 0.0118;$$

$$h^* = \frac{k}{d} Re_d P_r^{1/3} \frac{f}{8} = \frac{0.205}{0.08} 4.926 \times 10^6 (0.73^{0.333}) \frac{0.0118}{8} = 16766 \text{W/m}^2 \cdot \text{K}$$

$$r_b = \frac{h^*}{\rho_s C_p} \ln\left[1 + \frac{C_p}{C_s} \frac{(T_F - T_S)}{(T_S - T_i - \Delta H_s/C_s)}\right]$$
$$= \frac{16766}{1100(2083)} \ln\left[1 + \frac{2083}{2000} \cdot \frac{(2725 - 800)}{(800 - 288 - 0)}\right] = 0.01165 \,\text{m/s or } 1.17\,\text{cm/s}$$

Thus, the two models produce an identical result for fuel regression rate, in this case.

12.4. Looking at a hybrid rocket fuel's burning rate under mass flux and acceleration. Worst case is that there is sufficient oxidizer available for complete r_b augmentation due to a_n.

We will need to iterate between the two mechanisms of burning, given that each mechanism is dependent on the other mechanism as a base burning rate.

Effect of mass flux G:

From Problem 12.3(b),

$$h^* = \frac{k}{d} \mathrm{Re}_d \,\mathrm{Pr}^{1/3} \frac{f}{8} = \frac{0.205}{0.08} 4.926 \times 10^6 (0.73^{0.333}) \frac{0.0118}{8} = 16766 \,\text{W/m}^2\cdot\text{K},$$

remains constant.

Begin iteration, guess $r_b = 1.63 \times 0.01165$ m/s $= 0.019$ m/s:

$$h = \frac{\rho_s r_b C_p}{\exp\left(\frac{\rho_s r_b C_p}{h^*}\right) - 1} = \frac{1100(2083) r_b}{\exp\left(\frac{1100(2083) r_b}{16766}\right) - 1} = \frac{2.291 \times 10^6 r_b}{\exp(136.66 r_b) - 1} = 3506 \,\text{W/m}^2\cdot\text{K}$$

$$r_b = r_{o,a_n} + \frac{h(T_F - T_S)}{\rho_s C_s (T_S - T_i) - \rho_s \Delta H_S} = r_{o,a_n} + \frac{h(2725 - 800)}{1100(2000)(800 - 288) - 0}$$
$$= r_{o,a_n} + 1.709 \times 10^{-6} h = r_{o,a_n} + 0.006 = (0.019 - 0.006) + 0.006$$
$$= 0.013 + 0.006 = 0.019 \,\text{m/s},$$

tentatively. Need to check via remaining equations, to bring convergence to the solution. For now, $r_{o,G} = 0.006$ m/s.

Effect of normal acceleration a_n:

$$\delta_0 = \frac{k}{\rho_s r_o C_p} \ln\left[1 + \frac{C_p(T_F - T_S)}{C_S(T_S - T_i) - \Delta H_S}\right]$$
$$= \frac{0.205}{1100(r_{o,G})2083} \ln\left[1 + \frac{2083(1925)}{1100(512) - 0}\right] = \frac{1.874 \times 10^{-7}}{r_{o,G}}$$
$$= 3.123 \times 10^{-5} m$$

$$G_a = \frac{a_n p}{r_b} \frac{\delta_o}{RT_F} \frac{r_o}{r_b} = \frac{-4905(8.0 \times 10^6)}{r_b} \frac{1.874 \times 10^{-7}}{361.5(2725)r_{o,G}} \frac{r_{o,G}}{r_b} = -\frac{7.465 \times 10^{-3}}{r_b^2}$$
$$= -20.68 \, \text{kg/s m}^2$$

$$r_b = \frac{C_p(T_F - T_S)}{C_S(T_S - T_i) - \Delta H_S} \cdot \frac{r_b + G_a/\rho_s}{\exp\left[\frac{C_p \delta_o}{k}(\rho_s r_b + G_a)\right] - 1}$$

$$= 7.12 \cdot \frac{r_b - \dfrac{6.79 \times 10^{-6}}{r_b^2}}{\exp\left[\dfrac{2083(1.874 \times 10^{-7})}{0.205 r_{o,G}}\left(1100 r_b - \dfrac{7.465 \times 10^{-3}}{r_b^2}\right)\right] - 1}$$

$$= 7.12 \cdot \frac{r_b - \dfrac{6.79 \times 10^{-6}}{r_b^2}}{\exp\left[\dfrac{1.904 \times 10^{-3}}{r_{o,G}}\left(1100 r_b - \dfrac{7.465 \times 10^{-3}}{r_b^2}\right)\right] - 1} \approx 0.019 \, \text{m/s},$$

so guessed correctly on r_b

Thus, $\dfrac{r_b}{r_o} = \dfrac{0.019}{0.006} = 3.17$ as augmentation of burning rate due to radial vibration of 500 g.

Note how much mass-flux base burning was brought down by the vibration… from 0.01165 m/s down to 0.006 m/s (almost halved). Referencing the pre-vibration state, the augmentation ratio would in that case be 0.019/0.01165 = 1.63.

References

1. Trumpour AP, Greatrix DR, Karpynczyk J, Cherrington W (2010) Preliminary design of a university laboratory hybrid rocket engine. Can Aeronaut Space J 56:1–8
2. Chiaverini MJ (2007) Review of solid-fuel regression rate behavior in classical and nonclassical hybrid rocket motors. In: Chiaverini MJ, Kuo KK (eds) Fundamentals of hybrid rocket combustion and propulsion. AIAA, Reston (Virginia)
3. Greatrix DR, Gottlieb JJ (1987) Erosive burning model for composite-propellant rocket motors with large length-to-diameter ratios. Can Aeronaut Space J 33:133–142
4. Greatrix DR (2009) Regression rate estimation for standard-flow hybrid rocket engines. Aerosp Sci Technol 13:358–363
5. Sutton GP, Biblarz O (2001) Rocket propulsionRocket propulsion elements, 7th edn. Wiley, New York
6. Karabeyoglu MA, Ziliac G, Cantwell BJ, De Zilwa S, Castelluci, P (2003) Scale-up tests of high regression rate liquefying hybrid rocket fuels.In: Proceedings of 41st AIAA Aerospace Sciences Meeting, Reno (Nevada), January 6–9
7. Altman D, Humble RW (1995) Hybrid rocket propulsion systems. In: Humble RW, Henry GN, Larson WJ (eds) Space propulsion analysis and design. McGraw-Hill, New York

8. Greatrix DR, Gottlieb JJ, Constantinou T (1987) Quasi-steady analysis of the internal ballistics of solid-propellant rocket motors. Can Aeronaut Space J 33:61–70
9. Karabeyoglu MA, De Zilwa S, Cantwell BJ, Zilliac G (2005) Modeling of hybrid rocket low frequency instabilities. J Propul Power 21:1107–1116

Chapter 13
Air-Breathing Rocket Engines

13.1 Introduction

In some lower atmosphere applications (for part or all of a flight mission), utilizing outside air as a "free" source of oxidizer is a tempting proposition. Between a flight Mach number of about 2.5 and 5, ducting of outside air to the subsonic combustor section of a conventional ramjet [RJ; also referred to as an athodyd (*aerothermodynamic duct*)] provides adequate thrust for a number of applications at a specific impulse in the range of 1,500–2,000 s using a conventional fuel (here, I_{sp} is based on fuel consumption only… air as the oxidizer is free, since not in vehicle storage; [1]). A minimum flight Ma_∞ of approximately 2.5 is needed for sufficient aerodynamic ram compression (i.e. no active mechanical compression, thus no associated turbomachinery) in a system such as illustrated in Figs. 13.1 and 13.2. Above a Ma_∞ of about 5, the nominal starting point for hypersonic flight, thrust performance for RJs drop off due to stagnation/momentum losses enforced by the subsonic combustor (where the flow Mach number must be between say 0.4 and 0.6 for flame stability), and the maximum temperature obtainable in the combustor. A supersonic-flow combustor section in scramjets (SJs) allows for adequate thrust performance from Ma_∞ of about 4 up to about 16, depending on altitude and available air density for ram compression and the combustion process. To date, SJs are not too common with respect to practical usage, partly due to the need for a very refined design, and fuel that is capable of very rapid reaction rates to prevent blow-off at high-speed flow conditions (pyrophoric fuels that ignite in air almost instantly are useful, e.g. hydrogen [H_2] and ethylene [C_2H_4]). To partially compensate for these difficulties, dual-mode SJs combine both subsonic and supersonic combustion, to allow for practical performance (see Fig. 13.3). Shcramjets (shock-induced combustion ramjets) employ alternative means for combustion than conventional scramjets; a standing oblique detonation wave in the forward intake location is the principal approach used, as opposed to, say, a backward-facing step further downstream [2].

D. R. Greatrix, *Powered Flight*, DOI: 10.1007/978-1-4471-2485-6_13,
© Springer-Verlag London Limited 2012

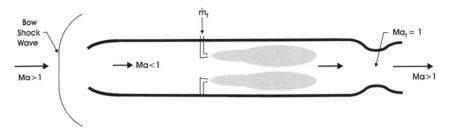

Fig. 13.1 Basic ramjet engine configuration

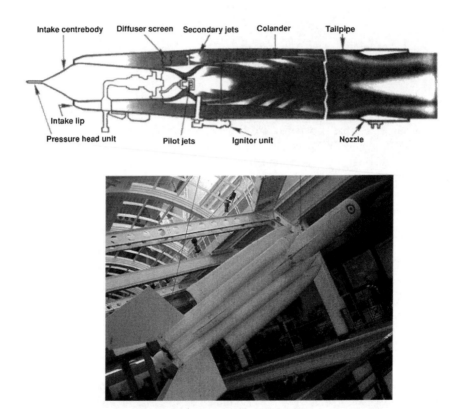

Fig. 13.2 Cutaway view of the Bristol Siddeley Thor ramjet engine used for the Bristol Bloodhound Mk. 1 surface-to-air missile (two Thors were used for cruise/sustain thrust, after being boosted to cruise by four strap-on solid rocket motors)

Please refer to the solution for Example problem 13.1 at the end of this chapter, for a completed sample ramjet engine performance analysis. A review of the cycle analysis done for turbojets in Chap. 6 will be helpful, in preparation for doing the ramjet problems at this chapter. For students learning this material as part of a course, I would encourage them to attempt the given example problem first, and then check the solution to see if any mistakes were made.

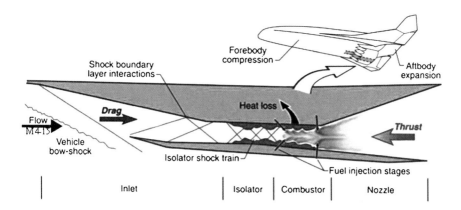

Fig. 13.3 Schematic diagram showing the operation of a dual-mode scramjet engine, employing an isolator channel (which acts to slow the flow entering the combustor, and separates combustion effects from the upstream inlet) and a diverging-duct/staged-injection combustor section to help control the local flow speed, allowing for combustion at both subsonic and supersonic flow conditions in the combustor, as required. Diagram courtesy of NASA Langley Research Center

As noted earlier Chap. 5, pulse detonation engines (PDEs) at some point may prove to be a competitive alternative to conventional RJs, at least in the sub-hypersonic flight regime ($2 < Ma_\infty < 5$). Thermodynamic cycle analyses illustrate that constant-volume combustors, such as the PDE (using a Humphrey cycle; pressure rises considerably during the transient wave-driven combustion process, as per Fig. 13.4), are more efficient than conventional constant-pressure combustors using the Brayton cycle. PDEs may potentially produce an I_{sp} of around 2,500 s using conventional hydrocarbon fuels, as compared to somewhat under 2000 s delivered from a conventional ramjet.

13.2 Performance Considerations

The ideal limit performance for aerodynamic ram compression in the air intake (going from ambient outside air pressure at station ∞ to the intake entrance at station 1, and from there through the intake/diffuser to the diffuser exit at station 2) is the isentropic (no non-isentropic shock waves being present ahead of the intake, or within the intake/diffuser) case, as follows:

$$\frac{p_2}{p_\infty} = \left\{ \frac{2 + (\gamma_a - 1)Ma_\infty^2}{2 + (\gamma_a - 1)Ma_2^2} \right\}^{\frac{\gamma_a}{\gamma_a - 1}} \tag{13.1}$$

where γ_a (for air) is nominally around 1.4, although at the elevated temperatures of higher compression levels, may drop in value. Of course, in practice, oblique and/

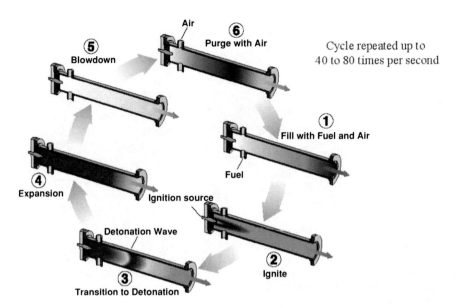

Fig. 13.4 Schematic diagram of operational cycle of a classical PDE. The quoted cycle time will depend in part on the length of the detonation tube, and the strength of the detonation waves being produced. Courtesy of DARPA

or normal shock waves may be present as part of the compression process in an actual intake, and thus the ram compression in practice may be significantly less than that predicted by the above equation. A normal (bow) shock ahead of the intake as the worst case would produce a static pressure ratio to the intake entrance as follows [3]:

$$\frac{p_1}{p_\infty} = \frac{1 + \gamma_a Ma_\infty^2}{1 + \gamma_a Ma_1^2} = \left(\frac{2\gamma_a}{\gamma_a + 1}\right)Ma_\infty^2 - \left(\frac{\gamma_a - 1}{\gamma_a + 1}\right) \tag{13.2}$$

and static temperature ratio:

$$\frac{T_1}{T_\infty} = \left(\frac{1 + \gamma_a Ma_\infty^2}{1 + \gamma_a Ma_1^2}\right)\frac{Ma_1}{Ma_\infty}\left[\frac{1 + \left(\frac{\gamma_a - 1}{2}\right)Ma_\infty^2}{1 + \left(\frac{\gamma_a - 1}{2}\right)Ma_1^2}\right]^{1/2} \tag{13.3}$$

where the approximate flow Mach number at or near the entrance, downstream of the normal shock, would be [3]:

$$Ma_1 = \left[\frac{Ma_\infty^2 + \dfrac{2}{\gamma_a - 1}}{\left(\dfrac{2\gamma_a}{\gamma_a - 1}\right)Ma_\infty^2 - 1}\right]^{1/2} \tag{13.4}$$

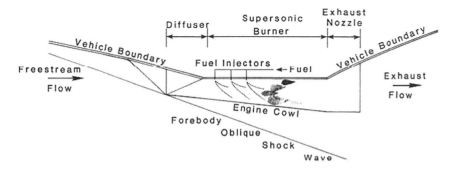

Fig. 13.5 Schematic diagram of a scramjet engine [4]. Reprinted with permission of the American Institute of Aeronautics and Astronautics

The peak adiabatic flame temperature for the fuel–air mixture will act to limit the ability of a ramjet to produce thrust above a certain flight Mach number. For example, a conventional jet fuel such as Jet A at stoichiometric conditions produces a flame temperature of around 2,500 K at the operational pressures under consideration here. Substituting this value as the stagnation temperature limit value in the isentropic flow relation for ideal compression at the diffuser exit (combustor inlet) for flight in the lower stratosphere ($T_\infty = 217\,\text{K}$) indicates the following:

$$T_{02,\text{lim}} = 2500 = T_\infty\left(1 + \left(\frac{\gamma_a - 1}{2}\right)Ma^2_{\infty,\text{lim}}\right) = 217(1 + 0.2Ma^2_{\infty,\text{lim}})$$

$$Ma_{\infty,\text{lim}} \approx 7.25$$

The above flight Mach number is reflective of the performance limitations expected for a conventional ramjet in going above approximately $Ma_\infty 5.0$, as noted earlier. Allowing the gas flow through the combustor to be supersonic, as done with scramjets (Fig. 13.5; [4]), keeps diffuser exit/combustor inlet static temperature T_2 substantially below the combustor exit static temperature T_4 at higher supersonic flight Mach numbers, and thus retain the ability to deliver thrust in turn for Ma_∞ values well above 10. In moving above $Ma_\infty 5.0$, one should be aware that real gas effects (which cause a deviation from ideal-gas behavior) will become progressively more significant, especially in those regions of highly shocked, high-temperature flow. Dissociation (molecular breakdown) and ionization (loss of electrons) of the air or gas downstream of strong shock wave fronts will at some point need to be accounted for in one's calculations, e.g., via the use of Mollier enthalpy diagrams [5].

13.3 Combined-Cycle Engine Technology

The above systems, RJs and SJs, by themselves are not normally referred to as rocket engines. Air-breathing rocket engines not only employ some aspects of RJ technology for at least some part of the flight mission, but also some rocket-based technology for other portions of the flight. This approach is an extension of that seen for earlier "variable-cycle" gas turbine engines that transform from a turbofan mode at lower flight speeds to a zero-bypass turbojet at higher flight speeds. "Multi-mode", "mixed-mode" or "combined-cycle" engines [5, 7] provide a single propulsion system for flight (at a relatively high I_{sp}, with intake of outside air), over a typically wide speed-altitude spectrum (e.g. the Aerojet Strutjet (Fig. 13.6; [8]), a combined-cycle rocket/scramjet/ramjet, proposed as a single-stage-to-orbit [SSTO] propulsion system for flight from $Ma_\infty 0.0$ at sea level to $Ma_\infty 8.0$ at 140 kft [43 km] altitude with air intake, and thereafter in non-air breathing rocket mode). Combined-cycle engines may employ an ejector approach for some part of the flight, simultaneously combining (mixing) and exhausting the hot streams of both cycles (ramjet/scramjet + rocket; see examples of Figs. 13.7, 13.8; [9]). Combined-cycle engines can be rocket-based (RBCC; rocket + ramjet/scramjet), or turbine-based (TBCC; gas turbine + ramjet/scramjet). Examples of both are discussed below.

At the lower speed aircraft regime, variable-cycle (TF to TJ), turboramjet or airturboramjet (ATRJ; TJ to RJ), and turbofan/ramjet (TF to RJ) engines have been proposed for supersonic civil transports. The rotating turbomachinery would be stopped in mid-flight and bypassed once in RJ mode, which requires no mechanical compressor (ram air compression at intake is sufficient). Conventional RJ-powered vehicles (missiles for the most part) have largely employed a separate SRM booster to go from low speed to an adequate supersonic flight Ma_∞ for RJ operation (with the booster either separated or jettisoned from the vehicle once burnt out, or the empty SRM combustion chamber retained or integral with the vehicle). In this regard, the integral rocket ramjet (IRR; see Fig. 13.9; [10]) uses the emptied SRM combustor as the RJ combustor at takeover, with the RJ fuel being liquid, solid or gel (gels are thicker/denser than liquids, with less sloshing… two advantages). One may also see the use of slurry fuels (insoluble solid particles immersed in a liquid mixture; e.g., boron particles immersed in kerosene, or aluminum powder immersed in water that is subsequently chilled to an ice-based mixture) to gain some advantages seen with gel fuels. A boost-phase nozzle insert may be ejected to provide a better cruise nozzle setting. The ducted rocket (DR) differs from the SFRJ in using a solid fuel/low oxidizer mix as the fuel-rich gas generator in more of an afterburner approach, allowing for lower flight speed operation versus a pure RJ, at the expense of a lower effective I_{sp} (Fig. 13.10; [10]). IRRs and DRs would typically be used for lower-cost, short-life applications such as one-flight-only missiles.

A typical I_{sp} for an air-breathing rocket engine such as an IRR operating in RJ mode during cruise may range from 1,000 to 2,000 s using conventional

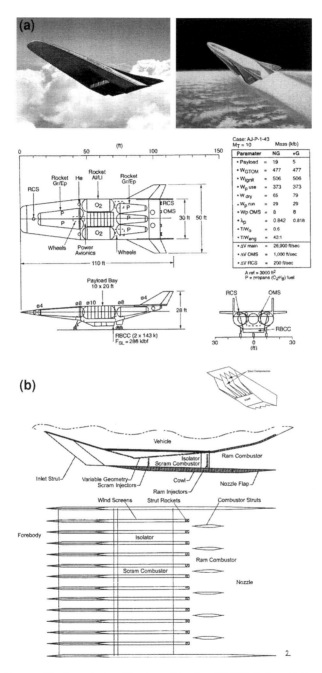

Fig. 13.6 a Proposed horizontal takeoff/horizontal landing (HTHL) flight vehicle employing two ventral Aerojet Strutjet engines, as part of the National AeroSpace Plane program. Cancelled circa 1994, the NASA X-30, to employ one Strutjet engine, was the first subscale prototype to be built (partially), but never flown. The large storage tanks contain liquid propane (as the fuel) and liquid oxygen (as the oxidizer, to be used when air intake is insufficient). Courtesy of NASA. **b** Aerojet Strutjet engine schematic diagram. Courtesy of NASA

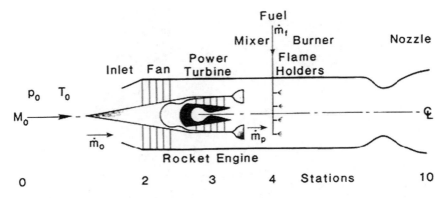

Fig. 13.7 Schematic diagram of a type of ejector ramjet (in this case, a turboramjet rocket, a.k.a., airturborocket, where the power turbine is driven by a liquid or solid fuel gas generator; [4]). Reprinted with permission of the American Institute of Aeronautics and Astronautics

Fig. 13.8 Schematic diagram of a type of ejector ramjet/scramjet (in this case, a ram rocket variant; [9]). Reprinted with permission of Cambridge University Press

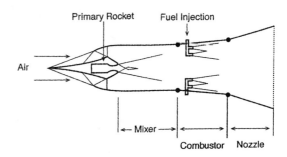

hydrocarbon fuel (see Fig. 13.11, for comparison of performance using standard hydrdocarbon fuels, versus fast-burning, high-energy hydrogen, which has a heat of reaction with air from 120 to 140 MJ/kg of H_2). For thrust performance under RJ/air induction mode, note the inclusion of the momentum drag term on net thrust delivered:

$$F = \dot{m}_e u_e - \dot{m}_{air} V_\infty + (p_e - p_\infty)A_e \qquad (13.5)$$

Given the parameter f is the fuel–air mixture ratio ($\dot{m}_F/\dot{m}_{air}$), one can note the following for the exiting mass flow:

$$\dot{m}_e = \dot{m}_{air} + \dot{m}_F = (1+f)\dot{m}_{air} \qquad (13.6)$$

For RJ operation, f is typically from 0.05 to 0.10 to produce a combustor temperature high enough to deliver the required thrust [no turbomachinery cooling concerns], and also depending on the fuel being used. Bypass IRRs allow for the ducting of some outside air to a mixing chamber (emptied of

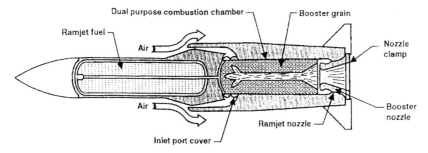

Liquid Fuel Ramjet

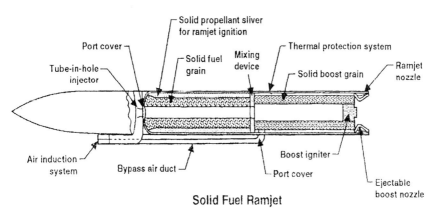

Solid Fuel Ramjet

Fig. 13.9 Schematic diagrams of a missile powered by an integral rocket ramjet engine (*above*, liquid fuel; *below*, solid fuel; [10]). The solid propellant grain is shown in place in the far right chamber; once emptied after the boost phase, the chamber becomes the aft combustor section for the sustainer ramjet engine stage using in these examples liquid fuel and solid fuel (*in the middle chamber*) that will react with incoming air. Diagrams reprinted with permission of the American Institute of Aeronautics and Astronautics

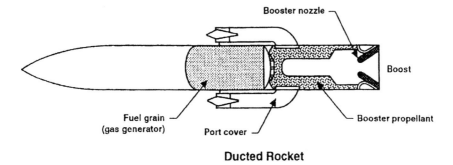

Ducted Rocket

Fig. 13.10 Schematic diagram of a ducted rocket [10]. Reprinted with permission of the American Institute of Aeronautics and Astronautics

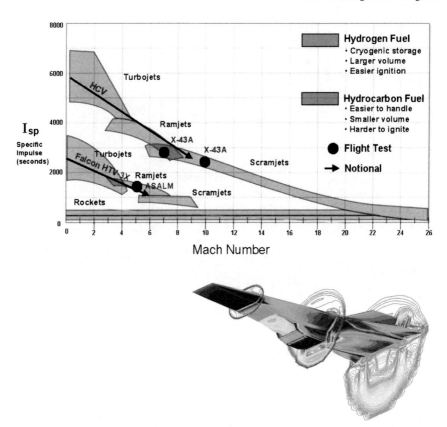

Fig. 13.11 Comparison of performance of various propulsion systems using conventional hydrocarbon fuels and hydrogen, as tied to vehicle flight Mach number. Both the Falcon HTV-3X and HCV are DARPA-proposed hypersonic vehicles for various applications. The X-43A Hyper-X is a NASA hypersonic technology demonstrator vehicle using a scramjet engine (*lower CFD diagram showing airflow contours around the vehicle at Ma$_\infty$ of* 7). The ASALM is the ramjet-powered Advanced Strategic Air-Launched Missile built by Martin Marietta and studied for cruise missile applications in the late 1970s, but never entered production. Courtesy of DARPA and NASA

SRM booster propellant) aft of the main combustor, to provide a more efficient combustion process (i.e. extend the combustion residence time). DRs commonly use the aft mixing approach as well, but if the aerodynamic ram compression is sufficiently high, the intake of air can take place ahead of the solid fuel grain for some portion of the flight (e.g. forward air intake closed at lower flight speeds, to prevent backflow). A dual-mode DR employs a combination of the conventional gas-generator approach with the higher-speed forward-ram approach.

Higher cost, longer-life applications for supersonic/hypersonic flight vehicles would more likely bring into consideration the use of gas turbine components for

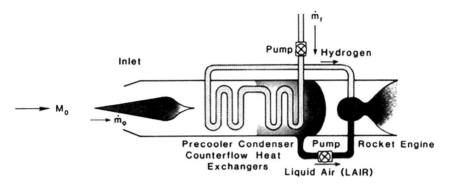

Fig. 13.12 Schematic diagram of a LACE propulsion system [4]. Reprinted with permission of the American Institute of Aeronautics and Astronautics

a combined-cycle engine, as is the case for the ATRJ noted above. For example, the airturborocket (ATR; see Fig. 13.7) uses a liquid or solid fuel-rich gas generator to provide a hot, high-pressure flow over a turbine for driving the forward compressor at lower flight speeds. The rotating machinery can be stopped/bypassed at higher supersonic speeds, and the system operated in ducted rocket mode. Moving into the high-end SSTO applications, the rocket/scramjet/ ramjet example was noted above. Another example is the liquid-air cycle engine (LACE; see Fig. 13.12) system proposed for the now-defunct HOTOL vehicle (horizontal takeoff and landing, SSTO). The engine had the proposed capability to compress and cool incoming air to liquefaction, and used in conjunction with on-board LH_2 and LO_2, to give high I_{sp} flight performance in moving up to orbital speed and altitude. A lower performance engine than the LACE is the cooled-air cycle engine (CACE), with a version called the *synergic air breathing engine* (SABRE) proposed for the Skylon launch vehicle (see Fig. 13.13). The SABRE engine is a combined-cycle rocket engine utilizing a LRE/TJ/RJ combination.

Fig. 13.14 illustrates the various constraints that can come into play when introducing supersonic and hypersonic commercial flight operations. The upper boundary is the required minimum flight vehicle dynamic pressure ($\frac{1}{2}\rho_\infty V_\infty^2$, e.g. for $Ma_\infty \sim 2$, $q_\infty \sim 8$ kPa at 24.4 km altitude, or 180 psf at 80 kft) needed for adequate air massflow and corresponding airbreathing engine operation (note: available aerodynamic lift may also be an issue). The lower boundary for supersonic and low hypersonic flight, a vehicle dynamic pressure of around 750 psf or 36 kPa, is the anticipated aerodynamic and aeroheating structural loading limit on the vehicle's airframe, given present technology. At higher hypersonic flight Mach numbers, $Ma_\infty > 5.5$ in the case of Fig. 13.14, the lower boundary becomes the allowed internal pressure loading limit on the intake duct structure, at that time (for the flight vehicles under study) established as being 100 psia or 690 kPa absolute. The upper and lower boundaries eventually meet at around a

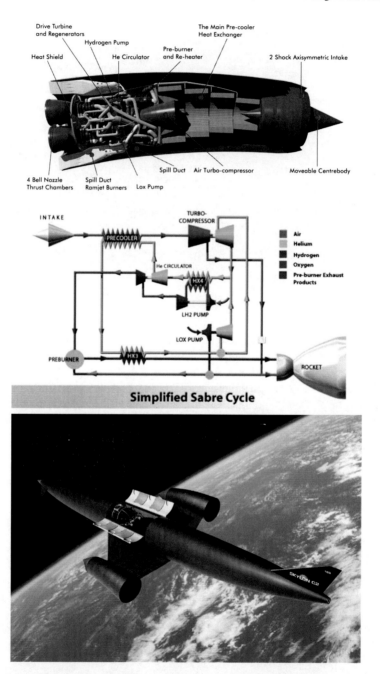

Fig. 13.13 Schematic diagrams of the proposed SABRE engine (above, one of two positioned port and starboard) proposed for usage by the Skylon single-stage-to-orbit flight vehicle (*lower illustration*). Courtesy of Reaction Engines Ltd

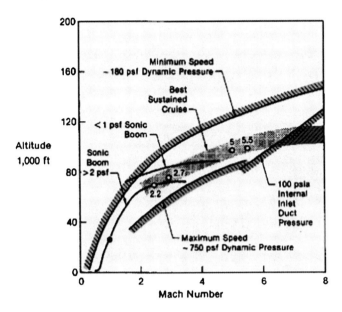

Fig. 13.14 Anticipated flight envelope, altitude vs. flight Mach number, for commercial supersonic and hypersonic flight vehicle operations. Sonic boom limits for sound pressure waves reaching the ground are also indicated. Graph from a McDonnell Douglas study, circa 1970s

flight Mach number of 9, and an altitude of 150 kft (46 km), the apparent closing of the allowed flight envelope, the so-called *hypersonic funnel*. Sonic boom limits for sound pressure waves produced by the flight vehicle moving through the air at that speed and altitude reaching the ground are also indicated in Fig. 13.14 (1 and 2 psf sound pressure limits, or 48 and 96 Pa). Obviously, the higher the flight vehicle can fly, the weaker the resulting sound waves that reach ground level. A newer hypersonic flight envelope chart, Fig. 13.15, shows the flight constraints on one currently proposed hypersonic vehicle design, the Lockheed Martin Falcon HTV-3X, and the expected flight corridor for future, bigger hypersonic flight vehicles (with a comparison to the Space Shuttle's performance limits in powered ascent to orbit, and in unpowered descent/re-entry from orbit). The HTV-3X has a right vertical boundary that is presumably a result of available engine thrust equaling aerodynamic drag in level flight at a flight Mach number of 6, or the aerodynamic structural/heating load limit in a dive at that Mach number.

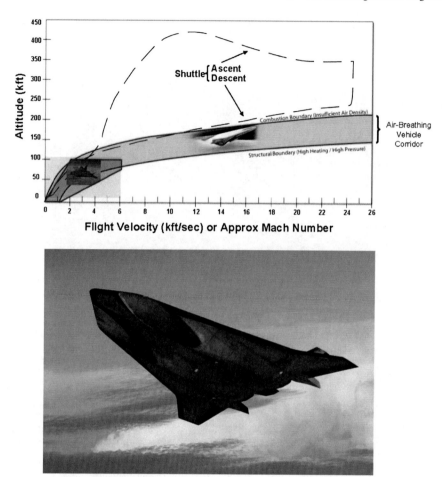

Fig. 13.15 Anticipated flight envelope, altitude versus flight Mach number, for commercial supersonic and hypersonic flight vehicle operations. The smaller closed envelope is for the Falcon HTV-3X (*"Blackswift"*) hypersonic technology demonstrator under study by Lockheed Martin and DARPA. The larger unclosed envelope (corridor), below the Space Shuttle's ascent/descent flight performance limits, is for future hypersonic vehicles with capabilities superseding those of the TBCC-powered HTV-3X. The lower diagram is an artist's conception of a possible HTV-3 variant. Courtesy of DARPA

13.4 Example Problems

13.1. Perform a design-point cycle analysis for a non-afterburning ramjet engine at its cruise Mach number of 2.8 and altitude of 12,200 m. With respect to the analysis, simply remove the compressor and turbine sections from the turbojet engine cycle, and assume ideal isentropic compression for the intake. Take note of the following design parameters:

diffuser, $\eta_d = 0.95, \gamma_d = 1.4; C_p$ entering burner $= 1{,}062$ J/(kg K)

burner, $\eta_b = 0.98, \gamma_b = 1.35, \pi_b = 0.94, q_R = 45$ MJ/kg of fuel, $T_{max} = 1{,}500$ K

nozzle, $\eta_n = 0.98, \gamma_n = 1.36$

Ascertain the specific thrust and TSFC. Compare these values to that seen for the TJ of Problem 6.2. Note: the fully expanded nozzle solution ("unchoked" case for TJ cycle analysis) is likely more applicable if one were using a convergent-divergent (CD) nozzle that is well-matched with the outside air pressure at cruise. Ramjets usually employ a CD nozzle.

13.2. Repeat the above analysis, but for a cruise Mach number of 3.3 at an altitude of 12,200 m. Ascertain specific thrust and TSFC, and compare to the previous result above.

13.3. Repeat the analysis of Problem 13.1, but for a $T_{max} = 2{,}500$ K.

13.4. Repeat the analysis of Problem 13.3, but for a $q_R = 140$ MJ/kg of fuel (H_2, as opposed to a conventional hydrocarbon fuel like kerosene).

13.5. Compare the compression effectiveness between an ideal isentropic compression process in going from a flow Mach number of 3.0 down to a flow Mach number of 0.475, in air, relative to the comparable process of passing through the normal shock section of a bow shock in front of a flight vehicle.

13.5 Solutions to Example Problems

13.1.

Looking at a ramjet engine, at a flight Mach no. Ma_∞ of 2.8, at 12,200 m altitude ($p_\infty = 18.75$ kPa, $T_\infty = 216.7$ K, $\rho_\infty = 0.30$ kg/m^3).

Starting from intake (diffuser):

$$T_{02} = T_\infty \left[1 + \tfrac{\gamma_d - 1}{2} Ma_\infty^2\right] = 216.7[1 + 0.2(2.8^2)] = 556K, \text{ at diffuser outlet}$$

$$p_{02} = p_\infty \left[1 + \eta_d(\tfrac{T_{02}}{T_\infty} - 1)\right]^{\frac{\gamma_d}{\gamma_d - 1}} = 18.75[1 + 0.97(556/216.7 - 1)]^{3.5} = 455.2 \text{ kPa},$$

at diffuser outlet.

No mechanical compressor, so $p_{03} = p_{02} = 455.2$ kPa, $T_{03} = T_{02} = 556$ K.

Moving through combustor (burner):

$$\bar{C}_{p,b} \approx \tfrac{C_{p,c} + C_{p,b,exit}}{2} = \tfrac{\gamma_b R_{air}}{\gamma_b - 1} = 1.35(287)/(1.35 - 1) = 1107 \text{ J/(kg K)}, \text{ so that at}$$

burner exit:

$$C_{p,b,\text{exit}} \approx 2\bar{C}_{p,b} - C_{p,b,\text{in}} = 2(1107) - 1062 = 1152 \text{ J}/(\text{kg K}).$$

$T_{04} = T_{\text{max}} = 1500 \text{ K}, p_{04} = \pi_b p_{03} = 0.94(455.2) = 427.9 \text{ kPa}$, at burner exit.

$$f = \frac{\dfrac{T_{04}}{T_{03}} - \dfrac{C_{p,b,\text{in}}}{C_{p,b,\text{exit}}}}{\dfrac{\eta_b q_R}{C_{p,b,\text{exit}} T_{03}} - \dfrac{T_{04}}{T_{03}}} = \frac{\dfrac{1500}{556} - \dfrac{1062}{1152}}{\dfrac{0.98(45 \times 10^6)}{1152(556)} - \dfrac{1500}{556}} = 0.0268, \text{ fuel–air ratio}$$

No turbine section in ramjet engine, and no afterburner in this case:

$T_{06} = T_{05} = T_{04} = 1{,}500 \text{ K}$, and $p_{06} = p_{05} = p_{04} = 427.9 \text{ kPa}$. Now, assume one is approaching subsonic converging nozzle exit (or nozzle throat, if a converging–diverging supersonic nozzle), need to check for choking (attaining sonic flow):

$$\frac{p_{06}}{p_\infty} = \frac{427.9}{18.75} = 22.8 \text{ and choking criterion } \left(\frac{\gamma_n + 1}{2}\right)^{\frac{\gamma_n}{\gamma_n - 1}} = 1.87 < \frac{p_{06}}{p_\infty}, \text{ so nozzle}$$

flow is *choked*.

$$C_{p,n} = \frac{\gamma_n R_{\text{air}}}{\gamma_n - 1} = 1084 \text{ J}/(\text{kg K}).$$

$$V_e = \sqrt{2\eta_n C_{p,n} T_{06}\left[\frac{\gamma_n - 1}{\gamma_n + 1}\right]} = \sqrt{2(0.98)1084(1500)[0.36/2.36]} = 697 \text{ m/s}$$

(note : $1{,}340$ m/s, if allowed to fully expand to local atmospheric pressure).

$T_7 = \dfrac{2}{\gamma_n + 1} T_{06} = 2/2.36 \cdot 1500 = 1271 \text{ K}$, exit nozzle static temperature at sonic flow condition.

$$p_7 = p_{06}\left[\frac{T_7}{T_{06}}\right]^{\frac{\gamma_n}{\gamma_n - 1}} = p_{06}\left[\frac{2}{\gamma_n + 1}\right]^{\frac{\gamma_n}{\gamma_n - 1}} = 427.9[2/2.36]^{3.778} = 229 \text{ kPa} = p_e > p_\infty.$$

For choked flow,

$$\frac{A_e}{\dot{m}} = \left[\frac{\gamma_n + 1}{2}\right]^{\frac{1}{\gamma_n - 1}}\left[\frac{\gamma_n + 1}{2\gamma_n R T_{06}}\right]^{0.5}\frac{R T_{06}}{p_{06}}$$

$$= \left[\frac{2.36}{2}\right]^{2.778}\left[\frac{2.36}{2(1.36)287(1500)}\right]^{0.5}\frac{287(1500)}{427900} = 0.00226 \text{ m}^2\text{s/kg}.$$

$V_\infty = a_\infty M a_\infty = \sqrt{\gamma_{\text{air}} R_{\text{air}} T_\infty}\, M a_\infty = 295(2.8) = 826 \text{ m/s}.$

For specific thrust,

$$\frac{F}{\dot{m}_a} = (1 + f)V_e - V_\infty + \frac{A_e}{\dot{m}}(p_e - p_\infty)(1 + f)$$

$$= (1 + 0.0268)697 - 826 + 0.00226(229000 - 18750)(1.0268)$$

$= 378$ (N s)/kg or 0.378 (kN s)/kg (if had been fully expanded, 0.547 (kN s)/kg), versus 0.76 for TJ of Problem No. 6.2.

$$\text{TSFC} = \frac{\dot{m}_f}{F} = \frac{f}{\frac{F}{\dot{m}_a}} = 0.0268/378 = 7.09 \times 10^{-5} \text{ (s N)} \text{ or } 0.256 \text{ kg/(h N)}$$

or 2.5 lb/(h lb), versus 1.2 lb/(h lb) for TJ of Problem No. 6.2

Fully expanded, TSFC $= 0.0268/547 = 4.9 \times 10^{-5}$ kg/(s N) or 0.177 kg/(h N) or 1.74 lb/(h lb).

13.2. Looking at a ramjet engine, at a flight Mach no. Ma_∞ of 3.3, at 12,200 m altitude ($p_\infty = 18.75$ kPa, $T_\infty = 216.7$ K, $\rho_\infty = 0.30$ kg/m^3).

Starting from intake (diffuser):

$$T_{02} = T_\infty \left[1 + \tfrac{\gamma_d - 1}{2} Ma_\infty^2 \right] = 216.7[1 + 0.2(3.3^2)] = 689 \text{ K, at diffuser outlet}$$

$$p_{02} = p_\infty \left[1 + \eta_d \left(\frac{T_{02}}{T_\infty} - 1 \right) \right]^{\frac{\gamma_d}{\gamma_d - 1}} = 18.75[1 + 0.97(689/216.7 - 1)]^{3.5} = 951.2 \text{ kPa, at}$$

diffuser outlet.

$\pi_d = 50.7$ for diffuser ram compression.

No mechanical compressor, so $p_{03} = p_{02} = 951.2$ kPa, $T_{03} = T_{02} = 689$ K.

Moving through combustor (burner):

$$\bar{C}_{p,b} \approx \frac{C_{p,c} + C_{p,b,\text{exit}}}{2} = \frac{\gamma_b R_{\text{air}}}{\gamma_b - 1} = 1.35(287)/(1.35 - 1) = 1107 \text{ J/(kg K), so that at}$$

burner exit:

$$C_{p,b,\text{exit}} \approx 2\bar{C}_{p,b} - C_{p,b,\text{in}} = 2(1107) - 1062 = 1152 \text{ J/(kg K)}.$$

$T_{04} = T_{\max} = 1500$ K, $p_{04} = \pi_b p_{03} = 0.94(951.2) = 894.1$ kPa, at burner exit.

$$f = \frac{\frac{T_{04}}{T_{03}} - \frac{C_{p,b,\text{in}}}{C_{p,b,\text{exit}}}}{\frac{\eta_b q_R}{C_{p,b,\text{exit}} T_{03}} - \frac{T_{04}}{T_{03}}} = \frac{\frac{1500}{689} - \frac{1062}{1152}}{\frac{0.98(45 \times 10^6)}{1152(689)} - \frac{1500}{689}} = 0.0235, \text{ fuel–air ratio}$$

No turbine section in ramjet engine, and no afterburner in this case:

$T_{06} = T_{05} = T_{04} = 1500$ K, and $p_{06} = p_{05} = p_{04} = 894.1$ kPa. Now, assume one is approaching subsonic converging nozzle exit (or nozzle throat, if a converging–diverging supersonic nozzle), need to check for choking (attaining sonic flow):

$$\frac{p_{06}}{p_\infty} = \frac{894.1}{18.75} = 47.7 \text{ and choking criterion } \left(\frac{\gamma_n + 1}{2} \right)^{\frac{\gamma_n}{\gamma_n - 1}} = 1.87 < \frac{p_{06}}{p_\infty}, \text{ so nozzle}$$

flow is *choked*.

$$C_{p,n} = \frac{\gamma_n R_{\text{air}}}{\gamma_n - 1} = 1084 \text{J/(kg K)}.$$

$$V_e = \sqrt{2\eta_n C_{p,n} T_{06} \left[\frac{\gamma_n - 1}{\gamma_n + 1} \right]} = \sqrt{2(0.98)1084(1500)[0.36/2.36]} = 697 \text{ m/s (note:1429 m/s, if}$$

allowed to fully expand to local atmospheric pressure).

$T_7 = \frac{2}{\gamma_n + 1} T_{06} = 2/2.36 \times 1500 = 1271$ K, exit nozzle static temperature at sonic flow condition.

$$p_7 = p_{06} \left[\frac{T_7}{T_{06}}\right]^{\frac{\gamma_n}{\gamma_n - 1}} = p_{06} \left[\frac{2}{\gamma_n + 1}\right]^{\frac{\gamma_n}{\gamma_n - 1}} = 894.1[2/2.36]^{3.778} = 478.4 \text{ kPa} = p_e > p_\infty.$$

For choked flow,

$$\frac{A_e}{\dot{m}} = \left[\frac{\gamma_n + 1}{2}\right]^{\frac{1}{\gamma_n - 1}} \left[\frac{\gamma_n + 1}{2\gamma_n RT_{06}}\right]^{0.5} \frac{RT_{06}}{p_{06}}$$

$$= \left[\frac{2.36}{2}\right]^{2.778} \left[\frac{2.36}{2(1.36)287(1500)}\right]^{0.5} \frac{287(1500)}{894100} = 0.001083 \text{ m}^2\text{s/kg}.$$

$$V_\infty = a_\infty Ma_\infty = \sqrt{\gamma_{air} R_{air} T_\infty} \, Ma_\infty = 295(3.3) = 973.5 \text{ m/s.} \qquad \text{For specific}$$
thrust,

$$\frac{F}{\dot{m}_a} = (1+f)V_e - V_\infty + \frac{A_e}{\dot{m}}(p_e - p_\infty)(1+f)$$

$$= (1 + 0.0235)697 - 973.5 + 0.001083(478400 - 18750)(1.0235)$$

$$= 249.4 \text{ (N s)/kg or } 0.249 \text{ (kN s)/kg (if had been fully-expanded, } 0.489 \text{ (kN s)/kg,}$$
versus 0.547 for fully expanded RJ of Problem No. 13.1.

$$\text{TSFC} = \frac{\dot{m}_f}{F} = \frac{f}{\frac{F}{\dot{m}_a}} = 0.0235/249.4$$

$$= 9.42 \times 10^{-5} \text{kg/(s N) or } 0.339 \text{ kg/(h N) or } 3.32 \text{ lb/(h lb)},$$

versus 1.74 lb/(h lb) for fully-expanded RJ of Problem No. 13.1.

Fully expanded, TSFC $= 0.0235/489 = 4.8 \times 10^{-5}$ kg/(s N) or 0.173 kg/(h N) or 1.70 lb/(h lb).

13.3. Looking at a ramjet engine, at a flight Mach no. Ma_∞ of 2.8, at 12,200 m altitude ($p_\infty = 18.75$ kPa, $T_\infty = 216.7$ K, $\rho_\infty = 0.30$ kg/m^3).

Starting from intake (diffuser):

$$T_{02} = T_\infty \left[1 + \tfrac{\gamma_d - 1}{2} Ma_\infty^2\right] = 216.7[1 + 0.2(2.8^2)] = 556 \text{ K, at diffuser outlet}$$

$$p_{02} = p_\infty [1 + \eta_d(\tfrac{T_{02}}{T_\infty} - 1)]^{\frac{\gamma_d}{\gamma_d - 1}} = 18.75[1 + 0.97(556/216.7 - 1)]^{3.5} = 455.2 \text{ kPa, at}$$
diffuser outlet

No mechanical compressor, so $p_{03} = p_{02} = 455.2$ kPa, $T_{03} = T_{02} = 556$ K.

Moving through combustor (burner):

$$\bar{C}_{p,b} \approx \frac{C_{p,c} + C_{p,b,exit}}{2} = \frac{\gamma_b R_{air}}{\gamma_b - 1} = 1.35(287)/(1.35 - 1) = 1107 \text{ J/(kg K), so that at}$$
burner exit:

$$C_{p,b,exit} \approx 2\bar{C}_{p,b} - C_{p,b,in} = 2(1107) - 1062 = 1152 \text{J/(kg K)}.$$

$T_{04} = T_{max} = 2500$ K , $p_{04} = \pi_b p_{03} = 0.94(455.2) = 427.9$ kPa, at burner exit.

$$f = \frac{\dfrac{T_{04}}{T_{03}} - \dfrac{C_{p,b,in}}{C_{p,b,exit}}}{\dfrac{\eta_b q_R}{C_{p,b,exit} T_{03}} - \dfrac{T_{04}}{T_{03}}} = \frac{\dfrac{2500}{556} - \dfrac{1062}{1152}}{\dfrac{1.0(45 \times 10^6)}{1152(556)} - \dfrac{2500}{556}} = 0.054, \text{ fuel–air ratio}$$

No turbine section in ramjet engine, and no afterburner in this case:
$T_{06} = T_{05} = T_{04} = 2{,}500$ K, and $p_{06} = p_{05} = p_{04} = 427.9$ kPa. Now, assume one is approaching subsonic converging nozzle exit (or nozzle throat, if a converging–diverging supersonic nozzle), need to check for choking (attaining sonic flow):

$$\frac{p_{06}}{p_\infty} = \frac{427.9}{18.75} = 22.8 \text{ and choking criterion } \left(\frac{\gamma_n + 1}{2}\right)^{\frac{\gamma_n}{\gamma_n - 1}} = 1.87 < \frac{p_{06}}{p_\infty}, \text{ so nozzle}$$

flow is *choked*.

$$C_{p,n} = \frac{\gamma_n R_{air}}{\gamma_n - 1} = 1084 \text{ J/(kg K)}.$$

$$V_e = \sqrt{2\eta_n C_{p,n} T_{06}\left[\frac{\gamma_n - 1}{\gamma_n + 1}\right]} = \sqrt{2(0.98)1084(2500)[0.36/2.36]} = 900 \text{ m/s (note : } 1{,}730 \text{ m/s,}$$

if allowed to fully expand to local atmospheric pressure).

$T_7 = \frac{2}{\gamma_n + 1} T_{06} = 2/2.36 \times 2500 = 2119$ K, exit nozzle static temperature at sonic flow condition.

$$p_7 = p_{06}\left[\frac{T_7}{T_{06}}\right]^{\frac{\gamma_n}{\gamma_n - 1}} = p_{06}\left[\frac{2}{\gamma_n + 1}\right]^{\frac{\gamma_n}{\gamma_n - 1}} = 427.9[2/2.36]^{3.778} = 229 \text{ kPa } = p_e > p_\infty.$$

For choked flow,

$$\frac{A_e}{\dot{m}} = \left[\frac{\gamma_n + 1}{2}\right]^{\frac{1}{\gamma_n - 1}}\left[\frac{\gamma_n + 1}{2\gamma_n RT_{06}}\right]^{0.5}\frac{RT_{06}}{p_{06}}$$

$$= \left[\frac{2.36}{2}\right]^{2.778}\left[\frac{2.36}{2(1.36)287(2500)}\right]^{0.5}\frac{287(2500)}{427900} = 0.00292 \text{ m}^2\text{s/kg}.$$

$V_\infty = a_\infty Ma_\infty = \sqrt{\gamma_{air} R_{air} T_\infty} \, Ma_\infty = 295(2.8) = 826$ m/s. For specific thrust,

$$\frac{F}{\dot{m}_a} = (1 + f)V_e - V_\infty + \frac{A_e}{\dot{m}}(p_e - p_\infty)(1 + f)$$

$$= (1 + 0.054)900 - 826 + 0.00292(229000 - 18750)(1.054)$$

$= 770$ (N s)/kg or 0.770 (kN s)/kg (if had been fully expanded, 0.997 (kN s)/kg), versus 0.547 for fully-expanded RJ of Problem No. 13.1.

TSFC $= \frac{\dot{m}_f}{F} = \frac{f}{\frac{F}{\dot{m}_a}} = $ $0.054/770 = 7.0 \times 10^{-5}$ kg/(s N) or 0.252 kg/(h N) or 2.47 lb/(h lb), versus 1.74 lb/(h lb) for fully-expanded RJ of Problem No. 13.1.

Fully expanded, TSFC $= 0.054/997 = 5.42 \times 10^{-5}$ kg/(s N) or 0.195 kg/(h N) or 1.91 lb/(h lb).

13.4. Looking at a ramjet engine, at a flight Mach No. Ma_∞ of 2.8, at 12,200 m altitude ($p_\infty = 18.75$ kPa, $T_\infty = 216.7$ K, $\rho_\infty = 0.30$ kg/m^3).

Starting from intake (diffuser):

$$T_{02} = T_\infty\left[1 + \frac{\gamma_d - 1}{2} Ma_\infty^2\right] = 216.7[1 + 0.2(2.8^2)] = 556 \text{ K, at diffuser outlet}$$

$$p_{02} = p_\infty \left[1 + \eta_d \left(\tfrac{T_{02}}{T_\infty} - 1\right)\right]^{\frac{\gamma_d}{\gamma_d - 1}} = 18.75[1 + 0.97(556/216.7 - 1)]^{3.5} = 455.2 \text{ kPa},$$

at diffuser outlet

No mechanical compressor, so $p_{03} = p_{02} = 455.2$ kPa, $T_{03} = T_{02} = 556$ K.

Moving through combustor (burner):

$$\bar{C}_{p,b} \approx \tfrac{C_{p,c} + C_{p,b,\text{exit}}}{2} = \tfrac{\gamma_b R_{\text{air}}}{\gamma_b - 1} = 1.35(287)/(1.35 - 1) = 1107 \ \text{J/(kg K)}, \text{ so that at}$$

burner exit:

$$C_{p,b,\text{exit}} \approx 2\bar{C}_{p,b} - C_{p,b,\text{in}} = 2(1107) - 1062 = 1152 \ \text{J/(kg K)}.$$

$T_{04} = T_{\max} = 2500$ K, $p_{04} = \pi_b p_{03} = 0.94(455.2) = 427.9$ kPa, at burner exit.

$$f = \frac{\dfrac{T_{04}}{T_{03}} - \dfrac{C_{p,b,\text{in}}}{C_{p,b,\text{exit}}}}{\dfrac{\eta_b q_R}{C_{p,b,\text{exit}} T_{03}} - \dfrac{T_{04}}{T_{03}}} = \frac{\dfrac{2500}{556} - \dfrac{1062}{1152}}{\dfrac{1.0(140 \times 10^6)}{1152(556)} - \dfrac{2500}{556}} = 0.0167, \text{ fuel–air ratio}$$

No turbine section in ramjet engine, and no afterburner in this case:

$T_{06} = T_{05} = T_{04} = 2{,}500$ K, and $p_{06} = p_{05} = p_{04} = 427.9$ kPa. Now, assume one is approaching subsonic converging nozzle exit (or nozzle throat, if a converging–diverging supersonic nozzle), need to check for choking (attaining sonic flow):

$$\frac{p_{06}}{p_\infty} = \frac{427.9}{18.75} = 22.8 \text{ and choking criterion } \left(\frac{\gamma_n + 1}{2}\right)^{\frac{\gamma_n}{\gamma_n - 1}} = 1.87 < \frac{p_{06}}{p_\infty} \text{ so nozzle}$$

flow is *choked*.

$$C_{p,n} = \frac{\gamma_n R_{\text{air}}}{\gamma_n - 1} = 1084 \ \text{J/(kg K)}.$$

$V_e = \sqrt{2\eta_n C_{p,n} T_{06} [\tfrac{\gamma_n - 1}{\gamma_n + 1}]} = \sqrt{2(0.98)1084(2500)[0.36/2.36]} = 900$ m/s (note : 1730 m/s, if allowed to fully expand to local atmospheric pressure).

$T_7 = \tfrac{2}{\gamma_n + 1} T_{06} = 2/2.36 \times 2{,}500 = 2{,}119$ K, exit nozzle static temperature at sonic flow condition.

$$p_7 = p_{06}\left[\frac{T_7}{T_{06}}\right]^{\frac{\gamma_n}{\gamma_n - 1}} = p_{06}\left[\frac{2}{\gamma_n + 1}\right]^{\frac{\gamma_n}{\gamma_n - 1}} = 427.9[2/2.36]^{3.778} = 229 \text{ kPa} = p_e > p_\infty. \qquad \text{For}$$

choked flow,

$$\begin{aligned}
\frac{A_e}{\dot{m}} &= \left[\frac{\gamma_n + 1}{2}\right]^{\frac{1}{\gamma_n - 1}} \left[\frac{\gamma_n + 1}{2\gamma_n R T_{06}}\right]^{0.5} \frac{R T_{06}}{p_{06}} \\
&= \left[\frac{2.36}{2}\right]^{2.778} \left[\frac{2.36}{2(1.36)287(2500)}\right]^{0.5} \frac{287(2500)}{427900} = 0.00292 \ \text{m}^2\text{s/kg}.
\end{aligned}$$

$V_\infty = a_\infty Ma_\infty = \sqrt{\gamma_{air} R_{air} T_\infty}\, Ma_\infty = 295(2.8) = 826$ m/s. For specific thrust,

$$\frac{F}{\dot{m}_a} = (1+f)V_e - V_\infty + \frac{A_e}{\dot{m}}(p_e - p_\infty)(1+f)$$

$$= (1 + 0.0167)900 - 826 + 0.00292(229000 - 18750)(1.0167)$$

$= 713$ (N s)/kg or 0.713 (kN s)/kg (if had been fully expanded, 0.933 (kN s)/kg, versus 0.547 for fully-expanded RJ of Problem No. 13.1.

TSFC $= \frac{\dot{m}_f}{F} = \frac{f}{\frac{F}{\dot{m}_a}} = 0.0167/713 = 2.34 \times 10^{-5}$ kg/(s N) or 0.084 kg/(h N) or 0.83 lb/(h lb), versus 1.74 lb/(h lb) for fully-expanded RJ of RJ of Problem No. 13.1 using conventional fuel.

Fully expanded, TSFC $= 0.0167/933 = 1.79 \times 10^{-5}$ kg/(s N) or 0.065 kg/(hr N) or 0.63 lb/(h b).

13.5. Isentropic compression:

$$\frac{p_2}{p_\infty} = \left\{\frac{2 + (\gamma_a - 1)Ma_\infty^2}{2 + (\gamma_a - 1)Ma_2^2}\right\}^{\frac{\gamma_a}{\gamma_a - 1}} = \left\{\frac{2 + 0.4(3^2)}{2 + 0.4(0.475^2)}\right\}^{3.5} = 31.47$$

Non-isentropic compression through a normal shock:

$$\frac{p_1}{p_\infty} = \frac{1 + \gamma_a Ma_\infty^2}{1 + \gamma_a Ma_1^2} = \left(\frac{2\gamma_a}{\gamma_a + 1}\right)Ma_\infty^2 - \left(\frac{\gamma_a - 1}{\gamma_a + 1}\right) = \left(\frac{2(1.4)}{1.4 + 1}\right)3^2 - \left(\frac{0.4}{2.4}\right)$$
$$= 10.33$$

$$Ma_1 = \left[\frac{Ma_\infty^2 + \frac{2}{\gamma_a - 1}}{\left(\frac{2\gamma_a}{\gamma_a - 1}\right)Ma_\infty^2 - 1}\right]^{1/2} = \left[\frac{3^2 + \frac{2}{0.4}}{\frac{2.8}{0.4} \times 3^2 - 1}\right]^{1/2} = 0.475, \text{ checks as same}$$

downstream subsonic flow Mach number as previous case.

So as a one-to-one comparison on static compression, compression through the normal shock section of the bow shock is approximately one-third as effective as the ideal isentropic compression case, for this example.

References

1. El-Sayed AF (2008) Aircraft propulsion and gas turbine engines. CRC Press, Boca Raton
2. Dudebout R, Sislian JP, Oppitz R (1998) Numerical simulation of hypersonic shock-induced combustion ramjets. J Propul Power 14:869–879
3. Fox RW, MacDonald AT (1978) Introduction to fluid mechanics, 2nd edn. Wiley, New York
4. Heiser WH, Pratt DT (1994) Hypersonic airbreathing propulsion. AIAA, Washington DC
5. John JEA (1984) Gas dynamics, 2nd edn. Prentice-Hall, Upper Saddle River
6. Farokhi S (2009) Aircraft propulsionAircraft propulsion. Wiley, New York

7. Forward RL (1995) Advanced propulsion systems. In: Humble RW, Henry GN, Larson WJ (eds) Space propulsionSpace propulsion analysis and design. McGraw-Hill, New York
8. Seibenhaar A, Bulman MJ (1995) The strutjet engine: the overlooked option for space launch. In: Proceedings 31st AIAA/ASME/SAE/ASEE Joint Propulsion Conference, San Diego, July 10–12
9. Segal C (2009) The scramjet engine—processes and characteristics. Cambridge University Press, Cambridge
10. Leisch S, Netzer DW (1996) Solid fuel ramjets. In: Jensen GE, Netzer DW (eds.) Tactical missile propulsion. AIAA, Reston (Virginia)

Chapter 14
Propulsion in Space: Beyond the Chemical Rocket

14.1 Introduction

As mankind moves beyond the confines of the Earth's atmosphere into space, the chemical rocket continues to play a key role in that venture. However, the vacuum and expanse of space does present an environment for innovation, where certain applications do allow for alternatives to the conventional chemical rocket approach for thrust delivery. As we will see, the physics of producing propulsive energy does not always require a combustion-based engine, and on occasion, the chemical rocket is not the best choice.

14.2 Electric Propulsion for Spaceflight Applications

With respect to more conventional systems used in space, satellites for example require frequent adjustments to their attitude and position via various approaches, including thrusters and non-propulsive devices like reaction wheels. At the very low end of performance, a pressurized gas like nitrogen in storage can be used as a straightforward cold-gas thruster (no combustion, but maybe some preheating before exhausting to get the performance up a bit). This can give a specific impulse from 200 s at the high-pressure end down to 50 s at the low end. This approach can be considered for short-life, low-cost, practical missions. For higher performance and longer-life missions, storable bipropellant (Fig. 14.1) or monopropellant (Fig. 14.2) liquid-propellant rocket engine (LRE) methods have been widely applied as low-thrust vernier thrusters, but electric propulsion (EP) techniques are becoming more popular with the advantages of higher specific impulse (I_{sp}) and lower system/propellant weights. Thrust levels with typical EP systems are currently quite low (on the order of μN's on the low end to low single-digit N's on the high end), so a lot more time may be required for a given flight maneuver than an LRE system (hours or days for an EP system, vs. seconds or minutes with a chemical system). For an EP system, a maneuver ΔV of 50 m/s per year may be a

D. R. Greatrix, *Powered Flight*, DOI: 10.1007/978-1-4471-2485-6_14,
© Springer-Verlag London Limited 2012

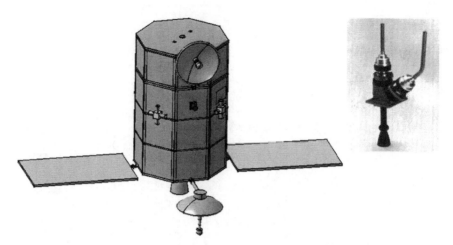

Fig. 14.1 Example diagram of a satellite using four-nozzle bipropellant thruster arrays positioned about the satellite's center of gravity, to provide attitude control. For newton-thrust-level applications such as the above, common bipropellant choices are nitrogen tetroxide/hydrazine (as for the example thruster shown at the *right*) and MON/MMH

typical requirement for a thruster, i.e., not a lot of kinetic energy imparted to the spacecraft.

The so-called ideal rocket equation (sometimes called Tsiolkovsky's equation, as introduced in his late ninetieth/early twentieth century space flight studies; [1]) for gravity-free, near-vacuum linear flight establishes the amount of propellant that must be consumed for a desired ("budgeted") vehicle velocity change ΔV:

$$\frac{m_{\text{final}}}{m_{\text{initial}}} = \exp\left(\frac{-\Delta V}{I_{\text{sp}} \cdot g_o}\right) = \exp\left(\frac{-\Delta V}{u_e}\right). \tag{14.1}$$

The parameter m_{initial} is the vehicle's mass at the start of the flight acceleration segment, and m_{final} is the vehicle's mass at the end of the flight segment, the difference of course being the amount of propellant consumed. The above equation can be presented in various formats in the literature, depending on the particular problem to be solved, including multi-staged flight.

The source of the electric power (e.g., solar radiation cells or receivers, batteries, chemical fuel cells, nuclear reactor, etc.) for an EP system is physically separate from the mechanism of gas/particle acceleration that produces the thrust (the acceleration may be via electric heating in simpler systems, or via electric and/or magnetic force fields in more sophisticated systems). A principal advantage in using an EP approach is that the electric energy added to the exhaust propellant (gas, or more discontinuous stream of atomic particles) can greatly increase its exit velocity, and therefore I_{sp} (thus more thrust produced for the same propellant flow/consumption rate). Consider some pertinent performance parameters for an EP system operating in the vacuum of space (and assuming ideal expansion of the exhaust jet):

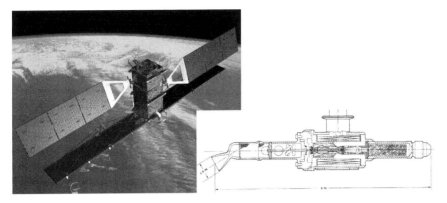

Fig. 14.2 The Canadian satellite RADARSAT-2 shown above is a collaboration between MDA and the Canadian Space Agency. The satellite uses six 1-N hydrazine monopropellant thrusters for attitude control (diagram courtesy of MDA and the Canadian Space Agency). *Lower right* diagram shows a schematic diagram of a hydrazine thruster (with a canted exhaust nozzle) in that thrust category

$$\frac{\text{kinetic power of jet exhaust}}{\text{thrust delivered}} = \frac{P_{\text{jet}}}{F} = \frac{\frac{1}{2}\dot{m}u_e^2}{\dot{m}u_e} = \frac{1}{2}u_e = \frac{1}{2}I_{\text{sp}}g_o \tag{14.2}$$

where in space,

$$I_{\text{sp}} \approx \frac{u_e}{g_o} \tag{14.3}$$

For thruster efficiency η_t, noting that P_{inp} is the input electric power (delivered via a current I and a voltage V):

$$\eta_t = \frac{P_{\text{jet}}}{P_{\text{inp}}} = \frac{P_{\text{jet}}}{IV} = \frac{\frac{1}{2}\dot{m}u_e^2}{IV} = \frac{FI_{\text{sp}}g_o}{2IV} \tag{14.4}$$

Typical values for η_t can range from as low as 20% to as high as 90% or more. The specific power α_s of an EP powerplant (excluding propellant mass) is defined by:

$$\alpha_s = \frac{P_{\text{inp}}}{m_{\text{pp}}}, \ \text{W/kg} \tag{14.5}$$

The parameter m_{pp} is the dry mass of the EP system. Current values for α_s range from 100 to 300 W/kg, with continuing efforts to push up these values. Figure 14.3 provides a chart showing the distribution of different EP systems in terms of delivered specific impulse and required electrical input power [1].

Current electric thrusters can be more or less grouped under three main categories:

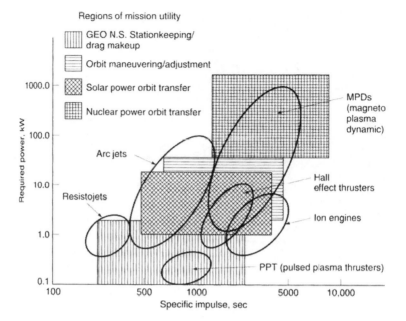

Fig. 14.3 Chart illustrating the performance ranges of various electric propulsion systems, given in terms of delivered specific impulse and required electrical input power [1]. Reprinted with permission of John Wiley & Sons, Inc.

1. **Electrothermal propulsion:** The propellant, which can originally be liquid or solid in storage, is heated electrically, and the resulting hot gas is then expanded and accelerated to supersonic/hypersonic velocities through an exhaust nozzle for a high u_e. *Resistojets* electrically heat high-resistance metal parts (e.g., tungsten), which in turn heat the propellant flowing over them. Among the oldest EP techniques to be put into practice, the specific impulse for a conventional resistojet may range from 200 to 300 s. Thrust levels can be on the order of newtons at the high end.

   Arcjets heat the propellant gas flow by passing it through an electric arc discharge produced between a central negatively-charged cathode and a surrounding coaxial positively charged anode that attracts the oncoming electrons (the anode may also act as the exhaust nozzle). Example liquid/gases used by arcjets include hydrazine, hydrogen and ammonia. While the surrounding structure may reach as high as 2200 K in temperature (close to the allowed structural limit), the temperature of the core gas within the arc may reach as high as 20,000 K. Specific impulse values for conventional arcjets can range from 400 up to 1,500 s, depending on the effectiveness in utilizing the energy from the electric arc. A diagram of a direct-current arcjet may be seen in Fig. 14.4 [2].

2. **Electrostatic/ion propulsion:** Positive ions (e.g., mercury, indium or cesium atoms stripped of one or more electrons) are generated from a propellant in

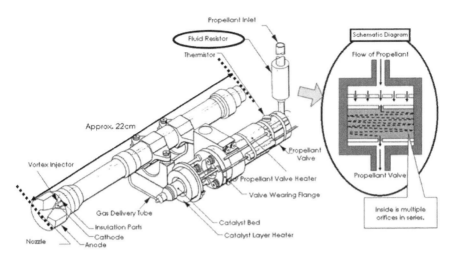

Fig. 14.4 Diagram of a direct-current (DC) arcjet thruster [2]. Hydrazine is used as the working propellant. Courtesy of the Japan Aerospace Exploration Agency

storage by various means, as the working medium to be used for thrust production (note: heavier ions benefit thrust delivery, as compared to lighter atomic particles or the very light electrons). Electron *b*ombardment of the propellant via electrons produced by a heated cathode is one *t*hruster technique (*EBT*; Fig. 14.5; [3]); contact with the propellant by a hot ionizer is an alternative technique (*ion contact* thruster); passage of the propellant in an appropriate format (e.g., as a colloid mixture, i.e., very small droplets of the working medium) through an intense electric field is a third technique for generating ions (*colloid* thruster, or if acting on individual ions rather than droplets, *FEEP*, *f*ield *e*mission *e*lectric *p*ropulsion thruster [4] as per Fig. 14.6); radio-frequency (RF) electromagnetic field ionization is the fourth (*RIT*, radio-frequency ion thruster; see Fig. 14.7; [5, 6]).

The resulting ions are accelerated to a high velocity by an electrostatic (electric) field established between the ion source and the accelerating cathode (negatively charged electrode; see Fig. 14.8 for a generic example [7, 8]). A FEEP thruster can have a specific impulse ranging from 1500 up to 5000 s. On the negative side, one should note that in the course of the positively charged ions exiting the system, they must be electrically neutralized, so as to keep the surrounding spacecraft structure electrically neutral for proper operation, by a second cathode that emits negatively charged electrons into the exhaust ion beam. This neutralization process tends to slow down the exhaust beam a bit, so acts to decrease the net thrust production.

3. **Electromagnetic propulsion:** In this category, the electrically heated propellant in storage produces a plasma (a hot, electrically neutral mixture of electrons, positive ions and neutral atoms) that is in turn accelerated to high velocity by various applications of electric and magnetic fields. Inert gases like

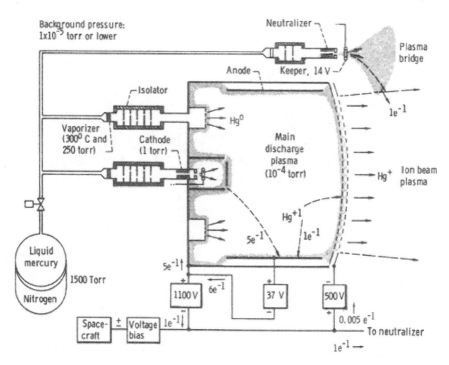

Fig. 14.5 Schematic diagram of an electron bombardment thruster, using mercury as the propellant [3]. Courtesy of NASA

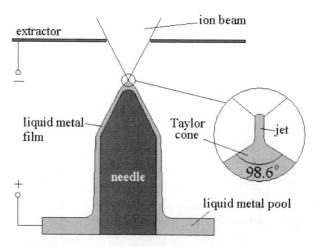

Fig. 14.6 Schematic diagram of indium-based FEEP thruster [4]. Courtesy of the European Space Agency

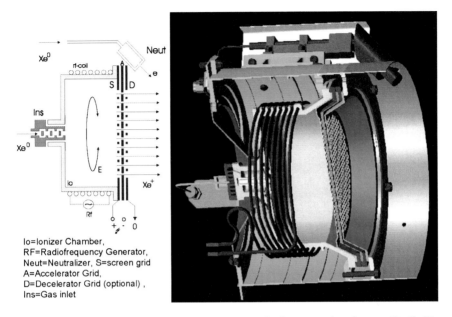

Io=Ionizer Chamber,
RF=Radiofrequency Generator,
Neut=Neutralizer, S=screen grid
A=Accelerator Grid,
D=Decelerator Grid (optional) ,
Ins=Gas inlet

Fig. 14.7 Cutaway schematic diagrams of 15 mN radio-frequency ion thruster [5, 6]. The thruster, developed by EADS Astrium, uses xenon as the propellant ionized by the 1 MHz radio-frequency coil, and delivers thrust at a nominal I_{sp} of 3,300 s. Reprinted with the permission of the Electric Rocket Propulsion Society

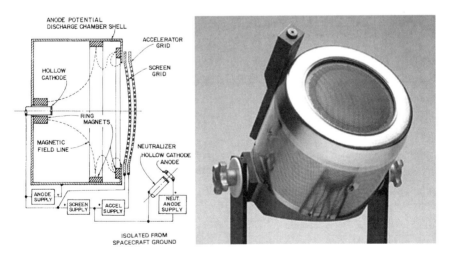

Fig. 14.8 Schematic diagram of generic direct-current ion thruster and associated operations [7]. *Right photo* of xenon-based 13 cm XIPS (Xenon Ion Propulsion System) ion thruster [8]; courtesy of NASA

Fig. 14.9 Vector orientation
diagram for EM thruster
operation

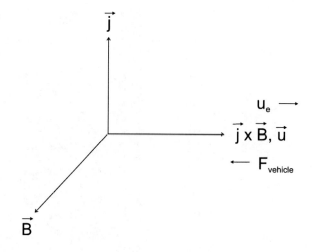

xenon or krypton may be used as the working propellant medium. Some of the
physics of EM propulsion will be discussed below.

The electrostatic (Coulomb) force component and the electromagnetic
(Lorentz) force component acting on a single charged particle is described by the
micro-scale force vector equation:

$$\vec{f} = q\vec{E} + q\vec{u} \times \vec{B}, \text{ N} \tag{14.6}$$

where q is the electric charge magnitude (+ or − in coulombs, C), $\vec{E}$ is the electric
field vector (strength, N/C or V/m), $\vec{u}$ is the velocity vector of the charged particle
(m/s) and $\vec{B}$ is the magnetic field vector (flux density in teslas, T, or Wb/m^2; note
that magnetic flux is measured in webers, Wb). In macroscopic terms, for a given
volume of plasma flow, the total force per unit volume due to electric and magnetic forces is given by:

$$\vec{F} = \rho_q\vec{E} + \vec{j} \times \vec{B}, \text{ N/m}^3 \tag{14.7}$$

where ρ_q is the net charge density (C/m^3) and $\vec{j}$ is the electric current vector density
[C/(s · m^2) or A/m^2]. Refer to Fig. 14.9 for the orientation of the various
parameters involved. Electrostatic thrusters rely on the net charge density (space–
charge density) of the first righthand (Coulomb) term to produce a force to
accelerate positively charged ions. Since the net charge density in plasmas goes to
zero, EM thrusters depend on the second (Lorentz) term on the righthand side for
thrust delivery.

The EM Lorentz force acting to accelerate the plasma is perpendicular to the
local electron current (as per Fig. 14.9, the current passing between the cathode
and anode), and also perpendicular to the local magnetic field (sometimes applied
by a permanent magnet, electromagnet or solenoid). In order to accelerate the

entire plasma (not just the positive ions), the plasma must have a high density. This results in a high number of atomic collisions with the surrounding engine casing that can eventually lead to structural failure, if not designed for. Another potential design constraint with EM systems is that the power requirement for producing the necessary plasma can be quite high. However, on the positive side of the ledger, thrust levels can be relatively high, on the order of tens of newtons on the high end, with a high I_{sp} accompanying this production.

As part of this third category, *Hall* or *stationary plasma thrusters* (*SPTs*) exploit the Hall effect (named after Edwin Hall) to strongly accelerate the plasma. A schematic diagram of a Hall thruster is provided in Fig. 14.10; [8–10]. In brief terms, the Hall effect is a strengthened axial electromagnetic force as the outcome of an axial electric field being in the presence of a perpendicular radial magnetic field, with moving/accelerating atoms ionized by the external-cathode-generated electrons moving transversely/spirally through the external exhaust flow, a second phenomenon resulting in conjunction from the Hall effect (i.e., the Hall current). Given a plasma is nominally electrically neutral with positively charged species countering negatively charged species in the flow, the cathode acting on the external exhaust beam is not primarily for neutralization in this case, but overall (through ionization and resulting electromagnet force generation) for beneficial thrust production. Hall thrusters presently have an I_{sp} ranging from 3,000 to 5,000 s, and existing systems have produced up to 3 N of thrust, which is quite substantial for an EP system.

Magneto-plasmadynamic (MPD) thrusters use an electric arc discharge (like an arcjet) to produce and accelerate the plasma, with the added impetus of an applied magnetic field reinforcing the self-induced magnetic field resulting from the plasma passing through the arc. The additional applied magnetic field may be generated via such techniques as a solenoid (where an electric current is passed through a spiral metal coil; an electromagnet would employ a metal core inside the coil). Specific impulse for MPDs can range from 1000 up to 8000 s, and thrust levels can be quite high, a number of newtons, if the required input power is available.

Pulsed plasma thrusters (*PPTs*) operate more on a pulsed discharge basis, versus the continous MPD approach, and may rely solely on the self-induced magnetic field to produce the electromagnetic force that accelerates the plasma to high speed. Solid Teflon has been a common choice for the propellant, with the plasma generated by interaction of the Teflon surface with the arc discharge. Radio-frequency energy, or laser energy, may also be used for plasma generation from the surface of various materials. Rather than as a block, some materials are implemented as a flexible roll of tape that can be slowly unwound as it is consumed. *Pulsed inductive thrusters* (*PITs*) apply a pulse-generated electric current through an applied magnetic field, which generates the axial electromagnetic force acting to accelerate the propellant plasma.

A final example of an EM/plasma propulsion system is the VASIMR® (*Variable Specific Impulse Magnetoplasma Rocket*) thruster/engine being proposed by the Ad Astra Rocket Company, as illustrated in Fig. 14.11. This approach [11] will

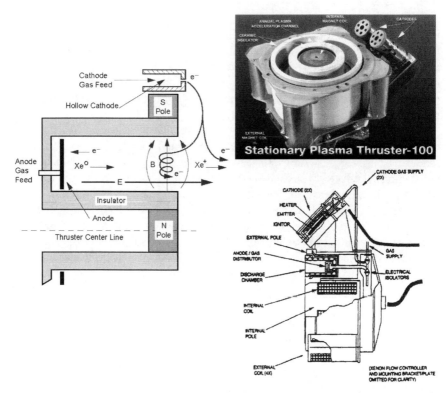

Fig. 14.10 Schematic diagram at *left* of Hall thruster and associated operations [8]; courtesy of NASA. *Right photo* and schematic diagram of SPT-100 Hall thruster [9, 10]; reprinted with permission of the American Institute of Aeronautics and Astronautics

require considerable electrical power to reach its full potential, a characteristic of EM systems, but the predicted performance is impressive.

Please refer to the solution for Example Problem 14.1 at the end of this chapter, for a completed sample electric propulsion system performance analysis. For students learning this material as part of a course, I would encourage them to attempt the given example problem first, and then check the solution to see if any mistakes were made.

14.3 Solar/Thermal Propulsion for Spaceflight

We see with electric propulsion systems the use of the Sun as an external power source for batteries, etc., that as a result can bring vehicle weight savings, for some spaceflight applications. Another means for harnessing solar energy is to optically concentrate infrared (IR) radiation from the Sun to heat a working fluid such as

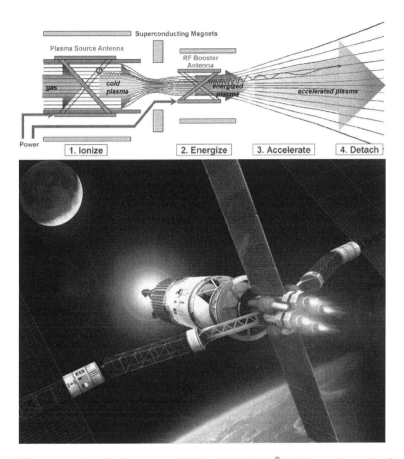

Fig. 14.11 *Upper* schematic diagram at of Ad Astra's VASIMR® EM/plasma thruster/engine. A two-step heating of the propellant gas/plasma via radio-frequency excitation brings it to very high temperatures (approaching 1,000,000 K). This impressive thermal energy level is converted downstream into kinetic energy, in accelerating the plasma in a strong applied magnetic field to a very high ultimate exit speed (ions reaching 50,000 m/s, which correlates to an I_{sp} of around 5000 s) [11]. Illustrations courtesy of Ad Astra Rocket Company

LH_2 (ideally, peak temperature that can be attained is that of the Sun's surface, about 2,400 K; note low molecular mass for diatomic hydrogen, 2 amu, thus producing superior I_{sp} via heating, to say a 9 amu gas produced via combustion of hydrogen and oxygen at a somewhat higher flame temperature). The Solar Orbit Transfer Vehicle (SOTV, as depicted in Fig. 14.12; [12]; for transfer from LEO to GEO) deploys an inflatable solar concentrator dish (mirrored) that focuses sunlight onto a graphite receiver that heats up to approximately 2,400 K. In "power" mode, the solar/thermal engine heats a thermionic converter (fuel cell) which produces electric power. In "propulsion" mode, LH_2 is passed through the graphite receiver, becoming a hot gas that is expanded out of a nozzle ($u_e \approx 4,000$–10,000 m/s to produce from 1 to 50 lb (4.5–225 N) of thrust, at an I_{sp} of 700–800 s.

Please refer to the solution for Example Problem 14.3 at the end of this chapter, for a completed sample solar/thermal rocket performance analysis.

14.4 Nuclear/Thermal Propulsion for Spaceflight

Instead of using solar energy, on-board nuclear energy sources (e.g., solid- or gas-core fission reactor, radioactive isotope decay source, or in the future, fusion reactor) can be used to deliver heat to the working fluid (again, typically LH$_2$ due to low molecular mass advantage), can in turn be expanded through a nozzle to a high exhaust velocity ($u_e \approx 6,000$–10,000 m/s, or more) at an I_{sp} for a solid-core fission engine from 600 to 900 s, and an I_{sp} from 1,500 to 6,000 s for a gas-core fission engine (substantially higher operating temperature). Nuclear rockets can deliver high thrust. The NERVA (Nuclear Engine for Rocket Vehicle Application; see Fig. 14.13; [13, 14]) solid graphite core fission engine prototype attained a peak thrust of 210,000 lb (935 kN), operating at an H$_2$ working fluid tempeature of 2500 K, and an I_{sp} of 850 s. A schematic diagram of a solid-core nuclear fission reactor based rocket engine is shown in Fig. 14.14; [15]. Higher I_{sp} plasma/gas-core fission approaches [16] have been proposed for the Earth-Mars flight mission, as a relatively attainable alternative to conventional chemical rockets (engine used from an Earth or Moon orbit starting point, to bypass potential environmental concerns). A schematic diagram of a gas-core nuclear fission reactor is shown in Fig. 14.15.

If storage volume is limited, a higher density liquid propellant like methane or ammonia may be a better choice than hydrogen for overall mission performance (e.g., total impulse, vs. specific impulse). A conventional solid-core nuclear rocket engine [15] will be comprised of a propellant tank, a radiation shield which protects the remaining vehicle from reactor radiation (the tank itself, and the propellant within [especially hydrogen], will also assist in this protection), a turbopump or pressure feed system for propellant delivery from (ideally) low-pressure storage, the nuclear reactor itself (comprised of the radioactive nuclear fuel core [typically enriched uranium compound like uranium oxide or uranium carbide], the moderator [typically carbide or graphite as fuel coating or separate element], control drums or rods and reflector [acting to keep the radiation energy within the nuclear reaction chamber; typically carbon compound], all of which are contained within a pressure vessel at a p_c typically ranging from 3 to 8 MPa), and finally the exhaust nozzle where the heated propellant gas exits after passing through flow channels at the periphery of the reactor. When inserted the nuclear reactor's control rods act to hinder ("poison") the nuclear reaction process, and conversely, when withdrawn, the reactor will ramp up in radiative heating performance. The conventional "slow" thermal fission method requires a moderator to slow down neutrons emitted from the nuclear fuel. The "fast fission" method

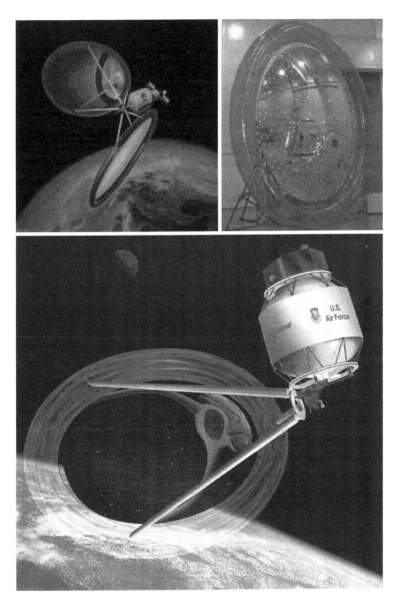

Fig. 14.12 Solar Orbit Transfer Vehicle illustrations at *upper left* and *below*, and an inflatable solar concentrator to be used by the SOTV on display at *upper right* [12]. Reprinted with permission of the American Institute of Aeronautics and Astronautics

used by the CERMET (*ceramic/metal* matrix used for nuclear core fuel; ceramic particles bonded into metal for high-temperature, high-strength applications [14]) engine designs may not require a moderator (given the metal acts to slow the

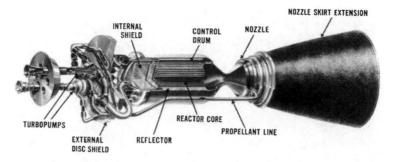

Fig. 14.13 NERVA engine diagram. Courtesy of NASA

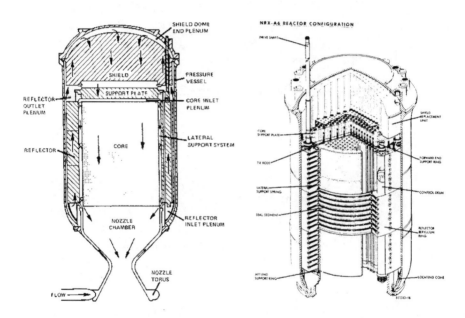

Fig. 14.14 Simple schematic diagram of nuclear/thermal rocket engine employing a solid-core (uranium fuel) fission reactor with hydrogen as the propellant being pumped through the nozzle walls for cooling first, then heated by the reactor, and exhausted out the nozzle, in a regenerative cooling approach [14, 15]. Diagrams courtesy of NASA

neutrons sufficiently in some cases). The bimodal engine proposed for the NASA NTR Transfer Vehicle employs a CERMET reactor approach (see Fig. 14.16). A particle-bed reactor employs the slow fission/moderator approach, but with superior surface area of contact between the nuclear fuel particles and the propellant flow, propellant temperatures can be higher (e.g., 3200 K, vs. conventional values like 2,400 K), thus providing a boost to delivered thrust and I_{sp}.

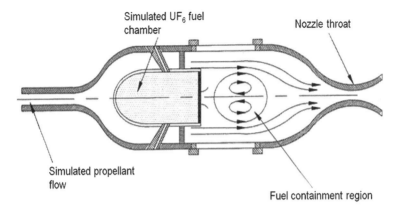

Fig. 14.15 Schematic diagram of nuclear/thermal rocket engine employing an open-cycle gas-core fission reactor. A gaseous uranium plasma fuel core is contained in the reactor by an induced toroidal vortex flow involving the injected hydrogen propellant [16]

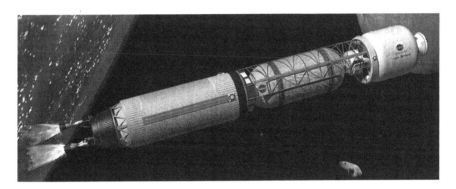

Fig. 14.16 Illustration of the proposed NASA NTR Transfer Vehicle for traveling to Mars, employing three 15,000 lbf nuclear-thermal rocket engines (engines have bimodal capability, meaning they can produce both thrust and electric power). Diagram courtesy of NASA

14.5 Unconventional Rocket Engine Concepts

Moving from the conventional or near-term, state-of-the-art technology realm (at least as far as practical present usage), let us look at a few examples of some more advanced rocket engine concepts. A first example, a nuclear/electric rocket (as opposed to a nuclear/thermal rocket), combines the advantages of electric propulsion (high I_{sp}) and nuclear reactor systems (on-board source of long-life energy). The actual thruster or rocket would use the applicable EP technique (present-day EP capability, low thrust delivery), with the electric power being delivered via a thermionic converter receiving heat from a nuclear reactor (present-day reactor would be fission-based, but later, possibly fusion-based). Certainly, an on-board nuclear power source would be required when radiation from the Sun is insufficient for a solar power approach, e.g., in an outer solar system or interstellar flight mission.

A nuclear-pulse rocket applies a repeated series of fission or fusion nuclear mini-explosions against a pusher-plate/shock-absorber attached to the rear of the spacecraft. This cyclic process imparts acceleration to the vehicle, and various studies (e.g., Project Orion and later, Project Daedalus [13], as per Fig. 14.17) have shown that this could be a viable candidate for interstellar flight missions. At this juncture, this approach would deliver a higher I_{sp} and thrust than any current EP system. Obviously, a considerable amount of radioactive waste would be left behind the spacecraft as it travels through space, but this is not a major concern in open space at some distance from Earth (protons in solar flares, and heavy ionized nuclei like Fe^{+26} in galactic cosmic rays can pose as much or more of a risk to the flight crew over long mission times).

An antimatter rocket uses the ultra-high energy release process of matter/antimatter annihilation to heat an appropriate propellant (expellant) gas for high-u_e exhaust. Antimatter is similar in some sense to matter, but its electrons have a positive charge (instead of negative), and its protons have a negative charge (instead of positive). When antimatter is brought together with an equivalent amount of normal matter, the ideal total energy release follows Einstein's famous equation (with a minor modification). In the case of 1 kg of antimatter, m, being used, the ideal energy released from the total amount of matter and antimatter is:

$$E = 2 \cdot mc^2 = 2 \cdot (1 \text{ kg})(3 \times 10^8 \text{ m/s})^2 = 1.8 \times 10^{17} \text{ J} \qquad (14.8)$$

The current ability to create and store antimatter is very limited, and shares many of the technical difficulties one is seeing with the development of contained/controlled nuclear fusion. Antimatter propulsion systems would be enhanced by directing energetic particles, resulting from the annihilation process at speeds approaching that of light, c, in a rearward direction. The use of a strong magnetic field may facilitate this directed energy beam approach, but system weights would typically need to be kept relatively low for this approach to be viable as a propulsion system for interstellar missions. A strong magnetic field may also aid in containing a high-temperature central expellant gas flow, while keeping the peripheral wall structure at a substantially lower temperature.

In some past unmanned spacecraft flights through our solar system, it has been demonstrated that the gravity induced by a planet (like Jupiter or Saturn) or star (in this case, the Sun) of substantial mass can be helpful in accelerating a spacecraft along a desired trajectory toward its ultimate destination. Some proposals for futuristic propulsion involve the use of antigravity, i.e., gravitomagnetic (non-Newtonian gravitational) forces that can act to counter conventional Newtonian gravity. For example, the acceleration of an item of substantial mass around a toroidal path (analogous to a magnetic coil) can produce an antigravitational field. However, the values involved are very high, and presently, unrealistic in terms of being attained for propulsive forces any time soon [17].

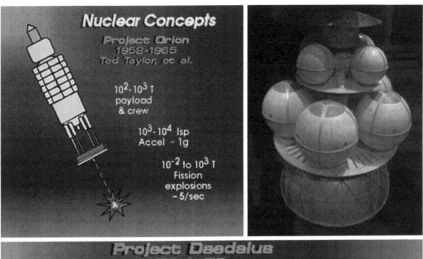

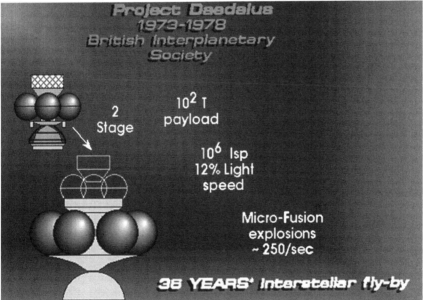

Fig. 14.17 NASA illustrations of the Projects Orion and Daedalus nuclear-pulse rocket concepts, with the Daedalus vehicle as a proposed means for flying by a nearby star (about 6 light-years away; *upper* retrorocket 2nd stage is for decelerating upon approaching star). *Right photo* of Daedalus scale model displayed at US Space and Rocket Center, Huntsville, Alabama

A ram-augmented interstellar rocket (RAIR [proposed by Alan Bond in 1974]; also called an interstellar ramjet [13]) uses a large forward scoop/intake to collect hydrogen atoms floating in open galactic space (say, around 1 hydrogen atom per cubic centimeter available). The scoop may be solid, or a mesh with a magnetic field [proposed by Robert Bussard in 1960 [13]; see Fig. 14.18]. The hydrogen atoms can

Fig. 14.18 Illustration of the
Bussard interstellar ramjet.
Courtesy of NASA

be used as both fuel (for a fission or fusion nuclear reactor onboard the vehicle) and
expellant. The efficiency of atomic collection would improve as vehicle speeds
increase to relativistic values (appreciable fraction of c, say >5–10%). The scoop
would by necessity have to be very light in weight, to be feasible. In addition, the
induced magnetic field as the collected, charged hydrogen atoms converge to the
central throat of the scoop may actually act to repel the incoming particles in a
bottleneck effect, rather than draw them through the engine. Some means for
countering this repulsion effect would be required, e.g., pulsing the forward scoop's
magnetic field.

Returning to the issue of flight at speeds approaching the speed of light c, which
in Albert Einstein's model must always be constant, one can note that Einstein's
energy-mass equivalence can be written in terms of mass at rest (rest mass, m_o)
and mass at some flight speed (relativistic mass, m):

$$E = mc^2 = \frac{m_o}{\sqrt{1 - \left(\frac{V}{c}\right)^2}} \cdot c^2 \qquad (14.9)$$

As a flight vehicle approaches c, the above equation suggests that the vehicle's
perceivable mass will begin to increase substantially, and then dramatically,
essentially toward an infinite mass at c. This is the road block commonly referred
to in discussions of future space travel, that one will need more and more propellant to accelerate closer to the speed of light, such that the required propellant
mass simply becomes infeasible at some fraction of c.

Since the rate of travel through Einstein's space–time frame of reference is a
constant for any object, then an object at some flight speed V must have its time
component slowed down, to compensate. The substantial slowing of time for
individuals in the relativistic flight vehicle is referred to as time dilation. Unlike
the mass issue raised above (a negative in terms of aerospace propulsion needs),
the slowing of time may allow for relativistic interstellar flights to be

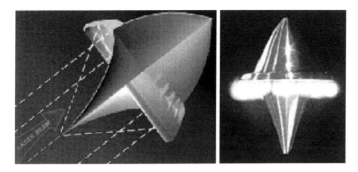

Fig. 14.19 Schematic diagram at *left* illustrating the NASA Laser Lightcraft vehicle concept (Feikema [18]). Diagram, and photo at *right*, courtesy of NASA Marshall Space Flight Center

accomplished (arrive at the destination star) by the same flight crew that started the flight mission from Earth. Of course, on the negative side, if one hoped to return to Earth to complete a two-way interstellar mission, it is with the expectation that one's family and friends will have aged substantially, when the still youthful flight crew is greeted.

Finally, associated with relativistic flight speeds, there is the third phenomenon of length contraction. Again, travel in space–time requires compensation, as one approaches c, in terms of the distance component (as well as the time component), and this will occur in terms of the flight vehicle contracting in length in the direction of flight, such that its length will be zero at the limit value of $V = c$.

14.6 Non-Rocket Concepts for Advanced Aerospace Transportation

Here, we will look at some advanced concepts for moving vehicles from point A to point B, that do not involve rocket propulsion. As a first example, there are a number of variations on the beamed-energy technique for moving the receiver vehicle through the atmosphere and/or space. The separate energy source that drives the propulsion of the vehicle can be heavy and relatively stationary (e.g., ground-based, but wheeled). The NASA Lightcraft (see Fig. 14.19; [18]) uses focused pulsed laser light from a ground source to instigate mini-explosions of heated air within the peripheral annular chamber at the top of the cone-shaped receiver vehicle, with the air exhaust downward through this hybrid aerospike nozzle arrangement propelling the vehicle upward.

On a larger scale, lightsails or solar sails require a large sail surface in order to be pushed by photonic pressure (impact and subsequent reflection of photons originating from the Sun, or alternatively, originating from a large space-based laser powered by solar energy [solar pumping approach]). To be feasible, the sails

Fig. 14.20 A large, 800 m-wide solar-sail-propelled vehicle (sailcraft) carrying a 850 kg payload, proposed by JPL/NASA in 1977 for a rendezvous mission with Halley's comet, for the 1986 passage by the Earth of the comet [19, 20]. The project ultimately did not come to fruition

must be very light in weight. An example diagram of a proposed large sailcraft pushed by solar radiation pressure is given in Fig. 14.20; [19, 20]. Along similar lines, a strong microwave beam can be used to propel a microwave sail. The sail in this case would be a fine metal grid. A magnetic sail (magsail) exploits the energy of the solar wind, whereby the sail's magnetic field acts to deflect the charged protons within the wind.

Catapult techniques also involve separating a large portion of the propulsive mechanism's mass and the flight vehicle itself. Light gas (low molecular mass, $\mathcal{M}$) guns and other gun or cannon approaches (e.g., Gerald Bull's High Altitude Research Program [HARP] rocket-assisted projectile system; [21]) employ a long gun barrel/tube of sufficient length to accelerate the vehicle, with high gas pressure pushing the craft along until exiting the muzzle at very high velocity. Gun approaches generally require the projectile to be able to withstand transient axial accelerations on the order of up to 100,000 g within the barrel/tube. Some electronics can presently be hardened to meet that requirement. Ram accelerators (see Fig. 14.21; [22]) as well require a considerable length of tube as the vehicle passes through a premixed fuel-oxidizer environment, whereby aft of the vehicle one has combustion and an associated high-pressure zone acting to push the vehicle forward. Electromagnetic rail guns (also known as coil guns, mass drivers or magnetic-coil accelerators) also require a substantial length of rail to permit adequate vehicle speeds upon exiting. Rail gun rails typically need to be supercooled to allow the superconducting electromagnets to function effectively. Magnetic levitation (MagLev) or magnetic lift techniques that aid in accelerating an item have similar design issues.

Tether techniques allow for movement of vehicles up or down in orbital altitude from the originating station on a thin, lightweight tether. This approach is commonly called a space elevator, especially associated with the most demanding case of the originating station being on the Earth's surface (see Fig. 14.22; Konstantin Tsiolkovsky in 1895 outlined some of the basic elements surrounding such an approach, albeit in terms of a "space tower" structure). Electrodynamic tethers

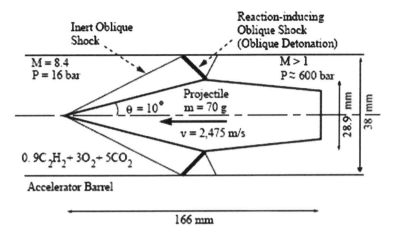

Fig. 14.21 Schematic diagram of a ram accelerator system (projectile/vehicle moving to the *left* through launch tube [accelerator barrel]) [22]. Reprinted with permission of the American Institute of Aeronautics and Astronautics

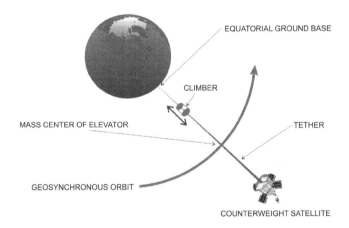

Fig. 14.22 Schematic diagram of a space elevator system using an ultra-lightweight tether

supply electric power through the conducting portion of the tether to enable onboard EP propulsion for vehicle movement along the tether. Tethers may also be used to sling vehicles into a new orbital or interplanetary flight path (example provided in Fig. 14.23; [23, 24]), given that the proper direction is chosen for the vehicle when being released from the long, rotating tether. The use of various tether techniques as discussed above become more feasible with the introduction of advanced lightweight materials like carbon nanotubes, that hypothetically can reach a strength-to-weight ratio approaching 600 times that of steel. Returning to the surface-based space elevator proposal, one needs to be aware of the inner and outer Van Allen radiation belts, which would be encountered on the way to GEO at an altitude of approximately 36,000 km; the relatively intense radiation from the

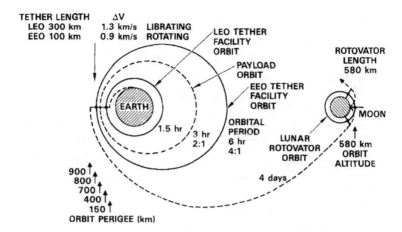

Fig. 14.23 LEO-lunar surface rotating tether transport system [24]. Two tethers in Earth orbit (one in *lower* circular LEO orbit, one in *higher* elliptical EEO orbit), and one lunar tether affixed to the Moon's surface, are employed in this proposal. The term "rotovator" refers to a rotating space elevator (tether) that in this example can collect items sent from Earth, and in turn sling items from the Moon back to Earth

energetic plasma, especially of the inner belt's energetic protons, could possibly have a significant disruptive effect on the various items (including humans) passing through. Present-day spacecraft that pass through the Van Allen belts are provided significant shielding to prevent any undue radiation damage.

In order to decelerate a spacecraft from interplanetary flight speed down to orbital or planetary entry/descent speeds, ideally without the use of retrorocket systems that can expend large amounts of propellant, there are a number of different aerodynamic braking (aerobraking) techniques that can be used to accomplish this action. They generally involve the deployment (e.g., via inflation) of large-surface-area, lightweight structures (e.g., bags) extending from the main vehicle. If no longer to be used, solar panels can double as aerobrakes. If the planet's atmospheric density is appreciable, the resulting enhanced pressure and viscous drag on the craft will bring its speed down. The aerobrakes may be ablative (material worn away from surface under high aeroheating) and expendable upon completion of the deceleration flight maneuver. Figure 14.24 illustrates a more conventional aerobraking technique, the use of parachutes to slow the final descent of a returning space vehicle. Parachutes may employ an airfoil shape, as a paraglider, to generate aerodynamic lift as well as drag, to gain forward distance while descending.

14.7 Stratospheric Balloons

While one would not typically see a discussion of balloons (or airships) in an aerospace propulsion book (except maybe at the beginning), balloons can, on occasion, be a competitive alternative to thrust-based flight systems for meeting a

Fig. 14.24 At *upper left*, splashdown imminent for Apollo 17 command module in the south Pacific Ocean, Dec. 19, 1972, a successful ending to a mission beginning Dec. 7 with a night-time launch of the Saturn V (C-5) rocket vehicle. Courtesy of NASA and US Space and Rocket Center

certain application. In terms of the exploration of space, it happens that large lighter-than-air balloons, through their buoyancy force, can take payloads to the "edge of space" [25], i.e., well above conventional passenger jet air traffic (commercial jets cruise around 9–12 km altitude, which is also the transitional region [tropopause] in going from the troposphere to the lower stratosphere, where air temperature becomes relatively constant at 217 K, until it begins to rise again with height in the upper stratosphere, above around 20 km). Balloons can go as high as 35 km (nominal air density of around 0.008 kg/m^3) to 50 km (nominal air density of around 0.001 kg/m^3) in altitude, depending on the payload mass. The air temperature will level off at around 270 K as the upper regions of the stratosphere are passed through (above around 47 km). Between 50 and 55 km, the stratosphere transitions through the stratopause into the mesosphere, where air temperature decreases with height. A meteor moving at high speed will likely burn up in this layer due to aerothermodynamic heating (only larger solid objects entering the Earth's atmosphere will ultimately reach the Earth's surface with any significant size). The mesosphere extends to about 80 km in altitude, and then transitions through the mesopause into the thermosphere. In the thermosphere, which can extend from 350 to 800 km in altitude depending on the local effects of solar radiation, and the air temperature nominally rises with altitude in this layer, there is no tangible atmosphere in terms of continuous air density above around 90 km (300 kft), with the gas molecules individually quite far apart at that point. The International Space Station [ISS] presently circles the Earth at about 350 km in altitude. Beyond the thermosphere is the exosphere, which is essentially made up of hydrogen and helium molecules separated by substantial distances, in general, as a precursor zone to entering open space.

In 2006, the Balloon-Borne Large Aperture Sub-Millimeter Telescope (BLAST; see Fig. 14.25; [26]) placed in a gondola, for a total payload of around two and half tonnes (2,500 kg), was carried to about 36 km altitude with a 1.2-million-cubic-meter-maximum-displacement helium balloon (total height when fully extended at high altitude, around that of a 33-storey building). By reaching that altitude, the far infrared/submillimeter-wavelength radiation telescope was able to clearly image the radiation from emerging new stars billions of light-years away. At the Earth's surface, the telescope would not have been able to do its job, since the intervening lower atmosphere would absorb the radiation in question, before reaching the telescope. For one and a half weeks, the balloon circled the Antarctic continent as the telescope collected the needed data, following a predictable wind pattern that allowed for a nominally safe ground recovery at the end of that period. The cost for this mission was significantly less than that required for a satellite-borne telescope, and the length of time required for imaging precluded the use of a high-altitude suborbital sounding rocket (a sounding rocket [SR] might provide up to twenty minutes of useful data-collection time during the upper end of its thirty-minute flight, with a cost of a single SR flight being comparable to the overall two-week mission cost of BLAST).

14.8 Example Problems

14.1. A spacecraft propulsion system engineer has acquired a competitor's FEEP thruster, as well as some specifications on its performance, but the data is incomplete:
Thrust rated at 0.1 N; specific impulse rated at $I_{sp} = 5,000$ s;
maximum current of 6 amperes; anticipated thruster efficiency $\eta_t \approx 60\%$

Fig. 14.25 A photo of the BLAST gondola in the *right foreground* (containing the telescope), and the helium balloon in the *left background* [26]. Photo by Prof. Mark Halpern, University of British Columbia

Based on this information, ascertain the remaining pertinent performance parameters, including estimated exhaust velocity (m/s), estimated voltage requirement (V), input electrical power (kW) and propellant feed rate (kg/s).

14.2. A spacecraft employs a solar/thermal propulsion system to move its position in orbit about the Earth. Via a solar concentrator, LH_2 is heated to a gas temperature of 2,400 K (you may assume the heated gas primarily remains diatomic), with a chamber pressure of 8.2 MPa. The nozzle throat diameter is 4 mm, while the nozzle exit diameter is 25.4 mm. Estimate the thrust and specific impulse of this engine (note $p_\infty \approx 0$ kPa).

14.3. A solar/thermal rocket engine has the following characteristics:
hydrogen gas, $\gamma = 1.4$; molecular weight, $\mathcal{M} = 2$ amu; $T_c = 2300$ K; chamber pressure at 11.3 MPa; nozzle throat dia. $= 2.54$ mm; nozzle exit dia. $= 25.4$ mm; engine/array dry mass of 2000 kg (i.e., excluding propellant mass);

A stationary plasma thruster (EP engine) has the following characteristics: $I_{sp} = 2800$ s; input power via solar-fed batteries, 20 kW; 75% thruster efficiency;
engine/array dry mass of 500 kg (i.e., excluding propellant mass)

For moving a client satellite (designate as a payload mass, of 3000 kg) from LEO to GEO (assume $p_\infty = 0$ atm), one needs a single propulsion system delivering a budgeted delta-V (ΔV) of 4,000 m/s. In this regard, you may use the ideal rocket equation (see book) as part of your calculations (i.e., linear space flight; neglect gravity, as well as drag).

By way of comparison for your customer, estimate the mission segment times required by the two alternative choices for carrier engine as noted above, for moving the customer package from LEO to GEO (the customer mentioned something about a 90-day limit).

14.4. For a three-stage rocket vehicle in space, establish that its nominal ideal linear flight performance can be described by:

$$\ell n\left[\left(\frac{m_{\text{final}}}{m_{\text{initial}}}\bigg|_1\right) \cdot \left(\frac{m_{\text{final}}}{m_{\text{initial}}}\bigg|_2\right) \cdot \left(\frac{m_{\text{final}}}{m_{\text{initial}}}\bigg|_3\right)\right] = -\frac{(\Delta V_{\text{total}})}{I_{\text{sp}} \cdot g_o}$$

assuming the same I_{sp} for each motor stage.

14.9 Solutions to Example Problems

14.1. Looking at a FEEP thruster (EP system). Performance assessed in vacuum of space.

$$u_e = g_o I_{\text{sp}} = 9.81(5000) = 49050 \,\text{m/s}$$

$$V = \frac{FI_{\text{sp}}g_o}{2I\eta_t} = \frac{0.1(5000)9.81}{2(6)0.6} = 681.3 \,\text{V}$$

$$P_{\text{inp}} = IV = 6(681.3) = 4088 \,\text{W} = 4.09 \,\text{kW}$$

$$\dot{m} = \frac{2\eta_t P_{\text{inp}}}{u_e^2} = \frac{2(0.6)(4088)}{49050^2} = 2.04 \times 10^{-6} \,\text{kg/s}$$

14.2. Looking at a solar/thermal rocket engine, at a flight altitude giving $p_\infty = 0$ kPa.

$R = \frac{\mathscr{R}}{\mathscr{M}} = 8314/2 = 4157$ J/kg · K, $\gamma = 1.4$ for diatomic gas like H_2

Nozzle throat area, $A_t = \frac{\pi d_t^2}{4} = \pi(0.004)^2/4 = 1.257 \times 10^{-5} \text{m}^2$;

Nozzle exit area, $A_e = \frac{\pi d_e^2}{4} = \pi(0.0254)^2/4 = 5.067 \times 10^{-4} \,\text{m}^2$

$$c* = \left[\frac{\gamma}{RT_f}\left(\frac{2}{\gamma+1}\right)^{\frac{\gamma+1}{\gamma-1}}\right]^{-1/2} = \left[1.4/[4157(2400)]\cdot(2/2.4)^{2.4/0.4}\right]^{-0.5} = 4613 \,\text{m/s}$$

$$\dot{m} = \frac{p_c A_t}{c*} = 8.2 \times 10^6(1.257 \times 10^{-5})/4613 = 0.0223 \,\text{kg/s}$$

$$C_{F,v} = \left[\frac{2\gamma^2}{\gamma-1}\left(\frac{2}{\gamma+1}\right)^{\frac{\gamma+1}{\gamma-1}}\right]^{1/2} = \left[2(1.4^2)/(0.4)\cdot(2/2.4)^{2.4/0.4}\right]^{0.5} = 1.812$$

Area-Mach No. relation, where Ma_t goes to unity:

$$\frac{A_t}{A_e} = \frac{Ma_e}{Ma_t}\left[\frac{2+(\gamma-1)Ma_t^2}{2+(\gamma-1)Ma_e^2}\right]^{\frac{\gamma+1}{2(\gamma-1)}} = Ma_e\left[\frac{2.4}{2+0.4Ma_e^2}\right]^{3.0} = 0.0248$$

Via iteration, find supersonic solution, $Ma_e = 5.6$. For nozzle exit pressure:

$$p_e = p_c\left[1 + \frac{\gamma-1}{2}Ma_e^2\right]^{\frac{-\gamma}{\gamma-1}} = 0.0079\ \text{MPa}.$$

$$F = C_{F,v}\left[1 - \left(\frac{p_e}{p_c}\right)^{\frac{\gamma-1}{\gamma}}\right]^{1/2}A_tp_c + (p_e - p_\infty)A_e$$

$$= 1.812\left[1 - (0.0079/8.2)^{0.286}\right]^{0.5}1.257 \times 10^{-5}(8.2 \times 10^6)$$

$$+ (7900 - 0)5.067 \times 10^{-4} = 173.5 + 4 = 177.5\ \text{N}$$

$$I_{sp} = \frac{F}{\dot{m}_e g_o} = \frac{177.5}{0.0223(9.81)} = 811.4\ \text{s in vacuum}$$

14.3. Looking at a solar/thermal engine and an SPT for space flight, $p\infty = 0$ kPa.
Solar:

$$\text{R} = \frac{\mathcal{R}}{M} = \frac{8314}{2} = 4157\ \text{J/kg} \cdot \text{K};\quad C_p = \frac{\gamma R}{\gamma-1} = \frac{1.4(4157)}{1.4-1} = 14550\ \text{J/kg} \cdot \text{K};$$

Nozzle throat area, $A_t = \frac{\pi d_t^2}{4} = \pi(0.00254)^2/4 = 5.067 \times 10^{-6}\ \text{m}^2$

Nozzle exit area, $A_e = \frac{\pi d_e^2}{4} = \pi(0.0254)^2/4 = 5.067 \times 10^{-4}\ \text{m}^2$;

Area-Mach No. relation, where Ma_t goes to unity:

$$\frac{A_t}{A_e} = \frac{Ma_e}{Ma_t}\left[\frac{2+(\gamma-1)Ma_t^2}{2+(\gamma-1)Ma_e^2}\right]^{\frac{\gamma+1}{2(\gamma-1)}} = Ma_e\left[\frac{2.4}{2+0.4Ma_e^2}\right]^3 = 0.01$$

Via iteration, find supersonic solution, $Ma_e = 6.9$. For nozzle exit pressure:

$$p_e = p_c\left[1 + \frac{\gamma-1}{2}Ma_e^2\right]^{\frac{-\gamma}{\gamma-1}} = 0.003\ \text{MPa}.$$

$$C_{F,v} = \left[\frac{2\gamma^2}{\gamma-1}\left(\frac{2}{\gamma+1}\right)^{\frac{\gamma+1}{\gamma-1}}\right]^{1/2} = 1.812;\quad c^* = \left[\frac{\gamma}{RT_f}\left(\frac{2}{\gamma+1}\right)^{\frac{\gamma+1}{\gamma-1}}\right]^{-1/2} = 4516\ \text{m/s}$$

$$F = C_{F,v}\left[1 - \left(\frac{p_e}{p_c}\right)^{\frac{\gamma-1}{\gamma}}\right]^{1/2} A_t p_c + (p_e - p_\infty)A_e$$

$$= 1.812\left[1 - (0.003/11.3)^{0.2857}\right]^{0.5} 5.067 \times 10^{-6}(11.3 \times 10^6)$$

$$+ (3000 - 0)5.067 \times 10^{-4}$$

$$= 98.7 + 1.52 = 100.2\,\text{N}$$

$$\dot{m} = \frac{p_c A_t}{c^*} = 11.3 \times 10^6(5.067 \times 10^{-6})/4516 = 0.0127\,\text{kg/s}$$

$$I_{sp} = \frac{F}{\dot{m}g_o} = \frac{100.2}{0.0127(9.81)} = 804.3\,\text{s}$$

Rocket equation:

$$\frac{m_{\text{fin}}}{m_{\text{ini}}} = \frac{m_{\text{PL}} + m_{\text{eng}}}{\Delta m_p + m_{\text{PL}} + m_{\text{eng}}} = \exp\left(\frac{-\Delta v}{I_{sp}g_o}\right)$$

$$\frac{3000 + 2000}{\Delta m_p + 3000 + 2000} = \exp\left(\frac{-4000}{804.3(9.81)}\right) = 0.6, \quad \Delta m_p = 3333\,\text{kg of propellant}$$

$$t = \frac{\Delta m_p}{\dot{m}} = \frac{3333}{0.0127} = 262441\,\text{s or 73 h or 3.04 days}$$

SPT:

$$\frac{m_{\text{PL}} + m_{\text{eng}}}{\Delta m_p + m_{\text{PL}} + m_{\text{eng}}} = \frac{3000 + 500}{\Delta m_p + 3000 + 500} = \exp\left(\frac{-\Delta v}{I_{sp}g_o}\right)$$

$$= \exp\left(\frac{-4000}{2800(9.81)}\right) = 0.865, \quad \text{so that:}$$

$$\Delta m_p = 546.3\,\text{kg of propellant}$$

$$F = \frac{\eta_t P_{\text{inp}}}{\frac{1}{2}I_{sp}g_o} = \frac{0.75(2000)}{0.5(2800)9.81} = 1.092\,\text{N}$$

$$\dot{m} = \frac{F}{I_{sp}g_o} = \frac{1.092}{2800(9.81)} = 3.98 \times 10^{-5}\,\text{kg/s}$$

$$t = \frac{\Delta m_p}{\dot{m}} = \frac{546.3}{3.98 \times 10^{-5}} = 1.373 \times 10^7\,\text{s or 158.9 days}$$

Since the customer wished to have the package moved within 90 days, then one must go with the solar-thermal rocket engine.

14.4. Single stage:

$$\frac{m_{\text{final}}}{m_{\text{initial}}} = \exp\left(\frac{-\Delta V}{I_{sp} \cdot g_o}\right) = \exp\left(\frac{-\Delta V}{u_e}\right)$$

$$\ln\left(\frac{m_{\text{final}}}{m_{\text{initial}}}\right) = \frac{-\Delta V}{I_{sp} \cdot g_o} = \frac{-\Delta V}{u_e}$$

Two stages (same I_{sp}):

$$\frac{m_{\text{final}}}{m_{\text{initial}}}\bigg|_1 + \frac{m_{\text{final}}}{m_{\text{initial}}}\bigg|_2 = \exp\left(\frac{-\Delta V_1}{I_{sp} \cdot g_o}\right) + \exp\left(\frac{-\Delta V_2}{I_{sp} \cdot g_o}\right)$$

$$\ln\left(\frac{m_{\text{final}}}{m_{\text{initial}}}\bigg|_1\right) + \ln\left(\frac{m_{\text{final}}}{m_{\text{initial}}}\bigg|_2\right) = \frac{-\Delta V_1}{I_{sp} \cdot g_o} + \frac{-\Delta V_2}{I_{sp} \cdot g_o} = -\frac{(\Delta V_1 + \Delta V_2)}{I_{sp} \cdot g_o}$$

$$\ln\left[\left(\frac{m_{\text{final}}}{m_{\text{initial}}}\bigg|_1\right) \cdot \left(\frac{m_{\text{final}}}{m_{\text{initial}}}\bigg|_2\right)\right] = -\frac{(\Delta V_1 + \Delta V_2)}{I_{sp} \cdot g_o}$$

Three stages (same I_{sp}):

$$\ln\left[\left(\frac{m_{\text{final}}}{m_{\text{initial}}}\bigg|_1\right) \cdot \left(\frac{m_{\text{final}}}{m_{\text{initial}}}\bigg|_2\right) \cdot \left(\frac{m_{\text{final}}}{m_{\text{initial}}}\bigg|_3\right)\right] = -\frac{(\Delta V_1 + \Delta V_2 + \Delta V_3)}{I_{sp} \cdot g_o} = -\frac{(\Delta V_{\text{total}})}{I_{sp} \cdot g_o}$$

References

1. Sutton GP, Biblarz O (2001) Rocket propulsion elements, 7th edn. Wiley, New York
2. Anonymous (2004) Situation of "Kodama" (data relay test satellite: DRTS). Press release, Japan Aerospace Exploration Agency (Public Affairs Department), 2 June 2004
3. Anonymous (1977) Ion propulsion for spacecraft. NASA Lewis Research Center, Cleveland
4. Steiger W, Genovese A, Tajmar M (2000) Micronewton indium FEEP thrusters. In: Proceedings of 3rd international conference of spacecraft propulsion, ESA SP-465, Cannes
5. Leiter HJ, Killinger R, Bassner H, et al. (2003) Development and performance of the advanced radio frequency ion thruster RIT-XT. In: Proceedings of 28th international electric propulsion conference, Toulouse, 17–21 Mar 2003
6. Bassner H, Killinger R, Leiter HJ, Müller J (2001) Development steps of the RF-ion thrusters RIT. In: Proceedings 27th international electric propulsion conference, Pasadena, 15–19 Oct 2001
7. Wilbur PJ, Rawlin VK, Beattie JR (1998) Ion thruster development trends and status in the United States. J Propul Power 14:708–715
8. Goebel DM, Katz I (2008) Fundamentals of electric propulsion: ion and Hall thrusters. Jet Propulsion Laboratory, California Institute of Technology, Pasadena
9. Pidgeon DL, Corey RL, Sauer B, Day ML (2006) Two years on-orbit performance of SPT-100 electric propulsion. In: Proceedings of 42nd AIAA/SAE/ASME/ASEE joint propulsion conference, Sacramento, 9–12 July 2006

10. Day ML, Maslennikov N, Randolph T, Rogers W (1996) SPT-100 subsystem qualification status. In: Proceedings of 32nd AIAA/SAE/ASME/ASEE joint propulsion confernece, Lake Buena Vista, 1–3 July 1996
11. Glover TW, Chang Diaz FR, Ilin AV, Vondra R (2007) Projected lunar cargo capabilities of high-power VASIMR propulsion. In: Proceedings of 30th international electric propulsion conference, Florence, 17–20 Sept 2007
12. Pearson JC, Gierow P, Lester D (1999) Near term in-space demonstration of an inflatable concentrator. In: Proceedings of 37th AIAA aerospace sciences meeting, Reno, 11–14 Jan 1999
13. Gatland KW, et al. (1981) The illustrated encyclopedia of space technology—a comprehensive history of space exploration. Crown Publishers, New York
14. Lawrence TJ, Witter JK, Humble RW (1995) Nuclear rocket propulsion systems. In: Humble RW, Henry GN, Larson WJ (eds) Space propulsion analysis & design. McGraw-Hill, New York
15. Buker JC, et al. (1967) NRX-A6 test prediction report. Report WANL-TME-1613, Westinghouse Electric Corporation, Astronuclear Laboratory; NASA-CR-148643
16. Frisbee RH (2003) Advanced space propulsion for the 21st century. J Propul Power 19:1129–1154
17. Davis EW (2004) Advanced propulsion study. Air Force Research Laboratory Report AFRL-PR-ED-TR-2004-0024, USAF
18. Feikema D (2000) Analysis of the laser propelled lightcraft vehicle. In: Proceedings of 31st AIAA plasmadynamics and lasers conference, Denver, 19–22 June 2000
19. Wie B (2005) Solar sailing kinetic energy interceptor (KEI) mission for impacting and deflecting near-Earth asteroids. In: Proceedings AIAA guidance, navigation and control conference, San Francisco, 15–18 August 2005
20. Wright JL (1992) Space sailing. Gordon and Breach, Philadelphia
21. Murphy CH, Bull GV, Boyer ED (1972) Gun-launched sounding rockets and projectiles. Ann NY Acad Sci 187:304–323
22. Grismer MJ, Powers JM (1995) Calculations for steady propagation of a generic ram accelerator configuration. J Propul Power 11:105–111
23. Forward RL (1995) Advanced propulsion systems. In: Humble RW, Henry GN, Larson WJ (eds.) Space propulsion analysis & design. McGraw- Hill, New York
24. Forward RL (1991) Tether transport from LEO to the lunar surface. In: Proceedings of 27th AIAA/SAE/ASME/ASEE joint propulsion conference, Sacramento, 24–26 June 1991
25. Jackson DD (1982) The epic of flight—the aeronauts. Time-Life, Chicago
26. Truch MDP, Ade PAR, Bock JJ et al (2009) The Balloon-Borne Large Aperture Submillimeter Telescope (BLAST) 2006: calibration and flight performance. Astrophys J 707:1723–1729

Appendix I
ICAO Standard Atmosphere

Altitude (km)	T_∞(K)	p_∞(Pa)	ρ_∞(kg/m)^3	a_∞(m/s)
0	288.155	101325	1.2250	340.294
1	281.651	89876	1.1117	336.435
2	275.154	79501	1.0066	332.532
3	268.659	70121	0.9093	328.583
4	262.166	61660	0.8194	324.589
5	255.676	54048	0.7364	320.545
6	249.187	47217	0.6601	316.452
7	242.700	41105	0.5900	312.306
8	236.215	35651	0.5258	308.105
9	229.733	30800	0.4671	303.848
10	223.252	26500	0.4135	299.532
11	216.774	22700	0.3648	295.154
12	216.650	19399	0.3119	295.069
13	216.650	16578	0.2666	295.069
14	216.650	14170	0.2279	295.069
15	216.650	12112	0.1948	295.069
16	216.650	10353	0.1665	295.069
17	216.650	8850	0.1423	295.069
18	216.650	7565	0.1217	295.069
19	216.650	6467	0.1040	295.069
20	216.650	5529	0.0889	295.069
21	217.581	4727	0.0757	295.703
22	218.574	4042	0.0645	296.377
23	219.567	3456	0.0550	297.049
24	220.560	2955	0.0469	297.720
25	221.552	2608	0.0401	298.389
26	222.544	2163	0.0343	299.056
27	223.536	1856	0.0293	299.722
28	224.527	1595	0.0251	300.386

(continued)

D. R. Greatrix, *Powered Flight*, DOI: 10.1007/978-1-4471-2485-6,
© Springer-Verlag London Limited 2012

(continued)

Altitude (km)	T_∞(K)	p_∞(Pa)	ρ_∞(kg/m)^3	a_∞(m/s)
29	225.518	1374	0.0215	301.048
30	226.509	1186	0.0184	301.709
31	227.5	1031	0.0158	302.4
32	228.5	889	0.0136	303.0
33	231.0	767	0.0116	304.7
34	233.8	663	0.00987	306.5
35	236.6	575	0.00846	308.3
36	239.3	499	0.00726	310.1
37	242.1	433	0.00624	311.9
38	244.9	377	0.00537	313.7
39	247.6	329	0.00463	315.4
40	250.4	287	0.00400	317.2
41	253.2	251	0.00346	318.9
42	255.9	220	0.00300	320.7
43	258.7	193	0.00260	322.4
44	261.5	170	0.00226	324.1
45	264.2	149	0.00197	325.8
46	267.0	131	0.00171	327.5
47	269.7	116	0.00150	329.2
48	270.7	102	0.00132	329.8
49	270.7	90.3	0.00116	329.8
50	270.7	79.8	0.00103	329.8
55	260.8	42.5	0.000568	323.7
60	247.1	22.0	0.000310	315.1
65	233.3	10.9	0.000163	306.2
70	219.6	5.22	0.0000828	297.1
75	208.4	2.39	0.0000399	289.4
80	198.7	1.05	0.0000185	282.6

Appendix II
Modeling the Atmosphere

In this appendix, we will review the various properties associated with the Earth's lower atmosphere, the environment in which conventional airplanes operate. As seen in various chapters of this book, some of these properties can play a significant role in influencing the flight performance of a given flight vehicle.

The troposphere refers to the lowest portion of the atmosphere, from sea level up to about 11 km in altitude. The lower stratosphere is above the troposphere (extending from 11 km to about 20 km), with the tropopause being the boundary between the two atmospheric layers. Conventional airplane flight will occur within these regions (the vast majority below 20 km in altitude). For preliminary design and calculations, one generally refers to the International Standard Atmosphere (ISA, produced by ICAO; [1, 2]) for atmospheric properties like air density ρ, temperature T, pressure p and sound speed a. One can look up values in a table (as in Appendix I), or alternatively, one can undertake calculations that generate values that closely correspond to the ISA numbers. By way of gaining some background, and potentially being useful for computer programs, let's look at some equations in this regard.

For the troposphere, the temperature ratio θ may be estimated as a function of altitude h (in metres, above sea level, ASL) via the so-called lapse-rate equation, as follows:

$$\theta = \frac{T}{T_{SL}} = 1 - 2.26 \times 10^{-5}h\,(\text{m}), \quad 0 < h < 11 \text{ km}$$

while for the lower stratosphere,

$$\theta = 0.752, \quad 11 \text{ km} < h < 20 \text{ km}$$

which corresponds to the static air temperature remaining constant at 216.6 K as one moves above the troposphere. Sea level reference outside air temperature T_{SL} is 288.2 K. For the troposphere, the density ratio σ may be found as a function of θ:

$$\sigma = \frac{\rho}{\rho_{SL}} = \theta^{4.256}, \quad 0 < h < 11 \text{ km}$$

D. R. Greatrix, *Powered Flight*, DOI: 10.1007/978-1-4471-2485-6,
© Springer-Verlag London Limited 2012

while for the lower stratosphere,

$$\sigma = 1.682 \exp[-1.577 \times 10^{-4} h \ (\text{m})], \quad 11 \ \text{km} < h < 20 \ \text{km}$$

noting that ρ_{SL} is 1.225 kg/m^3.

With respect to outside air pressure p, one can find the pressure ratio δ in the troposphere as follows:

$$\delta = \frac{p}{p_{SL}} = \theta^{5.256}, \quad 0 < h < 11 \ \text{km}$$

while for the lower stratosphere,

$$\delta = 1.265 \exp[-1.577 \times 10^{-4} h \ (\text{m})], \quad 11 \ \text{km} < h < 20 \ \text{km}$$

noting p_{SL} is 101325 Pa.

Moving from the lower stratosphere to the upper stratosphere, above around 20 km, the air temperature begins to rise again. It will rise from 217 K to a maximum of around 270 K at 47 km altitude, and hold that value until around 52 km in altitude, when the stratosphere transitions into the mesosphere. While the table in Appendix 1 goes from sea level to 80 km in altitude above sea level for a nominal atmospheric profile, as required, one can go to the literature for more thorough representations of the Earth's atmosphere, for different Earth locations (e.g., the North Pole, vs. the equator), and different times of the day or night or year (atmospheric conditions, depending on the particular layer, can change significantly as a function of these and other factors).

References

1. Hill PG, Peterson CR (1992) Mechanics and thermodynamics of propulsion, 2nd edn. Addison-Wesley, New York
2. Mattingly JD (1996) Elements of gas turbine propulsion. McGraw-Hill, New York

Appendix III
Unit Conversions

Length:

1 m = 3.28 ft, 1 ft = 0.305 m = 12 in, 1 km = 0.6214 statute miles = 0.54 nautical miles, 1 nautical mile (n.mi.) = 1.1508 statute miles = 6076 ft, 1 statute mile = 5280 ft = 1760 yards, 1 in = 2.54 cm = 0.0254 m, 1 angstrom (Å) = 1×10^{-8} cm

Volume:

1 ℓ (liter; also, L) = 1000 cm^3 = 0.001 m^3 = 0.0353 ft^3, 1 U.S. gal (gallon) = 0.134 ft^3 = 3.785 ℓ, 1 Imperial gal = 4.546 ℓ = 1.2 U.S. gal, 1 U.S. qt (quart, liquid) = 0.946 ℓ, 1 in^3 = 16.39 cc (cm^3), 1 pint (pt) = 0.5 qt, 1 gal = 4 qt = 8 pt, 1 U.S. qt (dry) = 1.164 U.S. qt (liquid)

Speed:

1 m/s = 3.28 ft/s = 1.944 knots (nautical miles per hour) = 3.6 km/hr = 2.237 mi/hr (mph), 1 ft/s = 0.3048 m/s = 60 ft/min = 3600 ft/hr = 0.682 mi/hr, 1 knot (kn or kt) = 1.69 ft/s = 0.515 m/s, 1 mph (statute mi/hr) = 0.447 m/s = 88 ft/min

Mass:

1 kg = 2.205 lbm = 0.0685 slugs, 1 lbm = 16 ounces (oz) = 0.454 kg = 0.031 slugs 1 tonne = 1 metric ton = 1000 kg = 2205 lbm, 1 ton (short) = 2000 lbm, 1 kg = 1000 g (grams; also, gm), 1 slug = 32.18 lbm (pounds, mass) = 14.594 kg, 1 ton (long) = 2240 lbm = 1.016 tonnes

Density:

1 kg/m^3 = 0.0624 lbm/ft^3 = 0.00194 $slugs/ft^3$ = 3.61 H 10^{-5} lbm/in^3; specific gravity (s.g.; ρ/ρ_{H_2O}) of water at standard conditions = 1.0, gasoline s.g. 0.72, kerosene s.g. 0.81, where ρ_{H_2O} at standard conditions = 1000 kg/m^3

D. R. Greatrix, *Powered Flight*, DOI: 10.1007/978-1-4471-2485-6,
© Springer-Verlag London Limited 2012

Force:

1 N (newton) = 0.2248 lbf, 1 kgf = 9.807 N, 1 dyne = 1×10^{-5} N
    1 lbf (pound, force) = 4.48 N

Pressure:

1 Pa (pascal) = 1 N/m^2 = 9.869×10^{-6} atm (atmospheres) = 0.02089 lbf/ft^2 = 1×10^{-5} bar, 1 Pa = 1.45×10^{-4} lbf/in^2 (psi), 1 atm = 101325 Pa = 14.69 psi – 1.013 bar = 2115 psf, 1 torr – 133.3 Pa, 1 atm = 406.8 inches of water = 76 cm of mercury (manometer)

Energy:

1 J (joule) = 0.7376 ft-lbf = 9.481×10^{-4} Btu (British thermal units) = 0.239 cal (calories), 1 Btu = 1055 J = 778 ft-lbf, 1 cal = 4.184 J, 1 erg = 1×10^{-7} J

Power:

1 W (watt) = 1 J/s = 0.7376 ft-lb/s = 1.341×10^{-3} hp (horsepower)
1 hp = 550 ft lb/s – 0.746 kW, 1 Btu/hr = 0.293 W

Temperature:

1 K = 9/5° R, °C = K – 273°, °F = °R –459°, 0° C = 32° F = 273 K = 491° R

Absolute (Dynamic) Viscosity (μ):

1 $N\text{-}s/m^2$ = 1 kg/m-s = 10 poise (P) = 10 g/cm-s = 0.0209 $lbf\text{-}s/ft^2$ = 0.0209 slug/ft-s, 1 lbm/ft-hr = 8.63×10^{-6} slug/ft-s = 4.13×10^{-4} kg/m-s, 1 P = 100 cP (centipoise), 1 lbm/ft-s = 1.488 $N\text{-}s/m^2$

Kinematic Viscosity (v):

1 m^2/s = 10.76 ft^2/s = 10000 stokes (St) = 10000 cm^2/s
1 St = 100 cSt (centistokes), 1 ft^2/s = 0.0929 m^2/s

Specific Heat:

1 J/kg-K = 2.388×10^{-4} Btu/lbm-°R = 5.98 ft-lbf/slug-°R

Thermal Conductivity:

1 W/m-K = 1.3374×10^{-5} Btu/s-in-°R = 0.0104 ft-lbf/s-in-°R = 0.1248 lbf/s-°R
1 Btu/hr-ft-°R = 1.731 W/m-K = 0.0833 Btu/hr-in-°R = 2.315×10^{-5} Btu/s-in-°R

Appendix IV
Acronyms

A&P	Airframe & powerplant (mechanic)
AAM	Air-to-air missile
AB	Afterburner; air breathing
ABL	Allegheny Ballistics Laboratory
ABM	Anti-ballistic missile
ABRE	Air-breathing rocket engine
ABS	Acrylonitrile butadiene styrene [plastic]; anti-skid braking system
AC	Alternating current; aircraft; air conditioning
ACOC	Air-cooled oil cooler
ACT	Active control technology
ADC	Analogue-to-digital converter (data acquisition system); air data computer
ADF	Automatic direction finder (radio navigation)
ADN	Ammonium dinitramide (oxidizer, solid propellant)
AEC	Atomic Energy Commission (U.S.)
AEO	All engines operative
AEW	Airborne early warning
AF	Activity factor (propeller)
AFCS	Automatic flight control system
AFM	Aircraft flight manual
AFOSR	Air Force Office of Scientific Research (U.S.)
AFR	Air-to-fuel ratio
AFRL	Air Force Research Laboratory (U.S.)
AGARD	NATO Advisory Group for Aerospace Research & Development; now called Nato Research & Technology Organization (RTO)
AGB	Accessory gearbox (engine)
AGL	Above ground level
AI	artificial intelligence
AIAA	American Institute of Aeronautics & Astronautics

D. R. Greatrix, *Powered Flight*, DOI: 10.1007/978-1-4471-2485-6,
© Springer-Verlag London Limited 2012

AIM	Aeronautical Information Manual
AJ	Jet primary nozzle area
AKM	Apogee kick motor (IUS rocket motor)
ALCM	Air-launched cruise missile
AME	Aircraft maintenance engineer
AMM	Anti-missile missile
AMU	Astronaut maneuvering unit (see EVA; also, MMU, manned); atomic mass unit (amu)
AN	Ammonium nitrate (crystalline oxidizer)
ANC	Active noise control
ANCP	Ammonium nitrate based composite propellant
ANSI	American National Standards Institute
AOM	Aircraft operating manual
AOS	Air & oil systems (engine)
AP	Ammonium perchlorate (crystalline oxidizer); automatic pilot
APCP	Ammonium perchlorate based composite propellant
APU	Auxiliary power unit (aircraft)
AR	Aspect ratio
ARC	Active rotor control (helicopter)
ARINC	Aeronautical Radio Inc. (more recently, avionics standards for suppliers)
ARO	Army Research Office (U.S.)
ASALM	Advanced Strategic Air-Launched (Cruise) Missile (U.S.)
ASAT	Anti-satellite (system(s))
ASE	Automatic stabilization equipment; aircraft survivability equipment; aeroservoelasticity
ASI	Agenzia Spaziale Italiana
ASL	Above sea level
ASM	American Society for Materials (originally, Metals); air-to-surface missile; anti-ship missile
ASME	American Society of Mechanical Engineers
ASTM	American Society for Testing Materials
ASW	Anti-submarine warfare (system(s))
ATC	Air traffic control; after top (dead) center [piston/crank position]
ATF	Altitude test facility
ATR	Air-turborocket
ATRJ	Air-turboramjet
AU	Astronomical unit (149,600,000 km = 1 AU = 1.58×10^{-5} light years = 4.85×10^{-6} parsecs)
AUW	All-up weight (aircraft; see TOGW)
AVEN	Axisymmetric vectoring exhaust nozzle
AWACS	Airborne warning & control system
BAMO	Bis-azidomethyloxetane (energetic material)
BATES	Ballistic test & evaluation system (SRM)

BDC	Bottom dead center (piston engine)
BECO	Booster engine cutoff (rocket)
BHP	Brake horsepower
BIS	British Interplanetary Society
BLAST	Balloon-Borne Large Aperture Sub-Millimeter Telescope
BLDC	Brushless direct-current (electric motor)
BMEP	Brake mean effective pressure (piston engine)
BOMARC	Boeing/(Univ. of) Michigan Aeronautical Research Center (IM-99 /CIM-10 anti-bomber surface-to-air missile)
BOV	Bleed-off valve (compressor air, gas turbine engine); blow-off valve, surge control
BP	Boiling point
BPN	Boron potassium nitrate (energetic material)
BPR	Bypass ratio (turbofan engine)
BRG	Bearing (engine shaft)
BSFC	Brake specific fuel consumption (engine)
BTC	Before top (dead) center (piston/crank position, degrees of crankshaft rotation)
BUC	Backup control
BVI	Blade vortex interaction (rotor noise/vibration)
C^3	Command, control & communication
C^3I	command, control, communication & intelligence
CAB	Cellulose acetyl butyrate (propellant)
CACE	Cooled air cycle engine (hypersonic flight)
CAS	Calibrated airspeed
CASI	Canadian Aeronautics & Space Institute
CATS	Cheap access to space
CCOC	Combustion chamber outer casing
CD	Convergent/divergent (nozzle)
CDA	Command & data acquisition
CDP	Component development program
CDR	Critical design review
CEP	Circular error probability
CERMET	Ceramic/metallic (structural matrix)
CEV	Crew Exploration Vehicle (NASA)
CF	Cooling fan (for cooling lubricant oil, engine); concentration factor (liquid mixtures)
CFD	Computational fluid dynamics (numerical flow analysis)
CFL	Courant, Friedrichs, Lewy (stability criterion, unsteady flow calculations)
CFM	Cubic feet per minute (volumetric flow rate); computational fluid mechanics
CFR	Code of Federal Regulations (U.S., EPA)
CG	Center of gravity

CH	Cutback height (for throttling engine back in takeoff climb for noise abatement)
CI	compression ignition (engine diesel engine); also, combustion intensity, combustion instability
CIVV	Compressor inlet variable vanes
CJ	Chapman-Jouguet (detonation wave equations)
CL-20	Low-smoke solid propellant (China Lake Compound #20, $C_6H_6N_6(NO_2)_6$; see GAP)
CLP	Combustor loading parameter
CPIA	Chemical Propulsion Information Agency (U.S.)
CMDB	Composite-modified double-base (solid propellant)
COTS	Commercially-available off-the-shelf; commercial orbital transportation system
CS	Contact surface (gasdynamic flow); control surface
CSD	Constant speed drive (gas turbine); Chemical Systems Division (United Technologies)
CV	Control volume; control valve
CVC	Constant-volume (Humphrey cycle) combustion
CTPB	Carboxyl-terminated polybutadiene (solid fuel)
DARPA	Defense Advanced Research Projects Agency (U.S.)
DB	Double-base (homogeneous solid propellant originally NC + NG)
DC	Direct current
DDT	Deflagration-to-detonation transition (PDE)
DED	Diethyl diphenyl (stabilizer)
DEEC	Digital electronic engine control
DELTA-V	Nominal velocity increment gained for a given flight phase under thrust
DETA	Diethylenetriamine (fuel)
DHS	Department of Homeland Security (U.S.)
DIGATEC	Digital gas turbine engine control
DL	Data link (antenna, telemetry)
DME	Distance measuring equipment
DN	Bearing speed capability index, (bearing bore) diameter × speed (of shaft rotation)
DNS	Direct Numerical Simulation (numerical viscous flow analysis)
DOD	Department of Defense (U.S.); also, domestic object damage (items coming loose within engine and impacting on internal engine structure)
DOT	Department of Transportation (U.S.)
DR	Ducted rocket (engine); degree of reaction (compressor, turbine; also, DoR, DOR)
DRO	Design requirements & objectives
DSO	Defense Sciences Office (U.S.)
DWG	Drawing (engineering)
EASA	European Aviation Safety Agency

EB	Erosive burning (SRM)
EBT	Electron bombardment thruster
EC	Ethyl centralite (plasticizer)
ECR	Engine compression ratio
ECU	Engine control unit
ED	Expansion-deflection (nozzle)
EDP	Engine development program
EDR	Engineering design review
EDS	Engine diagnostic system
EEO	Elliptical Earth orbit
EEC	Electronic engine control (gas turbine engine)
EED	Electroexplosive device
EFC	Electronic fuel control (gas turbine engine)
EFI	Electronic fuel injection
EFIS	Electronic flight instrument system
EGR	Exhaust gas recirculation (recycle)
EGT	Exhaust gas temperature
EHM	Engine health monitoring
EL	Elevation angle
ELINT	Electronic intelligence
ELT	Emergency locator transmitter
ELV	Expendable launch vehicle
EM	Electromagnetic; energetic material(s)
EMC	Electromagnetic compatibility
EMD	Engine model derivative
EME	Electromagnetic environment
EMI	Electromagnetic interference; electromagnetic induction
EMP	Electromagnetic pulse
EOD	Explosive ordnance disposal
EP	Electric propulsion
EPA	Environmental Protection Agency (U.S.)
EPN	Effective perceived noise (engine; EPNdB, decibels); EPNL, level
EPON	Epoxy resin
EPR	Engine pressure ratio (for engine thrust estimation)
ERAST	Environmental Research and Sensor Technology (NASA program)
EROPS	Extended range operations (large commercial aircraft)
ESA	European Space Agency
ESD	Electrostatic discharge
ESM	Electronic support measures
ESS	Environmental support system
ET	External tank (Space Shuttle main liquid propellant tank); extraterrestrial
ETOPS	Extended-range twin-engine operations (large commercial aircraft)

EVA	Extravehicular activity (e.g., "spacewalk" or "moonwalk")
EW	Empty weight; electronic warfare; early warning (system(s))
FAA	Federal Aviation Administration (U.S.)
FADEC	Full authority digital engine control; full authority digital electronic control; fully automated digital engine control
FANS	Favre-averaged Navier-Stokes (viscous compressible flow equations)
FAR	Federal Aviation Regulation (U.S.); also, fuel-to-air ratio
FAT	Fixed area turbojet
FBO	Fixed base operator (also, fixed base of operation; service provider at airport); fan blade off (engine test)
FBW	Fly by wire (electromechanical control, vs. conventional heavier hydraulic actuation)
FCC	Federal Communications Commission (U.S.); flight control computer
FCU	Fuel control unit (engine)
FE	Finite element (method; numerical structural analysis)
FEEP	Field emission electric propulsion (ion thruster)
FEM	Forced exhaust mixer; finite element method
FFT	Fast Fourier transform (signal processing)
FIR	Finite impulse response (filter, signal processing)
FLIR	Forward-looking infrared (imaging sensor)
FLOX-70	70% fluorine/30% oxygen (liquid oxidizer)
FM	figure of merit; frequency modulated
FOD	Foreign object damage (external items ingested by engine intake and impacting on internal structure of the engine)
FOV	Field of view
FP	Flash point (ignition threshold temperature for volatile mixture; see TCC); freezing point; flat plate (aerodynamics)
FR	Fineness ratio (length-to-diameter)
FS	Factor of safety
FSI	Fluid-structure interaction
FT	Free [power] turbine
FTS	Flight termination system
FWD	Forward
GA	General aviation (non-commercial, recreational)
GALCIT	Guggenheim Aeronautical Laboratory/California Institute of Technology (CalTech) Graduate Aerospace Laboratories/CIT, more recent name
GAP	Glycidyl azide polymer (solid homogeneous fuel)
GCS	Ground control station
GEO	Geostationary Earth orbit; geosynchronous Earth orbit (35800 km altitude)
GG	Gas generator (engine)
GHG	Greenhouse gas

GLONASS	Global Navigation Satellite System (Russia)
GLOW	Gross liftoff weight (rocket vehicle)
GNC	Guidance, navigation & control
GPH	Gallons per hour
GPS	Global Positioning System (navigation)
GSE	Ground support equipment
GTE	Gas turbine engine
GTO	Geosynchronous transfer orbit
GTS	Gas turbine starter
HAN	Hydroxyl ammonium nitrate (monopropellant)
HARP	High Altitude Research Program (gun-launched rocket projectile study)
HARV	High Alpha (Angle-of-Attack) Research Vehicle
HAZMAT	Hazardous materials (protection system, procedures)
HC	Hydrocarbon (exhaust product)
HCV	Hypersonic Cruise Vehicle (DARPA)
HDG	Heading (navigation)
HDPE	High density polyethylene
HEC	High energy composite (solid propellant with added high-explosive component like RDX)
HEDS	Human Exploration and Development of Space (NASA program)
HERO	Hazards of electromagnetic radiation to ordnance (pyrotechnics, igniters, etc.)
HHC	Higher harmonic control (rotor vibration)
HHV	Higher heating value (resulting from combustion, exhaust with water as liquid)
HiMAT	Highly Maneuverable Aircraft Technology (NASA program)
HIRF	High intensity radiation field (see HERO)
HOTOL	Horizontal takeoff & landing (aircraft, rocket vehicle)
HMX	Cyclotetramethylene-tetranitramine (solid homogeneous monopropellant)
HNF	Hydrazinum nitroformate (oxidizer)
HP	High pressure (engine)
HPC	High pressure compressor
HPT	High pressure turbine
HRE	Hybrid rocket engine (also, hybrid rocket motor, HRM)
HTC	Heat transfer coefficient
HTHL	Horizontal takeoff/horizontal landing (aircraft, rocket vehicle)
HTP	High test peroxide (oxidizer)
HTPB	Hydroxyl-terminated polybutadiene (solid rubber-based fuel)
HTV	Hypersonic technology vehicle
HUD	Head-up display (aircraft)
HVM	Hypervelocity missile
HWC	Hail water content (engine test)
HWIL	Hardware in the loop (test)

IAS	Indicated airspeed
IBC	Individual blade control (helicopter rotor)
IC	Internal combustion (engine); initial condition (analysis)
ICAO	International Civil Aviation Organization
ICBM	Intercontinental ballistic missile
ICO	Intermediate circular orbit (see MEO)
ICRPG	Interagency Chemical Rocket Propulsion Group (U.S.)
IDG	Integrated drive generator (engine electrical system)
IEEE	Institute of Electrical and Electronic Engineers
IFR	Instrument flight rules
IGV	Inlet guide vanes (engine compressor)
ILS	Instrument landing system
IM	Insensitive munition(s)
IMEP	Indicated mean effective pressure (piston engine)
IMI	Initial maintenance inspection
IMU	Inertial measurement unit
INS	Inertial navigation system
IO	Input/output (data acquisition)
IP	Intermediate pressure (engine); intellectual property (patents, etc.); indicated power (piston)
IPDP	Integrated product development process
IPT	Integrated product team
IR	Infrared (radiation)
IRBM	Intermediate range ballistic missile
IRFNA	Inhibited red fuming nitric acid (storable liquid oxidizer)
IRR	Integral rocket-ramjet
ISA	International Standard Atmosphere
ISBV	Interstage bleed valves (engine)
ISO	International Standards Organization (QC governing body)
ISP	Specific impulse
ISRO	Indian Space Research Organisation
ISS	International Space Station
ITAR	International Traffic in Arms Regulations (U.S.)
ITD	interturbine duct (from HP to LP section of turbine); integrated technology demonstrator
IUS	Inertial upper stage (for final orbital insertion phase)
IWFNA	Inhibited white fuming nitric acid (oxidizer)
JAA	Joint Aviation Authority (Europe)
JANNAF	Joint Army, Navy, NASA, Air Force (U.S.)
JATO	Jet-assisted takeoff (aircraft)
JAXA	Japan Aerospace Exploration Agency
JET	Jet engine exhaust (exit) temperature (see EGT)
JFS	Jet fuel starter (alternative to APU, for gas turbine starting)
JP-4	Jet propellant (#4, well-known kerosene-based aircraft fuel)
JPL	Jet Propulsion Laboratory (CalTech)

JPT	Jet pipe temperature
KE	kinetic energy
KD	Kantrowitz-Donaldson (supersonic diffuser pitot intake)
KDN	Potassium dinitramide (oxidizer)
KP	Potassium perchlorate
LACE	Liquid air cycle engine (hypersonic flight)
LANL	Los Alamos National Laboratory
LDA	Laser Doppler anemometry
LDPE	Low density polyethylene
LE	Leading edge
LEO	Low Earth orbit
LES	Large eddy simulation (numerical turbulent flow analysis)
LEV	Leading edge vortex
LFC	Laminar flow control
LFRJ	Liquid-fuel ramjet
LH	Liquid hydrogen (LH_2)
LHV	Lower heating value (resulting from combustion, exhaust water to vapor state by default, HV)
LHS	Lefthand side
LIDAR	Light detection and ranging (laser radar)
LLNL	Lawrence Livermore National Laboratory
LNG	Liquified natural gas (cryogenic methane)
LOF	Liftoff (also, LO)
LORAN	Long-range navigation (system)
LOS	Line of sight
LOT	Liftoff time
LOX	Liquid oxygen
LP	Low pressure (engine)
LPC	Low pressure compressor
LPG	Liquefied petroleum gas
LPT	Low pressure turbine
LRE	Liquid-propellant rocket engine
LRU	Line-replaceable unit
LS	Lifting surface (aerodynamics)
LTO	Low Earth transfer orbit; also, landing-takeoff (cycle)
LVDT	Linear variable differential transformer (linear displacement sensor)
MAC	Mean aerodynamic chord
MAP	Manifold air pressure (piston engine)
MAT	Material approval test
MATV	Multi-axis thrust-vectoring
Max-Q	Point in rocket vehicle's atmospheric flight of maximum dynamic pressure
MBT	Maximum brake torque
MCD	Minimum cost design

MDS	Material data sheet
MECO	Main engine cutoff (rocket)
MEMS	Micro-electromechanical system(s)
MEO	Medium Earth orbit (between LEO and GEO; 2000–35000 km altitude)
MEOP	Maximum expected operating pressure
METO	Maximum except takeoff (engine power setting)
MHD	Magnetohydrodynamic (flow within magnetic field)
MIL SPEC	Military specifications (U.S.), e.g., MIL-A-8660 "Airplane Strength & Rigidity: General Specifications"
MIL STD	Military design standards (U.S.), e.g., MIL-STD-100 "Engineering Drawing Practices"
MIMO	Multiple input/multiple output
MIRV	Multiple independently-targetable reentry vehicle (single ICBM, multiple warheads released)
MIT	Massachusetts Institute of Technology
MLE	Maximum Likelihood Estimation (system behavior identification technique); MMLE, Modified MLE
MMH	monomethylhydrazine (storable liquid fuel)
MMI	Man-machine interface
MMW	Millimeter-wave (radar)
MON	Mixed oxides of nitrogen (liquid oxidizer)
MOU	Memorandum of understanding
MP	Melting point
MPD	Magnetoplasmadynamic (flow, thruster)
MPS	Main propulsion system
MR	Mixture ratio (oxidizer-to-fuel); mass ratio (vehicle)
MRO	Marketing requirements & objectives; maintenance, repair & overhaul
MSL	Mean sea level (as an altitude reference)
MTBF	Mean time between failures
MTE	Mission task element
MTO	Maximum takeoff (thrust, power)
MTOW	Maximum takeoff weight (aircraft)
N1	LP engine shaft rotational speed
N2	HP engine shaft rotational speed
NACA	National Advisory Committee for Aeronautics (U.S.)
NAS	National Academy of Sciences (U.S.)
NASA	National Aeronautics & Space Administration (U.S.)
NASP	National AeroSpace Plane (NASA, X-30 hypersonic commercial air transport program)
NATO	North Atlantic Treaty Organization
NBC	Nuclear, biological, chemical (protection system)
NC	Nitrocellulose (solid monopropellant)
NDE/T	Non-destructive evaluation/testing, inspection for structural faults

NERVA	Nuclear Engine for Rocket Vehicle Applications
NFV	Non-free vortex (blade design, compressor or turbine)
NG	Nitroglycerine (liquid monopropellant/explosive; solid when combined with NC)
NGV	Nozzle guide vanes (engine turbine, entry)
NLF	Natural laminar flow
NMR	Nuclear magnetic resonance (imaging)
NOAA	National Oceanic & Atmospheric Administration (U.S.)
NOE	Nap of the Earth (low-altitude aircraft flight)
NORAD	North American Aerospace Defense Command (Canada/U.S agreement)
NOTAR	No tail rotor (helicopter)
NOTS	Naval Ordnance Test Station (U.S.)
NOX	Oxides of nitrogen (exhaust product)
NPR	Nozzle pressure ratio
NPSH	Net positive suction head (pump, cavitation limit)
NPSS	Numerical Propulsion System Simulation (program)
NQ	Nitroguanadine (energetic material; also, NGD)
NS	Navigation system normal shock (gasdynamic flow)
NSA	National Security Agency (U.S.)
NSF	National Science Foundation research funding & advisory program (U.S.)
NTO	Nitrogen tetroxide
NTR	Nuclear thermal rocket
NTSB	National Transportation Safety Board (U.S.)
NVC	Noise & vibration control
OAT	Outside air temperature
ODE	Ordinary differential equation
OEI	One engine inoperative
OEM	Original equipment manufacturer
OEW	Operating empty weight (for aircraft, with no payload or fuel); also, empty operating weight
OFR	Oxidizer-to-fuel ratio
OGE	Out of ground effect (aircraft)
OGV	Outlet guide vane
OHC	Overhead camshaft (piston engine)
OHV	Overhead valve (piston engine)
OL	Operating line (engine)
OM	operator's manual
OMS	Orbital maneuvering system (Space Shuttle)
ONERA	Office National d'Etudes et de Recherches Aérospatiales (France)
ONR	Office of Naval Research (U.S.)
OPR	overall pressure ratio (engine, in compression relative to outside ambient air)

OS	Oblique shock (gasdynamic flow)
OSHA	Occupational Safety & Health Administration (U.S.)
OTH	Over the horizon
OTRAG	Orbital Transport-und Raketen-Aktiengesellschaft (commercial space launch venture, using modular rocket engine bundling approach)
OXM	Oxamine (burning rate modifier)
PBAA	Polybutadiene acrylic acid (solid fuel)
PBAN	Polybutadiene acrylic-acid/acrylonitrile (solid fuel)
PCM	Pulse code modulation (analogue-to-digital data telemetry)
PDE	Pulse detonation engine; partial differential equation
PDF	Probability density function
PDN	Propanedial-dinitrate (energetic material)
PDPA	Phase Doppler particle anemometry
PDR	Preliminary design review
PE	Potential energy; polyethylene
PEG	Polyethylene glycol
PEM	Proton exchange membrane (electricity-generating hydrogen fuel cell)
PEO	Polar Earth orbit
PETN	Pentaerythritol tetranitrate (energetic material)
PLF	Payload fraction (of total vehicle weight)
PHM	Prognostic health management (engine, aircraft)
PI	Principal investigator
PID	Proportional-integral-derivative (control technique)
PIT	Pulsed inductive thruster
PJ	Pulsejet (engine)
PKM	Perigee kick motor (IUS rocket motor)
PLC	Programmable logic control (system); also, public limited company
PLF	Pressure loss factor
PMMA	Polymethyl methacrylate [plexiglass]
PPS	Pounds per second
PPT	Pulsed plasma thruster
PR	Pressure ratio (engine)
PS	Polysulfide
PSC	Performance-seeking control
PSIA	Pounds per square inch, absolute
PSIG	Pounds per square inch, gauge
PSLV	Polar Satellite [Space] Launch Vehicle (India)
PSRU	Propeller speed reduction unit
PU	Polyurethane
PVC	Polyvinyl chloride
QC	Quality control

QCPC	Quality clinic process chart (QC monitoring/improvement tool)
RADAR	Radio detection and ranging (radar)
RAE	Royal Aircraft Establishment (U.K.), now part of the Defence Research Agency
RAIR	Ram-augmented interstellar rocket
RAM	Radar-absorbent material (stealth technology)
RANS	Reynolds-averaged Navier-Stokes (viscous incompressible flow equations)
RAT	Ram air turbine
RATO	Rocket-assisted takeoff (aircraft)
RBCC	Rocket-based combined-cycle (engine)
R/C	Radio-controlled (aircraft); rate of climb
RCS	Reaction control system (Space Shuttle); radar cross-section (vehicle)
RCVV	Rear compressor variable vanes
RDE	Rotating detonation engine
RDTE	Research, development, test & evaluation/engineering
RDX	Cyclotrimethylene-trinitramine (solid homogeneous mono-propellant)
RF	Radio frequency
RFP	Request for proposal
RH	Rankine-Hugoniot (gasdynamic equations); also, righthand [side, RHS]
RIT	Radio-frequency ion thruster
RJ	Ramjet (engine); also, regional jet
RJ-1	Liquid kerosene-based hydrocarbon ramjet/rocket fuel
RKA	Russian Federal Space Agency
RLV	Reusable launch vehicle
RMS	Root mean square (of oscillating signal)
RP-1	Rocket Propellant No. 1, kerosene-based hydrocarbon storable liquid rocket fuel
RPG	Rocket-propelled grenade
RPM	Revolutions per minute (rotational speed)
RPS	Revolutions per second (rotational speed)
RPV	Remotely piloted vehicle (see UAV)
RSO	Range safety officer (flight tests, military test range)
RTA	Real-time analyzer (signal processing, sound)
RTG	Radioisotope thermoelectric generator
RTO	NATO Research & Technology Organization (formerly AGARD)
RVP	Reid vapor pressure (volatile fuel characteristic)
R & R	Refueling & reloading (aircraft operations); also, rendezvous & recovery
SAC	Strategic Air Command (U.S.)
SAE	Society of Automotive Engineers

SAM	Surface-to-air missile
SAR	Specific air range; search & rescue; synthetic aperture radar
SAS	Stability augmentation system
SBIR	Small Business Innovation Research funding program (U.S.)
SC	Supercharger
SCR	Selective catalytic reduction (for NOX reduction)
SCV	Surge control valve (gas turbine engine)
SFRJ	Solid-fuel ramjet
SGR	Specific ground range
SHARP	Stationary High Altitude Relay Platform (UAV project)
SHP	Shaft horsepower (brake horsepower, BHP; equivalent, ESHP)
SI	Système International [d'Unités] (for units metric system); also, spark ignition (engine)
SISO	Single input/single output
SIT	Spontaneous ignition temperature
SJ	Scramjet (engine; also SCRJ, for supersonic-combustion ramjet)
SLAT	Supersonic low-altitude target (drone)
SLCM	Ship-/submarine-launched cruise missile
SLF	Steady level flight
SLS	Sea level static (engine at zero flight speed)
SLTO	Sea level takeoff (aircraft)
SM	Surge margin; stall margin
SME	Society of Manufacturing Engineers
SN	Smoke number (exhaust product, SAE classification); also, signal-to-noise ratio
SOFC	Solid oxide fuel cell
SOP	Standard operating procedure(s)
SOT	Stator outlet temperature
SOTV	Solar Orbit Transfer Vehicle (NASA)
SOW	Statement of work
SOX	oxides of sulfur (exhaust product)
SPEC DECK	Collection (deck) of specifications file(s) (tabulating expected performance under different flight conditions, related data, etc.; commonly associated with air-breathing engines)
SPL	Sound pressure level (see EPN)
SPOT	Système Probatoire d'Observation de la Terre (satellite-based Earth imaging)
SPS	Secondary power system (aircraft power, excluding pro-pulsion)
SPT	Stationary plasma thruster
SR	Sounding rocket
SRB	Solid rocket booster (commonly associated with the two utilized by the Space Shuttle)
SRAM	Short-range attack missile
SRM	Solid-propellant rocket motor
SRR	Systems requirement review

SSL	Standard sea level atmospheric conditions
SSME	Space Shuttle Main Engine (high-thrust liquid-propellant rocket engine, with three SSMEs positioned at the rear of the Shuttle orbiter vehicle)
SSO	Sun synchronous orbit (near polar Earth orbit)
SST	Supersonic transport (aircraft)
SSTO	Single stage to orbit (rocket vehicle)
STEM	Shaped tube electrolytic machining (acid/electrochemical dril-ling of long holes in hard metals, e.g., for machining cooling channels in turbine blades and vanes); science, technology, engineering & mathematics
STIG	Steam-injected gas turbine (engine waste heat recovery technique)
STOVL	Short takeoff/vertical landing (aircraft)
STS	Space Transportation System (NASA Space Shuttle)
TACAN	Tactical air navigation
TAS	True airspeed
TBC	Thermal barrier coating
TBCC	Turbine-based combined cycle (engine)
TBO	Time between overhauls
TC	Type certificate (engine, aircraft certification); also, Transport Canada
TCC	Tag closed-cup [tester, for measurement of a liquid's flash point temperature]
TDC	Top dead center (piston engine)
TDF	Temperature distribution function
TDRSS	Tracking and data relay satellite system
TE	Trailing edge
TEMPEST	Transient Electromagnetic Pulse Emanation Standard; see HERO
TF	Turbofan (engine)
TGT	Turbine gas temperature
TIT	Turbine inlet temperature (also, TET, turbine entry temper-ature)
TJ	Turbojet (engine)
TM	Telemetry (data acquisition)
TMA	Trimethylamine (liquid fuel)
TNT	Trinitrotoluene (energetic material)
TO	Takeoff
TOG	Takeoff, ground roll
TOGW	Takeoff gross weight (aircraft)
TOP	Technical operating procedure
TOW	Tube-launched, optically-tracked, wire-guided (missile)
TP	Turboprop (engine); also, thrust power
TPM	Total productive maintenance, or total process management (pro-active method of QC)
TPS	Thermal protection system
TQM	Total quality management

TRU	Thrust reverser unit
TS	Turboshaft (engine); also, temperature-entropy (cycle diagram)
TSC	Turbosupercharger
TSFC	Thrust specific fuel consumption (engine)
TSO	Trial safety officer (engine firing or flight tests); also, technical standard order (minimum performance expected by transport authority)
TSS	Turbine engine, supersonic (flight)
TSTO	Two stage(s) to orbit (rocket vehicle)
TVC	Thrust vector control (engine)
UAV	Unmanned (uninhabited) air vehicle
UCAV	Unmanned combat air vehicle (for weapons delivery)
UCE	Usable cue environment (helicopter flight operations)
UDF	Unducted fan (engine)
UDMH	Unsymmetrical dimethylhydrazine (storable liquid rocket fuel)
UHB	Ultra-high bypass (engine)
UHC	Unburned hydrocarbons (exhaust product)
UHF	Ultra-high frequency
UHMWPE	Ultra-high molecular weight polyethylene
USAF	United States Air Force
UTM	Universal Transverse Mercator (Earth mapping model)
V-1	Vengeance [Weapon] No. 1 (German pulsejet-powered cruise missile, WWII)
V-2	Vengeance [Weapon] No. 2 (German ballistic missile, WWII; A-4)
VAB	Vehicle Assembly Building, NASA Kennedy Space Center, Florida (for Space Shuttle)
VASIMR®	Variable Specific Impulse Magnetoplasma Rocket
VFR	Visual flight rules
VHF	Very high frequency
VLA	Very light aircraft
VLF	Very low frequency
VOR	VHF omnidirectional range (radio navigation)
V/STOL	Vertical/short takeoff & landing (aircraft)
VTOL	Vertical takeoff & landing (aircraft)
WAC	Women's Army Corps (WAC-Corporal liquid-propellant sounding rocket, derived from MGM -5 Corporal tactical surface-to-surface nuclear missile)
WAR	Water air ratio
WAT	Weight-altitude-temperature chart for aircraft takeoff or landing maximum allowed weight
WEM	Whole engine model
WIG	Wing in ground-effect (near-surface flight)
WOT	Wide-open throttle setting (piston engine)
WSMR	White Sands Missile Range

WWII	World War Two (1939–1945)
ZS	Zinc-sulfur (solid propellants)
ZULU	Greenwich (United Kingdom) mean time (GMT)
1D	One-dimensional spatial framework (steady or unsteady)
2D	Two-dimensional (axisymmetric [cylindrical] or planar) spatial framework
2L	Two wave passes (1 left, 1 right) of acoustic chamber of length L, as operational cycle
3D	Three-dimensional (Cartesian [rectangular] or spherical) spatial framework
3DOF	Three degrees of freedom (point-mass dynamics in a two-dimensional plane, translation [two degrees] & rotation [one degree])
4L	Four wave passes (1 left, 1 right, 1 left, 1 right) of acoustic chamber of length L, as operational cycle
6DOF	Six degrees of freedom (three-dimensional point-mass dynam-ics, translation [three degrees] & rotation [three degrees])

Index

A

Acronyms, 493–509
Aerobraking, 478
Airbreathing rocket engines, 435, 440
  Aerojet Strutjet engine, 440–441
  air-turbo-rocket, 442
  combined cycle, 440, 445, 448
  ducted rocket, 440, 443
  fuel, 435, 437, 439–444
  integral rocket-ramjet, 440, 442–443
  Lockheed-Martin/DARPA Falcon HTV-3X vehicle, 447–448
  NASA X-30 flight vehicle, 441
  oxidizer, 435, 440–441
  flight performance envelope, 444, 447–448
  SABRE engine, 445–446
  Skylon flight vehicle, 445–446
  turbine-based combined cycle, 440, 448
Aircraft propulsion, 1, 29
  endurance, 30–37
  flight (mission) operations, 6, 30, 37, 104, 294, 445, 447–448
  range, 30–37
  vehicle aerodynamic drag, 29, 31–32
  vehicle aerodynamic lift, 23, 29, 31–32
Athodyd, 435
Atmosphere, 118, 487–490
  ISA profile, 51, 118, 487–490
  mesosphere, 480, 490
  stratosphere, 161, 439, 479–480, 489–490
  tropopause, 161, 479, 489
  troposphere, 161, 479, 489–490
Autogyro (gyrocopter), 83, 85

  Kellett K-2, 85
  McCulloch Super J-2, 85
Axial compressor, 180, 390
  bleed air, 156
  blowoff valves, 180, 186
  corrected mass flow, 178, 185–186
  corrected shaft speed, 178, 185–186
  degree of reaction, 183–184
  disk, 182, 186, 196, 204, 249
  inlet guide vane, 180, 182
  operating line, 185
  performance map, 185–186
  rotor, 181–186
  stage, 180–186
  stall, 185–186
  stator, 181–183, 186
  surge, 185–186
  surge margin, 185–186
Axial turbine, 152, 195, 392
  blade cooling, 196–197, 204
  blade materials, 204
  blisk, 204
  choking limit, 203–205
  degree of reaction, 198–200
  disk, 182, 186, 196, 204, 249
  flow coefficient, 202
  free vortex (blade), 203
  nozzle guide vane, 195
  operating line, 203
  performance map, 203–204
  pin fins, 197
  rotor, 195–204
  stage, 154, 195–202
  stage loading, 200–202

D. R. Greatrix, *Powered Flight*, DOI: 10.1007/978-1-4471-2485-6,
© Springer-Verlag London Limited 2012

Printed by Publishers' Graphics LLC
LMO130429.15.17.32